INTRODUCTION TO SPECTROSCOPY

A GUIDE FOR STUDENTS OF ORGANIC CHEMISTRY

Donald L. Pavia

Gary M. Lampman

George S. Kriz

Department of Chemistry
Western Washington University
Bellingham, Washington

HARCOURT COLLEGE PUBLISHERS

Fort Worth   Philadelphia   San Diego   New York   Orlando   Austin
San Antonio   Toronto   Montreal   London   Sydney   Tokyo

To All of Our "O-Spec" Students

Publisher: John Vondeling
Publisher: Emily Barrosse
Marketing Strategist: Pauline Mula
Developmental editor: Sandy Kiselica
Production Service: TSI Graphics
Production Manager: Charlene Catlett Squibb
Art Director: Paul Fry
Cover Designer: Jacqui LeFranc

PAVIA: INTRODUCTION TO SPECTROSCOPY, 3/e
ISBN 0-03-031961-7
Library of Congress Catalog Card Number: 00-030101

Address for domestic orders:
Harcourt College Publishers, 6277 Sea Harbor Drive, Orlando, FL 32887-6777
1-800-782-4479
e-mail   collegesales@harcourt.com

Address for international orders:
International Customer Service, Harcourt, Inc.
6277 Sea Harbor Drive, Orlando FL 32887-6777
(407) 345-3800
Fax (407) 345-4060
e-mail   hbintl@harcourt.com

Address for editorial correspondence:
Harcourt College Publishers, Public Ledger Building, Suite 1250, 150 S. Independence Mall West, Philadelphia, PA 19106-3412

Web Site Address
http://www.harcourtcollege.com

Printed in the United States of America

0 1 2 3 4 5 6 7 8 9   066   10 9 8 7 6 5 4 3 2 1

1191580

PREFACE

This is the third edition of our textbook in spectroscopy intended for students of organic chemistry. This textbook can serve as a supplement for the typical organic chemistry lecture textbook, and it can also be used as a "stand-alone" textbook for an advanced undergraduate course in spectroscopic methods of structure determination. This book is also a useful tool for students engaged in research. Our aim is not only to teach students to interpret spectra, but also to present basic theoretical concepts. As with the previous editions, we have tried to focus on the important aspects of each spectroscopic technique without dwelling excessively on theory or complex mathematical analyses.

This book is a continuing evolution of materials that we use in our own courses, both as a supplement to our organic chemistry lecture course series and also as the principal textbook in our upper division courses in spectroscopic methods and advanced NMR techniques. Explanations and examples that we have found to be effective in our courses as well as in our reviewers' courses have been incorporated into this edition.

New to This Edition

This third edition of *Introduction to Spectroscopy* contains some important changes. We have expanded our earlier chapter on advanced considerations in nuclear magnetic resonance spectroscopy into two new chapters. The first of these chapters (Chapter 5) focuses on spin coupling interactions, while the second new chapter (Chapter 6) discusses protons on heteroatoms, exchange phenomena, tautomerism, valence tautomerism, spin decoupling methods, and chemical shift reagents. Throughout the book we have also increased the number of 300-MHz spectra in order to provide spectra that compare more closely with those that students might obtain using modern instrumentation. We have provided expansions for many spectra so that students can see more clearly the multiplicity of the peaks. Hertz values have been included on the expansions to facilitate the determination of coupling constants. A section has been added in Chapter 5 that shows students how to extract the coupling constants in complex allylic systems.

We have included a discussion of some new techniques including NOE difference spectroscopy. We have also expanded our discussion of the information available from a DEPT experiment. We have added more tables of spectral data, and we have expanded the appendices to include more tables of chemical shift and coupling constant values.

We have added Spectral Analysis Boxes to the chapters on nuclear magnetic resonance and mass spectrometry. In the previous edition, these boxes were a feature used only in the chapter on infrared spectroscopy. We hope that these summary boxes will help students understand better how to use the information that the spectra contain. In addition, these boxes are part of a survey of NMR and mass spectral data *by functional group,* a new feature of this edition.

We have also added many new problems to each chapter. This book thus contains a great many problems, with a wide range of difficulty, ranging from relatively simple to more challenging. Students are provided answers to some of the problems, however, at the request of reviewers and instructors using the book, we have included some problems that are not answered at the back of the book. To make assignments easier for instructors, we have indicated the problems that have student answers with an asterisk (*). Answers to problems without student answers will be made available to qualified instructors *via* the World Wide Web. Chapter 9 *(Combined Structure Problems)* contains many new and challenging examples, most of them with 300-MHz spectra that include expansions. Problems making use of two-dimensional NMR data have been added to Chapter 10.

Acknowledgments

We want to express our deep gratitude to colleagues who have contributed significantly to this new edition. Our colleague, Prof. James R. Vyvyan (Western Washington University) made many valuable suggestions and provided several examples from his research. Charles Wandler, the instruments technician at Western Washington University did yeoman duty running spectra for us (running is his specialty!). Meagan Lindstrom, a chemistry undergraduate student, also ran many spectra. Special thanks are due to the students in our spectroscopy classes for the stimulation they provide us.

We wish to acknowledge the cooperation of Varian Associates and the Aldrich Chemical Company for their permission to use infrared and nuclear magnetic resonance spectra from their catalogues.

Many of our colleagues reviewed our manuscript and provided helpful criticism: Professors Jonathan Touster (Stanford University), Charles Garner (Baylor University), William F. Wood (Humboldt State University), and Stephen Branz (San Jose State University). Saunders College Publishing and TSI Graphics handled production of this textbook. Sandi Kiselica, our Developmental Editor, worked very hard keeping us up to the production schedule and providing us with a great deal of help. Tom Robinson at TSI Graphics capably handled the final production stages of the book.

Finally, once again we must thank our wives, Neva-Jean, Marian, and Carolyn, for the support and their patience. They deserve to have a break from dealing with piles of manuscript pages and spectra scattered over their dining room tables!

Don Pavia
Gary Lampman
George Kriz

CONTENTS

CHAPTER 3

NUCLEAR MAGNETIC RESONANCE SPECTROSCOPY

PART ONE: BASIC CONCEPTS   102

CHAPTER 4

NUCLEAR MAGNETIC RESONANCE SPECTROSCOPY

PART TWO: CARBON-13 SPECTRA, INCLUDING HETERONUCLEAR COUPLING WITH OTHER NUCLEI   167

CHAPTER 5

NUCLEAR MAGNETIC RESONANCE SPECTROSCOPY

PART THREE: SPIN-SPIN COUPLING   217

CHAPTER 6
NUCLEAR MAGNETIC RESONANCE SPECTROSCOPY
PART FOUR: OTHER TOPICS IN ONE-DIMENSIONAL NMR   306

CHAPTER 7
ULTRAVIOLET SPECTROSCOPY   353

CHAPTER 8

MASS SPECTROMETRY   390

CHAPTER 9

COMBINED STRUCTURE PROBLEMS   466

CHAPTER 10

NUCLEAR MAGNETIC RESONANCE SPECTROSCOPY
PART FIVE: ADVANCED NMR TECHNIQUES   526

ANSWERS TO SELECTED PROBLEMS   ANS-1

APPENDICES

INDEX   I-1

CHAPTER 1

MOLECULAR FORMULAS AND WHAT CAN BE LEARNED FROM THEM

Before attempting to deduce the structure of an unknown organic substance from an examination of its spectra, we can simplify the problem somewhat by examining the molecular formula of the substance. The purpose of this chapter is to describe how the molecular formula of a compound is determined and how structural information may be obtained from that formula. The early portion of this chapter reviews the classical analytical methods of determining a molecular formula. Many of these methods are still in routine use today, but the use of mass spectrometry has become a common alternative. A brief introduction to the use of the mass spectrometer is given at the end of this chapter (Section 1.6) and a more complete discussion may be found in Chapter 8.

1.1 ELEMENTAL ANALYSIS AND CALCULATIONS

The classical process of determining the molecular formula of a substance involves three steps. The first step is to perform a **qualitative elemental analysis** in order to find out what kinds of atoms are present in the molecule. The second step is to perform a **quantitative elemental analysis** in order to determine the relative numbers of the distinct kinds of atoms in the molecule. This analysis leads to an **empirical formula.** The third step is a **molecular mass determination** (or **molecular weight determination**) that, when combined with the empirical formula, shows the actual numbers of the distinct kinds of atoms. The result is the **molecular formula.**

Virtually all organic compounds contain carbon and hydrogen. In most cases it is not necessary to determine whether these elements are present; their presence is assumed. However, if it should be necessary to demonstrate that either carbon or hydrogen is present in a compound, that substance may be burned in the presence of oxygen. If combustion produces carbon dioxide, carbon must have been present in the unknown material; if combustion produces water, hydrogen atoms must have been present in the unknown.

It is important to note that there is no suitable direct method for determining the presence of oxygen in a substance. Consequently, the presence of oxygen is not demonstrated in a qualitative analysis.

Nitrogen, chlorine, bromine, iodine, and sulfur may be identified by tests similar to the **sodium fusion test.** To obtain the details of such tests, you should consult a textbook on qualitative organic analysis. The reference list at the end of this chapter includes examples of suitable textbooks.

To determine the precise amounts of carbon and hydrogen present in an unknown substance, a quantitative analysis is required. In practice, commercial laboratories frequently perform such analyses. The method of determining the amounts of carbon and hydrogen in a substance involves combustion to carbon dioxide and water. In a quantitative analysis, the carbon dioxide and water are collected and weighed. Methods are also available for the determination of the amounts of sulfur, nitrogen, and halogens that may be present. The mathematical methods for determining the percentage composition of an unknown substance from quantitative analysis for the elements should be familiar from general chemistry. For review, however, Table 1.1 summarizes the calculation of percentage composition, using an unknown organic compound as an example. Notice in the sample calculation that the percentage of oxygen in a sample is generally determined by difference.

TABLE 1.1
CALCULATION OF PERCENTAGE COMPOSITION FROM COMBUSTION DATA

$$C_xH_yO_z + \text{excess } O_2 \longrightarrow x\,CO_2 + y/2\,H_2O$$

9.83 mg   23.26 mg   9.52 mg

$$\text{millimoles } CO_2 = \frac{23.26 \text{ mg } CO_2}{44.01 \text{ mg/mmole}} = 0.5285 \text{ mmoles } CO_2$$

mmoles CO_2 = mmoles C in original sample

(0.5285 mmoles C)(12.01 mg/mmole C) = 6.35 mg C in original sample

$$\text{millimoles } H_2O = \frac{9.52 \text{ mg } H_2O}{18.02 \text{ mg/mmole}} = 0.528 \text{ mmoles } H_2O$$

$$(0.528 \text{ mmoles } H_2O)\left(\frac{2 \text{ mmoles H}}{1 \text{ mmole } H_2O}\right) = 1.056 \text{ mmoles H in original sample}$$

(1.056 mmoles H)(1.008 mg/mmole H) = 1.06 mg H in original sample

$$\% \text{ C} = \frac{6.35 \text{ mg C}}{9.83 \text{ mg sample}} \times 100 = 64.6\%$$

$$\% \text{ H} = \frac{1.06 \text{ mg H}}{9.83 \text{ mg sample}} \times 100 = 10.8\%$$

$$\% \text{ O} = 100 - (64.6 + 10.8) = 24.6\%$$

It is quite rare to find a research laboratory in which elemental analyses are performed by classical methods. Usually samples are sent to a commercial analytical laboratory that specializes in quantitative elemental analyses, or modern instrumentation is used. The commercial laboratory returns a report that lists the percentages of carbon and hydrogen in the sample. The report also includes the percentages of any other elements requested. In either case, the principles outlined in this section still apply, but the procedures tend to be too time consuming to be conducted in every organic chemistry laboratory.

Commercially available elemental analyzers are capable of determining simultaneously the percentages of carbon, hydrogen, and nitrogen in a compound. In these instruments, the sample is burned in a stream of oxygen. The gaseous products are converted to carbon dioxide, water, and nitrogen, which can be detected via gas chromatography, using thermal conductivity detectors. The precise amount of each gas produced in the combustion is determined by integration of the corresponding gas chromatography peaks. Some instruments are capable of performing oxygen analyses.

The percentage composition result may be used to calculate the empirical formula of the substance being studied. Note that the empirical formula expresses the correct **simplest whole-number ratios** of the elements in the test substance. The empirical formula may not be the true, or molecular, formula of the substance being examined. The molecular formula may be some multiple of the empirical formula. The mathematical methods used to obtain the empirical formula are familiar from general chemistry. As an illustration, however, a sample calculation, using the same unknown sample that was presented earlier, is shown in Table 1.2.

TABLE 1.2
CALCULATION OF EMPIRICAL FORMULA

Using a 100-g sample:

   64.6% of C = 64.6 g

   10.8% of H = 10.8 g

   24.6% of O $= \dfrac{24.6 \text{ g}}{100.0 \text{ g}}$

   moles C $= \dfrac{64.6 \text{ g}}{12.01 \text{ g/mole}} = 5.38$ moles C

   moles H $= \dfrac{10.8 \text{ g}}{1.008 \text{ g/mole}} = 10.7$ moles H

   moles O $= \dfrac{24.6 \text{ g}}{16.0 \text{ g/mole}} = 1.54$ moles O

giving the result

   $C_{5.38}H_{10.7}O_{1.54}$

Converting to the simplest ratio:

   $C_{\frac{5.38}{1.54}}H_{\frac{10.7}{1.54}}O_{\frac{1.54}{1.54}} = C_{3.49}H_{6.95}O_{1.00}$

which approximates

   $C_{3.50}H_{7.00}O_{1.00}$

or

   $C_7H_{14}O_2$

1.2 DETERMINATION OF MOLECULAR MASS

The next step in determining the molecular formula of a substance is to determine the weight of one mole of that substance. This may be accomplished in a variety of ways. Without knowledge of the molecular mass of the unknown, there is no way of determining whether the empirical formula, which is determined directly from elemental analysis, is the true formula of the substance or whether the empirical formula must be multiplied by some integral factor to obtain the molecular formula. In the example cited in Section 1.1, without knowledge of the molecular mass of the unknown, it is impossible to tell whether the molecular formula is $C_7H_{14}O_2$ or $C_{14}H_{28}O_4$.

In a modern laboratory, the molecular mass is determined through the use of mass spectrometry. The details of this method and the means of determining molecular mass can be found in Section 1.6 and Chapter 8, Section 8.4. This section reviews some classical methods of obtaining the same information.

An old method that is used occasionally is the **vapor density method.** In this method, a known volume of gas is weighed at a known temperature. After converting the volume of the gas to standard temperature and pressure, we can determine what fraction of a mole that volume represents. From that fraction, we can easily calculate the molecular mass of the substance.

Another method of determining the molecular mass of a substance is to measure the freezing-point depression of a solvent that is brought about when a known quantity of test substance is added. This is known as a **cryoscopic method.** Another method, which is used occasionally, is **vapor pressure osmometry,** in which the molecular weight of a substance is determined through an examination of the change in vapor pressure of a solvent when a test substance is dissolved in it.

If the unknown substance is a carboxylic acid, it may be titrated with a standardized solution of sodium hydroxide. By use of this procedure, a **neutralization equivalent** can be determined. The neutralization equivalent is identical to the equivalent weight of the acid. If the acid has only one carboxyl group, the neutralization equivalent and the molecular mass are identical. If the acid has more than one carboxyl group, the neutralization equivalent is equal to the molecular mass of the acid divided by the number of carboxyl groups. Many phenols, especially those substituted by electron-withdrawing groups, are sufficiently acidic to be titrated by this same method, as are sulfonic acids.

1.3 MOLECULAR FORMULAS

Once the molecular mass and the empirical formula are known, we may proceed directly to the **molecular formula.** Often the empirical formula weight and the molecular mass are the same. In such cases, the empirical formula is also the molecular formula. However, in many cases, the empirical formula weight is less than the molecular mass, and it is necessary to determine how many times the empirical formula weight can be divided into the molecular mass. The factor determined in this manner is the one by which the empirical formula must be multiplied in order to obtain the molecular formula.

Ethane provides a simple example. After quantitative element analysis, the empirical formula for ethane is found to be $CH_3.$ A molecular mass of 30 is determined. The empirical formula weight of ethane, 15, is half of the molecular mass, 30. Therefore, the molecular formula of ethane must be $2(CH_3)$ or $C_2H_6.$

For the sample unknown introduced earlier in this chapter, the empirical formula was found to be $C_7H_{14}O_2.$ The formula weight is 130. If we assume that the molecular mass of this substance was determined to be 130, we may conclude that the empirical formula and the molecular formula are identical and that the molecular formula must be $C_7H_{14}O_2.$

1.4 INDEX OF HYDROGEN DEFICIENCY

Frequently, a great deal can be learned about an unknown substance simply from a knowledge of its molecular formula. This information is based on the following general molecular formulas.

$$\left.\begin{array}{ll} \text{alkane} & C_nH_{2n+2} \\ \text{cycloalkane or alkene} & C_nH_{2n} \\ \text{alkyne} & C_nH_{2n-2} \end{array}\right\} \text{Difference of 2 hydrogens}$$

Notice that each time a ring or a π bond is introduced into a molecule, the number of hydrogens in the molecular formula is reduced by *two. For every triple bond (two π bonds) introduced into a molecule, the number of hydrogens in the molecular formula is reduced by four.*

When the molecular formula for a compound contains noncarbon or nonhydrogen elements, the ratio of carbon to hydrogen may change. Following are three simple rules that may be used to predict how this ratio will change.

1. To convert the formula of an open-chain, saturated hydrocarbon to a formula containing Group V elements (N, P, As, Sb, Bi), one additional hydrogen atom must be *added* to the molecular formula for each such Group V element present. In the following examples, each formula is correct for a two-carbon acyclic, saturated compound.

$$C_2H_6, \qquad C_2H_7N, \qquad C_2H_8N_2, \qquad C_2H_9N_3$$

2. To convert the formula of an open-chain, saturated hydrocarbon to a formula containing Group VI elements (O, S, Se, Te), *no change* in the number of hydrogens is required. In the following examples, each formula is correct for a two-carbon, acyclic, saturated compound.

$$C_2H_6, \qquad C_2H_6O, \qquad C_2H_6O_2, \qquad C_2H_6O_3$$

3. To convert the formula of an open-chain, saturated hydrocarbon to a formula containing Group VII elements (F, Cl, Br, I), one hydrogen must be *subtracted* from the molecular formula for each such Group VII element present. In the following examples, each formula is correct for a two-carbon, acyclic, saturated compound.

$$C_2H_6, \qquad C_2H_5F, \qquad C_2H_4F_2, \qquad C_2H_3F_3$$

The **index of hydrogen deficiency** (sometimes called the **unsaturation index**) is the number of π bonds and/or rings a molecule contains. It is determined from an examination of the molecular formula of an unknown substance and from a comparison of that formula with a formula for a corresponding acyclic, saturated compound. The difference in the numbers of hydrogens between these formulas, when divided by 2, gives the index of hydrogen deficiency.

The index of hydrogen deficiency can be very useful in structure determination problems. A great deal of information can be obtained about a molecule before a single spectrum is examined. For example, a compound with an index of **one** must have one double bond or one ring, but it cannot have both structural features. A quick examination of the infrared spectrum could confirm the presence of a double bond. If there were no double bond, the substance would have to be cyclic and saturated. A compound with an index of **two** could have a triple bond, or it could have two double bonds, or two rings, or one of each. Knowing the index of hydrogen deficiency of a substance, the chemist can proceed directly to the appropriate regions of the spectra to confirm the presence or absence of π bonds or rings. Benzene contains one ring and three "double bonds" and thus has an index of hydrogen deficiency of **four.** Any substance with an index of *four* or more may contain a benzenoid ring; a substance with an index less than *four* cannot contain such a ring.

To determine the index of hydrogen deficiency for a compound, apply the following steps.

1. Determine the formula for the saturated, acyclic hydrocarbon containing the same number of carbon atoms as the unknown substance.

2. Correct this formula for the nonhydrocarbon elements present in the unknown. Add one hydrogen atom for each Group V element present, and subtract one hydrogen atom for each Group VII element present.

3. Compare this formula with the molecular formula of the unknown. Determine the number of hydrogens by which the two formulas differ.

4. Divide the difference in the number of hydrogens by **two** to obtain the index of hydrogen deficiency. This equals the number of π bonds and/or rings in the structural formula of the unknown substance.

The following examples illustrate how the index of hydrogen deficiency is determined and how that information can be applied to the determination of a structure for an unknown substance.

■ EXAMPLE 1

The unknown substance introduced at the beginning of this chapter has the molecular formula $C_7H_{14}O_2$.

1. Using the general formula for a saturated, acyclic hydrocarbon (C_nH_{2n+2}, where $n = 7$), calculate the formula C_7H_{16}.

2. Correction for oxygens (no change in the number of hydrogens) gives the formula $C_7H_{16}O_2$.

3. This latter formula differs from that of the unknown by two hydrogens.

4. The index of hydrogen deficiency equals **one**. There must be one ring or one double bond in the unknown substance.

Having this information, the chemist can proceed immediately to the double-bond regions of the infrared spectrum. There she finds evidence for a carbon–oxygen double bond (carbonyl group). At this point the number of possible isomers which might include the unknown has been narrowed considerably. Further analysis of the spectral evidence leads to an identification of the unknown substance as **isopentyl acetate.**

$$CH_3-\overset{\overset{\displaystyle O}{\|}}{C}-O-CH_2-CH_2-\underset{\underset{\displaystyle CH_3}{|}}{CH}-CH_3$$

■ EXAMPLE 2

Nicotine has the molecular formula $C_{10}H_{14}N_2$.

1. The formula for a 10-carbon, saturated, acyclic hydrocarbon is $C_{10}H_{22}$.

2. Correction for the two nitrogens (add two hydrogens) gives the formula $C_{10}H_{24}N_2$.

3. This latter formula differs from that of nicotine by 10 hydrogens.

4. The index of hydrogen deficiency equals **five**. There must be some combination of five π bonds and/or rings in the molecule. Since the index is greater than *four*, a benzenoid ring could be included in the molecule.

Analysis of the spectrum quickly shows that a benzenoid ring is indeed present in nicotine. The spectral results indicate no other double bonds, suggesting that another ring, this one saturated, must be present in the molecule. More careful refinement of the spectral analysis leads to a structural formula for nicotine:

■ **EXAMPLE 3**

Chloral hydrate ("knockout drops") is found to have the molecular formula $C_2H_3Cl_3O_2$.

1. The formula for a two-carbon, saturated, acyclic hydrocarbon is C_2H_6.
2. Correction for oxygens (no additional hydrogens) gives the formula $C_2H_6O_2$.
3. Correction for chlorines (subtract three hydrogens) gives the formula $C_2H_3Cl_3O_2$.
4. This formula and the formula of chloral hydrate correspond exactly.
5. The index of hydrogen deficiency equals **zero.** Chloral hydrate cannot contain rings or double bonds.

Examination of the spectral results is limited to regions that correspond to singly bonded structural features. The correct structural formula for chloral hydrate follows. You can see that all of the bonds in the molecule are single bonds.

$$\underset{\displaystyle \overset{|}{OH}}{\overset{\displaystyle \overset{OH}{|}}{Cl_3C - C - H}}$$

1.5 THE RULE OF THIRTEEN

High-resolution mass spectrometry provides molecular mass information from which the user can determine the exact molecular formula directly. The discussion on exact mass determination in Chapter 8 will explain this process in detail. When such molar mass information is not available, however, it is often useful to be able to generate all the possible molecular formulas for a given mass. By applying other types of spectroscopic information, it may then be possible to distinguish among these possible formulas. A useful method for generating possible molecular formulas for a given molecular mass is the **Rule of Thirteen.**[1]

As a first step in the Rule of Thirteen, we generate a **base formula,** which contains only carbon and hydrogen. The base formula is found by dividing the molecular mass, M, by 13 (the mass of one carbon plus one hydrogen). This calculation provides a numerator, n, and a remainder, r.

$$\frac{M}{13} = n + \frac{r}{13}$$

The base formula thus becomes

$$C_nH_{n+r}$$

which is a combination of carbons and hydrogens that has the desired molecular mass, M.

The **index of hydrogen deficiency** (unsaturation index), U, that corresponds to the preceding formula is calculated easily by applying the relationship

$$U = \frac{(n - r + 2)}{2}$$

---

[1]Bright, J. W., and E. C. M. Chen, "Mass Spectral Interpretation Using the 'Rule of 13,'" *Journal of Chemical Education, 60* (1983): 557.

Of course, you can also calculate the index of hydrogen deficiency using the method shown in Section 1.4.

If we wish to derive a formula that includes other atoms besides carbon and hydrogen, then we must subtract the mass of a combination of carbons and hydrogens that equals the masses of the other atoms being included in the formula. For example, if we wish to convert the base formula to a new formula containing one oxygen atom, then we subtract one carbon and four hydrogens at the same time that we add one oxygen atom. Both changes involve a molecular mass equivalent of 16 ($O = CH_4 = 16$). Table 1.3 includes a number of C/H equivalents for replacement of carbon and hydrogen in the base formula by the most common elements likely to occur in an organic compound.[2]

To comprehend how the Rule of Thirteen might be applied, consider an unknown substance with a molecular mass of 94 amu. Application of the formula provides

$$\frac{94}{13} = 7 + \frac{3}{13}$$

According to the formula, $n = 7$ and $r = 3$. The base formula must be

$$C_7H_{10}$$

The index of hydrogen deficiency is

$$U = \frac{(7 - 3 + 2)}{2} = 3$$

TABLE 1.3
CARBON/HYDROGEN EQUIVALENTS FOR SOME COMMON ELEMENTS

Add Element	Subtract Equivalent	Add ΔU	Add Element	Subtract Equivalent	Add ΔU
C	H_{12}	7	^{35}Cl	C_2H_{11}	3
H_{12}	C	-7	^{79}Br	C_6H_7	-3
O	CH_4	1	^{79}Br	C_5H_{19}	4
O_2	C_2H_8	2	F	CH_7	2
O_3	C_3H_{12}	3	Si	C_2H_4	1
N	CH_2	$\frac{1}{2}$	P	C_2H_7	2
N_2	C_2H_4	1	I	C_9H_{19}	0
S	C_2H_8	2	I	$C_{10}H_7$	7

[2]In Table 1.3, the equivalents for chlorine and bromine are determined assuming that the isotopes are ^{35}Cl and ^{79}Br, respectively. Always use this assumption when applying this method.

A substance that fits this formula must contain some combination of three rings or multiple bonds. A possible structure might be

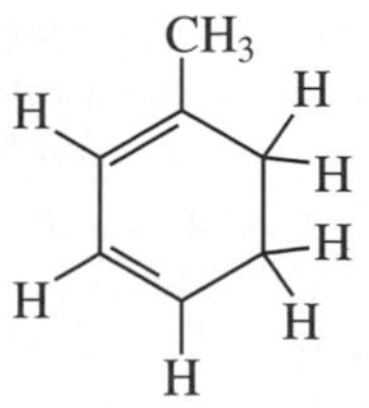

C_7H_{10}

$U = 3$

If we were interested in a substance that had the same molecular mass but that contained one oxygen atom, the molecular formula would become C_6H_6O. This formula is determined according to the following scheme.

1. Base formula $= C_7H_{10}$ $U = 3$

2. Add: $+ O$

3. Subtract: $- CH_4$

4. Change the value of U: $\Delta U = 1$

5. New formula $= C_6H_6O$

6. New index of hydrogen deficiency: $U = 4$

A possible substance that fits these data is

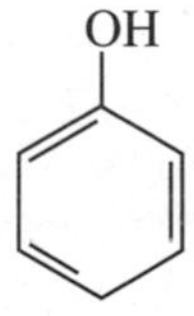

C_6H_6O

$U = 4$

There are additional possible molecular formulas that conform to a molecular mass of 94 amu:

$C_5H_2O_2$ $U = 5$ C_5H_2S $U = 5$

C_6H_8N $U = 3\frac{1}{2}$ CH_3Br $U = 0$

As the formula C_6H_8N shows, any formula that contains an even number of hydrogen atoms but an odd number of nitrogen atoms leads to a fractional value of U, an unlikely choice.

Any compound with a value of U less than zero (i.e., negative) is an impossible combination. Such a value is often an indicator that an oxygen or nitrogen atom must be present in the molecular formula.

When we calculate formulas using this method, if there are not enough hydrogens, we can subtract 1 carbon and add 12 hydrogens (and make the appropriate correction in U). This procedure works only if we obtain a positive value of U. Alternatively, if the value of U is greater than 7, we can obtain another potential molecular formula by adding 1 carbon and subtracting 12 hydrogens (and correcting U).

1.6 A QUICK LOOK AHEAD TO SIMPLE USES OF MASS SPECTRA

Chapter 8 contains a detailed discussion of the technique of **mass spectrometry.** See Sections 8.1–8.4 for applications of mass spectrometry to the problems of molecular formula determination. Briefly, the mass spectrometer is an instrument that subjects molecules to a high-energy beam of electrons. This beam of electrons converts molecules to positive ions by removing an electron. The stream of positively charged ions is accelerated along a curved path in a magnetic field. The radius of curvature of the path described by the ions depends upon the ratio of the mass of the ion to its charge (the e/m ratio). The ions strike a detector at positions that are determined by the radius of curvature of their paths. The number of ions with a particular mass-to-charge ratio is plotted as a function of that ratio.

The particle with the largest mass-to-charge ratio, assuming that the charge is 1, is the particle which represents the intact molecule with only one electron removed. This particle, called the **molecular ion** (see Chapter 8, Section 8.3), can be identified in the mass spectrum. From its position in the spectrum, its weight can be determined. Since the mass of the dislodged electron is so small, the mass of the molecular ion is essentially equal to the molecular mass of the original molecule. Thus, the mass spectrometer is an instrument capable of providing molecular mass information.

Virtually every element exists in nature in several isotopic forms. The natural abundance of each of these isotopes is known. Besides giving the mass of the molecular ion when each atom in the molecule is the most common isotope, the mass spectrum also gives peaks that correspond to that same molecule with heavier isotopes. The ratio of the intensity of the molecular ion peak to the intensities of the peaks corresponding to the heavier isotopes is determined by the natural abundance of each isotope. Because each type of molecule has a unique combination of atoms, and because each type of atom and its isotopes exist in a unique ratio in nature, the ratio of the intensity of the molecular ion peak to the intensities of the isotopic peaks can provide information about the number of each type of atom present in the molecule.

For example, the presence of bromine can be determined easily, because bromine causes a pattern of molecular ion peaks and isotope peaks that is easily identified. If we identify the mass of the molecular ion peak as M and the mass of the isotope peak that is two mass units heavier than the molecular ion as $M + 2$, then the ratio of the intensities of the M and $M + 2$ peaks will be approximately *one to one* when bromine is present (see Chapter 8, Section 8.5, for more details). When chlorine is present, the ratio of the intensities of the M and $M + 2$ peaks will be approximately *three to one*. These ratios reflect the natural abundances of the common isotopes of these elements. Thus, **isotope ratio studies** in mass spectrometry can be used to determine the molecular formula of a substance.

Another fact that can be used in determining the molecular formula is expressed as the **Nitrogen Rule.** This rule states that when the number of nitrogen atoms present in the molecule is odd, the

TABLE 1.4
PRECISE MASSES FOR SUBSTANCES OF MOLECULAR MASS EQUAL TO 44 AMU

Compound	Exact Mass (amu)
CO_2	43.9898
N_2O	44.0011
C_2H_4O	44.0262
C_3H_8	44.0626

molecular mass will be an odd number; when the number of nitrogen atoms present in the molecule is even (or zero), the molecular mass will be an even number. The Nitrogen Rule is explained further in Chapter 8, Section 8.4.

Since the advent of high-resolution mass spectrometers, it is also possible to use very precise mass determinations of molecular ion peaks to determine molecular formulas. When the atomic weights of the elements are determined very precisely, it is found that they do not have exactly integral values. Every isotopic mass is characterized by a small "mass defect," which is the amount by which the mass of the isotope differs from a perfectly integral mass number. The mass defect for every isotope of every element is unique. As a result, a precise mass determination can be used to determine the molecular formula of the sample substance, since every combination of atomic weights at a given nominal mass value will be unique when mass defects are considered. For example, each of the substances shown in Table 1.4 has a nominal mass of 44 amu. As can be seen from the table, their *exact masses*, obtained by adding exact atomic masses, are substantially different when measured to four decimal places.

PROBLEMS

*1. Researchers used a combustion method to analyze a compound used as an antiknock additive in gasoline. A 9.394-mg sample of the compound yielded 31.154 mg of carbon dioxide and 7.977 mg of water in the combustion.

   (a) Calculate the percentage composition of the compound.

   (b) Determine its empirical formula.

*2. The combustion of an 8.23-mg sample of unknown substance gave 9.62 mg CO_2 and 3.94 mg H_2O. Another sample, weighing 5.32 mg, gave 13.49 mg AgCl in a halogen analysis. Determine the percentage composition and empirical formula for this organic compound.

*3. An important amino acid has the percentage composition C 32.00%, H 6.71%, and N 18.66%. Calculate the empirical formula of this substance.

*4. A compound known to be a pain reliever had the empirical formula $C_9H_8O_4$. When a mixture of 5.02 mg of the unknown and 50.37 mg of camphor was prepared, the melting point of a portion of this mixture was determined. The observed melting point of the mixture was 156°C. What is the molecular mass of this substance?

*5. An unknown acid was titrated with 23.1 mL of 0.1 N sodium hydroxide. The weight of the acid was 120.8 mg. What is the equivalent weight of the acid?

*6. Determine the index of hydrogen deficiency for each of the following compounds.

   (a) C_8H_7NO (d) $C_5H_3ClN_4$

   (b) $C_3H_7NO_3$ (e) $C_{21}H_{22}N_2O_2$

   (c) $C_4H_4BrNO_2$

*7. A substance has the molecular formula C_4H_9N. Is there any likelihood that this material contains a triple bond? Explain your reasoning.

*8. (a) A researcher analyzed an unknown solid, extracted from the bark of spruce trees, to determine its percentage composition. An 11.32-mg sample was burned in a combustion apparatus. The carbon dioxide (24.87 mg) and water (5.82 mg) were collected and weighed. From the results of this analysis, calculate the percentage composition of the unknown solid.

   (b) Determine the empirical formula of the unknown solid.

(c) Through mass spectrometry, the molecular mass was found to be 420 g/mole. What is the molecular formula?

(d) How many aromatic rings could this compound contain?

*9. Calculate the molecular formulas for possible compounds with molecular masses of 136, using the Rule of Thirteen. You may assume that the only other atoms present in each molecule are carbon and hydrogen.

(a) A compound with two oxygen atoms

(b) A compound with two nitrogen atoms

(c) A compound with two nitrogen atoms and one oxygen atom

(d) A compound with five carbon atoms and four oxygen atoms

*10. An alkaloid was isolated from a common household beverage. The unknown alkaloid proved to have a molecular mass of 194. Using the Rule of Thirteen, determine a molecular formula and an index of hydrogen deficiency for the unknown. Alkaloids are naturally occurring organic substances that contain **nitrogen.** (*Hint:* There are four nitrogen atoms and two oxygen atoms in the molecular formula. The unknown is **caffeine.** Look up the structure of this substance in *The Merck Index* and confirm its molecular formula.)

*11. The Drug Enforcement Agency (DEA) confiscated a hallucinogenic substance during a drug raid. When the DEA chemists subjected the unknown hallucinogen to chemical analysis, they found that the substance had a molecular mass of 314. Elemental analysis revealed the presence of carbon and hydrogen only. Using the Rule of Thirteen, determine a molecular formula and an index of hydrogen deficiency for this substance. (*Hint:* The molecular formula of the unknown also contains two oxygen atoms. The unknown is **tetrahydrocannabinol,** the active constituent of marijuana. Look up the structure of tetrahydrocannabinol in *The Merck Index* and confirm its molecular formula.)

12. A carbohydrate was isolated from a sample of cow's milk. The substance was found to have a molecular mass of 342. The unknown carbohydrate can be hydrolyzed to form two isomeric compounds, each with a molecular mass of 180. Using the Rule of Thirteen, determine a molecular formula and an index of hydrogen deficiency for the unknown and for the hydrolysis products. (*Hint:* Begin by solving the molecular formula for the 180-amu hydrolysis products. These products have one oxygen atom for every carbon atom in the molecular formula. The unknown is **lactose.** Look up its structure in *The Merck Index* and confirm its molecular formula.)

---

*Answers are provided in the chapter, *Answers to Selected Problems*

REFERENCES

Budavari, S., ed. *The Merck Index,* 12th ed., Whitehouse Station, NJ: Merck & Co., 1996.

Pavia, D. L., Lampman, G. M., Kriz, G. S., and Engel, R. G., *Introduction to Organic Laboratory Techniques: A Small Scale Approach,* Philadelphia, PA: Saunders College Publishing, 1998.

Pavia, D. L., Lampman, G. M., Kriz, G. S., and Engel, R. G., *Introduction to Organic Laboratory Techniques: A Mi-* *croscale Approach,* 3rd ed., Philadelphia, PA: Saunders College Publishing, 1999.

Shriner, R. L., Hermann C. K. F., Morrill T. C., Curtin D. Y., and Fuson R. C., *The Systematic Identification of Organic Compounds,* 7th ed., New York, NY: John Wiley & Sons, 1998.

INFRARED SPECTROSCOPY

Almost any compound having covalent bonds, whether organic or inorganic, absorbs various frequencies of electromagnetic radiation in the infrared region of the electromagnetic spectrum. This region lies at wavelengths longer than those associated with visible light, which range from approximately 400 to 800 nm (1 nm $= 10^{-9}$ m), but lies at wavelengths shorter than those associated with microwaves, which are longer than 1 mm. For chemical purposes, we are interested in the **vibrational** portion of the infrared region. It includes radiation with wavelengths (λ) between 2.5 μm and 25 μm (1 μm $= 10^{-6}$ m). Although the more technically correct unit for wavelength in the infrared region of the spectrum is the micrometer (μm), you will often see the micron (μ) used on infrared spectra. Figure 2.1 illustrates the relationship of the infrared region to others included in the electromagnetic spectrum.

Figure 2.1 shows that the wavelength (λ) is inversely proportional to the frequency (ν) and is governed by the relationship $\nu = c/\lambda$, where c = speed of light. Also observe that the energy is directly proportional to the frequency: $E = h\nu$, where h = Planck's constant. From the latter equation, you can see qualitatively that the highest energy radiation corresponds to the X-ray region of the spectrum, where the energy may be great enough to break bonds in molecules. At the other end of the electromagnetic spectrum, radiofrequencies have very low energies, only enough to cause nuclear or electronic spin transitions within molecules—that is, nuclear magnetic resonance (NMR) or electron spin resonance (ESR), respectively.

Table 2.1 summarizes the regions of the spectrum and the types of energy transitions observed there. Several of these regions, including the infrared, give vital information about the structures of organic molecules. Nuclear magnetic resonance, which occurs in the radiofrequency part of the spectrum, is discussed in Chapters 3, 4, 5, 6 and 10 whereas ultraviolet and visible spectroscopy are described in Chapter 7.

Most chemists refer to the radiation in the vibrational infrared region of the electromagnetic spectrum in terms of a unit called a **wavenumber** ($\overline{\nu}$), rather than wavelength (μ or μm). Wavenumbers are expressed as reciprocal centimeters (cm^{-1}) and are easily computed by taking the reciprocal of

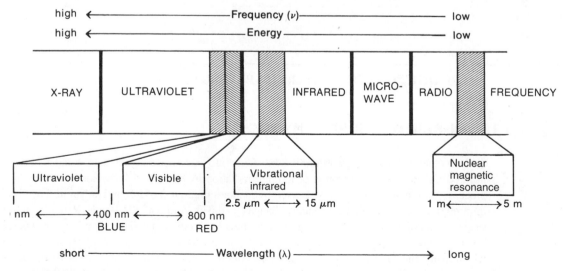

FIGURE 2.1 A portion of the electromagnetic spectrum, showing the relationship of the vibrational infrared to other types of radiation.

TABLE 2.1
TYPES OF ENERGY TRANSITIONS IN EACH REGION
OF THE ELECTROMAGNETIC SPECTRUM

Region of Spectrum	Energy Transitions
X-rays	Bond breaking
Ultraviolet/visible	Electronic
Infrared	Vibrational
Microwave	Rotational
Radiofrequencies	Nuclear spin (nuclear magnetic resonance)
	Electronic spin (electron spin resonance)

the wavelength expressed in centimeters. Convert a wavenumber to a frequency (ν) by multiplying it by the speed of light (expressed in centimeters per second).

$$\bar{\nu}\,(\text{cm}^{-1}) = \frac{1}{\lambda\,(\text{cm})} \qquad \nu\,(\text{Hz}) = \bar{\nu}c = \frac{c\,(\text{cm/sec})}{\lambda\,(\text{cm})}$$

The main reason chemists prefer to use wavenumbers as units is that they are directly proportional to energy (*a higher wavenumber corresponds to a higher energy*). Thus, in terms of wavenumbers, the vibrational infrared extends from about 4000 to 400 cm^{-1}. This range corresponds to wavelengths of 2.5 to 25 μm. *We will use wavenumber units exclusively in this textbook.* You may encounter wavelength values in older literature. Convert wavelengths (μ or μm) to wavenumbers (cm^{-1}) by using the following relationships.

$$\text{cm}^{-1} = \frac{1}{(\mu\text{m})} \times 10{,}000 \qquad \text{and} \qquad \mu\text{m} = \frac{1}{(\text{cm}^{-1})} \times 10{,}000$$

INTRODUCTION TO INFRARED SPECTROSCOPY

2.1 THE INFRARED ABSORPTION PROCESS

As with other types of energy absorption, molecules are excited to a higher energy state when they absorb infrared radiation. The absorption of infrared radiation is, like other absorption processes, a quantized process. A molecule absorbs only selected frequencies (energies) of infrared radiation. The absorption of infrared radiation corresponds to energy changes on the order of 8 to 40 kJ/mole. Radiation in this energy range corresponds to the range encompassing the stretching and bending vibrational frequencies of the bonds in most covalent molecules. In the absorption process, those frequencies of infrared radiation which match the natural vibrational frequencies of the molecule in question are absorbed, and the energy absorbed serves to increase the **amplitude** of the vibrational motions of the bonds in the molecule. Note, however, that not all bonds in a molecule are capable of absorbing infrared energy, even if the frequency of the radiation exactly matches that of the bond motion. Only those bonds that have a **dipole moment** that changes as a function of time are capable of absorbing infrared radiation. Symmetric bonds, such as those of H_2 or Cl_2, do not absorb infrared radiation. A bond must present an electrical dipole that is changing at the same frequency as the incoming radiation in order for energy to be transferred. The changing electrical dipole of the bond can then couple with the sinusoidally changing electromagnetic field of the incoming radiation. Thus, a symmetric bond which has identical or nearly identical groups on each end will not absorb in the infrared. For the purposes of an organic chemist, the bonds most likely to be affected by this restraint are those of symmetric or pseudosymmetric alkenes (C=C) and alkynes (C≡C).

$$CH_3\diagdown \atop CH_3 \diagup C=C \diagdown ^{CH_3} \atop _{CH_3} \qquad\qquad CH_3-CH_2 \diagdown \atop CH_3 \diagup C=C \diagdown ^{CH_3} \atop _{CH_3}$$

$$CH_3-C≡C-CH_3 \qquad\qquad CH_3-CH_2-C≡C-CH_3$$

Symmetric   Pseudosymmetric

2.2 USES OF THE INFRARED SPECTRUM

Since every type of bond has a different natural frequency of vibration, and since two of the same type of bond in two different compounds are in two slightly different environments, no two molecules of different structure have exactly the same infrared absorption pattern, or **infrared spectrum.** Although some of the frequencies absorbed in the two cases might be the same, in no case of two different molecules will their infrared spectra (the patterns of absorption) be identical. Thus, the infrared spectrum can be used for molecules much as a fingerprint can be used for humans. By comparing the infrared spectra of two substances thought to be identical, you can establish whether they are, in fact, identical. If their infrared spectra coincide peak for peak (absorption for absorption), in most cases the two substances will be identical.

A second and more important use of the infrared spectrum is to determine structural information about a molecule. The absorptions of each type of bond (N−H, C−H, O−H, C−X, C=O, C−O, C−C, C=C, C≡C, C≡N, and so on) are regularly found only in certain small portions of the vibrational infrared region. A small range of absorption can be defined for each type of bond. Outside this range, absorptions are normally due to some other type of bond. For instance, any absorption in the range 3000 ± 150 cm^{-1} is almost always due to the presence of a C−H bond in the molecule; an absorption in the range $1715 \pm$ 100 cm^{-1} is normally due to the presence of a C=O bond (carbonyl group) in the molecule. The same type of range applies to each type of bond. Figure 2.2 illustrates schematically how these are spread out over the vibrational infrared. Try to fix this general scheme in your mind for future convenience.

FREQUENCY (cm^{-1})

4000		2500	2000	1800	1650	1550	650
O−H	C−H	C≡C	VERY FEW BANDS	C=O	C=N		C−C1 C−O
		C≡N					C−N
N−H						C=C	C−C
		X=C=Y (C,O,N,S)				N=O N=O	

2.5	4	5	5.5	6.1	6.5	15.4

WAVELENGTH (μ)

FIGURE 2.2 The approximate regions where various common types of bonds absorb (stretching vibrations only; bending, twisting, and other types of bond vibrations have been omitted for clarity).

2.3 THE MODES OF STRETCHING AND BENDING

The simplest types, or **modes,** of vibrational motion in a molecule that are **infrared active**—that is, which give rise to absorptions—are the stretching and bending modes.

C—H

Stretching   Bending

However, other, more complex types of stretching and bending are also active. The following illustrations of the normal modes of vibration for a methylene group introduce several terms. In general, asymmetric stretching vibrations occur at higher frequencies than symmetric stretching vibrations; also, stretching vibrations occur at higher frequencies than bending vibrations. The terms **scissoring, rocking, wagging,** and **twisting** are commonly used in the literature to describe the origins of infrared bands.

In any group of three or more atoms, at least two of which are identical, there are *two* modes of stretching: symmetric and asymmetric. Examples of such groupings are $-CH_3$, $-CH_2-$ (see page 17), $-NO_2$, $-NH_2$, and anhydrides. The methyl group gives rise to a symmetric stretching vibration at about 2872 cm^{-1} and an asymmetric stretch at about 2962 cm^{-1}. The anhydride functional group gives two absorptions in the C=O region because of the asymmetric and symmetric modes of stretch. A similar phenomenon occurs in the amino group, where a primary amine (NH_2) usually has two absorptions in the N—H stretch region while a secondary amine (R_2NH) has one absorption peak. Amides exhibit similar bands. There are two strong N=O stretch peaks for a nitro group, with the symmetric stretch appearing at about 1350 cm^{-1} and the asymmetric stretch appearing at about 1550 cm^{-1}.

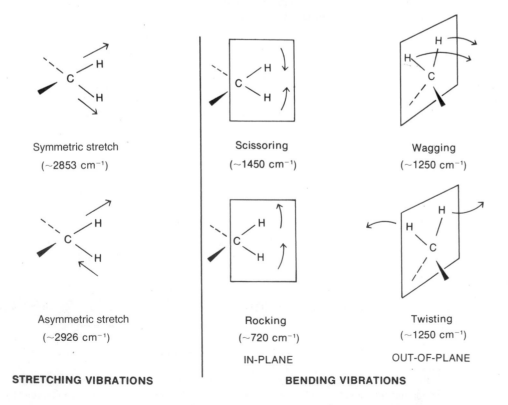

Symmetric stretch
(~2853 cm^{-1})

Scissoring
(~1450 cm^{-1})

Wagging
(~1250 cm^{-1})

Asymmetric stretch
(~2926 cm^{-1})

Rocking
(~720 cm^{-1})

Twisting
(~1250 cm^{-1})

IN-PLANE

OUT-OF-PLANE

STRETCHING VIBRATIONS   **BENDING VIBRATIONS**

	Symmetric Stretch	**Asymmetric Stretch**
Methyl		

~2872 cm⁻¹ ~2962 cm⁻¹

Anhydride

~1760 cm⁻¹ ~1800 cm⁻¹

Amino

~3300 cm⁻¹ ~3400 cm⁻¹

Nitro

~1350 cm⁻¹ ~1550 cm⁻¹

The vibrations we have been discussing are called **fundamental absorptions.** They arise from excitation from the ground state to the lowest-energy excited state. Usually the spectrum is complicated because of the presence of weak overtone, combination, and difference bands. **Overtones** result from excitation from the ground state to higher energy states, which correspond to integral multiples of the frequency of the fundamental (ν). For example, you might observe weak overtone bands at $2\overline{\nu}$, $3\overline{\nu}$, ... Any kind of physical vibration generates overtones. If you pluck a string on a cello, the string vibrates with a fundamental frequency. However, less intense vibrations are also set up at several overtone frequencies. An absorption in the infrared at 500 cm⁻¹ may well have an accompanying peak of lower intensity at 1000 cm⁻¹—an overtone.

When two vibrational frequencies ($\overline{\nu}_1$ and $\overline{\nu}_2$) in a molecule couple to give rise to a vibration of a new frequency within the molecule, and when such a vibration is infrared active, it is called a **combination band.** This band is the sum of the two interacting bands ($\overline{\nu}_{\text{comb}} = \overline{\nu}_1 + \overline{\nu}_2$). Not all possible combinations occur. The rules that govern which combinations are allowed are beyond the scope of our discussion here.

Difference bands are similar to combination bands. The observed frequency in this case results from the difference between the two interacting bands ($\nu_{\text{diff}} = \overline{\nu}_1 - \overline{\nu}_2$).

One can calculate overtone, combination, and difference bands by directly manipulating frequencies in wavenumbers via multiplication, addition, and subtraction, respectively. When a fundamental vibration couples with an overtone or combination band, the coupled vibration is called **Fermi resonance.** Again, only certain combinations are allowed. Fermi resonance is often observed in carbonyl compounds.

Although rotational frequencies of the whole molecule are not infrared active, they often couple with the stretching and bending vibrations in the molecule to give additional fine structure to these absorptions, thus further complicating the spectrum. One of the reasons a band is broad rather than sharp in the infrared spectrum is rotational coupling, which may lead to a considerable amount of unresolved fine structure.

2.4 BOND PROPERTIES AND ABSORPTION TRENDS

Let us now consider how bond strength and the masses of the bonded atoms affect the infrared absorption frequency. For the sake of simplicity, we will restrict the discussion to a simple heteronuclear diatomic molecule (two *different* atoms) and its stretching vibration.

A diatomic molecule can be considered as two vibrating masses connected by a spring. The bond distance continually changes, but an equilibrium or average bond distance can be defined. Whenever the spring is stretched or compressed beyond this equilibrium distance, the potential energy of the system increases.

As for any harmonic oscillator, when a bond vibrates, its energy of vibration is continually and periodically changing from kinetic to potential energy and back again. The total amount of energy is proportional to the frequency of the vibration,

$$E_{osc} \propto h\nu_{osc}$$

which for a harmonic oscillator is determined by the force constant (K) of the spring, or its stiffness, and the masses (m_1 and m_2) of the two bonded atoms. The natural frequency of vibration of a bond is given by the equation

$$\bar{\nu} = \frac{1}{2\pi c} \sqrt{\frac{K}{\mu}}$$

which is derived from Hooke's Law for vibrating springs. The **reduced mass,** μ, of the system is given by

$$\mu = \frac{m_1 m_2}{m_1 + m_2}$$

K is a constant that varies from one bond to another. As a first approximation, the force constants for triple bonds are three times those of single bonds, whereas the force constants for double bonds are twice those of single bonds.

Two things should be noticeable immediately. One is that stronger bonds have a larger force constant K and vibrate at higher frequencies than weaker bonds. The second is that bonds between atoms of higher masses (larger reduced mass, μ) vibrate at lower frequencies than bonds between lighter atoms.

In general, triple bonds are stronger than double or single bonds between the same two atoms and have higher frequencies of vibration (higher wavenumbers):

$C\equiv C$	$C=C$	$C-C$
2150 cm^{-1}	1650 cm^{-1}	1200 cm^{-1}

$$\longleftarrow$$
Increasing K

The C—H stretch occurs at about 3000 cm^{-1}. As the atom bonded to carbon increases in mass, the reduced mass (μ) increases, and the frequency of vibration decreases (wavenumbers get smaller):

$C-H$	$C-C$	$C-O$	$C-Cl$	$C-Br$	$C-I$
3000 cm^{-1}	1200 cm^{-1}	1100 cm^{-1}	750 cm^{-1}	600 cm^{-1}	500 cm^{-1}

$$\longrightarrow$$
Increasing μ

Bending motions occur at lower energy (lower frequency) than the typical stretching motions because of the lower value for the bending force constant K.

C—H stretching   C—H bending

~3000 cm^{-1} ~1340 cm^{-1}

Hybridization affects the force constant K, also. Bonds are stronger in the order $sp > sp^2 > sp^3$, and the observed frequencies of C—H vibration illustrate this nicely.

sp sp^2 sp^3

$\equiv$C—H   =C—H   —C—H

3300 cm^{-1} 3100 cm^{-1} 2900 cm^{-1}

Resonance also affects the strength and length of a bond and hence its force constant K. Thus, whereas a normal ketone has its C=O stretching vibration at 1715 cm^{-1}, a ketone that is conjugated with a C=C double bond absorbs at a lower frequency, near 1675 to 1680 cm^{-1}. That is because resonance lengthens the C=O bond distance and gives it more single-bond character:

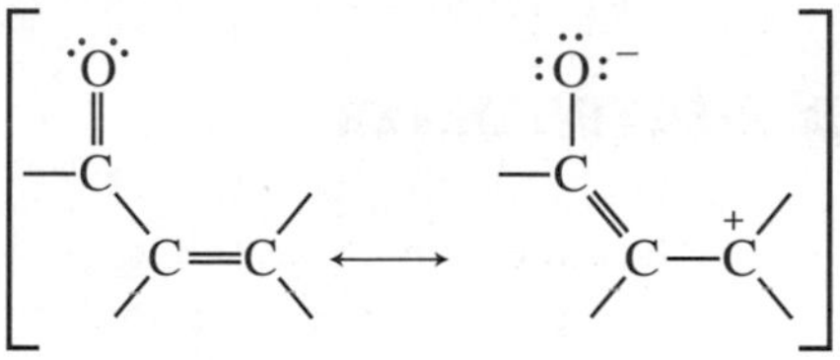

Resonance has the effect of reducing the force constant K, and the absorption moves to a lower frequency.

The Hooke's Law expression given earlier may be transformed into a very useful equation as follows:

$$\overline{v} = \frac{1}{2\pi c} \sqrt{\frac{K}{\mu}}$$

$\overline{v}$ = frequency in cm^{-1}

c = velocity of light = 3×10^{10} cm/sec

K = force constant in dynes/cm

$\mu = \dfrac{m_1 m_2}{m_1 + m_2}$,   masses of atoms in grams,

or $\dfrac{M_1 M_2}{(M_1 + M_2)(6.02 \times 10^{23})}$,   masses of atoms in amu

Removing Avogadro's number (6.02×10^{23}) from the denominator of the reduced mass expression (μ) by taking its square root, we obtain the expression

$$\overline{v} = \frac{7.76 \times 10^{11}}{2\pi c} \sqrt{\frac{K}{\mu}}$$

A new expression is obtained by inserting the actual values of π and c:

$$\overline{\nu}(\text{cm}^{-1}) = 4.12\sqrt{\frac{K}{\mu}}$$

$$\mu = \frac{M_1 M_2}{M_1 + M_2}, \quad \text{where } M_1 \text{ and } M_2 \text{ are atomic weights}$$

$$K = \text{force constant in dynes/cm [1 dyne} = 1.020 \times 10^{-3} \text{ g]}$$

This equation may be used to calculate the approximate position of a band in the infrared spectrum by assuming that K for single, double, and triple bonds is 5, 10, and 15×10^5 dynes/cm, respectively. Table 2.2 gives a few examples. Notice that excellent agreement is obtained with the experimental values given in the table. However, experimental and calculated values may vary considerably owing to resonance, hybridization, and other effects which operate in organic molecules. Nevertheless, good *qualitative* values are obtained by such calculations.

2.5 THE INFRARED SPECTROMETER

The instrument that determines the absorption spectrum for a compound is called an **infrared spectrometer** or, more precisely, a **spectrophotometer.** Two types of infrared spectrometers are in common use in the organic laboratory: dispersive and Fourier transform (FT) instruments. Both of these types of instruments provide spectra of compounds in the common range of 4000 to 400 cm^{-1}. Although the two provide nearly identical spectra for a given compound, FT infrared spectrometers provide the infrared spectrum much more rapidly than the dispersive instruments.

A.   Dispersive Infrared Spectrometers

Figure 2.3 schematically illustrates the components of a simple dispersive infrared spectrometer. The instrument produces a beam of infrared radiation from a hot wire and, by means of mirrors, divides it into two parallel beams of equal-intensity radiation. The sample is placed in one beam, and the other beam is used as a reference. The beams then pass into the **monochromator,** which disperses each into a continuous spectrum of frequencies of infrared light. The monochromator consists of a rapidly rotating sector (beam chopper) that passes the two beams alternately to a diffraction grating (a prism in older instruments). The slowly rotating diffraction grating varies the frequency or wavelength of radiation reaching the thermocouple detector. The detector senses the ratio between the intensities of the reference and sample beams. In this way the detector determines which frequencies have been absorbed by the sample and which frequencies are unaffected by the light passing through the sample. After the signal from the detector is amplified, the recorder draws the resulting spectrum of the sample on a chart. It is important to realize that the spectrum is recorded as the frequency of infrared radiation changes by rotation of the diffraction grating. Dispersive instruments are said to record a spectrum in the **frequency domain.**

TABLE 2.2
CALCULATION OF STRETCHING
FREQUENCIES FOR DIFFERENT TYPES
OF BONDS

C=C bond:

$$\bar{v} = 4.12\sqrt{\frac{K}{\mu}}$$

$K = 10 \times 10^5$ dynes/cm

$$\mu = \frac{M_C M_C}{M_C + M_C} = \frac{(12)(12)}{12 + 12} = 6$$

$$\bar{v} = 4.12\sqrt{\frac{10 \times 10^5}{6}} = 1682 \text{ cm}^{-1} \text{ (calculated)}$$

$\bar{v} = 1650 \text{ cm}^{-1}$ (experimental)

C—H bond:

$$\bar{v} = 4.12\sqrt{\frac{K}{\mu}}$$

$K = 5 \times 10^5$ dynes/cm

$$\mu = \frac{M_C M_H}{M_C + M_H} = \frac{(12)(1)}{12 + 1} = 0.923$$

$$\bar{v} = 4.12\sqrt{\frac{5 \times 10^5}{0.923}} = 3032 \text{ cm}^{-1} \text{ (calculated)}$$

$\bar{v} = 3000 \text{ cm}^{-1}$ (experimental)

C—D bond:

$$\bar{v} = 4.12\sqrt{\frac{K}{\mu}}$$

$K = 5 \times 10^5$ dynes/cm

$$\mu = \frac{M_C M_D}{M_C + M_D} = \frac{(12)(2)}{12 + 2} = 1.71$$

$$\bar{v} = 4.12\sqrt{\frac{5 \times 10^5}{1.71}} = 2228 \text{ cm}^{-1} \text{ (calculated)}$$

$\bar{v} = 2206 \text{ cm}^{-1}$ (experimental)

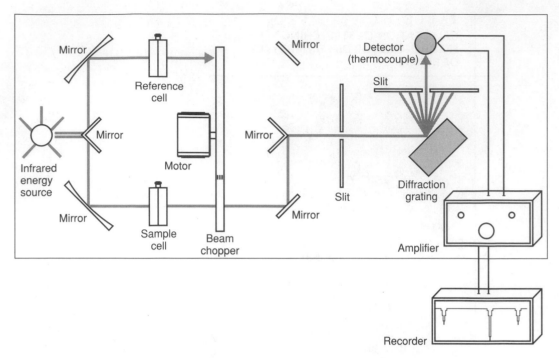

FIGURE 2.3 A schematic diagram of a dispersive infrared spectrometer.

Note that it is customary to plot frequency (wavenumber, cm^{-1}) versus light transmitted, not light absorbed. This is recorded as **percent transmittance (%T)** because the detector records the ratio of the intensities of the two beams, and

$$\text{percent transmittance} = \frac{I_s}{I_r} \times 100$$

where I_s is the intensity of the sample beam and I_r is the intensity of the reference beam. In many parts of the spectrum the transmittance is nearly 100%, meaning that the sample is nearly transparent to radiation of that frequency (does not absorb it). Maximum absorption is thus represented by a *minimum* on the chart. Even so, the absorption is traditionally called a **peak.**

The chemist often obtains the spectrum of a compound by dissolving it in a solvent (Section 2.6). The solution is then placed in the **sample beam** while pure solvent is placed in the **reference beam** in an identical cell. The instrument automatically "subtracts" the spectrum of the solvent from that of the sample. The instrument also cancels out the effects of the infrared-active atmospheric gases, carbon dioxide and water vapor, from the spectrum of the sample (they are present in both beams). This convenience feature is the reason most dispersive infrared spectrometers are double-beam (sample + reference) instruments that measure intensity ratios; since the solvent absorbs in both beams, it is in both terms of the ratio I_s / I_r and cancels out. If a pure liquid is analyzed (no solvent), the compound is placed in the sample beam and nothing is inserted into the reference beam. When the spectrum of the liquid is obtained, the effects of the atmospheric gases are automatically canceled since they are present in both beams.

B.   Fourier Transform Spectrometers

The most modern infrared spectrometers (spectrophotometers) operate on a different principle. The design of the optical pathway produces a pattern called an **interferogram.** The interferogram is a complex signal, but its wave-like pattern contains all the frequencies that make up the infrared spectrum. An interferogram is essentially a plot of intensity versus time (a **time-domain spectrum**). However, a chemist is more interested in a spectrum that is a plot of intensity versus frequency (a **frequency-domain spectrum**). A mathematical operation known as a **Fourier transform (FT)** can separate the individual absorption frequencies from the interferogram, producing a spectrum virtually identical to that obtained with a dispersive spectrometer. This type of instrument is known as a **Fourier transform infrared spectrometer,** or **FT-IR.**[1] The advantage of an FT-IR instrument is that it acquires the interferogram in less than a second. It is thus possible to collect dozens of interferograms of the same sample and accumulate them in the memory of a computer. When a Fourier transform is performed on the sum of the accumulated interferograms, a spectrum with a better signal-to-noise ratio can be plotted. An FT-IR instrument is therefore capable of greater speed and greater sensitivity than a dispersion instrument.

Computer-interfaced FT-IR instruments operate in a single-beam mode. To obtain a spectrum of a compound, the chemist first obtains an interferogram of the "background," which consists of the infrared-active atmospheric gases, carbon dioxide and water vapor (oxygen and nitrogen are not infrared active). The interferogram is subjected to a Fourier transform, which yields the spectrum of the background. Then the chemist places the compound (sample) into the beam and obtains the spectrum resulting from the Fourier transform of the interferogram. This spectrum contains absorption bands for *both the compound and the background*. The computer software automatically subtracts the spectrum of the background from the sample spectrum, yielding the spectrum of the compound being analyzed. The subtracted spectrum is essentially identical to that obtained from a traditional double-beam dispersive instrument.

2.6 PREPARATION OF SAMPLES FOR INFRARED SPECTROSCOPY

To determine the infrared spectrum of a compound, one must place the compound in a sample holder, or cell. In infrared spectroscopy this immediately poses a problem. Glass and plastics absorb strongly throughout the infrared region of the spectrum. Cells must be constructed of ionic substances—typically sodium chloride or potassium bromide. Potassium bromide plates are more expensive than sodium chloride plates and have the advantage of usefulness in the range of 4000 to 400 cm^{-1}. Sodium chloride plates are used widely because of their relatively low cost. The practical range for their use in spectroscopy extends from 4000 to 650 cm^{-1}. Sodium chloride begins to absorb at 650 cm^{-1}, and any bands with frequencies less than this value will not be observed. Since few important bands appear below 650 cm^{-1}, sodium chloride plates are in most common use for routine infrared spectroscopy.

---

[1] The principles of interferometry and the operation of an FT-IR instrument are explained in two articles by W. D. Perkins: "Fourier Transform–Infrared Spectroscopy, Part 1: Instrumentation," *Journal of Chemical Education, 63* (January 1986): A5–A10, and "Fourier Transform–Infrared Spectroscopy, Part 2: Advantages of FT-IR," *Journal of Chemical Education, 64* (November 1987): A269–271.

Liquids. A drop of a liquid organic compound is placed between a pair of polished sodium chloride or potassium bromide plates, referred to as **salt plates.** When the plates are squeezed gently, a thin liquid film forms between them. A spectrum determined by this method is referred to as a **neat** spectrum since no solvent is used. Salt plates break easily and are water soluble. Organic compounds analyzed by this technique must be free of water. The pair of plates is inserted into a holder which fits into the spectrometer.

Solids. There are at least three common methods for preparing a solid sample for spectroscopy. The first method involves mixing the finely ground solid sample with powdered potassium bromide and pressing the mixture under high pressure. Under pressure, the potassium bromide melts and seals the compound into a matrix. The result is a **KBr pellet** which can be inserted into a holder in the spectrometer. The main disadvantage of this method is that potassium bromide absorbs water, which may interfere with the spectrum that is obtained. If a good pellet is prepared, the spectrum obtained will have no interfering bands since potassium bromide is transparent down to 400 cm^{-1}.

The second method, a **Nujol mull,** involves grinding the compound with mineral oil (Nujol) to create a suspension of the finely ground sample dispersed in the mineral oil. The thick suspension is placed between salt plates. The main disadvantage of this method is that the mineral oil obscures bands that may be present in the analyzed compound. Nujol bands appear at 2924, 1462, and 1377 cm^{-1} (p. 30).

The third common method used with solids is to dissolve the organic compound in a solvent, most commonly carbon tetrachloride (CCl_4). Again, as was the case with mineral oil, some regions of the spectrum are obscured by bands in the solvent. Although it is possible to cancel out the solvent from the spectrum by computer or instrumental techniques, the region around 785 cm^{-1} is often obscured by the strong C—Cl stretch that occurs there.

2.7 WHAT TO LOOK FOR WHEN EXAMINING INFRARED SPECTRA

An infrared spectrometer determines the positions and relative sizes of all the absorptions, or peaks, in the infrared region and plots them on a piece of paper. This plot of absorption intensity versus wavenumber (or sometimes wavelength) is referred to as the **infrared spectrum** of the compound. Figure 2.4 shows a typical infrared spectrum, that of 3-methyl-2-butanone. The spectrum exhibits at least two strongly absorbing peaks at about 3000 and 1715 cm^{-1} for the C—H and C=O stretching frequencies, respectively.

The strong absorption at 1715 cm^{-1} that corresponds to the carbonyl group (C=O) is quite intense. In addition to the characteristic position of absorption, the *shape* and *intensity* of this peak are also unique to the C=O bond. This is true for almost every type of absorption peak; both shape and intensity characteristics can be described, and these characteristics often enable the chemist to distinguish the peak in potentially confusing situations. For instance, to some extent C=O and C=C bonds absorb in the same region of the infrared spectrum:

$$C=O \quad 1850-1630 \text{ cm}^{-1}$$

$$C=C \quad 1680-1620 \text{ cm}^{-1}$$

However, the C=O bond is a strong absorber, whereas the C=C bond generally absorbs only weakly (Fig. 2.5). Hence, trained observers would not interpret a strong peak at 1670 cm^{-1} to be a C=C double bond, nor would they interpret a weak absorption at this frequency to be due to a carbonyl group.

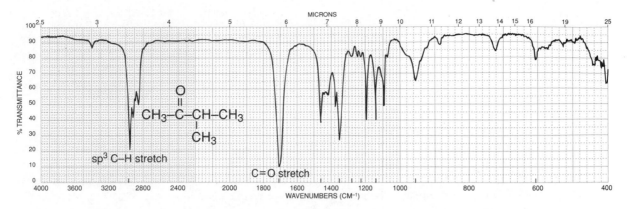

FIGURE 2.4 The infrared spectrum of 3-methyl-2-butanone (neat liquid, KBr plates).

The shape and fine structure of a peak often give clues to its identity, as well. Thus, although the N—H and O—H regions overlap,

$$O–H \qquad 3650–3200 \text{ cm}^{-1}$$

$$N–H \qquad 3500–3300 \text{ cm}^{-1}$$

the N—H absorption usually has one or two *sharp* absorption bands of lower intensity, whereas O—H, when it is in the N—H region, usually gives a *broad* absorption peak. Also, primary amines give *two* absorptions in this region, whereas alcohols as pure liquids give only one (Fig. 2.6). Figure 2.6 also shows typical patterns for the C—H stretching frequencies at about 3000 cm^{-1}.

Therefore, while you study the sample spectra in the pages that follow, take notice of shapes and intensities. They are as important as the frequency at which an absorption occurs, and the eye must be trained to recognize these features. Often, when reading the literature of organic chemistry, you will find absorptions referred to as strong (s), medium (m), weak (w), broad, or sharp. The author is trying to convey some idea of what the peak looks like without actually drawing the spectrum.

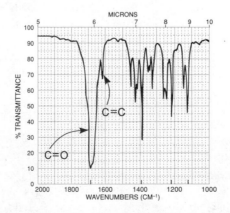

FIGURE 2.5 A comparison of the intensities of the C=O and C=C absorption bands.

TABLE 2.3
A SIMPLIFIED CORRELATION CHART

	Type of Vibration		Frequency (cm^{-1})	Intensity	Page Reference
C—H	Alkanes	(stretch)	3000–2850	s	29
	—CH$_3$	(bend)	1450 and 1375	m	
	—CH$_2$—	(bend)	1465	m	
	Alkenes	(stretch)	3100–3000	m	31
		(out-of-plane bend)	1000–650	s	
	Aromatics	(stretch)	3150–3050	s	41
		(out-of-plane bend)	900–690	s	
	Alkyne	(stretch)	ca. 3300	s	33
	Aldehyde		2900–2800	w	54
			2800–2700	w	
C—C	Alkane		Not interpretatively useful		
C=C	Alkene		1680–1600	m–w	31
	Aromatic		1600 and 1475	m–w	41
C≡C	Alkyne		2250–2100	m–w	33
C=O	Aldehyde		1740–1720	s	54
	Ketone		1725–1705	s	56
	Carboxylic acid		1725–1700	s	60
	Ester		1750–1730	s	62
	Amide		1680–1630	s	68
	Anhydride		1810 and 1760	s	71
	Acid chloride		1800	s	70
C—O	Alcohols, ethers, esters, carboxylic acids, anhydrides		1300–1000	s	45, 48, 60, 62, and 71
O—H	Alcohols, phenols				
	Free		3650–3600	m	47
	H-bonded		3400–3200	m	47
	Carboxylic acids		3400–2400	m	61
N—H	Primary and secondary amines and amides				
	(stretch)		3500–3100	m	72
	(bend)		1640–1550	m–s	72
C—N	Amines		1350–1000	m–s	72
C=N	Imines and oximes		1690–1640	w–s	75
C≡N	Nitriles		2260–2240	m	75
X=C=Y	Allenes, ketenes, isocyanates, isothiocyanates		2270–1940	m–s	75
N=O	Nitro (R—NO$_2$)		1550 and 1350	s	77
S—H	Mercaptans		2550	w	79
S=O	Sulfoxides		1050	s	79
	Sulfones, sulfonyl chlorides, sulfates, sulfonamides		1375–1300 and 1350–1140	s	80
C—X	Fluoride		1400–1000	s	83
	Chloride		785–540	s	83
	Bromide, iodide		<667	s	83

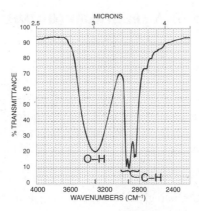
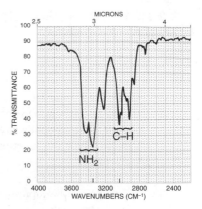

FIGURE 2.6 A comparison of the shapes of the absorption bands for the O—H and N—H groups.

2.8 CORRELATION CHARTS AND TABLES

To extract structural information from infrared spectra, you must be familiar with the frequencies at which various functional groups absorb. You may consult **infrared correlation tables,** which provide as much information as is known about where the various functional groups absorb. The references listed at the end of this chapter contain extensive series of correlation tables. Sometimes the absorption information is presented in the form of a chart called a **correlation chart.** Table 2.3 is a simplified correlation table; a more detailed chart appears in Appendix 1.

The mass of data in Table 2.3 looks as though it may be difficult to assimilate. However, it is really quite easy if you start simply and then slowly increase your familiarity with and ability to interpret the finer details of an infrared spectrum. You can do this most easily by first establishing the broad visual patterns of Figure 2.2 quite firmly in mind. Then, as a second step, memorize a "typical absorption value"—a single number that can be used as a pivotal value—for each of the functional groups in this pattern. For example, start with a simple aliphatic ketone as a model for all typical carbonyl compounds. The typical aliphatic ketone has a carbonyl absorption of about 1715 ± 10 cm^{-1}. Without worrying about the variation, memorize 1715 cm^{-1} as the base value for carbonyl absorption. Then, more slowly, familiarize yourself with the extent of the carbonyl range and the visual pattern showing where the different kinds of carbonyl groups appear throughout this region. See, for instance, Section 2.14 (p. 51), which gives typical values for the various types of carbonyl compounds. Also learn how factors such as ring size (when the functional group is contained in a ring) and conjugation affect the base values (i.e., in which direction the values are shifted). Learn the trends, always keeping the memorized base value (1715 cm^{-1}) in mind. As a beginning, it might prove useful to memorize the base values for this approach, given in Table 2.4. Notice that there are only eight of them.

TABLE 2.4
BASE VALUES FOR ABSORPTIONS OF BONDS

O—H	3400 cm^{-1}	C≡C	2150 cm^{-1}
N—H	3400	C=O	1715
C—H	3000	C=C	1650
C≡N	2250	C—O	1100

2.9 HOW TO APPROACH THE ANALYSIS OF A SPECTRUM (OR WHAT YOU CAN TELL AT A GLANCE)

When analyzing the spectrum of an unknown, concentrate your first efforts on determining the presence (or absence) of a few major functional groups. The C=O, O—H, N—H, C—O, C=C, C≡C, C≡N, and NO_2 peaks are the most conspicuous and give immediate structural information if they are present. Do not try to make a detailed analysis of the C—H absorptions near 3000 cm^{-1}; almost all compounds have these absorptions. Do not worry about subtleties of the exact environment in which the functional group is found. Following is a major checklist of the important gross features.

1. Is a carbonyl group present? The C=O group gives rise to a strong absorption in the region 1820–1660 cm^{-1}. The peak is often the strongest in the spectrum and of medium width. You can't miss it.

2. If C=O is present, check the following types (if it is absent, go to 3).

ACIDS	Is O—H also present?
	• *Broad* absorption near 3400–2400 cm^{-1} (usually overlaps C—H).
AMIDES	Is N—H also present?
	• Medium absorption near 3400 cm^{-1}; sometimes a double peak with equivalent halves.
ESTERS	Is C—O also present?
	• Strong-intensity absorptions near 1300–1000 cm^{-1}.
ANHYDRIDES	*Two* C=O absorptions near 1810 and 1760 cm^{-1}.
ALDEHYDES	Is aldehyde C—H present?
	• Two weak absorptions near 2850 and 2750 cm^{-1} on right side of the aliphatic C—H absorptions.
KETONES	The preceding five choices have been eliminated.

3. If C=O is absent:

ALCOHOLS, PHENOLS	Check for O—H.
	• *Broad* absorption near 3400–3300 cm^{-1}.
	• Confirm this by finding C—O near 1300–1000 cm^{-1}.
AMINES	Check for N—H.
	• Medium absorption(s) near 3400 cm^{-1}.
ETHERS	Check for C—O near 1300–1000 cm^{-1} (and absence of O—H near 3400 cm^{-1}).

4. Double bonds and/or aromatic rings

    • C=C is a weak absorption near 1650 cm.

    • Medium to strong absorptions in the region 1600–1450 cm^{-1}; these often imply an aromatic ring.

    • Confirm the double bond or aromatic ring by consulting the C—H region; aromatic and vinyl C—H occurs to the left of 3000 cm^{-1} (aliphatic C—H occurs to the right of this value).

5. Triple bonds

- $C \equiv N$ is a medium, sharp absorption near 2250 cm^{-1}.
- $C \equiv C$ is a weak, sharp absorption near 2150 cm^{-1}.
- Check also for acetylenic C—H near 3300 cm^{-1}.

6. Nitro groups

- Two strong absorptions at 1600–1530 cm^{-1} and 1390–1300 cm^{-1}.

7. Hydrocarbons

- None of the preceding is found.
- Major absorptions are in C—H region near 3000^{-1}.
- Very simple spectrum; the only other absorptions appear near 1460 and 1375 cm^{-1}.

The beginning student should resist the idea of trying to assign or interpret *every* peak in the spectrum. You simply will not be able to do it. Concentrate first on learning these *major* peaks and recognizing their presence or absence. This is best done by carefully studying the illustrative spectra in the sections that follow.

A SURVEY OF THE IMPORTANT FUNCTIONAL GROUPS, WITH EXAMPLES

The following sections describe the behaviors of important functional groups toward infrared radiation. These sections are organized as follows.

1. The *basic* information about the functional group or type of vibration is abstracted and placed in a **Spectral Analysis Box,** where it may be consulted easily.

2. Examples of spectra follow the basic section. The *major* absorptions of diagnostic value are indicated on each spectrum.

3. Following the spectral examples, a discussion section provides details about the functional groups and other information that may be of use in identifying organic compounds. You may choose to omit this section on your first reading of the material.

2.10 HYDROCARBONS: ALKANES, ALKENES, AND ALKYNES

A. Alkanes

Alkanes show very few absorption bands in the infrared spectrum. They yield four or more C—H stretching peaks near 3000 cm^{-1} plus CH$_2$ and CH$_3$ bending peaks in the range 1475–1365 cm^{-1}.

SPECTRAL ANALYSIS BOX

ALKANES

The spectrum is usually simple, with few peaks.

C—H   Stretch occurs around 3000 cm^{-1}.

In alkanes (except strained ring compounds), sp^3 C—H absorption always occurs at frequencies less than 3000 cm^{-1} (3000–2840 cm^{-1}).

If a compound has vinylic, aromatic, acetylenic, or cyclopropyl hydrogens, the C—H absorption is greater than 3000 cm^{-1}. These compounds have sp^2 and sp hybridizations (see Sections 2.10B and 2.10C).

CH_2 Methylene groups have a characteristic bending absorption of approximately 1465 cm^{-1}.

CH_3 Methyl groups have a characteristic bending absorption of approximately 1375 cm^{-1}.

CH_2 The bending (rocking) motion associated with four or more CH_2 groups in an open chain occurs at about 720 cm^{-1} (called a long-chain band).

C—C   Stretch not interpretatively useful; many weak peaks.

Examples: decane (Fig. 2.7), mineral oil (Fig. 2.8), and cyclohexane (Fig. 2.9).

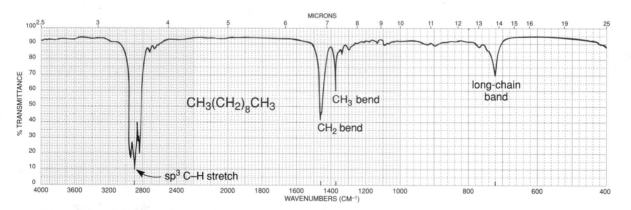

FIGURE 2.7 The infrared spectrum of decane (neat liquid, KBr plates).

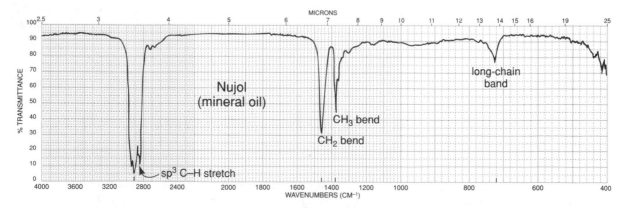

FIGURE 2.8 The infrared spectrum of mineral oil (neat liquid, KBr plates).

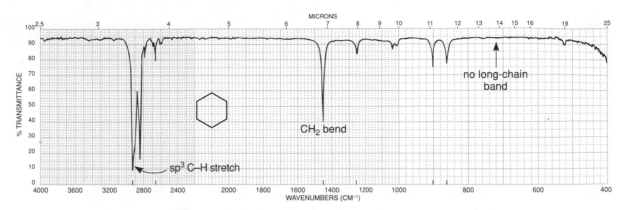

F I G U R E 2 . 9 The infrared spectrum of cyclohexane (neat liquid, KBr plates).

B. Alkenes

Alkenes show many more peaks than alkanes. The principal peaks of diagnostic value are the C—H stretching peaks for the sp^2 carbon at values greater than 3000 cm^{-1}, along with C—H peaks for the sp^3 carbon atoms appearing below that value. Also prominent are the out-of-plane bending peaks that appear in the range 1000–650 cm^{-1}. For unsymmetrical compounds, you should expect to see the C=C stretching peak near 1650 cm^{-1}.

SPECTRAL ANALYSIS BOX

ALKENES

=C—H   Stretch for sp^2 C—H occurs at values greater than 3000 cm^{-1}.

(3095–3010 cm^{-1})

=C—H   Out-of-plane (oop) bending occurs in the range 1000–650 cm^{-1}.

These bands can be used to determine the degree of substitution on the double bond (see discussion).

C=C   Stretch occurs at 1660–1600 cm^{-1}; often conjugation moves C=C stretch to lower frequencies and increases the intensity.

Symmetrically substituted bonds (e.g., 2,3-dimethyl-2-butene) do not absorb in the infrared (no dipole change).

Symmetrically disubstituted (*trans*) double bonds are often vanishingly weak in absorption; *cis* are stronger.

Examples: 1-hexene (Fig. 2.10), cyclohexene (Fig. 2.11), *cis*-2-pentene (Fig. 2.12), and *trans*-2-pentene (Fig. 2.13).

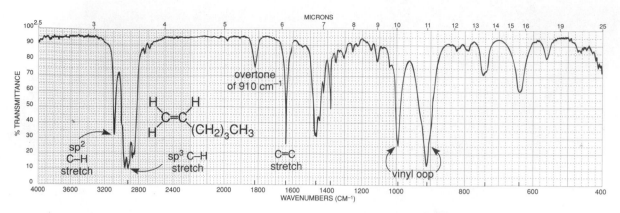

FIGURE 2.10 The infrared spectrum of 1-hexene (neat liquid, KBr plates).

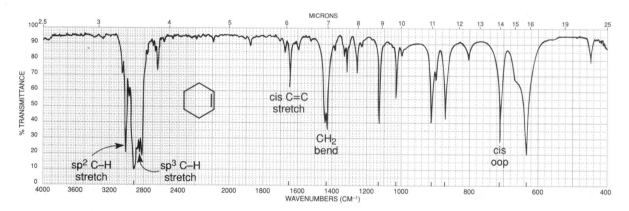

FIGURE 2.11 The infrared spectrum of cyclohexene (neat liquid, KBr plates).

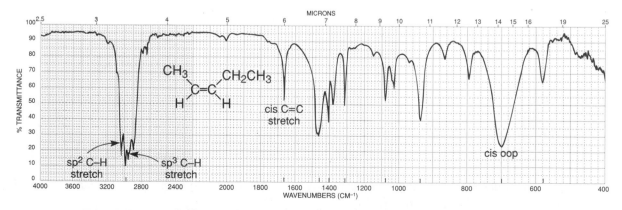

FIGURE 2.12 The infrared spectrum of *cis*-2-pentene (neat liquid, KBr plates).

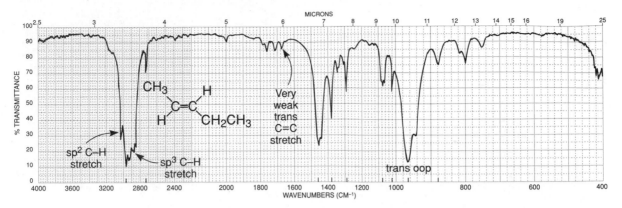

FIGURE 2.13 The infrared spectrum of *trans*-2-pentene (neat liquid, KBr plates).

C. Alkynes

Terminal alkynes will show a prominent peak at about 3300 cm^{-1} for the *sp*-hybridized C—H. A C≡C will also be a prominent feature in the spectrum for the terminal alkyne, appearing at about 2150 cm^{-1}. The alkyl chain will show C—H stretching frequencies for the *sp^3* carbon atoms. Other features include the bending bands for CH$_2$ and CH$_3$ groups. Nonterminal alkynes will not show the C—H band at 3300 cm^{-1}. The C≡C at 2150 cm^{-1} will be very weak or absent from the spectrum.

SPECTRAL ANALYSIS BOX

ALKYNES

≡C—H   Stretch for *sp* C—H usually occurs near 3300 cm^{-1}.

C≡C   Stretch occurs near 2150 cm^{-1}; conjugation moves stretch to lower frequency. Disubstituted or symmetrically substituted triple bonds give either no absorption or weak absorption.

Examples: 1-octyne (Fig. 2.14) and 4-octyne (Fig. 2.15).

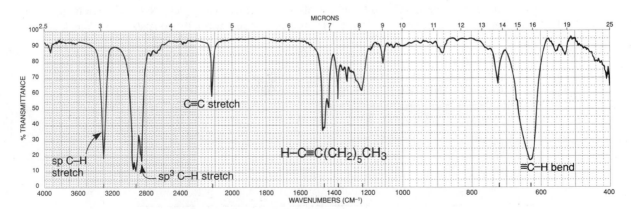

FIGURE 2.14 The infrared spectrum of 1-octyne (neat liquid, KBr plates).

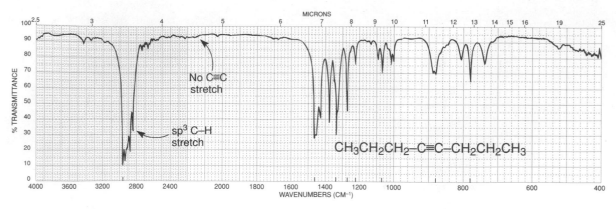

FIGURE 2.15 The infrared spectrum of 4-octyne (neat liquid, KBr plates).

DISCUSSION SECTION

C—H Stretch Region

The C—H stretching and bending regions are two of the most difficult regions to interpret in infrared spectra. The C—H stretching region, which ranges from 3300 to 2750 cm^{-1}, is generally the more useful of the two. As discussed in Section 2.4, the frequency of the absorption of C—H bonds is a function mostly of the type of hybridization that is attributed to the bond. The sp-$1s$ C—H bond present in acetylenic compounds is stronger than the sp^2-$1s$ bond present in C=C double-bond compounds (vinyl compounds). This strength results in a larger vibrational force constant and a higher frequency of vibration. Likewise, the sp^2-$1s$ C—H absorption in vinyl compounds occurs at a higher frequency than the sp^3-$1s$ C—H absorption in saturated aliphatic compounds. Table 2.5 gives some physical constants for various C—H bonds involving sp-, sp^2-, and sp^3-hybridized carbon.

As Table 2.5 demonstrates, the frequency at which the C—H absorption occurs indicates the type of carbon to which the hydrogen is attached. Figure 2.16 shows the entire C—H stretching region. Except for the aldehyde hydrogen, an absorption frequency of less than 3000 cm^{-1} usually implies a saturated compound (only sp^3-$1s$ hydrogens). An absorption frequency higher than 3000 cm^{-1} but not above about 3150 cm^{-1} usually implies aromatic or vinyl hydrogens. However, cyclopropyl C—H bonds, which have extra s character because of the need to put more p character into the ring C—C bonds to reduce angle distortion, also give rise to absorption in the region of 3100 cm^{-1}. Cyclopropyl hydrogens can easily be distinguished from aromatic hydrogens or vinyl hydrogens by cross-reference to the C=C and C—H out-of-plane regions. The aldehyde C—H stretch appears at lower frequencies than the saturated C—H absorptions and normally consists of two weak absorp-

TABLE 2.5
PHYSICAL CONSTANTS FOR sp-, sp^2-, AND sp^3-HYBRIDIZED CARBON AND THE RESULTING C—H ABSORPTION VALUES

Bond	≡C—H	=C—H	—C—H
Type	sp-$1s$	sp^2-$1s$	sp^3-$1s$
Length	1.08 Å	1.10 Å	1.12 Å
Strength	506 kJ	444 kJ	422 kJ
IR frequency	3300 cm^{-1}	~3100 cm^{-1}	~2900 cm^{-1}

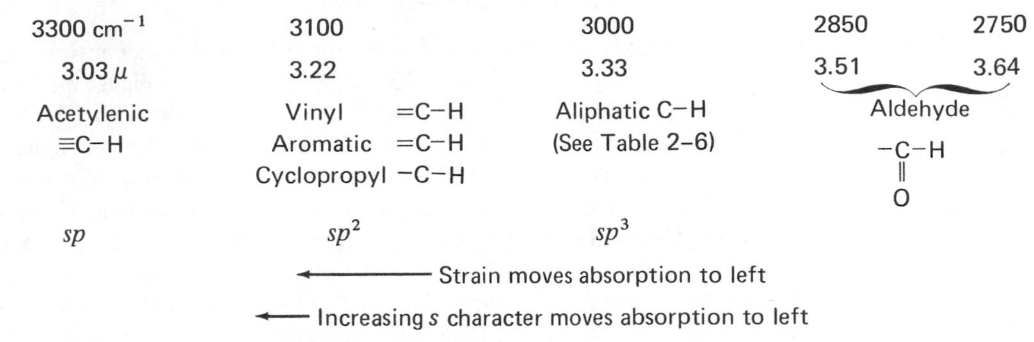

3300 cm^{-1}	3100	3000	2850	2750
3.03 μ	3.22	3.33	3.51	3.64
Acetylenic	Vinyl =C—H	Aliphatic C—H	Aldehyde	
≡C—H	Aromatic =C—H	(See Table 2-6)		
	Cyclopropyl —C—H		$-\overset{\displaystyle\parallel}{\underset{\displaystyle O}{C}}-H$	
sp	sp^2	sp^3		

———— Strain moves absorption to left

←———— Increasing s character moves absorption to left

FIGURE 2.16 The C—H stretch region.

tions at about 2850 and 2750 cm^{-1}. The 2850-cm^{-1} band usually appears as a shoulder on the saturated C—H absorption bands. The band at 2750 cm^{-1} is rather weak and may be missed in an examination of the spectrum. However, it appears at lower frequencies than aliphatic sp^3 C—H bands. If you are attempting to identify an aldehyde, look for this pair of weak but very diagnostic bands for the aldehyde C—H stretch.

Table 2.6 lists the sp^3-hybridized C—H stretching vibrations for methyl, methylene, and methine. The tertiary C—H (methine hydrogen) gives only one weak C—H stretch absorption, usually near 2890 cm^{-1}. Methylene hydrogens (—CH$_2$—), however, give rise to two C—H stretching bands, representing the symmetric (sym) and asymmetric (asym) stretching modes of the group. In effect, the 2890-cm^{-1} methine absorption is split into two bands at 2926 cm^{-1} (asym) and 2853 cm^{-1} (sym). The asymmetric mode generates a larger dipole moment and is of greater intensity than the symmetric mode. The splitting of the 2890-cm^{-1} methine absorption is larger in the case of a methyl group. Peaks appear at about 2962 and 2872 cm^{-1}. Section 2.3 showed the asymmetric and symmetric stretching modes for methylene and methyl.

Since several bands may appear in the C—H stretch region, it is probably a good idea to decide only whether the absorptions are acetylenic (3300 cm^{-1}), vinylic or aromatic (>3000 cm^{-1}), aliphatic (<3000 cm^{-1}), or aldehydic (2850 and 2750 cm^{-1}). Further interpretation of C—H stretching vibrations may not be worth extended effort. The C—H *bending vibrations* are often more useful for determining whether methyl or methylene groups are present in a molecule.

TABLE 2.6
STRETCHING VIBRATIONS FOR VARIOUS sp^3-HYBRIDIZED C—H BONDS

Group		Stretching Vibration (cm^{-1})	
		Asymmetric	*Symmetric*
Methyl	CH$_3$—	2962	2872
Methylene	—CH$_2$—	2926	2853
Methine	$-\overset{\displaystyle\mid}{\underset{\displaystyle H}{C}}-$	2890 Very weak	

C—H Bending Vibrations for Methyl and Methylene

The presence of methyl and methylene groups, when not obscured by other absorptions, may be determined by analyzing the region from 1465 to 1370 cm^{-1}. As shown in Figure 2.17, the band due to CH$_2$ scissoring usually occurs at 1465 cm^{-1}. One of the bending modes for CH$_3$ usually absorbs strongly near 1375 cm^{-1}. These two bands can often be used to detect methylene and methyl groups, respectively. Furthermore, the 1375-cm^{-1} methyl band is usually split into *two* peaks of nearly equal intensity (symmetric and asymmetric modes) if a geminal dimethyl group is present. This doublet is often observed in compounds with isopropyl groups. A *tert*-butyl group results in an even wider splitting of the 1375-cm^{-1} band into two peaks. The 1370-cm^{-1} band is more intense than the 1390-cm^{-1} one. Figure 2.18 shows the expected patterns for the isopropyl and *tert*-butyl groups. Note that some variation from these idealized patterns may occur. Nuclear magnetic resonance spectroscopy may be used to confirm the presence of these groups. In cyclic hydrocarbons, which do not have attached methyl groups, the 1375-cm^{-1} band is missing, as can be seen in the spectrum of cyclohexane (Fig. 2.9). Finally, a rocking band (Section 2.3) appears near 720 cm^{-1} for long-chain alkanes of four carbons or more (see Fig. 2.7).

C=C Stretching Vibrations

Simple Alkyl-Substituted Alkenes. The C=C stretching frequency usually appears between 1670 and 1640 cm^{-1} for simple noncyclic (acyclic) alkenes. The C=C frequencies increase as alkyl groups are added to a double bond. For example, simple monosubstituted alkenes yield values near 1640 cm^{-1}, 1,1-disubstituted alkenes absorb at about 1650 cm^{-1}, and tri- and tetrasubstituted alkenes absorb near 1670 cm^{-1}. *trans*-Disubstituted alkenes absorb at higher frequencies

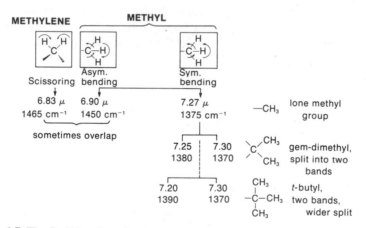

FIGURE 2.17 The C—H bending vibrations for methyl and methylene groups.

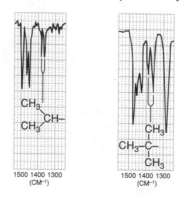

FIGURE 2.18 C—H bending patterns for the isopropyl and *tert*-butyl groups.

(1670 cm^{-1}) than *cis*-disubstituted alkenes (1658 cm^{-1}). Unfortunately, the C=C group has a rather weak intensity, certainly much weaker than a typical C=O group. In many cases, such as in tetrasubstituted alkenes, the double bond may be so weak that it is not observed at all. Recall from Section 2.1 that if the attached groups are arranged symmetrically, no change in dipole moment occurs during stretching, and hence no infrared absorption is observed. *cis*-Alkenes, which have less symmetry than *trans*-alkenes, generally absorb more strongly than the latter. Double bonds in rings, because they are often symmetric or nearly so, absorb more weakly than those not contained in rings. Terminal double bonds in monosubstituted alkenes generally have stronger absorption.

Conjugation Effects. Conjugation of a C=C double bond with either a carbonyl group or another double bond provides the multiple bond with more single-bond character (through resonance, as the following example shows), a lower force constant K, and thus a lower frequency of vibration. For example, the vinyl double bond in styrene gives an absorption at 1630 cm^{-1}.

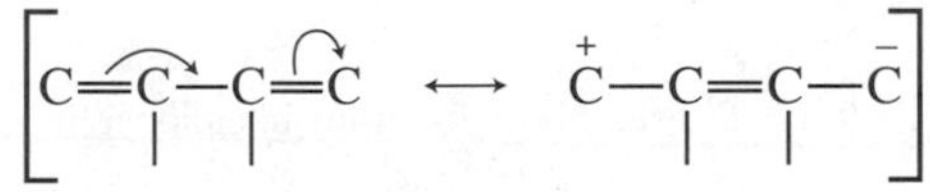

With several double bonds, the number of C=C absorptions often corresponds to the number of conjugated double bonds. An example of this correspondence is found in 1,3-pentadiene, where absorptions are observed at 1600 cm^{-1} and 1650 cm^{-1}. In the exception to the rule, butadiene gives only one band near 1600 cm^{-1}. If the double bond is conjugated with a carbonyl group, the C=C absorption shifts to a lower frequency and is also intensified by the strong dipole of the carbonyl group. Often, two closely spaced C=C absorption peaks are observed for these conjugated systems resulting from two possible conformations.

Ring-Size Effects with Internal Double Bonds. The absorption frequency of *internal (endo)* double bonds in cyclic compounds is very sensitive to ring size. As shown in Figure 2.19, the absorption frequency decreases as the internal angle decreases, until it reaches a minimum at 90° in cyclobutene. The frequency increases again for cyclopropene when the angle drops to 60°. This initially unexpected increase in frequency occurs because the C=C vibration in cyclopropene is strongly coupled to the attached C—C single-bond vibration. When the attached C—C bonds are perpendicular to the C=C axis, as in cyclobutene, their vibrational mode is orthogonal to that of the C=C bond (i.e., on a different axis) and does not couple. When the angle is greater than 90° (120° in the following example), the C—C single-bond stretching vibration can be resolved into two components, one of which is coincident with the direction of the C=C stretch. In the diagram, components **a** and **b** of the C—C stretching vector are shown. Since component **a** is in line with the C=C stretching vector, the C—C and C=C bonds are coupled, leading to a higher frequency of absorption. A similar pattern exists for cyclopropene, which has an angle less than 90°.

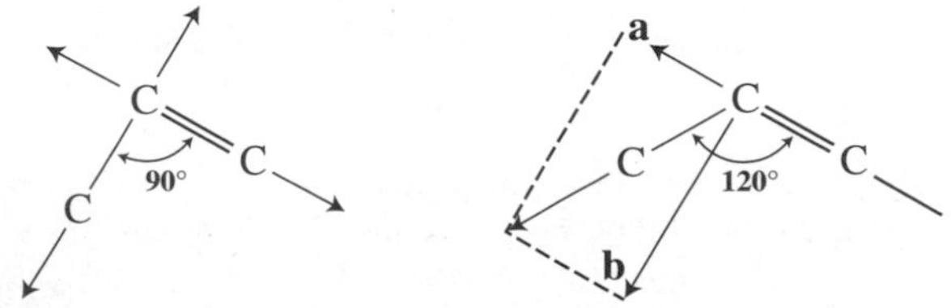

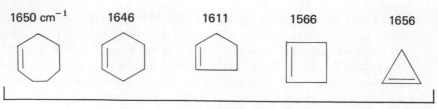

1650 cm^{-1} 1646   1611   1566   1656

Endo double bonds

(a) Strain moves the peak to the right.
    Anomaly: Cyclopropene

(b) If an endo double bond is at a ring fusion, the absorption moves to the right an amount equivalent to the change that would occur if one carbon were removed from the ring.

e.g.:   ~1611 cm^{-1}

FIGURE 2.19 C=C stretching vibrations in endocyclic systems.

Significant increases in the frequency of the absorption of a double bond contained in a ring are observed when one or two alkyl groups are attached directly to the double bond. The increases are most dramatic for small rings, especially cyclopropenes. For example, Figure 2.20 shows that the base value of 1656 cm^{-1} for cyclopropene increases to about 1788 cm^{-1} when one alkyl group is attached to the double bond; with two alkyl groups the value increases to

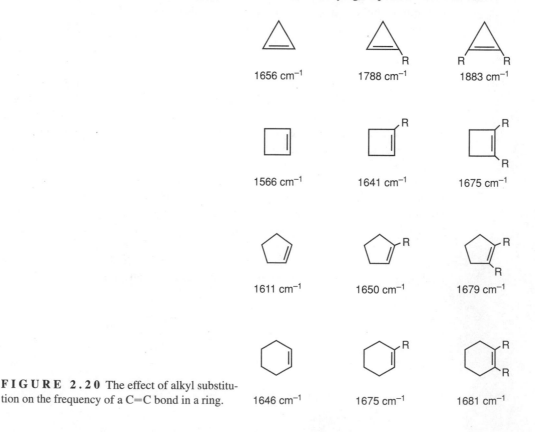

1656 cm^{-1} 1788 cm^{-1} 1883 cm^{-1}

1566 cm^{-1} 1641 cm^{-1} 1675 cm^{-1}

1611 cm^{-1} 1650 cm^{-1} 1679 cm^{-1}

1646 cm^{-1} 1675 cm^{-1} 1681 cm^{-1}

FIGURE 2.20 The effect of alkyl substitution on the frequency of a C=C bond in a ring.

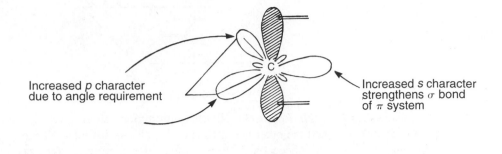

Increased *p* character
due to angle requirement

Increased *s* character
strengthens σ bond
of π system

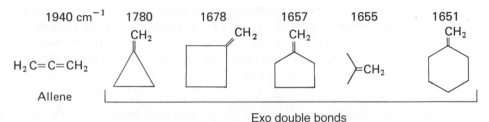

1940 cm^{-1} 1780   1678   1657   1655   1651

H$_2$C=C=CH$_2$

Allene

Exo double bonds

(a) Strain moves the peak to the left.

(b) Ring fusion moves the absorption to the left.

FIGURE 2.21 C=C stretching vibrations in exocyclic systems.

about 1883 cm^{-1}. The figure shows additional examples. It is important to realize that the ring size must be determined before the illustrated rules are applied. Notice, for example, that the double bonds in the 1,2-dialkylcyclopentene and 1,2-dialkylcyclohexene absorb at nearly the same value.

Ring-Size Effects with External Double Bonds. *External (exo)* double bonds give an increase in absorption frequency with decreasing ring size, as shown in Figure 2.21. Allene is included in the figure because it is an extreme example of an *exo* double-bond absorption. Smaller rings require the use of more *p* character to make the C—C bonds form the requisite small angles (recall the trend: $sp = 180°$, $sp^2 = 120°$, $sp^3 = 109°$, $sp^{>3} = <109°$). This removes *p* character from the sigma bond of the double bond but gives it more *s* character, thus strengthening and stiffening the double bond. The force constant K is then increased, and the absorption frequency increases.

C—H Bending Vibrations for Alkenes

The C—H bonds in alkenes can vibrate by bending both in-plane and out-of-plane when they absorb infrared radiation. The scissoring in-plane vibration for terminal alkenes occurs at about 1415 cm^{-1}. This band appears at this value as a medium to weak absorption for both monosubstituted and 1,1-disubstituted alkenes.

The most valuable information for alkenes is obtained from analysis of the C—H out-of-plane region of the spectrum, which extends from 1000 to 650 cm^{-1}. These bands are frequently the strongest peaks in the spectrum. The number of absorptions and their positions in the spectrum can be used to indicate the substitution pattern on the double bond.

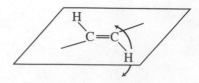

C—H out-of-plane bending

Monosubstituted Double Bonds (Vinyl). This substitution pattern gives rise to two strong bands, one near 990 cm^{-1} and the other near 910 cm^{-1} for alkyl-substituted alkenes. An overtone of the 910-cm^{-1} band usually appears at 1820 cm^{-1} and helps confirm the presence of the vinyl group. The 910-cm^{-1} band is shifted to a lower frequency, as low as 810 cm^{-1}, when a group attached to the double bond can release electrons by a resonance effect (Cl, F, OR). The 910-cm^{-1} group shifts to a higher frequency, as high as 960, when the group withdraws electrons by a resonance effect (C=O, C≡N). The use of the out-of-plane vibrations to confirm the monosubstituted structure is considered very reliable. The absence of these bands almost certainly indicates that this structural feature is not present within the molecule.

cis- and trans-1,2-Disubstituted Double Bonds. A *cis* arrangement about a double bond gives one strong band near 700 cm^{-1}, while a *trans* double bond absorbs near 970 cm^{-1}. This kind of information can be valuable in the assignment of stereochemistry about the double bond (see Figs. 2.12 and 2.13).

1,1-Disubstituted Double Bonds. One strong band near 890 cm^{-1} is obtained for a *gem*-dialkyl-substituted double bond. When electron-releasing or electron-withdrawing groups are attached to the double bond, shifts similar to that just given for monosubstituted double bonds are observed.

Trisubstituted Double Bonds. One medium-intensity band near 815 cm^{-1} is obtained.

Tetrasubstituted Double Bonds. These alkenes do not give any absorption in this region because of the absence of a hydrogen atom on the double bond. In addition, the C=C stretching vibration is very weak (or absent) at about 1670 cm^{-1} in these highly substituted systems.

Figure 2.22 shows the C—H out-of-plane bending vibrations for substituted alkenes, together with the frequency ranges.

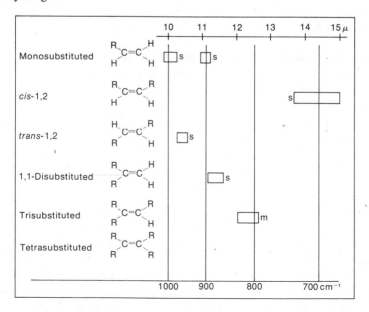

F I G U R E   2 . 2 2 The C—H out-of-plane bending vibrations for substituted alkenes.

2.11 AROMATIC RINGS

Aromatic compounds show a number of absorption bands in the infrared spectrum, many of which are not of diagnostic value. The C—H stretching peaks for the sp^2 carbon appear at values greater than 3000 cm^{-1}. Since C—H stretching bands for alkenes appear in the same range, it may be difficult to use the C—H stretching bands to differentiate between alkenes and aromatic compounds. However, the C=C stretching bands for aromatic rings usually appear between 1600 and 1450 cm^{-1} outside the usual range where the C=C appears for alkenes (1650 cm^{-1}). Also prominent are the out-of-plane bending peaks that appear in the range 900–690 cm^{-1}, which, along with weak overtone bands at 2000–1667 cm^{-1}, can be used to assign substitution on the ring.

SPECTRAL ANALYSIS BOX

AROMATIC RINGS

=C—H	Stretch for sp^2 C—H occurs at values greater than 3000 cm^{-1} (3050–3010 cm^{-1}).
=C—H	Out-of-plane (oop) bending occurs at 900–690 cm^{-1}. These bands can be used with great utility to assign the ring substitution pattern (see discussion).
C=C	Ring stretch absorptions often occur in pairs at 1600 cm^{-1} and 1475 cm^{-1}.

Overtone/combination bands appear between 2000 and 1667 cm^{-1}. These *weak* absorptions can be used to assign the ring substitution pattern (see discussion).

Examples: toluene (Fig. 2.23), *ortho*-diethylbenzene (Fig. 2.24), *meta*-diethylbenzene (Fig. 2.25), *para*-diethylbenzene (Fig. 2.26), and styrene (Fig. 2.27).

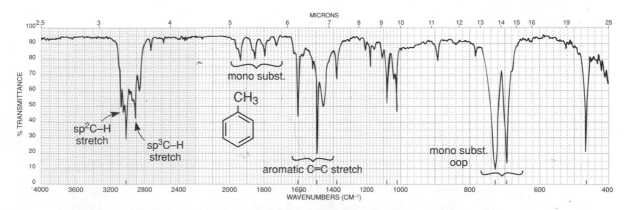

FIGURE 2.23 The infrared spectrum of toluene (neat liquid, KBr plates).

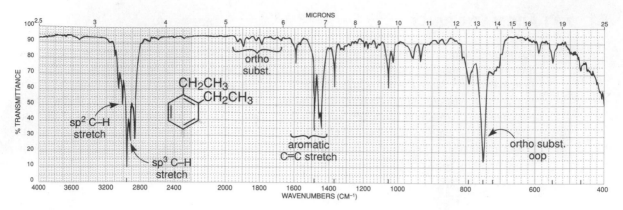

FIGURE 2.24 The infrared spectrum of *ortho*-diethylbenzene (neat liquid, KBr plates).

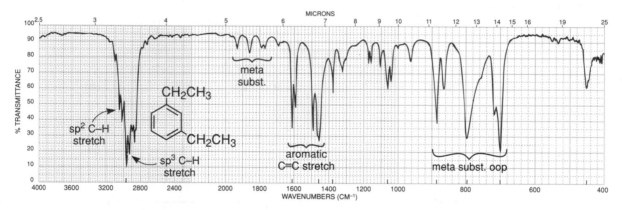

FIGURE 2.25 The infrared spectrum of *meta*-diethylbenzene (neat liquid, KBr plates).

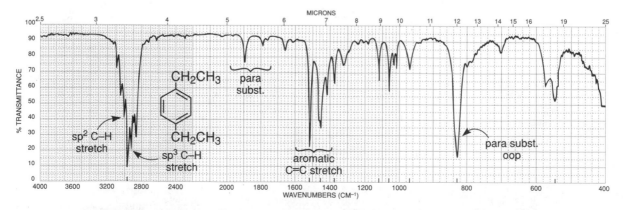

FIGURE 2.26 The infrared spectrum of *para*-diethylbenzene (neat liquid, KBr plates).

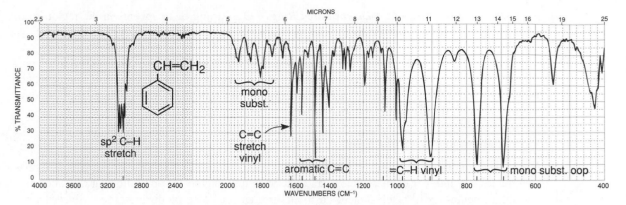

FIGURE 2.27 The infrared spectrum of styrene (neat liquid, KBr plates).

DISCUSSION SECTION

C—H Bending Vibrations

The in-plane C—H bending vibrations occur between 1300 and 1000 cm^{-1}. However, these bands are rarely useful because they overlap other, stronger absorptions which occur in this region.

The out-of-plane C—H bending vibrations, which appear between 900 and 690 cm^{-1}, are far more useful than the in-plane bands. These extremely intense absorptions, resulting from strong coupling with adjacent hydrogen atoms, can be used to assign the positions of substituents on the aromatic ring. The assignment of structure based upon these out-of-plane bending vibrations is most reliable for alkyl-, alkoxy-, halo-, amino-, or carbonyl-substituted aromatic compounds. Aromatic nitro compounds, derivatives of aromatic carboxylic acids, and derivatives of sulfonic acids often lead to unsatisfactory interpretation.

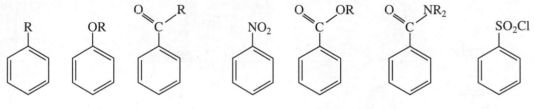

Reliable interpretation   Unreliable interpretation

Monosubstituted Rings. This substitution pattern always gives a strong absorption near 690 cm^{-1}. If this band is absent, no monosubstituted ring is present. A second strong band usually appears near 750 cm^{-1}. When the spectrum is taken in a halocarbon solvent, the 690-cm^{-1} band may be obscured by the strong C—X stretch absorptions. The typical two-peak monosubstitution pattern appears in the spectra of toluene (Fig. 2.23) and styrene (Fig. 2.27). In addition, the spectrum of styrene shows a pair of bands for the vinyl out-of-plane bending modes.

ortho-Disubstituted Rings (1,2-Disubstituted Rings). One strong band near 750 cm^{-1} is obtained. This pattern is seen in the spectrum of *ortho*-diethylbenzene (Fig. 2.24).

meta-Disubstituted Rings (1,3-Disubstituted Rings). This substitution pattern gives the 690-cm^{-1} band plus one near 780 cm^{-1}. A third band of medium intensity is often found near 880 cm^{-1}. This pattern is seen in the spectrum of *meta*-diethylbenzene (Fig. 2.25).

para-Disubstituted Rings (1,4-Disubstituted Rings). One strong band appears in the region from 800 to 850 cm^{-1}. This pattern is seen in the spectrum of *para*-diethylbenzene (Fig. 2.26).

Figure 2.28a shows the C—H out-of-plane bending vibrations for the common substitution patterns already given plus some others, together with the frequency ranges. Note that the bands appearing in the 720- to 667-cm^{-1} region actually result from C=C out-of-plane ring bending vibrations rather than from C—H out-of-plane bending (shaded boxes).

Combinations and Overtone Bands

Many *weak* combination and overtone absorptions appear between 2000 and 1667 cm^{-1}. The relative shapes and number of these peaks can be used to tell whether an aromatic ring is mono-, di-, tri-, tetra-, penta-, or hexasubstituted. Positional isomers can also be distinguished. Since the absorptions are weak, these bands are best observed by using neat liquids or concentrated solutions. If the compound has a high-frequency carbonyl group, this absorption will overlap the weak overtone bands so that no useful information can be obtained from the analysis of the region.

Figure 2.28b shows the various patterns obtained in this region. The monosubstitution pattern that appears in the spectra of toluene (Fig. 2.23) and styrene (Fig. 2.27) is particularly useful and helps to confirm the out-of-plane data given in the preceding section. Likewise, the *ortho-, meta-,*

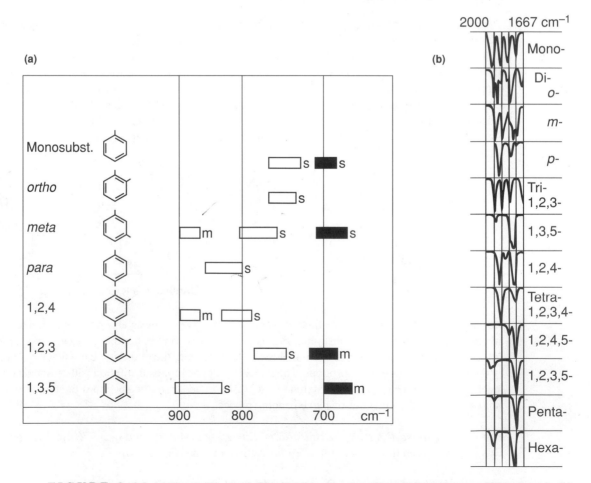

FIGURE 2.28 (a) The C—H out-of-plane bending vibrations for substituted benzenoid compounds. (b) The 2000- to 1667-cm^{-1} region for substituted benzenoid compounds (from Dyer, John R., *Applications of Absorption Spectroscopy of Organic Compounds,* Prentice–Hall, Englewood Cliffs, N.J., 1965).

and *para*-disubstituted patterns may be consistent with the out-of-plane bending vibrations discussed earlier. The spectra of *ortho*-diethylbenzene (Fig. 2.24), *meta*-diethylbenzene (Fig. 2.25), and *para*-diethylbenzene (Fig. 2.26) each shows bands in *both* the 2000- to 1667-cm^{-1} and 900- to 690-cm^{-1} regions, consistent with their structures. Note, however, that the out-of-plane vibrations are generally more useful for diagnostic purposes.

2.12 ALCOHOLS AND PHENOLS

Alcohols and phenols will show strong and broad hydrogen bonded O—H stretching bands centering between 3400 and 3300 cm^{-1}. In solution, it will also be possible to observe a free O—H stretching band at about 3600 cm^{-1} (sharp and weaker) to the left of the hydrogen bonded O—H peak. In addition, a C—O stretching band will appear in the spectrum at 1260–1000 cm^{-1}.

SPECTRAL ANALYSIS BOX

ALCOHOLS AND PHENOLS

O—H   The free O—H stretch is a *sharp* peak at 3650–3600 cm^{-1}. This band appears in combination with the hydrogen-bonded O—H peak when the alcohol is dissolved in a solvent (see discussion).

The hydrogen-bonded O—H band is a *broad* peak at 3400–3300 cm^{-1}. This band is usually the only one present in an alcohol that has not been dissolved in a solvent (neat liquid). When the alcohol is dissolved in a solvent, the free O—H and hydrogen-bonded O—H bands are present together, with the relatively weak free O—H on the left (see discussion).

C—O—H   Bending appears as a broad and weak peak at 1440–1220 cm^{-1} often obscured by the CH$_3$ bendings.

C—O   Stretching vibration usually occurs in the range 1260–1000 cm^{-1}. This band can be used to assign a primary, secondary, or tertiary structure to an alcohol (see discussion).

Examples: The hydrogen-bonded O—H stretch is present in the pure liquid (neat) samples of 1-hexanol (Fig. 2.29), 2-butanol (Fig. 2.30), and *para*-cresol (Fig. 2.31).

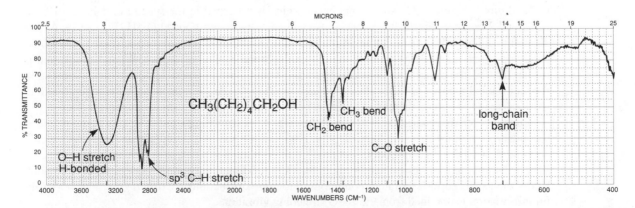

FIGURE 2.29 The infrared spectrum of 1-hexanol (neat liquid, KBr plates).

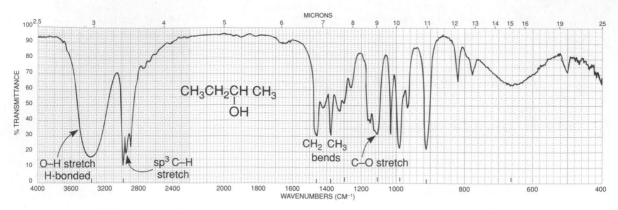

FIGURE 2.30 The infrared spectrum of 2-butanol (neat liquid, KBr plates).

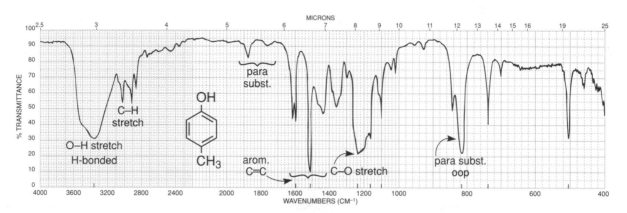

FIGURE 2.31 The infrared spectrum of *para*-cresol (neat liquid, KBr plates).

DISCUSSION SECTION

O—H Stretching Vibrations

When alcohols or phenols are determined as pure (neat) liquid films, as is common practice, a broad O—H stretching vibration is obtained for intermolecular hydrogen bonding in the range from 3400 to 3300 cm^{-1}. Figure 2.32a shows this band, which is observed in the spectra of 1-hexanol (Fig. 2.29) and 2-butanol (Fig. 2.30). Phenols also show the hydrogen-bonded O—H (Fig. 2.31). As the alcohol is diluted with carbon tetrachloride, a sharp "free" (non-hydrogen-bonded) O—H stretching band appears at about 3600 cm^{-1}, to the left of the broad band (Fig. 2.32b). When the solution is further diluted, the broad intermolecular hydrogen-bonded band is reduced considerably, leaving as the major band the free O—H stretching absorption (Fig. 2.32c). Intermolecular hydrogen bonding weakens the O—H bond, thereby shifting the band to lower frequency (lower energy).

Some workers have used the position of the free O—H stretching band to help assign a primary, secondary, or tertiary structure to an alcohol. For example, the free stretch occurs near 3640, 3630, 3620, and 3610 cm^{-1} for primary, secondary, and tertiary alcohols and for phenols, respectively. These absorptions can be analyzed only if the O—H region is expanded and carefully calibrated. Under the usual routine laboratory conditions, these fine distinctions are of little use. Far more useful information is obtained from the C—O stretching vibrations.

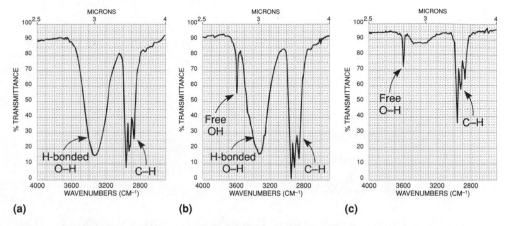

FIGURE 2.32 The O—H stretch region. (a) Hydrogen-bonded O—H only (neat liquid). (b) Free and hydrogen-bonded O—H (dilute solution). (c) Free and hydrogen-bonded O—H (very dilute solution).

Intramolecular hydrogen bonding, present in *ortho*-carbonyl-substituted phenols, usually shifts the broad O—H band to a lower frequency. For example, the O—H band is centered at about 3200 cm^{-1} in the neat spectrum of methyl salicylate, while O—H bands from normal phenols are centered at about 3350 cm^{-1}. The intramolecular hydrogen-bonded band does not change its position significantly even at high dilution, because the internal bonding is not altered by a change in concentration.

Methyl salicylate

Although phenols often have broader O—H bands than alcohols, it is difficult to assign a structure based upon this absorption; use the aromatic C=C region and the C—O stretching vibration (to be discussed shortly) to assign a phenolic structure. Finally, the O—H stretching vibrations in carboxylic acids also occur in this region. They may easily be distinguished from alcohols and phenols by the presence of a *very broad* band extending from 3400 to 2400 cm^{-1} *and* the presence of a carbonyl absorption (see Section 2.14D).

C—O—H Bending Vibrations

This bending vibration is coupled to H—C—H bending vibrations to yield some weak and broad peaks in the 1440–1220 cm^{-1} region. These broad peaks are difficult to observe because they are usually located under the more strongly absorbing CH$_3$ bending peaks at 1375 cm^{-1} (see Fig. 2.29).

C—O Stretching Vibrations

The strong C—O single-bond stretching vibrations are observed in the range from 1260 to 1000 cm^{-1}. Since the C—O absorptions are coupled with the adjacent C—C stretching vibrations, the position of the band may be used to assign a primary, secondary, or tertiary structure to an alcohol or to determine whether a phenolic compound is present. Table 2.7 gives the expected absorption bands for the C—O stretching vibrations in alcohols and phenols. For comparison, the O—H stretch values are also tabulated.

TABLE 2.7
C−O AND O−H STRETCHING VIBRATIONS IN ALCOHOLS AND PHENOLS

Compound	C−O Stretch (cm^{-1})	O−H Stretch (cm^{-1})
Phenols	1220	3610
3° Alcohols (saturated)	1150	3620
2° Alcohols (saturated)	1100	3630
1° Alcohols (saturated)	1050	3640

(C−O Stretch: *Decrease* ↓; O−H Stretch: *Increase* ↓)

Unsaturation on adjacent carbons or a cyclic structure lowers the frequency of C−O absorption. 2° examples:

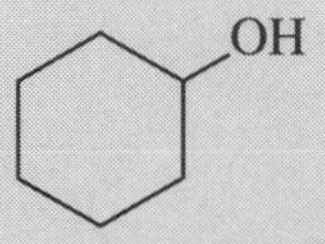

1100 → 1070 cm^{-1} 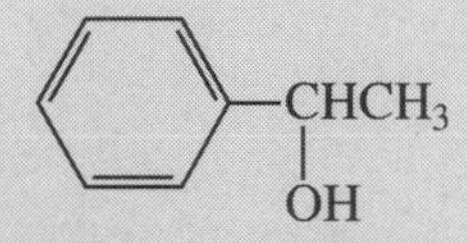   1100 → 1070 cm^{-1} 1100 → 1060 cm^{-1}

1° examples:

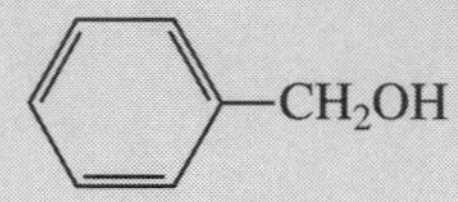

1050 → 1017 cm^{-1} 1050 → 1030 cm^{-1}

The spectrum of 1-hexanol, a primary alcohol, has its C−O absorption at 1058 cm^{-1} (Fig. 2.29), whereas that of 2-butanol, a secondary alcohol, has its C−O absorption at 1109 cm^{-1} (Fig. 2.30). Thus, both alcohols have their C−O bands near the expected values given in Table 2.7. Phenols give a C−O absorption at about 1220 cm^{-1} because of conjugation of the oxygen with the ring, which shifts the band to higher energy (more double-bond character). In addition to this band, an O–H in-plane bending absorption is usually found near 1360 cm^{-1} for neat samples of phenols. This latter band is also found in alcohols determined as neat (undiluted) liquids. It usually overlaps the C−H bending vibration for the methyl group at 1375 cm^{-1}.

The numbers in Table 2.7 should be considered *base* values. These C−O absorptions are shifted to lower frequencies when unsaturation is present on adjacent carbon atoms or when the O−H is attached to a ring. Shifts of 30 to 40 cm^{-1} from the base values are common, as seen in some selected examples in Table 2.7.

2.13 ETHERS

Ethers show at least one C−O band in the range 1300–1000 cm^{-1}. Simple aliphatic ethers can be distinguished from alkanes by the presence of the C−O band. In all other respects, the spectra of simple ethers look very similar to those of alkanes. Aromatic ethers, epoxides, and acetals are discussed in this section.

SPECTRAL ANALYSIS BOX

ETHERS

C—O   The most prominent band is that due to C—O stretch, 1300–1000 cm^{-1}. Absence of C=O and O—H is required to ensure that C—O stretch is not due to an ester or an alcohol. Phenyl alkyl ethers give two strong bands at about 1250 and 1040 cm^{-1}, while aliphatic ethers give one strong band at about 1120 cm^{-1}.

Examples: dibutyl ether (Fig. 2.33) and anisole (Fig. 2.34).

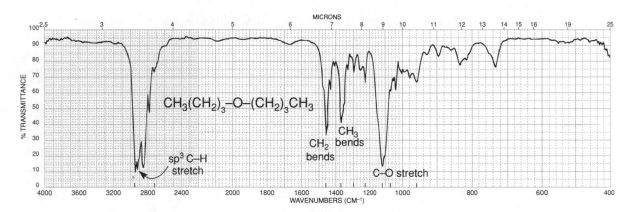

FIGURE 2.33 The infrared spectrum of dibutyl ether (neat liquid, KBr plates).

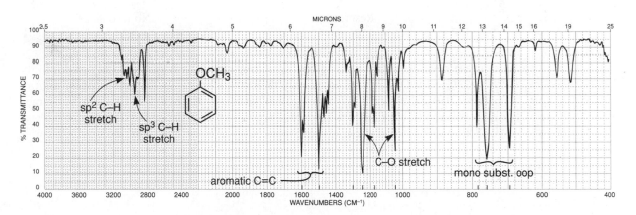

FIGURE 2.34 The infrared spectrum of anisole (neat liquid, KBr plates).

DISCUSSION SECTION

Ethers and related compounds such as epoxides, acetals, and ketals give rise to C—O—C stretching absorptions in the range from 1300 to 1000 cm^{-1}. Alcohols and esters also give strong C—O absorptions in this region, and these latter possibilities must be eliminated by observing the absence of bands in the O—H stretch region (Section 2.12) and in the C=O stretch region (Section 2.14), respectively. Ethers are generally encountered more often than epoxides, acetals, and ketals.

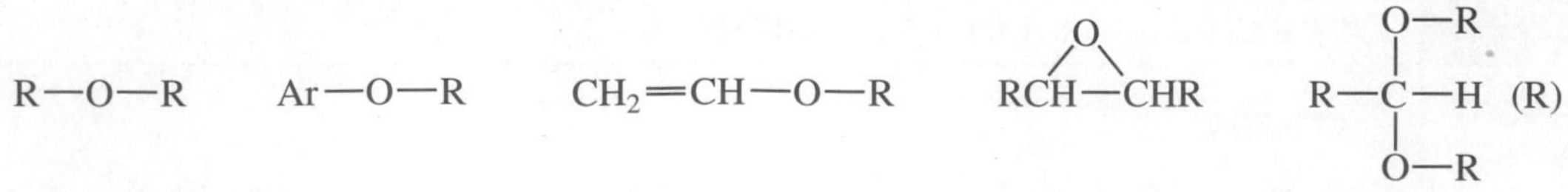

R—O—R   Ar—O—R   CH$_2$=CH—O—R   RCH—CHR   R—C—H (R)

(epoxide: O bridging RCH—CHR)

(acetal: O—R above and O—R below R—C—H)

Dialkyl   Aryl ethers   Vinyl ethers   Epoxides   Acetals
ethers   (Ketals)

Dialkyl Ethers. The asymmetric C—O—C stretching vibration leads to a single strong absorption that appears at about 1120 cm^{-1}, as seen in the spectrum of dibutyl ether (Fig. 2.33). The symmetric stretching band at about 850 cm^{-1} is usually very weak. The asymmetric C—O—C absorption also occurs at about 1120 cm^{-1} for a six-membered ring containing oxygen.

Aryl and Vinyl Ethers. Aryl alkyl ethers give rise to *two* strong bands: an asymmetric C—O—C stretch near 1250 cm^{-1} and a symmetric stretch near 1040 cm^{-1}, as seen in the spectrum of anisole (Fig. 2.34). Vinyl alkyl ethers also give two bands: one strong band assigned to an asymmetric stretching vibration at about 1220 cm^{-1} and one very weak band due to a symmetric stretch at about 850 cm^{-1}.

The shift in the asymmetric stretching frequencies in aryl and vinyl ethers to values higher than were found in dialkyl ethers can be explained through resonance. For example, the C—O band in vinyl alkyl ethers is shifted to a higher frequency (1220 cm^{-1}) because of the increased double-bond character, which strengthens the bond. In dialkyl ethers the absorption occurs at 1120 cm^{-1}. In addition, because resonance increases the polar character of the C=C double bond, the band at about 1640 cm^{-1} is considerably stronger than in normal C=C absorption (Section 2.10B).

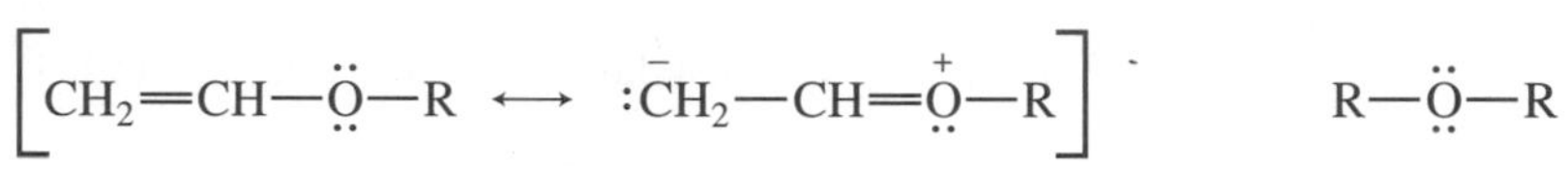

$$\left[\text{CH}_2\!=\!\text{CH}\!-\!\ddot{\text{O}}\!-\!\text{R} \longleftrightarrow \ :\!\bar{\text{C}}\text{H}_2\!-\!\text{CH}\!=\!\overset{+}{\ddot{\text{O}}}\!-\!\text{R}\right] \qquad\qquad \text{R}\!-\!\ddot{\text{O}}\!-\!\text{R}$$

Resonance   No resonance
1220 cm^{-1} 1120 cm^{-1}

Epoxides. These small-ring compounds give a *weak* ring-stretching band (breathing mode) in the range 1280–1230 cm^{-1}. Of more importance are the two *strong* ring deformation bands, one that appears between 950 and 815 cm^{-1} (asymmetric) and the other between 880 and 750 cm^{-1} (symmetric). For monosubstituted epoxides, this latter band appears in the upper end of the range, often near 835 cm^{-1}. Disubstituted epoxides have absorption in the lower end of the range, closer to 775 cm^{-1}.

Acetals and Ketals. Molecules that contain kctal or acetal linkages often give *four or five strong bands,* respectively, in the region from 1200 to 1020 cm^{-1}. These bands are often unresolved.

2.14 CARBONYL COMPOUNDS

The carbonyl group is present in aldehydes, ketones, acids, esters, amides, acid chlorides, and anhydrides. This group absorbs strongly in the range from 1850 to 1650 cm^{-1} because of its large change in dipole moment. Since the C=O stretching frequency is sensitive to attached atoms, the common functional groups already mentioned absorb at characteristic values. Figure 2.35 provides the normal base values for the C=O stretching vibrations of the various functional groups. The C=O frequency of a ketone, which is approximately in the middle of the range, is usually considered the reference point for comparisons of these values.

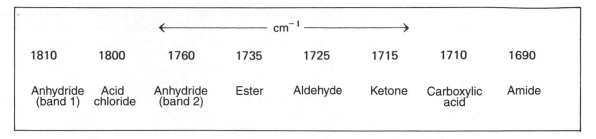

F I G U R E 2 . 3 5 Normal base values for the C=O stretching vibrations for carbonyl groups.

The range of values given in Figure 2.35 may be explained through the use of electron-withdrawing effects (inductive effects), resonance effects, and hydrogen bonding. The first two effects operate in opposite ways to influence the C=O stretching frequency. First, an electronegative element may tend to draw in the electrons between the carbon and oxygen atoms through its electron-withdrawing effect, so that the C=O bond becomes somewhat stronger. A higher-frequency (higher-energy) absorption results. Since oxygen is more electronegative than carbon, this effect dominates in an ester to raise the C=O frequency above that of a ketone. Second, a resonance effect may be observed when the unpaired electrons on a nitrogen atom conjugate with the carbonyl group, resulting in increased single-bond character and a lowering of the C=O absorption frequency. This second effect is observed in an amide. Since nitrogen is less electronegative than an oxygen atom, it can more easily accommodate a positive charge. The resonance structure shown here introduces single-bond character into the C=O group and thereby lowers the absorption frequency below that of a ketone.

Ester   Amide

Electron-withdrawing effect raises   Resonance effect lowers C=O frequency
C=O frequency

In acid chlorides, the highly electronegative halogen atom strengthens the C=O bond through an enhanced inductive effect and shifts the frequency to values even higher than are found in esters. Anhydrides are likewise shifted to frequencies higher than are found in esters because of a concentration of electronegative oxygen atoms. In addition, anhydrides give two absorption bands that are due to symmetric and asymmetric stretching vibrations (Section 2.3).

A carboxylic acid exists in monomeric form in very dilute solution, and it absorbs at about 1760 cm^{-1} because of the electron-withdrawing effect just discussed. However, acids in concentrated solution, in the form of neat liquid, or in the solid state (KBr pellet and Nujol) tend to dimerize via hydrogen bonding. This dimerization weakens the C=O bond and lowers the stretching force constant K, resulting in a lowering of the carbonyl frequency of saturated acids to about 1710 cm^{-1}.

Ketones absorb at a lower frequency than aldehydes because of their additional alkyl group, which is electron donating (compared to H) and supplies electrons to the C=O bond. This electron-releasing effect weakens the C=O bond in the ketone and lowers the force constant and the absorption frequency.

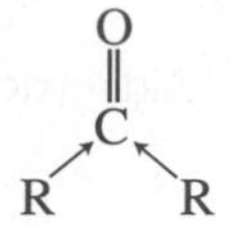

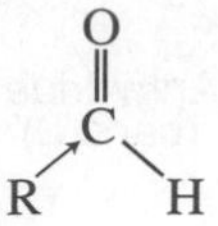

A. Factors That Influence the C=O Stretching Vibration

Conjugation Effects. The introduction of a C=C bond adjacent to a carbonyl group results in delocalization of the π electrons in the C=O and C=C bonds. This conjugation increases the single-bond character of the C=O and C=C bonds in the resonance hybrid and hence lowers their force constants, resulting in a lowering of the frequencies of carbonyl and double-bond absorption. Conjugation with triple bonds also shows this effect.

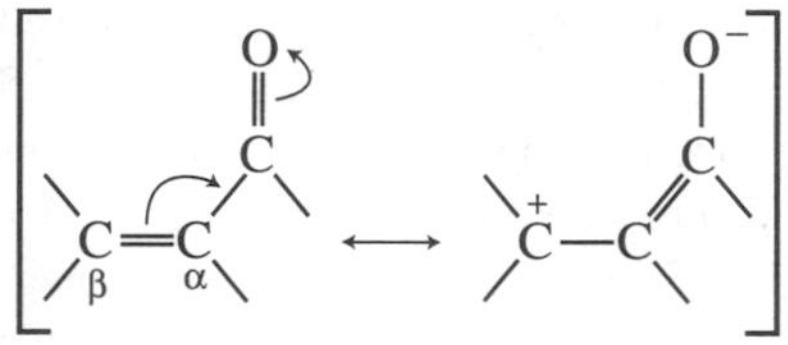

Generally, the introduction of an α,β double bond in a carbonyl compound results in a 25- to 45-cm^{-1} lowering of the C=O frequency from the base value given in Figure 2.35. A similar lowering occurs when an adjacent aryl group is introduced. Further addition of unsaturation (γ,δ) results in a further shift to lower frequency, but only by about 15 cm^{-1} more. In addition, the C=C absorption shifts from its "normal" value, about 1650 cm^{-1}, to a lower-frequency value of about 1640 cm^{-1}, and the C=C absorption is greatly intensified. Often two closely spaced C=O absorption peaks are observed for these conjugated systems, resulting from two possible conformations, the *s-cis* and *s-trans*. The *s-cis* conformation absorbs at a frequency higher than the *s-trans* conformation. In some cases, the C=O absorption is broadened rather than split into the doublet.

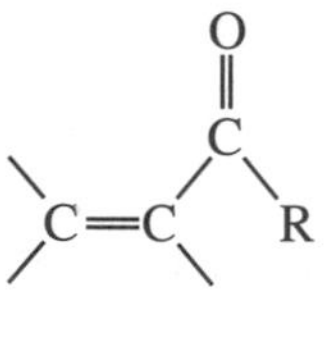

s-cis 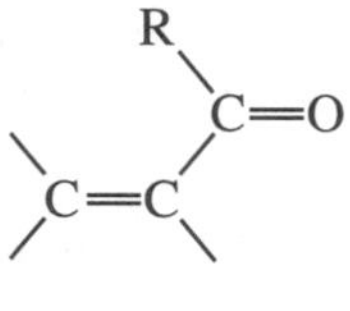   *s-trans*

The following examples show the effects of conjugation on the C=O frequency.

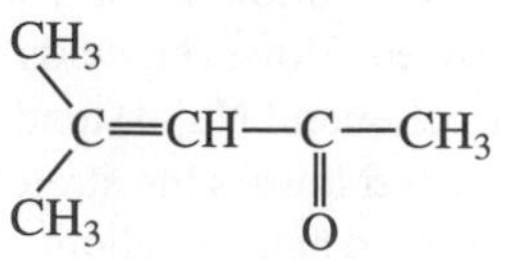

| α,β-Unsaturated ketone | Aryl-substituted aldehyde | Aryl-substituted acid |
| 1715 → 1690 cm^{-1} | 1725 → 1700 cm^{-1} | 1710 → 1680 cm^{-1} |

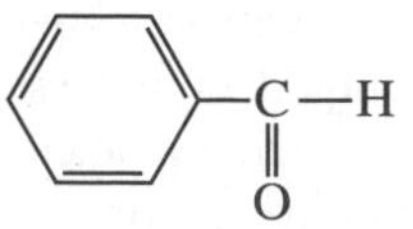

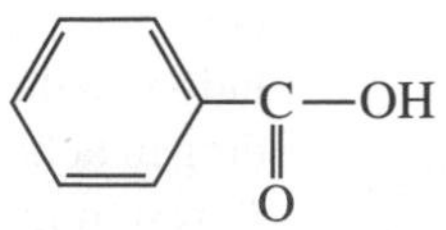

Conjugation does not reduce the C=O frequency in amides. The introduction of α,β unsaturation causes an *increase in frequency* from the base value given in Figure 2.35. Apparently, the introduction of

sp^2-hybridized carbon atoms removes electron density from the carbonyl group and strengthens the bond instead of interacting by resonance as in other carbonyl examples. Since the parent amide group is already highly stabilized (see p. 51), the introduction of the C=C unsaturation does not overcome this resonance.

Ring-Size Effects. Six-membered rings with carbonyl groups are unstrained and absorb at about the values given in Figure 2.35. Decreasing the ring size *increases the frequency* of the C=O absorption for the reasons discussed in Section 2.10 (C=C stretching vibrations and exocyclic double bonds; p. 38). All of the functional groups listed in Figure 2.35, which can form rings, give increased frequencies of absorption with increased angle strain. For ketones and esters, there is often a 30-cm^{-1} increase in frequency for each carbon removed from the unstrained six-membered ring values. Some examples are:

Cyclic ketone	Cyclic ketone	Cyclic ester	Cyclic amide
$1715 \rightarrow 1745$ cm^{-1}	$1715 \rightarrow 1780$ cm^{-1}	(lactone)	(lactam)
		$1735 \rightarrow 1770$ cm^{-1}	$1690 \rightarrow 1705$ cm^{-1}

In ketones, larger rings have frequencies that range from nearly the same value as in cyclohexanone (1715 cm^{-1}) to values slightly less than 1715 cm^{-1}. For example, cycloheptanone absorbs at about 1705 cm^{-1}.

α-Substitution Effects. When the carbon next to the carbonyl is substituted with a chlorine (or other halogen) atom, the carbonyl band shifts to a *higher frequency*. The electron-withdrawing effect removes electrons from the carbon of the C=O bond. This removal is compensated for by a tightening of the π bond (shortening), which increases the force constant and leads to an increase in the absorption frequency. This effect holds for all carbonyl compounds.

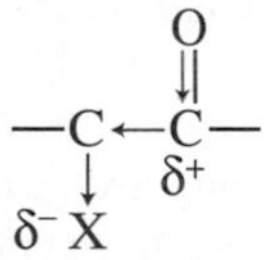

In ketones, two bands result from the substitution of an adjacent chlorine atom. One arises from the conformation in which the chlorine is rotated next to the carbonyl, and the other is due to the conformation in which the chlorine is away from the group. When the chlorine is next to the carbonyl, nonbonded electrons on the oxygen atom are repelled, resulting in a stronger bond and a higher absorption frequency. Information of this kind can be used to establish a structure in rigid ring systems, such as in the following examples.

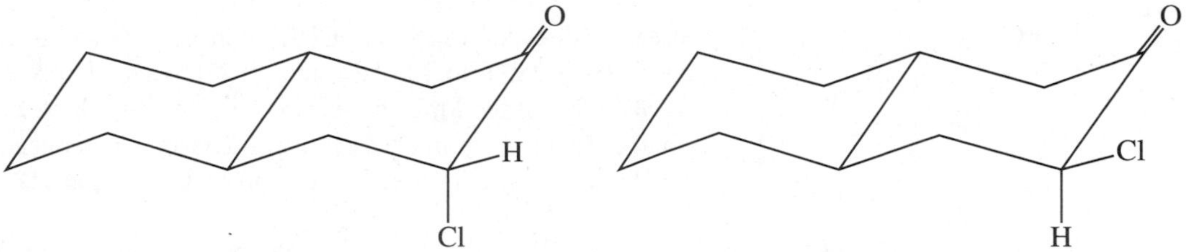

Axial chlorine	Equatorial chlorine
~1725 cm^{-1}	~1750 cm^{-1}

Hydrogen-Bonding Effects. Hydrogen bonding to a carbonyl group lengthens the C=O bond and lowers the stretching force constant K, resulting in a *lowering* of the absorption frequency. Examples of this effect are the decrease in the C=O frequency of the carboxylic acid dimer (p. 51) and the lowering of the ester C=O frequency in methyl salicylate caused by intramolecular hydrogen bonding:

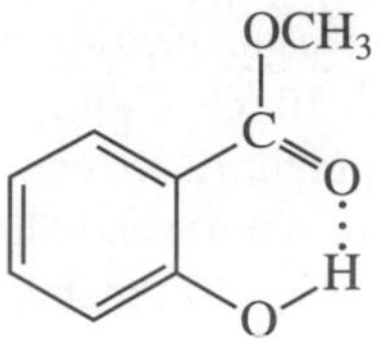

Methyl salicylate
1680 cm^{-1}

B.   Aldehydes

Aldehydes show a very strong band for the carbonyl group (C=O) that appears in the range of 1740–1725 cm^{-1} for simple aliphatic aldehydes. This band is shifted to lower frequencies with conjugation to a C=C or phenyl group. A very important doublet can be observed in the C—H stretch region for the aldehyde C—H near 2850 and 2750 cm^{-1}. The presence of this doublet allows aldehydes to be distinguished from other carbonyl-containing compounds.

SPECTRAL ANALYSIS BOX

ALDEHYDES

C=O

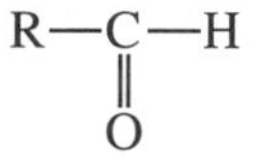

R—C—H (with =O)	C=O stretch appears in range 1740–1725 cm^{-1} for normal aliphatic aldehydes.
C=C—C—H (with =O)	Conjugation of C=O with α,β C=C; 1700–1680 cm^{-1} for C=O and 1640 cm^{-1} for C=C.
Ar—C—H (with =O)	Conjugation of C=O with phenyl; 1700–1660 cm^{-1} for C=O and 1600–1450 cm^{-1} for ring.
Ar—C=C—C—H (with =O)	Longer conjugated system; 1680 cm^{-1} for C=O.
C—H	Stretch, aldehyde hydrogen (—CHO), consists of a pair of *weak* bands, one at 2860–2800 cm^{-1} and the other at 2760–2700 cm^{-1}. It is easier to see the band at the lower frequency because it is not obscured by the usual C—H bands from the alkyl chain. The higher-frequency aldehyde C—H stretch is often buried in the aliphatic C—H bands.

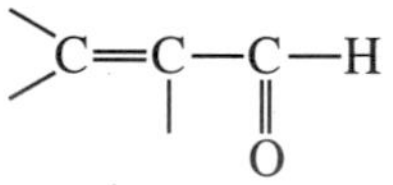

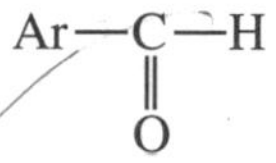

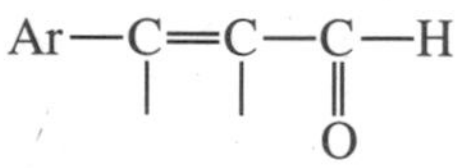

Examples: nonanal (Fig. 2.36), crotonaldehyde (Fig. 2.37), and benzaldehyde (Fig. 2.38).

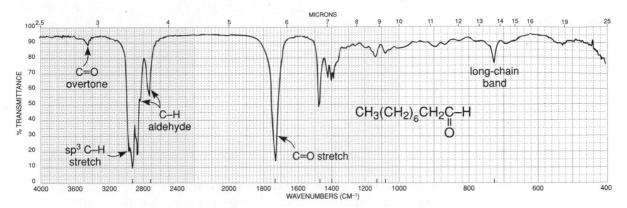

FIGURE 2.36 The infrared spectrum of nonanal (neat liquid, KBr plates).

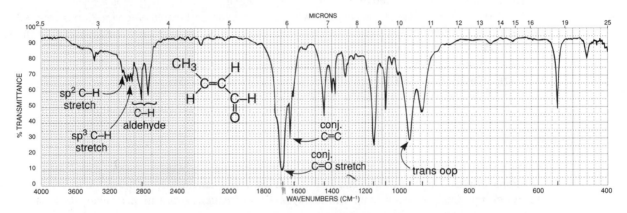

FIGURE 2.37 The infrared spectrum of crotonaldehyde (neat liquid, KBr plates).

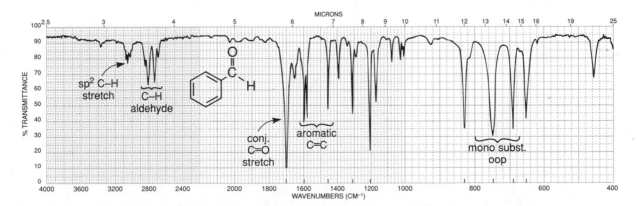

FIGURE 2.38 The infrared spectrum of benzaldehyde (neat liquid, KBr plates).

DISCUSSION SECTION

The spectrum of nonanal (Fig. 2.36) exhibits the normal aldehyde stretching frequency at 1725 cm^{-1}. Since the positions of these absorptions are not very different from those of ketones, it may not be easy to distinguish between aldehydes and ketones on this basis. Conjugation of the carbonyl group with an aryl or an α,β double bond shifts the normal C=O stretching band to a lower frequency (1700–1680 cm^{-1}), as predicted in Section 2.14A (Conjugation Effects). This effect is seen in crotonaldehyde (Fig. 2.37), which has α,β unsaturation, and in benzaldehyde (Fig. 2.38), in which an aryl group is attached directly to the carbonyl group. Halogenation on the α carbon leads to an increased frequency for the carbonyl group (p. 53).

The C—H stretching vibrations found in aldehydes (—CHO) at about 2750 cm^{-1} and 2850 cm^{-1} are extremely important for distinguishing between ketones and aldehydes. Typical ranges for the pairs of C—H bands are 2860–2800 cm^{-1} and 2760–2700 cm^{-1}. The band at 2750 cm^{-1} is probably the more useful of the pair because it appears in a region where other C—H absorptions (CH$_3$, CH$_2$, and so on) are absent. The 2850-cm^{-1} band overlaps other C—H bands and is not as easy to see (see nonanal, Fig. 2.36). If the 2750-cm^{-1} band is present together with the proper C=O absorption value, an aldehyde functional group is almost certainly indicated.

The doublet that is observed in the range 2860–2700 cm^{-1} for an aldehyde is a result of *Fermi* resonance (p. 17). The second band appears when the aldehyde C—H *stretching* vibration is coupled with the first overtone of the medium intensity aldehyde C—H *bending* vibration appearing in the range 1400–1350 cm^{-1}.

The medium-intensity absorption in nonanal (Fig. 2.36) at 1460 cm^{-1} is due to the scissoring (bending) vibration of the CH$_2$ group next to the carbonyl group. Methylene groups often absorb more strongly when they are attached directly to a carbonyl group.

C.   Ketones

Ketones show a very strong band for the C=O group that appears in the range of 1720–1708 cm^{-1} for simple aliphatic ketones. This band is shifted to lower frequencies with conjugation to a C=C or phenyl group. An α-halogen atom will shift the C=O frequency to a higher value. Ring strain moves the absorption to a higher frequency in cyclic ketones.

SPECTRAL ANALYSIS BOX

KETONES

C=O

R—C—R ‖ O	C=O stretch appears in range 1720–1708 cm^{-1} for normal aliphatic ketones.
C=C—C—R ‖ O	Conjugation of C=O with α,β C=C; 1700–1675 cm^{-1} for C=O and 1644–1617 cm^{-1} for C=C.
Ar—C—R ‖ O	Conjugation of C=O with phenyl; 1700–1680 cm^{-1} for C=O and 1600–1450 cm^{-1} for ring.

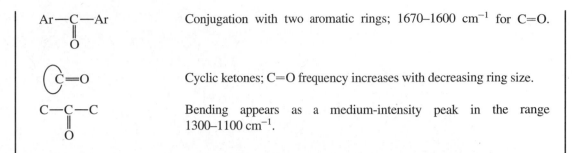

Ar—C—Ar   Conjugation with two aromatic rings; 1670–1600 cm^{-1} for C=O.
   ||
   O

C=O   Cyclic ketones; C=O frequency increases with decreasing ring size.

C—C—C   Bending appears as a medium-intensity peak in the range
   || 1300–1100 cm^{-1}.
   O

Examples: 3-methyl-2-butanone (Fig. 2.4), mesityl oxide (Fig. 2.39), acetophenone (Fig. 2.40), cyclopentanone (Fig. 2.41), and 2,4-pentanedione (Fig. 2.42).

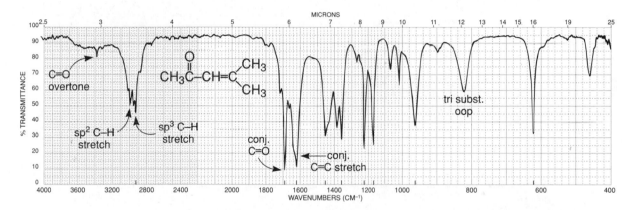

FIGURE 2.39 The infrared spectrum of mesityl oxide (neat liquid, KBr plates).

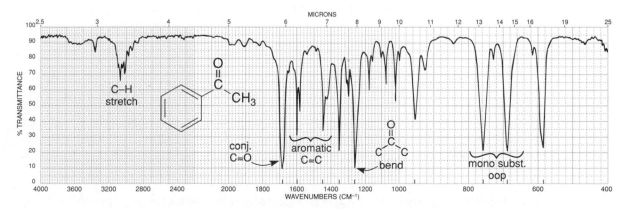

FIGURE 2.40 The infrared spectrum of acetophenone (neat liquid, KBr plates).

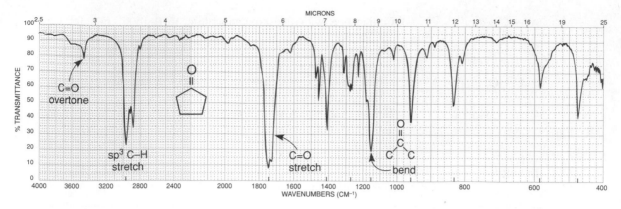

F I G U R E 2.41 The infrared spectrum of cyclopentanone (neat liquid, KBr plates).

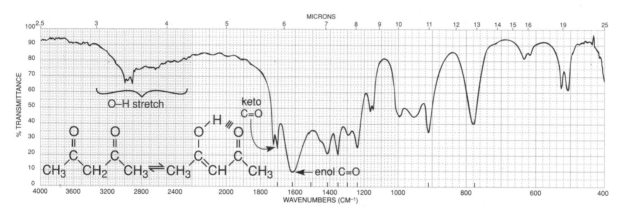

F I G U R E 2.42 The infrared spectrum of 2,4-pentanedione (neat liquid, KBr plates).

DISCUSSION SECTION

Normal C=O Bands. The spectrum of 3-methyl-2-butanone (Fig. 2.4) exhibits a normal, or unconjugated, ketone stretching frequency at 1715 cm^{-1}. A very weak overtone band from the C=O (1715 cm^{-1}) appears at twice the frequency of the C=O absorption (3430 cm^{-1}). Small bands of this type should not be confused with O–H absorptions, which also appear near this value. The O–H stretching absorptions are *much more intense.*

Conjugation Effects. Conjugation of the carbonyl group with an aryl or an α,β double bond shifts the normal C=O stretching band (1715 cm^{-1}) to a lower frequency (1700–1675 cm^{-1}), as predicted in Section 2.14A (p. 52). Rotational isomers may lead to a splitting or broadening of the carbonyl band (p. 52). The effect of conjugation on the C=O band is seen in mesityl oxide (Fig. 2.39), which has α,β unsaturation, and in acetophenone (Fig. 2.40), in which an aryl group is attached to the carbonyl group. Both exhibit C=O shifts to lower frequencies. Figure 2.43 presents some typical C=O stretching vibrations, which demonstrate the influence of conjugation.

Cyclic Ketones (Ring Strain). Figure 2.44 provides some values for the C=O absorptions for cyclic ketones. Note that ring strain shifts the absorption values to a higher frequency, as was predicted in

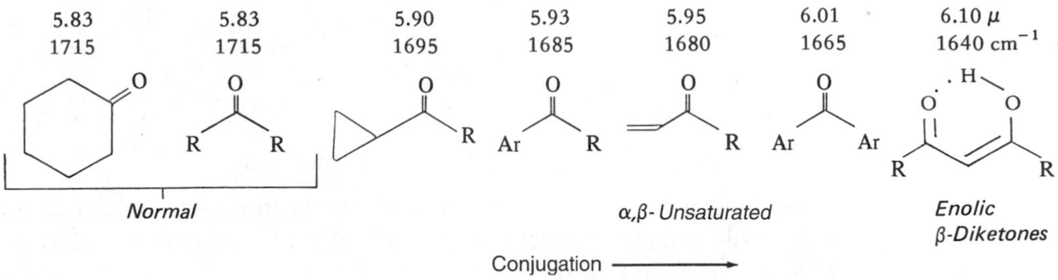

FIGURE 2.43 The C=O stretching vibrations in conjugated ketones.

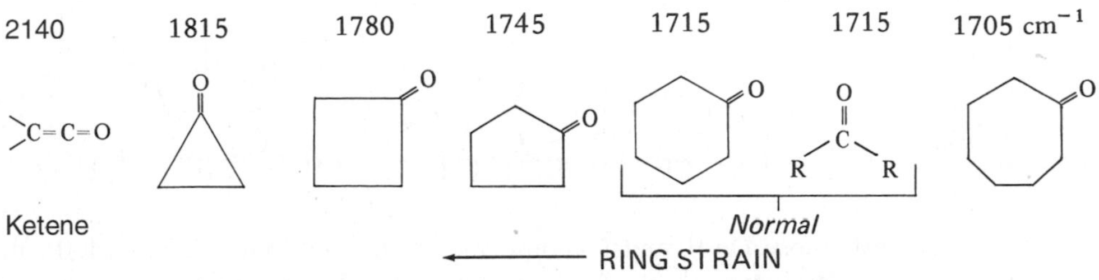

FIGURE 2.44 The C=O stretching vibrations for cyclic ketones and ketene.

Section 2.13A (p. 53). Ketene is included in Figure 2.44 because it is an extreme example of an *exo* double-bond absorption (see p. 38). The *s* character in the C=O group increases as the ring size decreases, until it reaches a maximum value that is found in the *sp*-hybridized carbonyl carbon in ketene. The spectrum of cyclopentanone (Fig. 2.41) shows how ring strain increases the frequency of the carbonyl group.

α-Diketones (1,2-Diketones). Unconjugated diketones that have the two carbonyl groups adjacent to each other show one strong absorption peak at about 1716 cm^{-1}. If the two carbonyl groups are conjugated with aromatic rings, the absorption is shifted to a lower frequency value, about 1680 cm^{-1}. In the latter case, a narrowly spaced doublet rather than a single peak may be observed, due to symmetric and asymmetric absorptions.

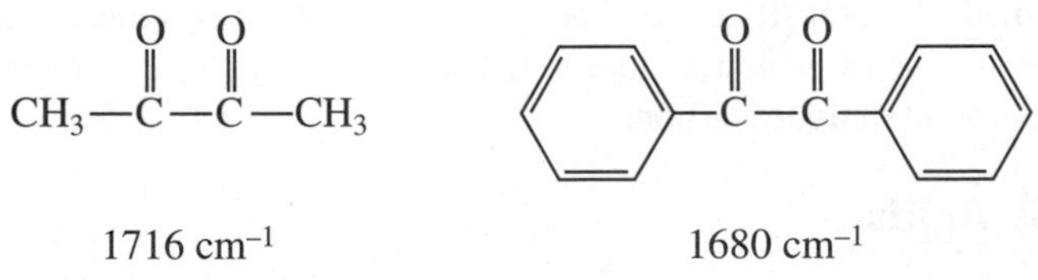

β-Diketones (1,3-Diketones). Diketones with carbonyl groups located 1,3 with respect to each other may yield a more complicated pattern than those observed for most ketones (2,4-pentane-dione, Fig. 2.42). These β-diketones often exhibit tautomerization, which yields an equilibrium mixture of enol and keto tautomers. Since many β-diketones contain large amounts of the enol form, you may observe carbonyl peaks for both the enol and keto tautomers.

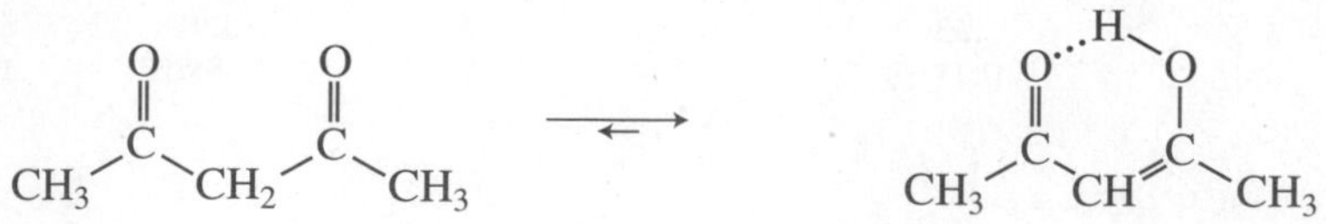

Keto tautomer

C=O doublet
1723 cm^{-1} (symmetric stretch)
1706 cm^{-1} (asymmetric stretch)

Enol tautomer

C=O (hydrogen-bonded), 1622 cm^{-1}
O—H (hydrogen-bonded), 3200–2400 cm^{-1}

The carbonyl group in the enol form appearing at about 1622 cm^{-1} is substantially shifted and intensified in comparison to the normal ketone value, 1715 cm^{-1}. The shift is a result of internal hydrogen bonding, as discussed in Section 2.14A (p. 54). Resonance, however, also contributes to the lowering of the carbonyl frequency in the enol form. This effect introduces single-bond character into the enol form.

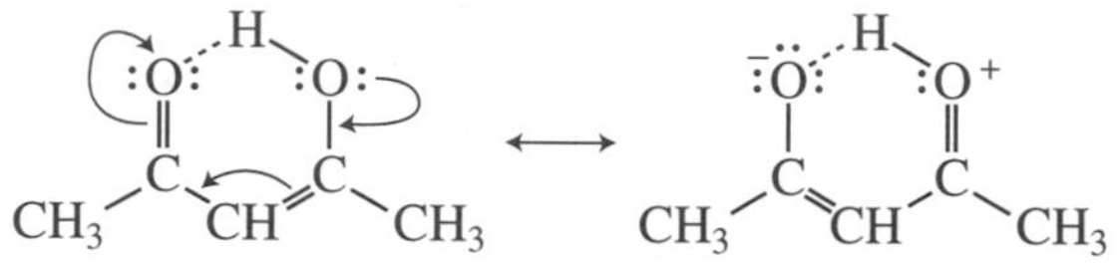

A weak, broad O—H stretch is observed for the enol form at 3200–2400 cm^{-1}. Since the keto form is also present, a doublet for the asymmetric and symmetric stretching frequencies is observed for the two carbonyl groups (Fig. 2.42). The relative intensities of the enol and keto carbonyl absorptions depend on the percentages present at equilibrium. Hydrogen-bonded carbonyl groups in enol forms are often observed in the region 1640–1570 cm^{-1}. The keto forms generally appear as doublets in the range from 1730 to 1695 cm^{-1}.

α-Haloketones. Substitution of a halogen atom on the α carbon shifts the carbonyl absorption peak to a higher frequency, as discussed in Section 2.14A (p. 53). Similar shifts occur with other electron-withdrawing groups, such as an alkoxy group (—O—CH$_3$). For example, the carbonyl group in chloroacetone appears at 1750 cm^{-1}, whereas that in methoxyacetone appears at 1731 cm^{-1}. When the more electronegative fluorine atom is attached, the frequency shifts to an even higher value, 1781 cm^{-1}, in fluoroacetone.

Bending Modes. A medium to strong absorption occurs in the range from 1300 to 1100 cm^{-1} for coupled stretching and bending vibrations in the C—CO—C group of ketones. Aliphatic ketones absorb to the right in this range (1220 to 1100 cm^{-1}), as seen in the spectrum of 3-methyl-2-butanone (Fig. 2.4), where a band appears at about 1180 cm^{-1}. Aromatic ketones absorb to the left in this range (1300 to 1220 cm^{-1}), as seen in the spectrum of acetophenone (Fig. 2.40), where a band appears at about 1260 cm^{-1}.

A medium-intensity band appears for a methyl group adjacent to a carbonyl at about 1370 cm^{-1}, for the symmetric bending vibration. These methyl groups absorb with greater intensity than methyl groups found in hydrocarbons.

D. Carboxylic Acids

Carboxylic acids show a very strong band for the C=O group that appears in the range of 1730–1700 cm^{-1} for simple aliphatic carboxylic acids in the *dimeric* form (p. 51). This band is shifted to lower frequencies with conjugation to a C=C or phenyl group. The O—H stretch appears in the spectrum as a *very broad* band extending from 3400 to 2400 cm^{-1}. This broad band centers on about 3000 cm^{-1}, and partially obscures the C—H stretching bands. If this very broad O—H stretch band is seen, along with a C=O peak, it almost certainly indicates the compound is a carboxylic acid.

SPECTRAL ANALYSIS BOX

CARBOXYLIC ACIDS

O—H   Stretch, usually *very broad* (strongly H-bonded), occurs at 3400–2400 cm^{-1} and often overlaps the C—H absorptions.

C=O   Stretch, broad, occurs at 1730–1700 cm^{-1}. Conjugation moves the absorption to a lower frequency.

C—O   Stretch occurs in the range 1320–1210 cm^{-1}, medium intensity.

Examples: isobutyric acid (Fig. 2.45) and benzoic acid (Fig. 2.46).

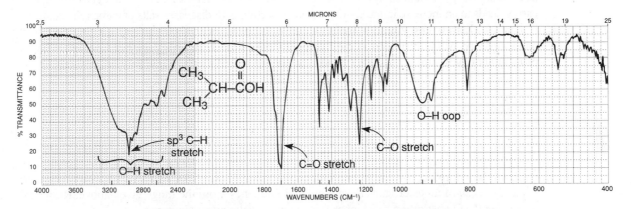

FIGURE 2.45 The infrared spectrum of isobutyric acid (neat liquid, KBr plates).

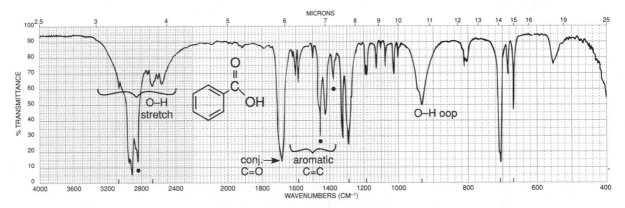

FIGURE 2.46 The infrared spectrum of benzoic acid (Nujol mull, KBr plates). Dots indicate the Nujol (mineral oil) absorption bands (see Fig. 2.8).

DISCUSSION SECTION

The most characteristic feature in the spectrum of a carboxylic acid is the *extremely broad* O—H absorption occurring in the region from 3400 to 2400 cm^{-1}. This band is attributed to the strong hydrogen bonding present in the dimer, which was discussed in the introduction to Section 2.14 (p. 51). The absorption often obscures the C—H stretching vibrations that occur in the same region. If this broad hydrogen-bonded band is present *together with* the proper C=O absorption value, a carboxylic acid is almost certainly indicated. Figures 2.45 and 2.46 show the spectra of an aliphatic carboxylic acid and an aromatic carboxylic acid, respectively.

The carbonyl stretching absorption, which occurs at about 1730 to 1700 cm^{-1} for the dimer, is usually broader and more intense than that present in an aldehyde or a ketone. For most acids, when the acid is diluted with a solvent, the C=O absorption appears between 1760 and 1730 cm^{-1} for the monomer. However, the monomer is not often seen experimentally, since it is usually easier to run the spectrum as a neat liquid. Under these conditions, as well as in a potassium bromide pellet or a Nujol mull, the dimer exists. It should be noted that some acids exist as dimers even at high dilution. Conjugation with a C=C or aryl group usually shifts the absorption band to a lower frequency, as predicted in Section 2.14A (p. 52) and as shown in the spectrum of benzoic acid (Fig. 2.46). Halogenation on the α carbon leads to an increase in the C=O frequency. Section 2.18 discusses salts of carboxylic acids.

The C—O stretching vibration for acids (dimer) appears near 1260 cm^{-1} as a medium-intensity band. A broad band, attributed to the hydrogen-bonded O—H out-of-plane bending vibration, appears at about 930 cm^{-1}. This latter band is usually of low to medium intensity.

E.   Esters

Esters show a very strong band for the C=O group that appears in the range of 1750–1735 cm^{-1} for simple aliphatic esters. The C=O band is shifted to lower frequencies when it is conjugated to a C=C or phenyl group. On the other hand, conjugation of a C=C or phenyl group with the *single-bonded oxygen* of an ester leads to an increased frequency from the range given above. Ring strain moves the C=O absorption to a higher frequency in cyclic esters (lactones).

SPECTRAL ANALYSIS BOX

ESTERS

C=O

R—C—O—R
$\parallel$
O
  — C=O stretch appears in range 1750–1735 cm^{-1} for normal aliphatic esters.

C=C—C—O—R
$\parallel$
O
  — Conjugation of C=O with α,β C=C; 1740–1715 cm^{-1} for C=O and 1640–1625 cm^{-1} for C=C (two bands for some C=C, *cis and trans,* p. 52.)

Ar—C—O—R
$\parallel$
O
  — Conjugation of C=O with phenyl; 1740–1715 cm^{-1} for C=O and 1600–1450 cm^{-1} for ring.

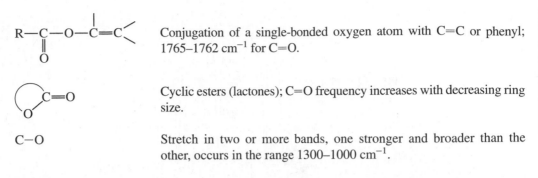

Conjugation of a single-bonded oxygen atom with C=C or phenyl; 1765–1762 cm⁻¹ for C=O.

Cyclic esters (lactones); C=O frequency increases with decreasing ring size.

C—O Stretch in two or more bands, one stronger and broader than the other, occurs in the range 1300–1000 cm⁻¹.

Examples: ethyl butyrate (Fig. 2.47), methyl methacrylate (Fig. 2.48), vinyl acetate (Fig. 2.49), methyl benzoate (Fig. 2.50), and methyl salicylate (Fig. 2.51).

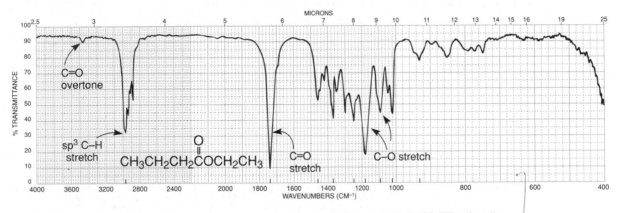

FIGURE 2.47 The infrared spectrum of ethyl butyrate (neat liquid, KBr plates).

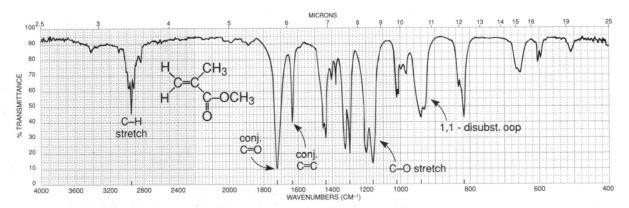

FIGURE 2.48 The infrared spectrum of methyl methacrylate (neat liquid, KBr plates).

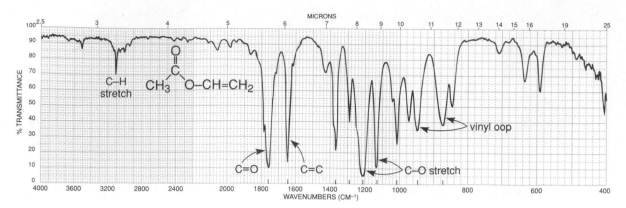

FIGURE 2.49 The infrared spectrum of vinyl acetate (neat liquid, KBr plates).

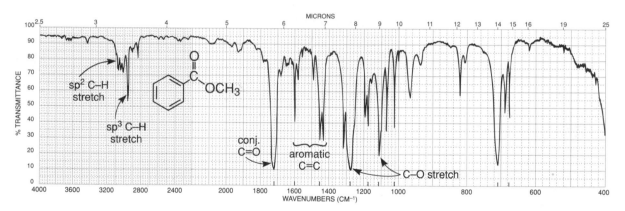

FIGURE 2.50 The infrared spectrum of methyl benzoate (neat liquid, KBr plates).

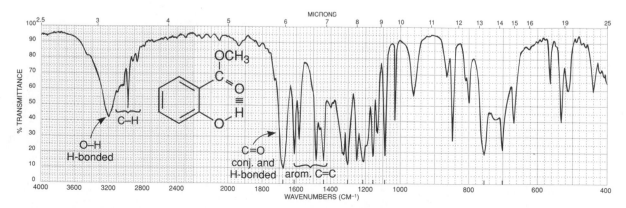

FIGURE 2.51 The infrared spectrum of methyl salicylate (neat liquid, KBr plates).

DISCUSSION SECTION

General Features of Esters. The two most characteristic features in the spectrum of a normal ester are the strong C=O, which appears in the range from 1750 to 1735 cm^{-1}, and C—O stretching absorptions, which appear in the range from 1300 to 1000 cm^{-1}. Although some ester carbonyl groups may appear in the same general area as ketones, one can usually eliminate ketones from consideration by observing the *strong* and *broad* C—O stretching vibrations that appear in a region (1300 to 1000 cm^{-1}) where ketonic absorptions appear as weaker and narrower bands. For example, compare the spectrum of a ketone, mesityl oxide (Fig. 2.39) with that of an ester, ethyl butyrate (Fig. 2.47) in the 1300- to 1000-cm^{-1} region. Ethyl butyrate (Fig. 2.47) shows the typical C=O stretching vibration at about 1738 cm^{-1}.

Conjugation with a Carbonyl Group (α,β Unsaturation or Aryl Substitution). The C=O stretching vibrations are shifted by about 15 to 25 cm^{-1} to lower frequencies with α,β unsaturation or aryl substitution, as predicted in Section 2.14A (Conjugation Effects, p. 52). The spectra of both methyl methacrylate (Fig. 2.48) and methyl benzoate (Fig. 2.50) show the C=O absorption shift from the position in a normal ester, ethyl butyrate (Fig. 2.47). Also notice that the C=C absorption band at 1630 cm^{-1} in methyl methacrylate has been intensified over what is obtained with a nonconjugated double bond (Section 2.10B).

| Ethyl butyrate | Methyl methacrylate | Methyl benzoate |
| 1738 cm^{-1} | 1725 cm^{-1} | 1724 cm^{-1} |

Conjugation with the Ester Single-Bonded Oxygen. Conjugation involving the single-bonded oxygen shifts the C=O vibrations to higher frequencies. Apparently, the conjugation interferes with possible resonance with the carbonyl group, leading to an increase in the absorption frequency for the C=O band.

In the spectrum of vinyl acetate (Fig. 2.49), the C=O band appears at 1762 cm^{-1}, an increase of 25 cm^{-1} above a normal ester. Notice that the C=C absorption intensity is increased in a manner similar to the pattern obtained with vinyl ethers (Section 2.13). The substitution of an aryl group on the oxygen would exhibit a similar pattern.

| Ethyl butyrate | Vinyl acetate | Phenyl acetate |
| 1738 cm^{-1} | 1762 cm^{-1} | 1765 cm^{-1} |

Figure 2.52 shows the general effect of α,β unsaturation or aryl substitution and conjugation with oxygen on the C=O vibrations.

Hydrogen-Bonding Effects. When intramolecular (internal) hydrogen bonding is present, the C=O is shifted to a lower frequency, as predicted in Section 2.14A (p. 54) and shown in the spectrum of methyl salicylate (Fig. 2.51).

Methyl salicylate
1680 cm^{-1}

Cyclic Esters (Lactones). The C=O vibrations are shifted to higher frequencies with decreasing ring size, as predicted in Section 2.14A (p. 53). The unstrained, six-membered cyclic ester δ-valerolactone absorbs at about the same value as a noncyclic ester (1735 cm^{-1}). Because of increased angle strain, γ-butyrolactone absorbs at about 35 cm^{-1} higher than δ-valerolactone.

δ-Valerolactone
1735 cm^{-1}

γ-Butyrolactone
1770 cm^{-1}

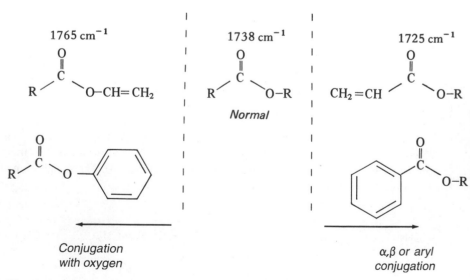

FIGURE 2.52 The effect of α,β unsaturation or aryl substitution and conjugation with oxygen on the C=O vibrations in noncyclic (acyclic) esters.

Table 2.8 presents some typical lactones together with their C=O stretching absorption values. Inspection of these values reveals the influence of ring size, conjugation with a carbonyl group, and conjugation with the single-bond oxygen.

α-Halo Effects. Halogenation on the $α$ carbon leads to an increase in the C=O frequency.

α-Keto Esters. In principle, one should see two carbonyl groups for a compound with "ketone" and "ester" functional groups. Usually one sees a shoulder on the main absorption band near 1735 cm^{-1} or a single broadened absorption band.

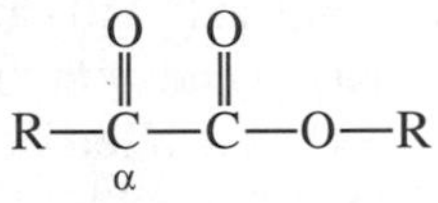

$$R-\overset{\overset{\displaystyle O}{\|}}{\underset{\alpha}{C}}-\overset{\overset{\displaystyle O}{\|}}{C}-O-R$$

β-Keto Esters. Although this class of compounds exhibits tautomerization like that observed in $β$-diketones (p. 59), less evidence exists for the enol form because $β$-keto esters do not enolize to as great an extent. $β$-Keto esters exhibit a *strong-intensity* doublet for the two carbonyl groups at about 1720 and 1740 cm^{-1} in the "keto" tautomer, presumably for the ketone and ester C=O groups. Evidence for the *weak-intensity* C=O band in the "enol" tautomer (often a doublet) appears at about 1650 cm^{-1}. Because of the low concentration of the enol tautomer, one generally cannot observe the broad O—H stretch that was observed in $β$-diketones.

TABLE 2.8

EFFECTS OF RING SIZE, $α,β$ UNSATURATION, AND CONJUGATION WITH OXYGEN ON THE C=O VIBRATIONS IN LACTONES

Ring-Size Effects (cm^{-1})	$α,β$ Conjugation (cm^{-1})	Conjugation with Oxygen (cm^{-1})
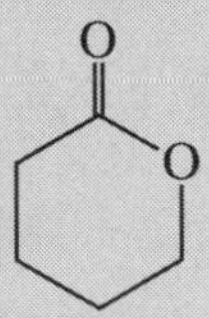 1735	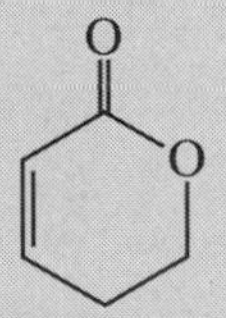 1725	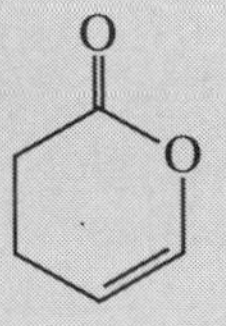 1760
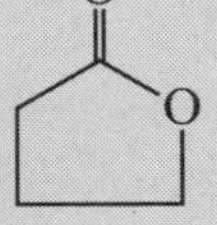 1770	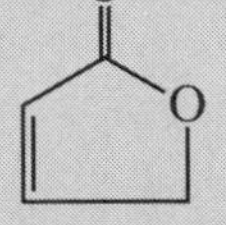 1750	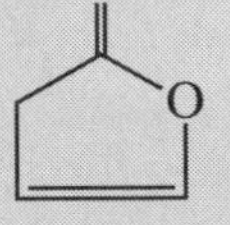 1800
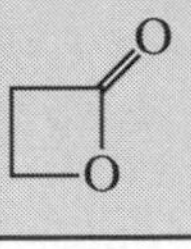 1820		

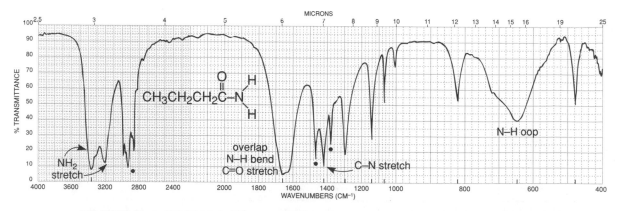

Keto tautomer   Enol tautomer

C—O Stretching Vibrations in Esters. Two (or more) bands appear for the C—O stretching vibrations in esters in the range from 1300 to 1000 cm^{-1}. Generally, the C—O stretch next to the carbonyl group (the "acid" side) of the ester is one of the strongest and broadest bands in the spectrum. This absorption appears between 1300 and 1150 cm^{-1} for most common esters; esters of aromatic acids absorb nearer the higher-frequency end of this range, and esters of saturated acids absorb nearer the lower-frequency end. The C—O stretch for the "alcohol" part of the ester may appear as a weaker band in the range from 1150 to 1000 cm^{-1}. In analyzing the 1300- to 1000-cm^{-1} region to confirm an ester functional group, do not worry about fine details. It is usually sufficient to find at least one very strong and broad absorption to help identify the compound as an ester.

F.   Amides

Amides show a very strong band for the C=O group that appears in the range of 1680–1630 cm^{-1}. The N—H stretch is observed in the range of 3475–3150 cm^{-1}. Unsubstitued (primary) amides, R—CO—NH$_2$, show two bands in the N—H region while monosubstituted (secondary) amides, R—CO—NH—R, show only one band. The presence of N—H bands plus an unusually low value for the C=O would suggest the presence of an amide functional group. Disubstituted (tertiary) amides, R—CO—NR$_2$, will show the C=O in the range of 1680–1630 cm^{-1}, but will not show an N—H stretch.

SPECTRAL ANALYSIS BOX

AMIDES

C=O   Stretch occurs at approximately 1680–1630 cm^{-1}.

N—H   Stretch in primary amides (—NH$_2$) gives two bands near 3350 and 3180 cm^{-1}. Secondary amides have one band (—NH) at about 3300 cm^{-1}.

N—H   Bending occurs around 1640–1550 cm^{-1} for primary and secondary amides.

Examples: propionamide (Fig. 2.53) and *N*-methylacetamide (Fig. 2.54).

FIGURE 2.53 The infrared spectrum of propionamide (Nujol mull, KBr plates). Dots indicate the Nujol (mineral oil) absorption bands (see Fig. 2.8).

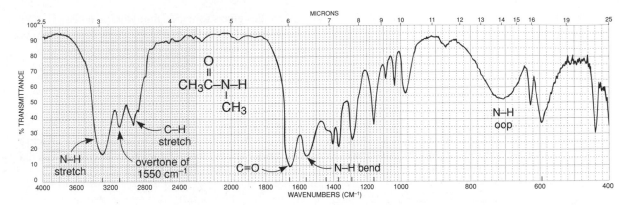

FIGURE 2.54 The infrared spectrum of *N*-methylacetamide (neat liquid, KBr plates).

DISCUSSION SECTION

Carbonyl Absorption in Amides. Primary and secondary amides in the solid phase (potassium bromide pellet or Nujol) have broad C=O absorptions in the range from 1680 to 1630 cm^{-1}. The C=O band partially overlaps the N—H bending band which appears in the range 1640–1620 cm^{-1}, making the C=O band appear as a doublet. In very dilute solution, the band appears at about 1690 cm^{-1}. This effect is similar to that observed for carboxylic acids, in which hydrogen bonding reduces the frequency in the solid state or in concentrated solution. Tertiary amides, which cannot form hydrogen bonds, have C=O frequencies that are not influenced by the physical state and absorb in about the same range as do primary and secondary amides (1680–1630 cm^{-1}).

Primary amide   Secondary amide   Tertiary amide

Cyclic amides (lactams) give the expected increase in C=O frequency for decreasing ring size, as shown for lactones in Table 2.8.

~1660 cm^{-1} ~1705 cm^{-1} ~1745 cm^{-1}

N—H and C—N Stretching Bands. A pair of fairly strong N—H stretching bands appears at about 3350 cm^{-1} and 3180 cm^{-1} for a primary amide in the solid state (KBr or Nujol). The 3350- and 3180-cm^{-1} bands result from the asymmetric and symmetric vibrations, respectively (Section 2.3). Figure 2.53 shows an example, the spectrum of propionamide. In the solid state, secondary amides and lactams give one band at about 3300 cm^{-1}. A weaker band may appear at about 3100 cm^{-1} in secondary amides; it is attributed to a Fermi resonance overtone of the 1550-cm^{-1} band. A C—N stretching band appears at about 1400 cm^{-1} for primary amides.

N—H Bending Bands. In the solid state, primary amides give strong bending vibrational bands in the range from 1640 to 1620 cm^{-1}. They often nearly overlap the C=O stretching bands. Primary amides give other bending bands at about 1125 cm^{-1} and a very broad band in the range from 750 to 600 cm^{-1}. Secondary amides give relatively strong bending bands at about 1550 cm^{-1}; these are attributed to a combination of a C—N stretching band and an N—H bending band.

G. Acid Chlorides

Acid chlorides show a very strong band for the C=O group that appears in the range of 1810–1775 cm^{-1} for aliphatic acid chlorides. Acid chloride and anhydrides are the most common functional groups that have a C=O appearing at such a high frequency. Conjugation lowers the frequency.

SPECTRAL ANALYSIS BOX

ACID CHLORIDES

C=O   Stretch occurs in the range 1810–1775 cm^{-1} in unconjugated chlorides. Conjugation lowers the frequency to 1780–1760 cm^{-1}.

C—Cl   Stretch occurs in the range 730–550 cm^{-1}.

Examples: acetyl chloride (Fig. 2.55) and benzoyl chloride (Fig. 2.56).

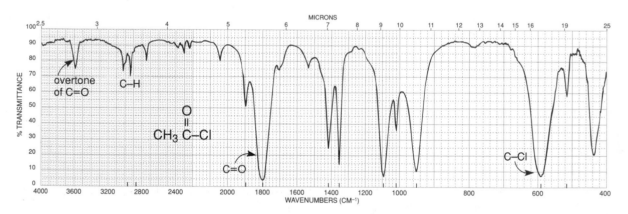

FIGURE 2.55 The infrared spectrum of acetyl chloride (neat liquid, KBr plates).

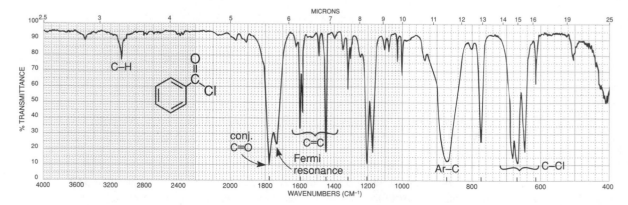

FIGURE 2.56 The infrared spectrum of benzoyl chloride (neat liquid, KBr plates).

DISCUSSION SECTION

C=O Stretching Vibrations. By far the most common acid halides, and the only ones discussed in this book, are acid chlorides. The strong carbonyl absorption appears at a characteristically high frequency of about 1800 cm^{-1} for saturated acid chlorides. Figure 2.55 shows the spectrum of acetyl chloride. Conjugated acid chlorides absorb at a lower frequency (1780 to 1760 cm^{-1}), as predicted in Section 2.14A (p. 52). Figure 2.56 shows an example of an aryl-substituted acid chloride, benzoyl chloride. In this spectrum the main absorption occurs at 1774 cm^{-1}, but a weak shoulder appears on the higher-frequency side of the C=O band (about 1810 cm^{-1}). The shoulder is probably the result of an overtone of a strong band in the 1000- to 900-cm^{-1} range. A weak band is also seen at about 1900 cm^{-1} in the spectrum of acetyl chloride (Fig. 2.55). Sometimes this overtone band is relatively strong.

In some aromatic acid chlorides one may observe another rather strong band, often on the lower-frequency side of the C=O band, which makes the C=O appear as a doublet. This band, which appears in the spectrum of benzoyl chloride (Fig. 2.56) at about 1730 cm^{-1}, is probably a Fermi resonance band originating from an interaction of the C=O vibration, with an overtone of a strong band for aryl-C stretch often appearing in the range from 900 to 800 cm^{-1}. When a fundamental vibration couples with an overtone or combination band, the coupled vibration is called **Fermi resonance.** The Fermi resonance band may also appear on the higher-frequency side of the C=O in many aromatic acid chlorides. This type of interaction can lead to splitting in other carbonyl compounds, as well.

C—Cl Stretching Vibrations. These bands, which appear in the range from 730 to 550 cm^{-1}, are best observed if KBr plates or cells are used. One strong C—Cl band appears in the spectrum of acetyl chloride. In most other acid chlorides, one may observe as many as four bands, due to the many conformations that are possible.

H. Anhydrides

Anhydrides show two strong bands for the C=O groups. Simple alkyl-substituted anhydrides generally give bands near 1820 and 1750 cm^{-1}. Anhydrides and acid chlorides are the most common functional groups that have a C=O peak appearing at such a high frequency. Conjugation shifts each of the bands to lower frequencies (about 30 cm^{-1} each). Simple five-membered ring anhydrides have bands at near 1860 and 1780 cm^{-1}.

SPECTRAL ANALYSIS BOX

ANHYDRIDES

C=O   Stretch always has two bands, 1830–1800 cm^{-1} and 1775–1740 cm^{-1}, with variable relative intensity. Conjugation moves the absorption to a lower frequency. Ring strain (cyclic anhydrides) moves the absorptions to a higher frequency.

C—O   Stretch (multiple bands) occurs in the range 1300–900 cm^{-1}.

Example: propionic anhydride (Fig. 2.57).

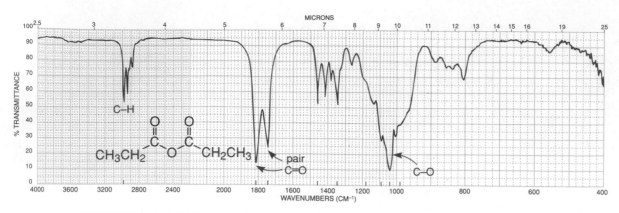

FIGURE 2.57 The infrared spectrum of propionic anhydride (neat liquid, KBr plates).

DISCUSSION SECTION

The characteristic pattern for noncyclic and saturated anhydrides is the appearance of *two strong bands,* not necessarily of equal intensities, in the regions from 1830 to 1800 cm^{-1} and from 1775 to 1740 cm^{-1}. The two bands result from asymmetric and symmetric stretch (Section 2.3). Conjugation shifts the absorption to a lower frequency, while cyclization (ring strain) shifts the absorption to a higher frequency. The *strong* and *broad* C—O stretching vibrations occur in the region from 1300 to 900 cm^{-1}. Figure 2.57 shows the spectrum of propionic anhydride.

2.15 AMINES

Primary amines, R—NH$_2$, show two N—H stretching bands in the range 3500–3300 cm^{-1}, whereas secondary amines, R$_2$N—H, show only one band in that region. Tertiary amines will not show an N—H stretch. Because of these features, it is easy to differentiate among primary, secondary, and tertiary amines by inspection of the N—H stretch region.

SPECTRAL ANALYSIS BOX

AMINES

N—H	Stretch occurs in the range 3500–3300 cm^{-1}. Primary amines have two bands. Secondary amines have one band: a vanishingly weak one for aliphatic compounds and a stronger one for aromatic secondary amines. Tertiary amines have no N—H stretch.
N—H	Bend in primary amines results in a broad band in the range 1640–1560 cm^{-1}. Secondary amines absorb near 1500 cm^{-1}.
N—H	Out-of-plane bending absorption can sometimes be observed near 800 cm^{-1}.
C—N	Stretch occurs in the range 1350–1000 cm^{-1}.

Examples: butylamine (Fig. 2.58), dibutylamine (Fig. 2.59), tributylamine (Fig. 2.60), and *N*-methylaniline (Fig. 2.61).

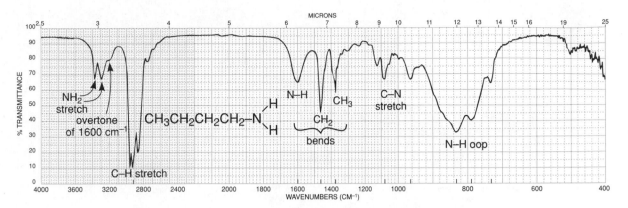

FIGURE 2.58 The infrared spectrum of butylamine (neat liquid, KBr plates).

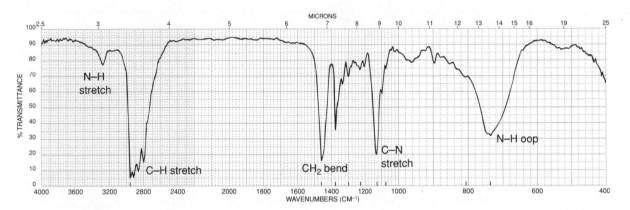

FIGURE 2.59 The infrared spectrum of dibutylamine (neat liquid, KBr plates).

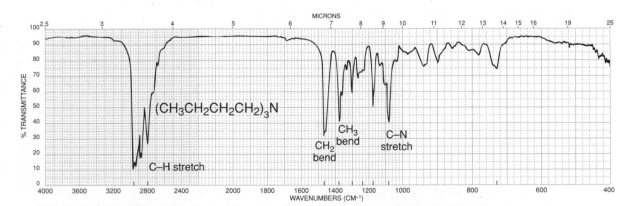

FIGURE 2.60 The infrared spectrum of tributylamine (neat liquid, KBr plates).

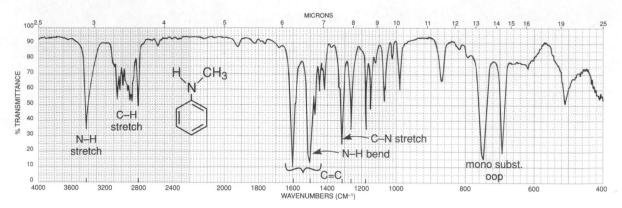

FIGURE 2.61 The infrared spectrum of *N*-methylaniline (neat liquid, KBr plates).

DISCUSSION SECTION

The N—H stretching vibrations occur in the range from 3500 to 3300 cm^{-1}. In neat liquid samples, the N—H bands are often weaker and sharper than an O—H band (see Fig. 2.6). Amines may sometimes be differentiated from alcohols on that basis. Primary amines, determined as neat liquids (hydrogen bonded), give *two bands* at about 3400 and 3300 cm^{-1}. The higher-frequency band in the pair is due to the asymmetric vibration, whereas the lower-frequency band results from a symmetric vibration (Section 2.3). In dilute solution, the two free N—H stretching vibrations are shifted to higher frequencies. Figure 2.58 shows the spectrum of an aliphatic primary amine. A low-intensity shoulder appears at about 3200 cm^{-1} on the low-frequency side of the symmetric N—H stretching band. This low-intensity band has been attributed to an overtone of the N—H *bending* vibration that appears near 1600 cm^{-1}. The 3200-cm^{-1} shoulder has been enhanced by a Fermi resonance interaction with the symmetric N—H stretching band near 3300 cm^{-1}. The overtone band is often even more pronounced in aromatic primary amines.

Aliphatic secondary amines determined as neat liquids give *one band* in the N—H stretching region at about 3300 cm^{-1}, but the band is often vanishingly weak. On the other hand, an aromatic secondary amine gives a stronger N—H band near 3400 cm^{-1}. Figures 2.59 and 2.61 are the spectra of an aliphatic secondary amine and an aromatic secondary amine, respectively. Tertiary amines do not absorb in this region, as shown in Figure 2.60.

In primary amines, the N—H bending mode (scissoring) appears as a medium- to strong-intensity (broad) band in the range from 1640 to 1560 cm^{-1}. In aromatic secondary amines, the band shifts to a lower frequency and appears near 1500 cm^{-1}. However, in aliphatic secondary amines the N—H bending vibration is very weak and usually is not observed. The N—H vibrations in aromatic compounds often overlap the aromatic C=C ring absorptions, which also appear in this region. An out-of-plane N—H bending vibration appears as a broad band near 800 cm^{-1} for primary and secondary amines. These bands appear in the spectra of compounds determined as neat liquids and are seen most easily in aliphatic amines (Figs. 2.58 and 2.59).

The C—N stretching absorption occurs in the region from 1350 to 1000 cm^{-1} as a medium to strong band for all amines. Aliphatic amines absorb from 1250 to 1000 cm^{-1}, whereas aromatic amines absorb from 1350 to 1250 cm^{-1}. The C—N absorption occurs at a higher frequency in aromatic amines because resonance increases the double-bond character between the ring and the attached nitrogen atom.

2.16 NITRILES, ISOCYANATES, ISOTHIOCYANATES, AND IMINES

Nitriles, isocyanates, and isothiocyanates all have sp-hybridized carbon atoms similar to the $C \equiv C$ bond. They absorb in the region 2100–2270 cm^{-1}. On the other hand the $C = N$ bond of an imine has an sp^2 carbon atom. Imines and similar compounds absorb near where double bonds appear, 1690–1640 cm^{-1}.

SPECTRAL ANALYSIS BOX

NITRILES   R—C≡N

—C≡N   Stretch is a medium-intensity, sharp absorption near 2250 cm^{-1}. Conjugation with double bonds or aromatic rings moves the absorption to a lower frequency.

Examples: butyronitrile (Fig. 2.62) and benzonitrile (Fig. 2.63).

ISOCYANATES   R—N=C=O

—N=C=O   Stretch in an isocyanate gives a broad, intense absorption near 2270 cm^{-1}.

Example: benzyl isocyanate (Fig. 2.64).

ISOTHIOCYANATES   R—N=C=S

—N=C=S   Stretch in an isothiocyanate gives one or two broad, intense absorptions centering near 2125 cm^{-1}.

IMINES   R₂C=N—R

—C=N—   Stretch in an imine, oxime, and so on gives a variable-intensity absorption in the range 1690–1640 cm^{-1}.

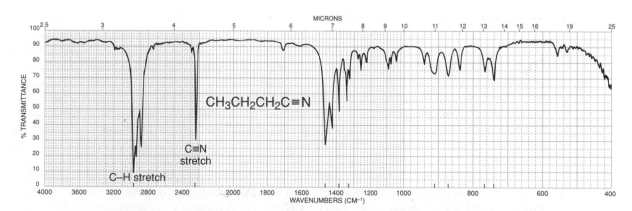

FIGURE 2.62 The infrared spectrum of butyronitrile (neat liquid, KBr plates).

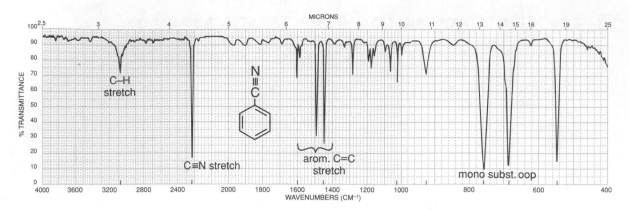

FIGURE 2.63 The infrared spectrum of benzonitrile (neat liquid, KBr plates).

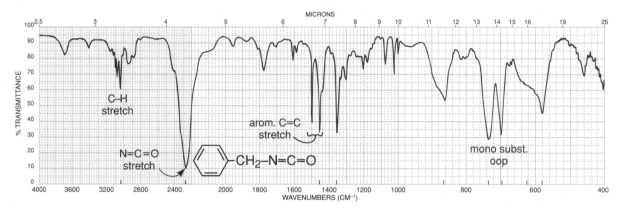

FIGURE 2.64 The infrared spectrum of benzyl isocyanate (neat liquid, KBr plates).

DISCUSSION SECTION

***sp*-Hybridized Carbon.** The C≡N group in a nitrile gives a medium-intensity, sharp band in the triple-bond region of the spectrum (2270 to 2210 cm^{-1}). The C≡C bond, which absorbs near this region (2150 cm^{-1}), usually gives a weaker and broader band unless it is at the end of the chain. Aliphatic nitriles absorb at about 2250 cm^{-1} whereas their aromatic counterparts absorb at lower frequencies, near 2230 cm^{-1}. Figures 2.62 and 2.63 are the spectra of an aliphatic nitrile and an aromatic nitrile, respectively. Aromatic nitriles absorb at lower frequencies with increased intensity because of conjugation of the triple bond with the ring. Isocyanates also contain an *sp*-hybridized carbon atom (R−N=C=O). This class of compounds gives a broad, intense band at about 2270 cm^{-1} (Fig. 2.64).

***sp*^{2}-Hybridized Carbon.** The C=N bond absorbs in about the same range as a C=C bond. Although the C=N band varies in intensity from compound to compound, it usually is more intense than that obtained from the C=C bond. An oxime (R−CH=N−O−H) gives a C=N absorption in the range from 1690 to 1640 cm^{-1} and a broad O−H absorption between 3650 and 2600 cm^{-1}. An imine (R−CH=N−R) gives a C=N absorption in the range from 1690 to 1650 cm^{-1}.

2.17 NITRO COMPOUNDS

Nitro compounds show two strong bands in the infrared spectrum. One appears near 1550 cm⁻¹ and the other near 1350 cm⁻¹. Although these two bands may partially overlap the aromatic ring region, 1600–1450 cm⁻¹, it is usually easy to see the NO_2 peaks.

SPECTRAL ANALYSIS BOX

NITRO COMPOUNDS

Aliphatic nitro compounds: asymmetric stretch (strong), 1600–1530 cm⁻¹; symmetric stretch (medium), 1390–1300 cm⁻¹.

Aromatic nitro compounds (conjugated): asymmetric stretch (strong), 1550–1490 cm⁻¹; symmetric stretch (strong), 1355–1315 cm⁻¹.

Examples: 1-nitrohexane (Fig. 2.65) and nitrobenzene (Fig. 2.66).

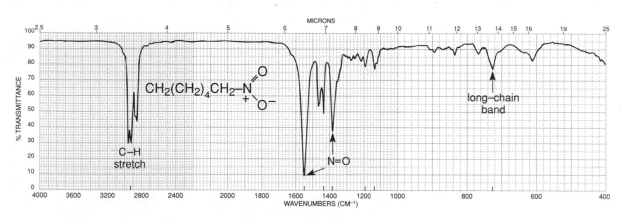

FIGURE 2.65 The infrared spectrum of 1-nitrohexane (neat liquid, KBr plates).

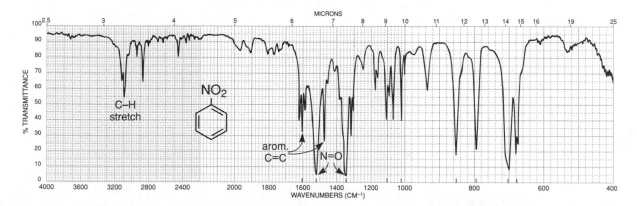

FIGURE 2.66 The infrared spectrum of nitrobenzene (neat liquid, KBr plates).

DISCUSSION SECTION

The nitro group (NO_2) gives two strong bands in the infrared spectrum. In aliphatic nitro compounds, the asymmetric stretching vibration occurs in the range from 1600 to 1530 cm^{-1}, and the symmetric stretching band appears between 1390 and 1300 cm^{-1}. An aliphatic nitro compound—for example, 1-nitrohexane (Fig. 2.65)—absorbs at about 1550 and 1380 cm^{-1}. Normally, its lower-frequency band is less intense than its higher-frequency band. In contrast with aliphatic nitro compounds, aromatic compounds give two bands of nearly equal intensity. Conjugation of a nitro group with an aromatic ring shifts the bands to lower frequencies: 1550–1490 cm^{-1} and 1355–1315 cm^{-1}. For example, nitrobenzene (Fig. 2.66) absorbs strongly at 1525 and 1350 cm^{-1}. The nitroso group (R—N=O) gives only one strong band that appears in the range from 1600 to 1500 cm^{-1}.

2.18 CARBOXYLATE SALTS, AMINE SALTS, AND AMINO ACIDS

This section covers compounds with ionic bonds. Included here are carboxylate salts, amine salts, and amino acids. Amino acids are included in this section because of their zwitterionic nature.

SPECTRAL ANALYSIS BOX

CARBOXYLATE SALTS $R—\overset{\overset{O}{\|}}{C}—O^-$ Na^+

Asymmetric stretch (strong) occurs near 1600 cm^{-1}; symmetric stretch (strong) occurs near 1400 cm^{-1}.

Frequency of C=O absorption is lowered from the value found for the parent carboxylic acid because of resonance (more single-bond character).

AMINE SALTS NH_4^+ RNH_3^+ $R_2NH_2^+$ R_3NH^+

N—H   Stretch (broad) occurs at 3300–2600 cm^{-1}. The ammonium ion absorbs to the left in this range, while the tertiary amine salt absorbs to the right. Primary and secondary amine salts absorb in the middle of the range, 3100–2700 cm^{-1}. A broad band often appears near 2100 cm^{-1}.

N—H   Bend (strong) occurs at 1610–1500 cm^{-1}. Primary (two bands) is asymmetric at 1610 cm^{-1}, symmetric at 1500 cm^{-1}. Secondary absorbs in the range 1610–1550 cm^{-1}. Tertiary absorbs only weakly.

AMINO ACIDS

$$R—\underset{\underset{NH_2}{|}}{CH}—\overset{\overset{O}{\|}}{C}—OH \longrightarrow R—\underset{\underset{{}^+NH_3}{|}}{CH}—\overset{\overset{O}{\|}}{C}—O^-$$

These compounds exist as zwitterions (internal salts) and exhibit spectra that are combinations of carboxylate and primary amine salts. Amino acids show NH_3^+ stretch (very broad), N—H bend (asymmetric/symmetric), and COO$^-$ stretch (asymmetric/symmetric).

Example: leucine (Fig. 2.67).

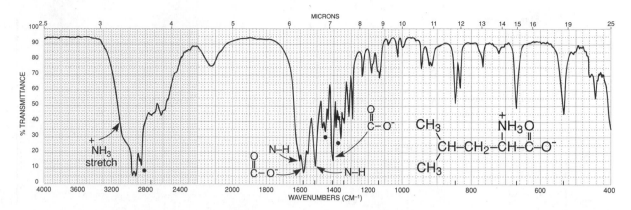

FIGURE 2.67 The infrared spectrum of leucine (Nujol mull, KBr plates). Dots indicate the Nujol (mineral oil) absorption bands (see Fig. 2.8).

2.19 SULFUR COMPOUNDS

Infrared spectral data for sulfur-containing compounds are covered in this section. Included here are single-bonded compounds (mercaptans or thiols and sulfides). Double-bonded S=O compounds are also included in this section.

SPECTRAL ANALYSIS BOX

MERCAPTANS R—S—H

S—H   Stretch, one weak band, occurs near 2550 cm^{-1} and virtually confirms the presence of this group, since few other absorptions appear here.

Example: benzenethiol (Fig. 2.68).

SULFIDES R—S—R

Little useful information is obtained from the infrared spectrum.

SULFOXIDES R—S—R
$$\underset{O}{\overset{||}{}}$$

S=O   Stretch, one strong band, occurs near 1050 cm^{-1}.

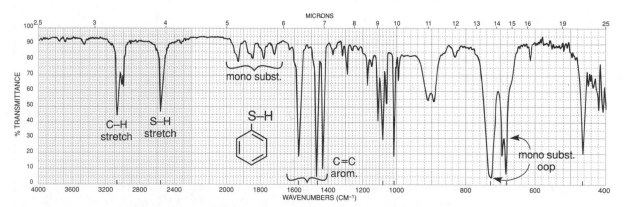

FIGURE 2.68 The infrared spectrum of benzenethiol (neat liquid, KBr plates).

SULFONES

S=O   Asymmetric stretch (strong) occurs at 1300 cm^{-1}, symmetric stretch (strong) at 1150 cm^{-1}.

SULFONYL CHLORIDES

S=O   Asymmetric stretch (strong) occurs at 1375 cm^{-1}, symmetric stretch (strong) at 1185 cm^{-1}.

Example: benzenesulfonyl chloride (Fig. 2.69).

SULFONATES

S=O   Asymmetric stretch (strong) occurs at 1350 cm^{-1}, symmetric stretch (strong) at 1175 cm^{-1}.

S—O   Stretch, several strong bands, occurs in the range 1000–750 cm^{-1}.

Example: methyl *p*-toluenesulfonate (Fig. 2.70).

SULFONAMIDES
(Solid State)

 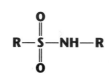

S=O   Asymmetric stretch (strong) occurs at 1325 cm^{-1}, symmetric stretch (strong) at 1140 cm^{-1}.

N—H   Primary stretch occurs at 3350 and 3250 cm^{-1}; secondary stretch occurs at 3250 cm^{-1}; bend occurs at 1550 cm^{-1}.

Example: benzenesulfonamide (Fig. 2.71).

SULFONIC ACIDS
(Anhydrous)

S=O   Asymmetric stretch (strong) occurs at 1350 cm^{-1}, symmetric stretch (strong) at 1150 cm^{-1}.

S—O   Stretch (strong) occurs at 650 cm^{-1}.

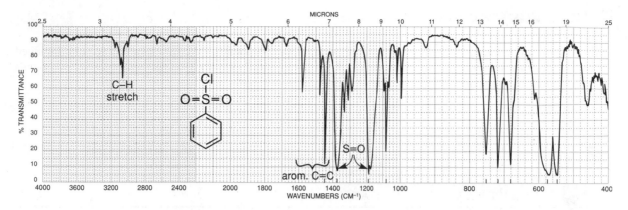

FIGURE 2.69 The infrared spectrum of benzenesulfonyl chloride (neat liquid, KBr plates).

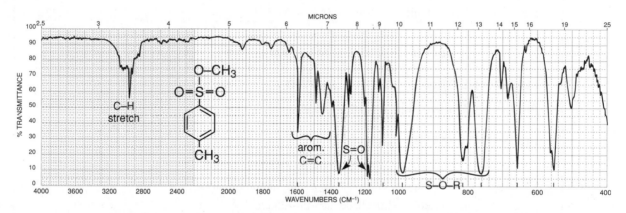

FIGURE 2.70 The infrared spectrum of methyl *p*-toluenesulfonate (neat liquid, KBr plates).

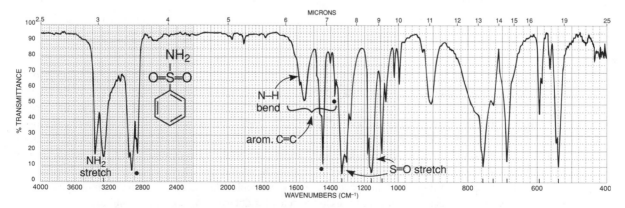

FIGURE 2.71 The infrared spectrum of benzenesulfonamide (Nujol mull, KBr plates). Dots indicate the Nujol (mineral oil) absorption bands (see Fig. 2.8).

2.20 PHOSPHORUS COMPOUNDS

Infrared spectral data for phosphorus-containing compounds are covered in this section. Included here are single-bonded compounds (P—H, P—R, and P—O—R). Double bonded P=O compounds are also included in this section.

SPECTRAL ANALYSIS BOX

PHOSPHINES   RPH$_2$ R$_2$PH

P—H	Stretch, one strong, sharp band, at 2320–2270 cm^{-1}.
PH$_2$	Bend, medium bands, at 1090–1075 cm^{-1} and 840–810 cm^{-1}.
P—H	Bend, medium band, at 990–885 cm^{-1}.
P—CH$_3$	Bend, medium bands, at 1450–1395 cm^{-1} and 1346–1255 cm^{-1}.
P—CH$_2$—	Bend, medium band, at 1440–1400 cm^{-1}.

PHOSPHINE OXIDES   R$_3$P=O   Ar$_3$P=O

P=O	Stretch, one very strong band, at 1210–1140 cm^{-1}.

PHOSPHATE ESTERS   (RO)$_3$P=O

P=O	Stretch, one very strong band, at 1300–1240 cm^{-1}.
R—O	Stretch, one or two strong bands, at 1088–920 cm^{-1}.
P—O	Stretch, medium band, at 845–725 cm^{-1}.

2.21 ALKYL AND ARYL HALIDES

Infrared spectral data for halogen-containing compounds are covered in this section. It is difficult to determine the presence or the absence of a halide in a compound via infrared spectroscopy. There are several reasons for this problem. First, the C—X absorption occurs at very low frequencies, to the extreme right of the spectrum, where a number of other bands appear (fingerprint). Second, the sodium chloride plates or cells that are often used obscure the region where halogens absorb (these plates are transparent only above 650 cm^{-1}). Other inorganic salts, most commonly KBr, can be used to extend the region down to 400 cm^{-1}. Mass spectral methods (Sections 8.5B and 8.6N) provide more reliable information for this class of compounds. The spectra of carbon tetrachloride and chloroform are shown in this section. These solvents are often used to dissolve solids for determining spectra in solution.

SPECTRAL ANALYSIS BOX

FLUORIDES R—F

C—F   Stretch (strong) at 1400–1000 cm⁻¹. Monofluoroalkanes absorb at the lower-frequency end of this range, while polyfluoroalkanes give multiple strong bands in the range 1350–1100 cm⁻¹. Aryl fluorides absorb between 1250 and 1100 cm⁻¹.

CHLORIDES R—Cl

C—Cl   Stretch (strong) in aliphatic chlorides occurs in the range 785–540 cm⁻¹. Primary chlorides absorb at the upper end of this range, while tertiary chlorides absorb near the lower end. Two or more bands may be observed, due to the different conformations which are possible.

Multiple substitution on a single-carbon atom results in an intense absorption at the upper-frequency end of this range: CH_2Cl_2 (739 cm⁻¹), $HCCl_3$ (759 cm⁻¹), and CCl_4 (785 cm⁻¹). Aryl chlorides absorb between 1100 and 1035 cm⁻¹.

CH_2—Cl   Bend (wagging) at 1300–1230 cm⁻¹.

Examples: carbon tetrachloride (Fig. 2.72) and chloroform (Fig. 2.73).

BROMIDES R—Br

C—Br   Stretch (strong) in aliphatic bromides occurs at 650–510 cm⁻¹, out of the range of routine spectroscopy using NaCl plates or cells. The trends indicated for aliphatic chlorides hold for bromides. Aryl bromides absorb between 1075 and 1030 cm⁻¹.

CH_2—Br   Bend (wagging) at 1250–1190 cm⁻¹.

IODIDES R—I

C—I   Stretch (strong) in aliphatic iodides occurs at 600–485 cm⁻¹, out of the range of routine spectroscopy using NaCl plates or cells. The trends indicated for aliphatic chlorides hold for iodides.

CH_2—I   Bend (wagging) at 1200–1150 cm⁻¹.

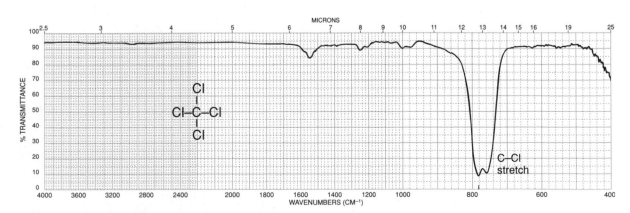

FIGURE 2.72 The infrared spectrum of carbon tetrachloride (neat liquid, KBr plates).

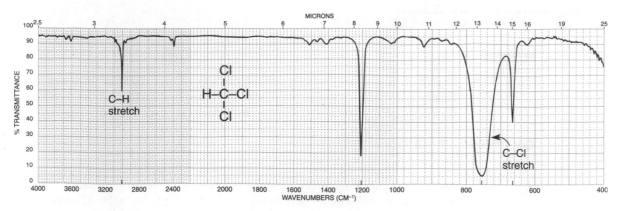

FIGURE 2.73 The infrared spectrum of chloroform (neat liquid, KBr plates).

PROBLEMS

When a molecular formula is given, it is advisable to calculate an index of hydrogen deficiency (Section 1.4). The index often gives useful information about the functional group or groups that may be present in the molecule.

*1. In each of the following parts, a molecular formula is given. Deduce the structure that is consistent with the infrared spectrum. There may be more than one possible answer.

   (a) C_3H_3Cl

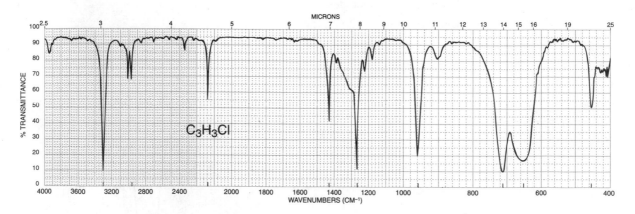

(b) $C_{10}H_{14}$

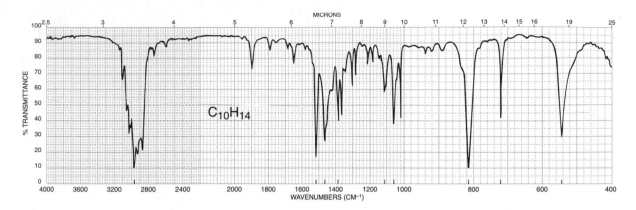

(c) C_7H_9N

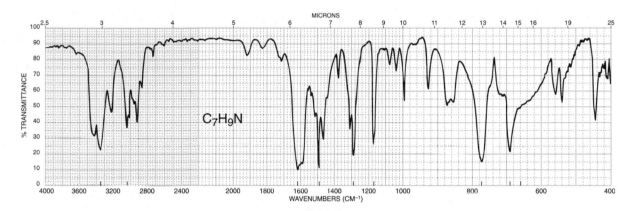

(d) C_7H_8O

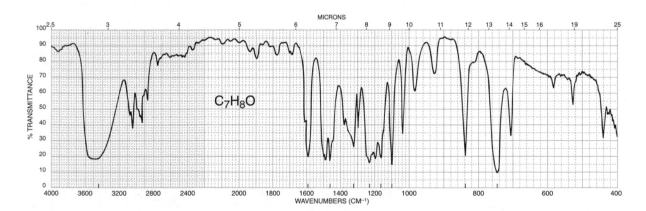

(e) $C_8H_{11}N$

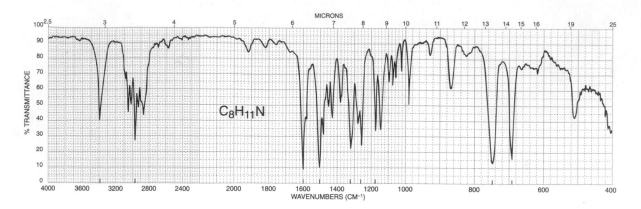

(f) C_7H_7Cl

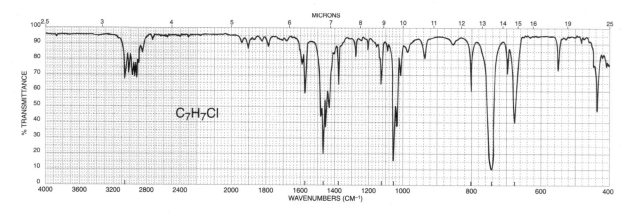

(g) $C_3H_5O_2Cl$

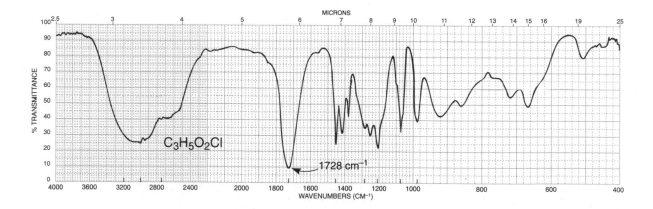

(h) C$_5$H$_{12}$O

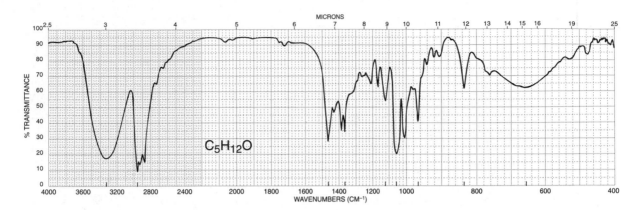

(i) C$_6$H$_{10}$O

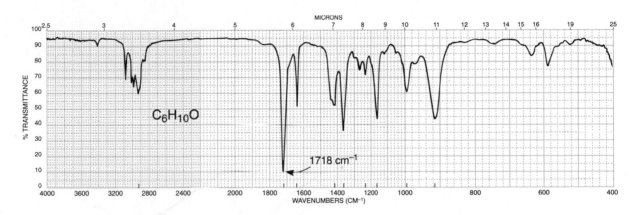

(j) C$_{10}$H$_{12}$ (two six-membered rings)

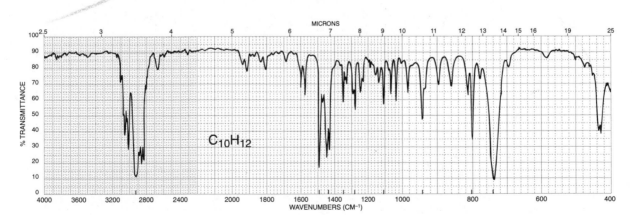

(k) $C_5H_{10}N_2$

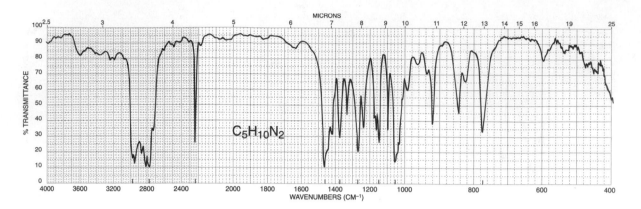

(l) C_4H_8O

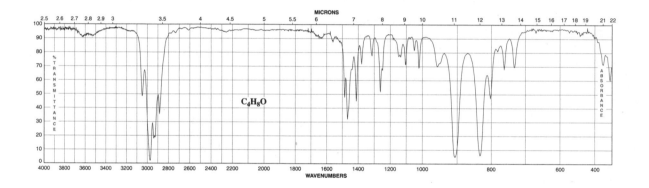

*2. Ants emit tiny amounts of chemicals called alarm pheromones to warn other ants (of the same species) of the presence of an enemy. Several of the components of the pheromone in one species have been identified, and two of their structures follow. Which compound has the infrared spectrum shown?

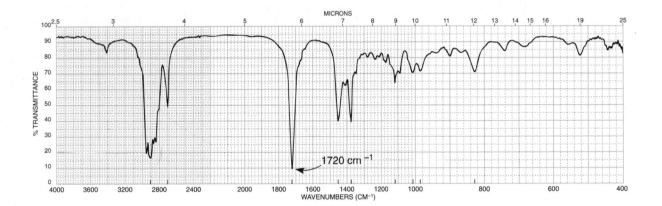

Citral   Citronellal

*3. The main constituent of cinnamon oil has the formula C_9H_8O. From the following infrared spectrum, deduce the structure of this component.

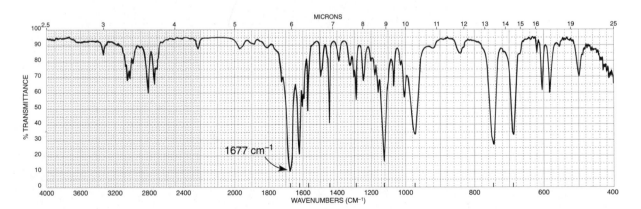

*4. The infrared spectra of *cis*- and *trans*-3-hexen-1-ol follow. Assign a structure to each.

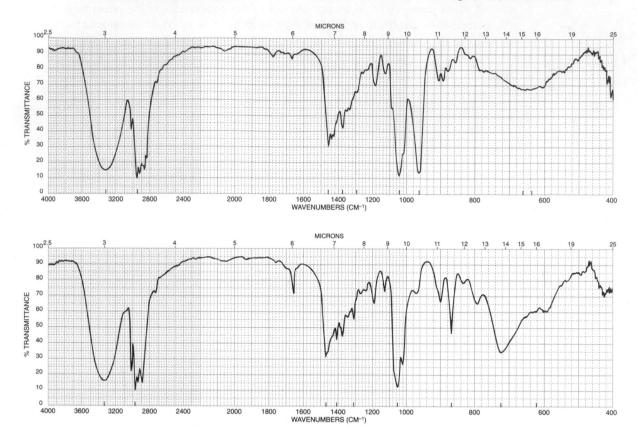

5. In each part, choose the structure that best fits the infrared spectrum shown.

*(a)

A

B

C

D

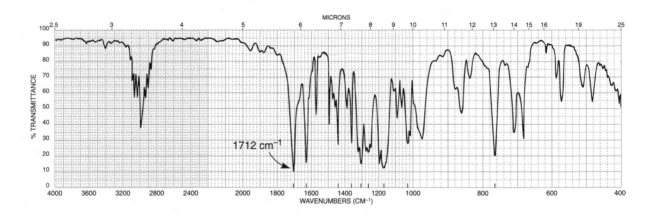

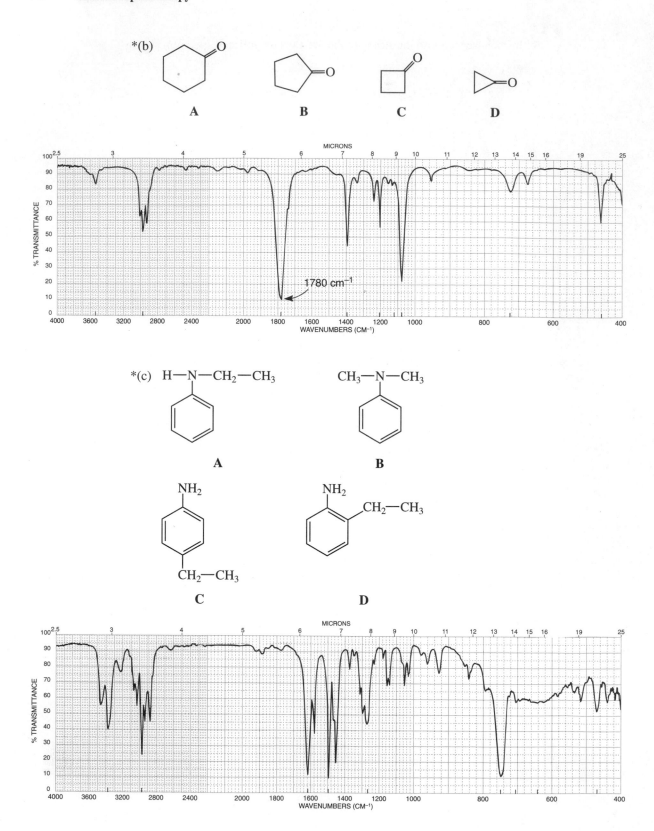

*(d)

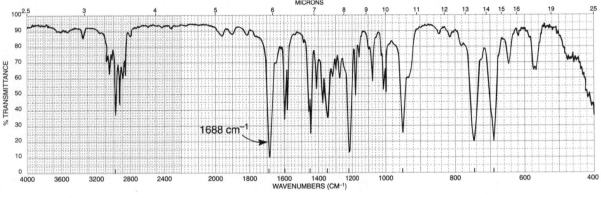

A

B

C

D

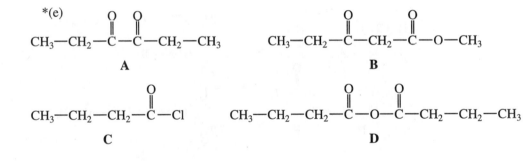

1688 cm⁻¹

*(e)

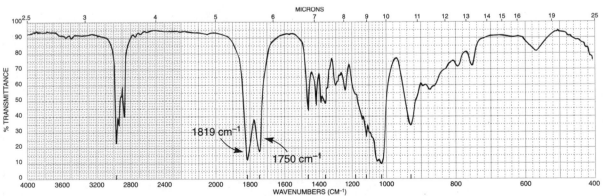

A

B

C

D

1819 cm⁻¹

1750 cm⁻¹

(f)

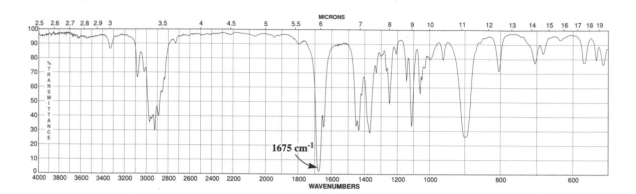

A   B   C   D

(g)

A

B

C

D

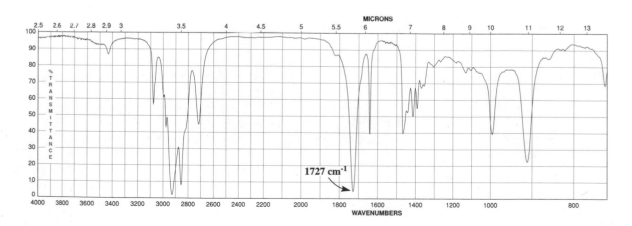

(h)

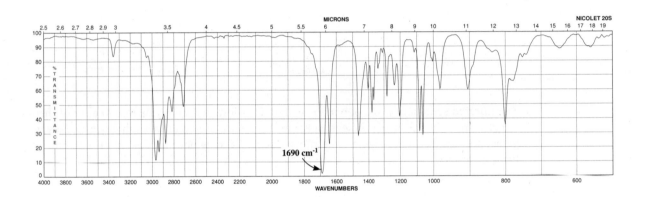

(i)

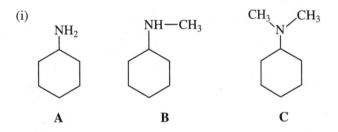

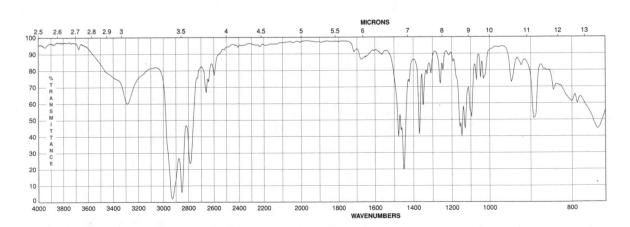

(j) CH$_3$—CH$_2$—CH$_2$—S—CH$_2$—CH$_2$—CH$_3$ CH$_3$(CH$_2$)$_4$CH$_2$—S—H

A   **B**

CH$_3$(CH$_2$)$_4$CH$_2$—O—H   CH$_3$—CH$_2$—CH$_2$—O—CH$_2$—CH$_2$—CH$_3$

C   **D**

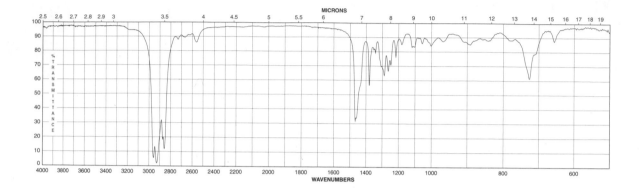

*6. The infrared spectra of some polymeric materials follow. Assign a structure to each of them, selected from the following choices: polyamide (Nylon), poly(methyl methacrylate), polyethylene, polystyrene, and poly(acrylonitrile-styrene). You may need to look up the structures of these materials.

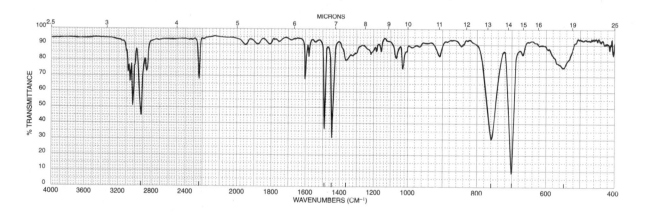

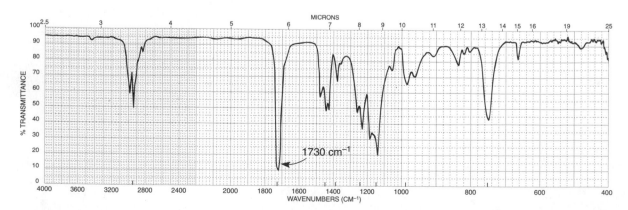

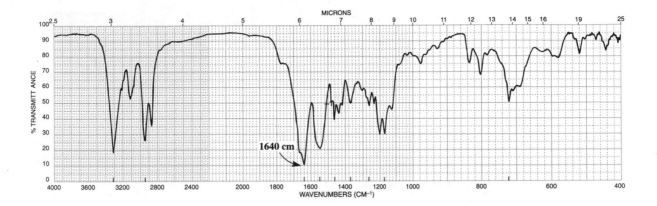

7. Assign a structure to each of the spectra shown. Choose from among the following 5-carbon esters.

$$CH_3-CH_2-\overset{\overset{\displaystyle O}{\|}}{C}-O-CH=CH_2 \qquad CH_3-CH_2-CH_2-O-\overset{\overset{\displaystyle O}{\|}}{C}-CH_3$$

$$CH_3-\overset{\overset{\displaystyle O}{\|}}{C}-O-CH_2-CH=CH_2 \qquad CH_2=CH-\overset{\overset{\displaystyle O}{\|}}{C}-O-CH_2-CH_3$$

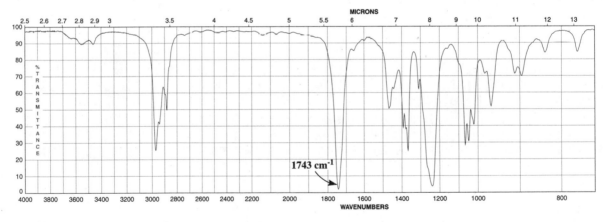

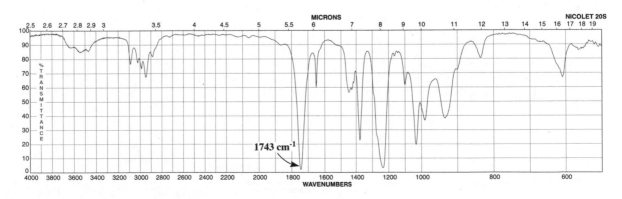

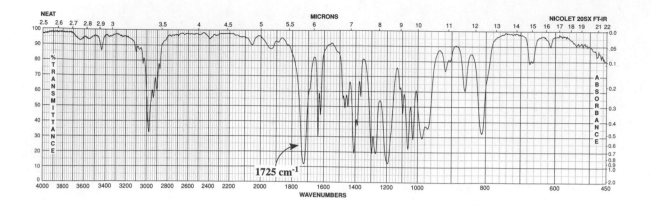

8. Assign a structure to each of the following three spectra. The structures are shown here.

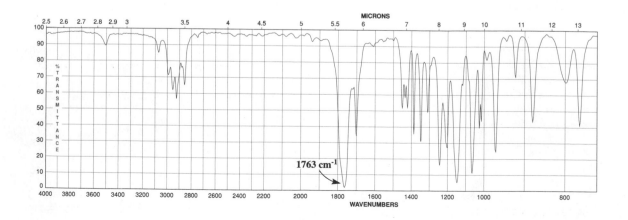

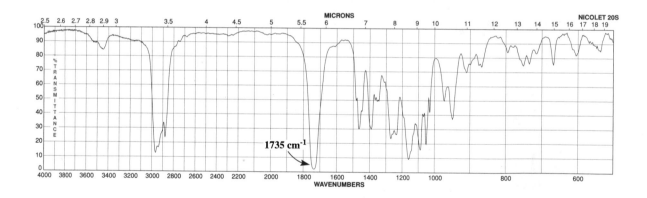

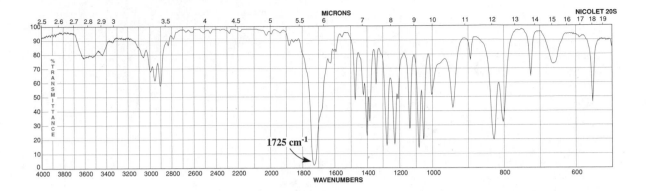

9. Assign a structure to each of the following three spectra. The structures are shown here.

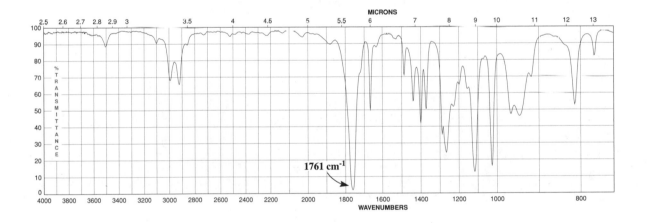

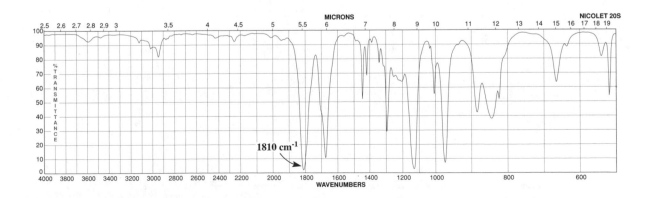

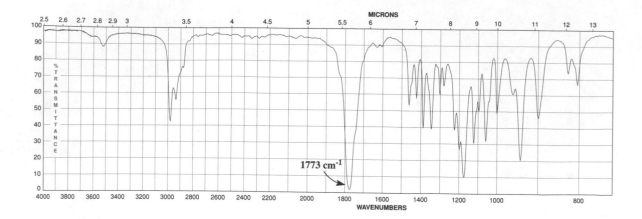

10. Assign a structure to each of the spectra shown. Choose from among the following 5-carbon alcohols.

$$\underset{CH_3}{\overset{CH_3}{\diagdown}}C=CH-CH_2OH$$

$$CH_2=C\underset{CH_2CH_2OH}{\overset{CH_3}{\diagup}}$$

$$CH_3CH_2CH_2CH_2CH_2OH \qquad CH_2=CH-CH_2CH_2CH_2OH$$

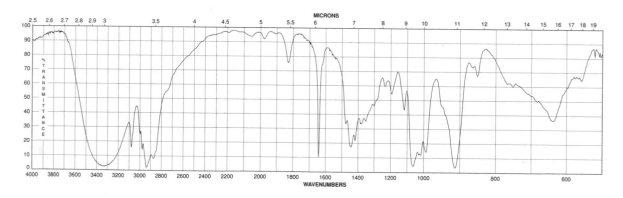

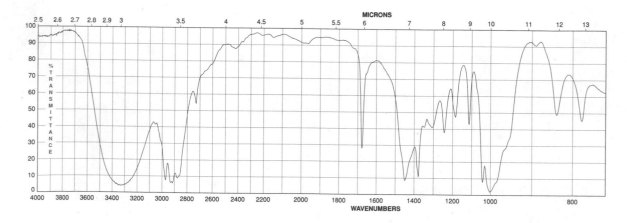

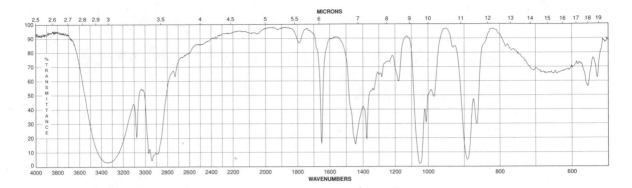

11. Substitution of an amino group on the *para* position of acetophenone shifts the C=O frequency from about 1685 to 1652 cm^{-1}, whereas a nitro group attached to the *para* position yields a C=O frequency of 1693 cm^{-1}. Explain the shift for each substituent from the 1685 cm^{-1} base value for acetophenone.

REFERENCES

Books and Compilations of Spectra

Bellamy, L. J., *The Infrared Spectra of Complex Molecules,* 3d ed., John Wiley, New York, 1975.

Colthrup, N. B., L. H. Daly, and S. E. Wiberley, *Introduction to Infrared and Raman Spectroscopy,* 3rd ed., Academic Press, New York, 1990.

Lin-Vien, D., N. B. Colthrup, W. G. Fateley, and J. G. Grasselli, *The Handbook of Infrared and Raman Characteristic Frequencies of Organic Molecules,* Academic Press, New York, 1991.

Nakanishi, K., and P. H. Solomon, *Infrared Absorption Spectroscopy,* Holden–Day, San Francisco, 1977.

Pouchert, C. J., *Aldrich Library of FT-IR Spectra,* Aldrich Chemical Co., Milwaukee, WI, 1985 (1st edition) and 1997 (2nd edition).

Pretsch, E., T. Clerc, J. Seibl, and W. Simon, *Tables of Spectral Data for Structure Determination of Organic Compounds,* 2nd ed., Springer-Verlag, Berlin and New York, 1989. Translated from the German by K. Biemann.

Sadtler Standard Spectra, Sadtler Research Laboratories Division, Bio-Rad Laboratories, Inc., 3316 Spring Garden Street, Philadelphia, PA 19104-2596. Numerous FT-IR search libraries are available for computers.

Silverstein, R. M., and F. X. Webster, *Spectrometric Identification of Organic Compounds,* 6th ed., John Wiley, New York, 1998.

Szymanski, H. A., *Interpreted Infrared Spectra,* Vols. 1–3, Plenum Press, New York, 1964–1967.

Computer Programs that Teach Spectroscopy

Clough, F. W., "Introduction to Spectroscopy," Version 2.0 for MS-DOS and Macintosh, Trinity Software, 607 Tenney Mtn. Highway, Suite 215, Plymouth, NH 03264; www.trinitysoftware.com

"IR Tutor," John Wiley & Sons, 1 Wiley Drive, Somerset, NJ 08875-1272.

Pavia, D. L., "Spectral Interpretation," MS-DOS version, Trinity Software, 607 Tenney Mtn. Highway, Suite 215, Plymouth, NH 03264; www.trinitysoftware.com

Schatz, P. F., "Spectrabook I and II and Spectradeck I and II," MS-DOS and Macintosh versions, Falcon Software, One Hollis Street, Wellesley, MA 02482; *www.falconsoftware.com*

Web sites

http://www.dq.fct.unl.pt/qoa/jas/ir.html

This site lists a number of resources for infrared spectroscopy, including databases, tutorials, problems, and theory.

http://www.aist.go.jp/RIODB/SDBS/menu-e.html

Integrated Spectral DataBase System for Organic Compounds, National Institute of Materials and Chemical Research, Tsukuba, Ibaraki 305-8565, Japan. This database includes infrared, mass spectra, and NMR data (proton and carbon-13) for a number of compounds.

http://webbook.nist.gov/chemistry/

The National Institute of Standards and Technology (NIST) has developed the WebBook. This site includes gas phase infrared spectra and mass spectral data for compounds.

http://www.chem.ucla.edu/~webnmr/index.html

UCLA Department of Chemistry and Biochemistry in connection with Cambridge University Isotope Laboratories maintains a website, WebSpecta, which provides NMR and IR spectroscopy problems for students to interpret. They provide links to other sites with problems for students to solve.

CHAPTER 3

NUCLEAR MAGNETIC RESONANCE SPECTROSCOPY
Part One: Basic Concepts

Nuclear magnetic resonance (NMR) is a spectroscopic method that is even more important to the organic chemist than infrared spectroscopy. Many nuclei may be studied by NMR techniques, but hydrogen and carbon are most commonly available. Whereas infrared spectroscopy reveals the types of functional groups present in a molecule, NMR gives information about the number of magnetically distinct atoms of the type being studied. When hydrogen nuclei (protons) are studied, for instance, one can determine the number of each of the distinct types of hydrogen nuclei as well as obtain information regarding the nature of the immediate environment of each type. Similar information can be determined for the carbon nuclei. The combination of IR and NMR data is often sufficient to determine completely the structure of an unknown molecule.

3.1 NUCLEAR SPIN STATES

Many atomic nuclei have a property called **spin:** the nuclei behave as if they were spinning. In fact, any atomic nucleus that possesses either *odd* mass, *odd* atomic number, or both has a quantized **spin angular momentum** and a magnetic moment. The more common nuclei that possess spin include ^1_1H, ^2_1H, $^{13}_6C$, $^{14}_7N$, $^{17}_8O$, and $^{19}_9F$. Notice that the nuclei of the ordinary (most abundant) isotopes of carbon and oxygen, $^{12}_6C$ and $^{16}_8O$, are not included among those with the spin property. However, the nucleus of the ordinary hydrogen atom, the proton, does have spin. For each nucleus with spin, the number of allowed spin states it may adopt is quantized and is determined by its nuclear spin quantum number, I. For each nucleus, the number I is a physical constant, and there are $2I + 1$ allowed spin states with integral differences ranging from $+I$ to $-I$. The individual spin states fit into the sequence

$$+I, (I - 1), \ldots, (-I + 1), -I \qquad \text{Equation 3.1}$$

For instance, a proton (hydrogen nucleus) has the spin quantum number $I = \frac{1}{2}$ and has two allowed spin states $[2(\frac{1}{2}) + 1 = 2]$ for its nucleus: $-\frac{1}{2}$ and $+\frac{1}{2}$. For the chlorine nucleus, $I = \frac{3}{2}$ and there are four allowed spin states $[2(\frac{3}{2}) + 1 = 4]$: $-\frac{3}{2}, -\frac{1}{2}, +\frac{1}{2}$, and $+\frac{3}{2}$. Table 3.1 gives the spin quantum numbers of several nuclei.

TABLE 3.1
SPIN QUANTUM NUMBERS OF SOME COMMON NUCLEI

Element	^1_1H	^2_1H	$^{12}_6C$	$^{13}_6C$	$^{14}_7N$	$^{16}_8O$	$^{17}_8O$	$^{19}_9F$	$^{31}_{15}P$	$^{35}_{17}Cl$
Nuclear spin quantum number	$\frac{1}{2}$	1	0	$\frac{1}{2}$	1	0	$\frac{5}{2}$	$\frac{1}{2}$	$\frac{1}{2}$	$\frac{3}{2}$
Number of spin states	2	3	0	2	3	0	6	2	2	4

In the absence of an applied magnetic field, all the spin states of a given nucleus are of equivalent energy (degenerate), and in a collection of atoms, all of the spin states should be almost equally populated, with the same number of atoms having each of the allowed spins.

3.2 NUCLEAR MAGNETIC MOMENTS

Spin states are not of equivalent energy in an applied magnetic field, because the nucleus is a charged particle, and any moving charge generates a magnetic field of its own. Thus, the nucleus has a magnetic moment, μ, generated by its charge and spin. A hydrogen nucleus may have a clockwise $(+\frac{1}{2})$ or counterclockwise $(-\frac{1}{2})$ spin, and the nuclear magnetic moments (μ) in the two cases are pointed in opposite directions. In an applied magnetic field, all protons have their magnetic moments either aligned with the field or opposed to it. Figure 3.1 illustrates these two situations.

Hydrogen nuclei can adopt only one or the other of these orientations with respect to the applied field. The spin state $+\frac{1}{2}$ is of lower energy since it is aligned with the field, while the spin state $-\frac{1}{2}$ is of higher energy since it is opposed to the applied field. This should be intuitively obvious to anyone who thinks a little about the two situations depicted in Figure 3.2, involving magnets. The aligned configuration of magnets is stable (low energy). However, where the magnets are opposed (not aligned), the center magnet is repelled out of its current (high-energy) orientation. If the central

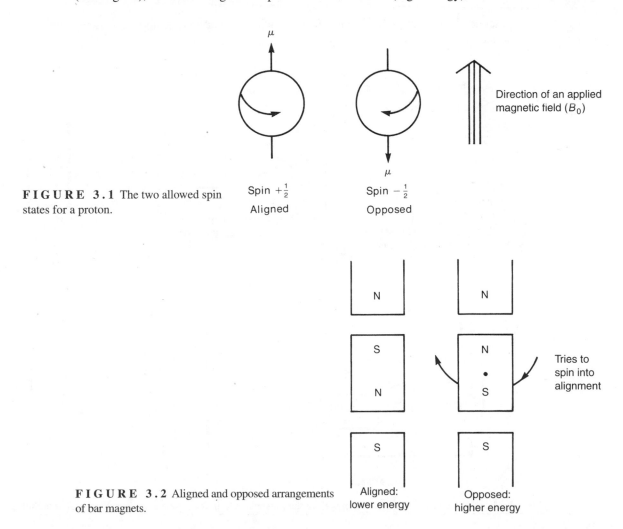

FIGURE 3.1 The two allowed spin states for a proton.

Spin $+\frac{1}{2}$
Aligned

Spin $-\frac{1}{2}$
Opposed

Direction of an applied magnetic field (B_0)

FIGURE 3.2 Aligned and opposed arrangements of bar magnets.

Aligned: lower energy

Opposed: higher energy

Tries to spin into alignment

magnet were placed on a pivot, it would spontaneously spin around the pivot into alignment (low energy). Hence, as an external magnetic field is applied, the degenerate spin states split into two states of unequal energy, as shown in Figure 3.3.

In the case of a chlorine nucleus, there are four energy levels, as shown in Figure 3.4. The $+\frac{3}{2}$ and $-\frac{3}{2}$ spin states are aligned with the applied field and opposed to the applied field, respectively. The $+\frac{1}{2}$ and $-\frac{1}{2}$ spin states have intermediate orientations, as indicated by the vector diagram on the right in Figure 3.4.

3.3 ABSORPTION OF ENERGY

The nuclear magnetic resonance phenomenon occurs when nuclei aligned with an applied field are induced to absorb energy and change their spin orientation with respect to the applied field. Figure 3.5 illustrates this process for a hydrogen nucleus.

The energy absorption is a quantized process, and the energy absorbed must equal the energy difference between the two states involved.

$$E_{absorbed} = (E_{-\frac{1}{2}\,state} - E_{+\frac{1}{2}\,state}) = h\upsilon \qquad \textbf{Equation 3.2}$$

In practice, this energy difference is a function of the strength of the applied magnetic field, B_0, as illustrated in Figure 3.6.

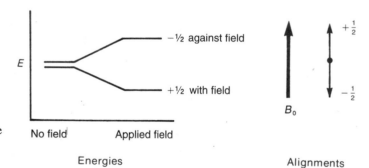

F I G U R E 3 . 3 The spin states of a proton in the absence and in the presence of an applied magnetic field.

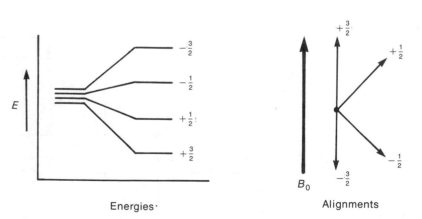

F I G U R E 3 . 4 The spin states of a chlorine atom both in the presence and in the absence of an applied magnetic field.

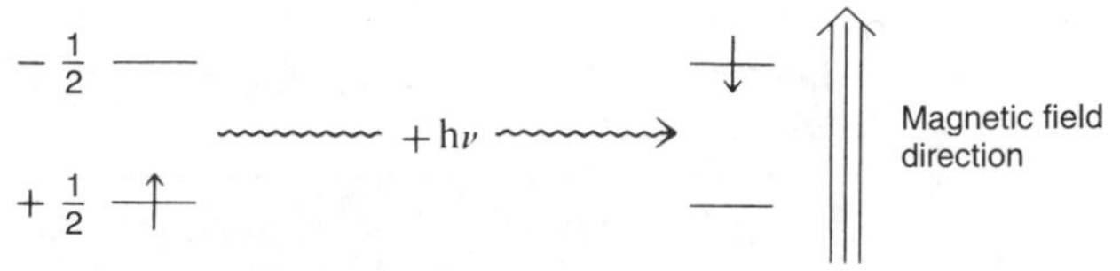

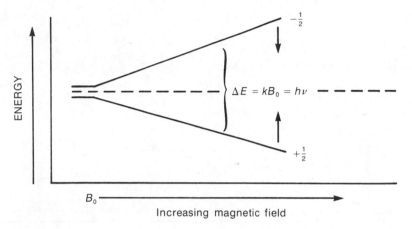

FIGURE 3.5 The NMR absorption process for a proton.

FIGURE 3.6 The spin-state energy separation as a function of the strength of the applied magnetic field, B_0.

The stronger the applied magnetic field, the greater the energy difference between the possible spin states:

$$\Delta E = f(B_0)$$ **Equation 3.3**

The magnitude of the energy-level separation also depends on the particular nucleus involved. Each nucleus (hydrogen, chlorine, and so on) has a different ratio of magnetic moment to angular momentum, since each has different charge and mass. This ratio, called the **magnetogyric ratio, γ,** is a constant for each nucleus and determines the energy dependence on the magnetic field:

$$\Delta E = f(\gamma B_0) = h\nu$$ **Equation 3.4**

Since the angular momentum of the nucleus is quantized in units of $h/2\pi$, the final equation takes the form

$$\Delta E = \gamma \left(\frac{h}{2\pi}\right) B_0 = h\nu$$ **Equation 3.5**

Solving for the frequency of the absorbed energy,

$$\upsilon = \left(\frac{\gamma}{2\pi}\right) B_0$$ **Equation 3.6**

If the correct value of γ for the proton is substituted, one finds that an unshielded proton should absorb radiation of frequency 42.6 MHz in a field of strength 1 Tesla (10,000 Gauss) or radiation of frequency 60.0 MHz in a field of strength 1.41 Tesla (14,100 Gauss). Table 3.2 shows the field strengths and frequencies at which several nuclei have resonance (i.e., absorb energy and make transitions).

Although many nuclei are capable of exhibiting magnetic resonance, the organic chemist is mainly interested in proton and carbon resonances. This chapter emphasizes hydrogen. Chapter 4

TABLE 3.2

FREQUENCIES AND FIELD STRENGTHS AT WHICH SELECTED NUCLEI HAVE THEIR NUCLEAR RESONANCES

Isotope	Natural Abundance (%)	Field Strength, B_0 (Tesla[a])	Frequency, ν (MHz)	Magnetogyric Ratio, γ (radians/Tesla)
^1H	99.98	1.00	42.6	267.53
		1.41	60.0	
		2.35	100.0	
		4.70	200.0	
		7.05	300.0	
^2H	0.0156	1.00	6.5	41.1
^{13}C	1.108	1.00	10.7	67.28
		1.41	15.1	
		2.35	25.0	
		4.70	50.0	
		7.05	75.0	
^{19}F	100.0	1.00	40.0	251.7
^{31}P	100.0	1.00	17.2	108.3

[a] 1 Tesla = 10,000 Gauss.

will discuss nuclei other than hydrogen—for example, carbon-13, fluorine-19, phosphorus-31, and deuterium (hydrogen-2).

For a proton (the nucleus of a hydrogen atom), if the applied magnetic field has a strength of approximately 1.41 Tesla, the difference in energy between the two spin states of the proton is about 2.39×10^{-5} kJ/mole. Radiation with a frequency of about 60 MHz (60,000,000 Hz), which lies in the radiofrequency region of the electromagnetic spectrum, corresponds to this energy difference. Other nuclei have both larger and smaller energy differences between spin states than do hydrogen nuclei. The earliest nuclear magnetic resonance spectrometers applied a variable magnetic field with a range of strengths near 1.41 Tesla and supplied a constant radiofrequency radiation of 60 MHz. They effectively induced transitions only among proton (hydrogen) spin states in a molecule and were not useful for other nuclei. Separate instruments were required to observe transitions in the nuclei of other elements, such as carbon and phosphorus. The newer, more expensive Fourier transform instruments (Section 3.7B), which are in common use today, are equipped to observe the nuclei of several different elements in a single instrument. Instruments operating at frequencies of 200 and 300 MHz are now quite common, and instruments with frequencies up to 600 MHz are found in the larger research universities.

3.4 THE MECHANISM OF ABSORPTION (RESONANCE)

To understand the nature of a nuclear spin transition, the analogy of a child's spinning top is useful. Protons absorb energy because they begin to precess in an applied magnetic field. The phenomenon of precession is similar to that of a spinning top. Owing to the influence of the earth's gravitational field, the top begins to "wobble," or precess, about its axis (Fig. 3.7a). A spinning nucleus behaves in a similar fashion under the influence of an applied magnetic field (Fig. 3.7b).

When the magnetic field is applied, the nucleus begins to precess about its own axis of spin with angular frequency ω, which is sometimes called its **Larmor frequency.** The frequency at

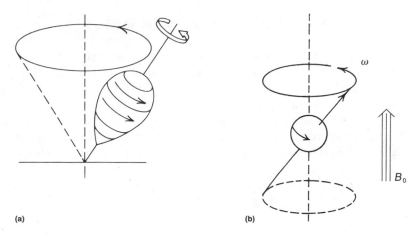

FIGURE 3.7 (a) A top precessing in the earth's gravitational field; (b) the precession of a spinning nucleus resulting from the influence of an applied magnetic field.

which a proton precesses is directly proportional to the strength of the applied magnetic field; the stronger the applied field, the higher the rate (angular frequency, ω) of precession. For a proton, if the applied field is 1.41 Tesla (14,100 Gauss), the frequency of precession is approximately 60 MHz.

Since the nucleus has a charge, the precession generates an oscillating electric field of the same frequency. If radiofrequency waves of this frequency are supplied to the precessing proton, the energy can be absorbed. That is, when the frequency of the oscillating electric field component of the incoming radiation just matches the frequency of the electric field generated by the precessing nucleus, the two fields can couple, and energy can be transferred from the incoming radiation to the nucleus, thus causing a spin change. This condition is called **resonance,** and the nucleus is said to have resonance with the incoming electromagnetic wave. Figure 3.8 schematically illustrates the resonance process.

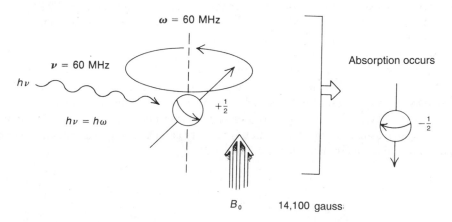

FIGURE 3.8 The nuclear magnetic resonance process; absorption occurs when $\upsilon = \omega$.

▨ 3.5 POPULATION DENSITIES OF NUCLEAR SPIN STATES

For a proton, if the applied magnetic field has a strength of approximately 1.41 Tesla, resonance occurs at about 60 MHz, and, using $\Delta E = h\nu$, we can calculate that the difference in energy between the two spin states of the proton is about 2.39×10^{-5} kJ/mole. Thermal energy resulting from room temperature is sufficient to populate both of these energy levels, since the energy separation between the two levels is small. There is, however, a slight excess of nuclei in the lower-energy spin state. The magnitude of this difference can be calculated using the Boltzmann distribution equations. Equation 3.7 gives the Boltzmann ratio of nuclear spins in the upper and lower levels.

$$\frac{N_{\text{upper}}}{N_{\text{lower}}} = e^{-\Delta E/kT} = e^{-h\nu/kT}$$ **Equation 3.7**

$$h = 6.624 \times 10^{-34}\,\text{J·sec}$$

$$k = 1.380 \times 10^{-23}\,\text{J/K · molecule}$$

$$T = \text{absolute temperature (K)}$$

where ΔE is the energy difference between the upper and lower energy states and k is the molecular (not molar) gas constant. Since $\Delta E = h\nu$, the second form of the equation is derived, where ν is the operating frequency of the instrument and h is Planck's constant.

Using Equation 3.7, one can calculate that at 298 K (25°C), for an instrument operating at 60 MHz there are 1,000,009 nuclei in the lower (favored) spin state for every 1,000,000 that occupy the upper spin state:

$$\frac{N_{\text{upper}}}{N_{\text{lower}}} = 0.999991 = \frac{1,000,000}{1,000,009}$$

In other words, in approximately 2 million nuclei, there are only 9 more nuclei in the lower spin state. Let us call this number (9) the **excess population** (Fig. 3.9).

The excess nuclei are the ones that allow us to observe resonance. When the 60-MHz radiation is applied, it not only induces transitions upward but also stimulates transitions downward. If the populations of the upper and lower states become exactly equal, we observe no net signal. This situation is called **saturation.** One must be careful to avoid saturation when performing an NMR experiment. Saturation is achieved quickly if the power of the radiofrequency (RF) signal is too high. Therefore, the very small excess of nuclei in the lower spin state is quite important to NMR spectroscopy, and we can see that very sensitive NMR instrumentation is required to detect the signal.

If we increase the operating frequency of the NMR instrument, the energy difference between the two states increases (see Fig. 3.6), which causes an increase in this excess. Table 3.3 shows how the excess increases with operating frequency. It also clearly shows why modern instrumentation has been designed with increasingly higher operating frequencies. The sensitivity of the instrument is increased, and the resonance signals are stronger, because more nuclei can undergo transition at higher frequency. Before the advent of higher-field instruments it was very difficult to observe less sensitive nuclei such as carbon-13, which is not very abundant (1.1%) and has a detection frequency much lower than that of hydrogen (see Table 3.2).

Population

_____ N

$N = 1,000,000$
Excess = 9

FIGURE 3.9 The excess population of nuclei in the lower spin state at 60 MHz.

_____ $N + 9$

TABLE 3.3
**VARIATION OF ¹H EXCESS NUCLEI
WITH OPERATING FREQUENCY**

Frequency (MHz)	Excess Nuclei
20	3
40	6
60	9
80	12
100	16
200	32
300	48
600	96

3.6 THE CHEMICAL SHIFT AND SHIELDING

Nuclear magnetic resonance has great utility because not all protons in a molecule have resonance at the same frequency. This variability is due to the fact that the protons in a molecule are surrounded by electrons and exist in slightly different electronic environments from one another. The valence-shell electron densities vary from one proton to another. The protons are **shielded** by the electrons that surround them. In an applied magnetic field, the valence electrons of the protons are caused to circulate. This circulation, called a **local diamagnetic current,** generates a counter magnetic field which opposes the applied magnetic field. Figure 3.10 illustrates this effect, which is called **diamagnetic shielding** or **diamagnetic anisotropy.**

Circulation of electrons around a nucleus can be viewed as being similar to the flow of an electric current in an electric wire. From physics, we know that the flow of a current through a wire induces a magnetic field. In an atom, the local diamagnetic current generates a secondary, induced magnetic field which has a direction opposite that of the applied magnetic field.

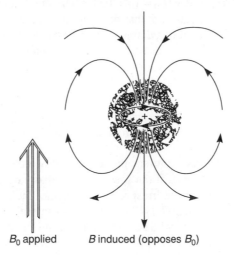

B_0 applied B induced (opposes B_0)

FIGURE 3.10 Diamagnetic anisotropy—the diamagnetic shielding of a nucleus caused by the circulation of valence electrons.

As a result of diamagnetic anisotropy, each proton in a molecule is shielded from the applied magnetic field to an extent that depends on the electron density surrounding it. The greater the electron density around a nucleus, the greater the induced counter field that opposes the applied field. The counter field that shields a nucleus diminishes the net applied magnetic field that the nucleus experiences. As a result, the nucleus precesses at a lower frequency. This means that it also absorbs radiofrequency radiation at this lower frequency. Each proton in a molecule is in a slightly different chemical environment and consequently has a slightly different amount of electronic shielding, which results in a slightly different resonance frequency.

These differences in resonance frequency are very small. For instance, the difference between the resonance frequencies of the protons in chloromethane and those in fluoromethane is only 72 Hz when the applied field is 1.41 Tesla. Since the radiation used to induce proton spin transitions at that magnetic field strength is of a frequency near 60 MHz, the difference between chloromethane and fluoromethane represents a change in frequency of only slightly more than one part per million! It is very difficult to measure exact frequencies to that precision; hence, no attempt is made to measure the exact resonance frequency of any proton. Instead, a reference compound is placed in the solution of the substance to be measured, and the resonance frequency of each proton in the sample is measured relative to the resonance frequency of the protons of the reference substance. In other words, the frequency *difference* is measured directly. The standard reference substance that is used universally is **tetramethylsilane, $(CH_3)_4Si$,** also called **TMS.** This compound was chosen initially because the protons of its methyl groups are more shielded than those of most other known compounds. At that time, no compounds which had better-shielded hydrogens than TMS were known, and it was assumed that TMS would be a good reference substance since it would mark one end of the range. Thus, when another compound is measured, the resonances of its protons are reported in terms of how far (in Hertz) they are shifted from those of TMS.

The shift from TMS for a given proton depends on the strength of the applied magnetic field. In an applied field of 1.41 Tesla the resonance of a proton is approximately 60 MHz, whereas in an applied field of 2.35 Tesla (23,500 Gauss) the resonance appears at approximately 100 MHz. The ratio of the resonance frequencies is the same as the ratio of the two field strengths:

$$\frac{100 \text{ MHz}}{60 \text{ MHz}} = \frac{2.35 \text{ Tesla}}{1.41 \text{ Tesla}} = \frac{23,500 \text{ Gauss}}{14,100 \text{ Gauss}} = \frac{5}{3}$$

Hence, for a given proton, the shift (in Hertz) from TMS is $\frac{5}{3}$ larger in the 100-MHz range ($B_0 = 2.35$ Tesla) than in the 60-MHz range ($B_0 = 1.41$ Tesla). This can be confusing for workers trying to compare data if they have spectrometers that differ in the strength of the applied magnetic field. The confusion is easily overcome if one defines a new parameter that is independent of field strength—for instance, by dividing the shift in Hertz of a given proton by the frequency in megahertz of the spectrometer with which the shift value was obtained. In this manner, a field-independent measure called the **chemical shift** (δ) is obtained

$$\delta = \frac{(\text{shift in Hz})}{(\text{spectrometer frequency in MHz})} \qquad \textbf{Equation 3.8}$$

The chemical shift in δ units expresses the amount by which a proton resonance is shifted from TMS, in parts per million (ppm), of the spectrometer's basic operating frequency. Values of δ for a given proton are always the same irrespective of whether the measurement was made at 60 MHz ($B_0 = 1.41$ Tesla) or at 100 MHz ($B_0 = 2.35$ Tesla). For instance, at 60 MHz the shift of the protons in CH_3Br is 162 Hz from TMS, while at 100 MHz the shift is 270 Hz. However, both of these correspond to the same value of δ (2.70 ppm):

$$\delta = \frac{162 \text{ Hz}}{60 \text{ MHz}} = \frac{270 \text{ Hz}}{100 \text{ MHz}} = 2.70 \text{ ppm}$$

By agreement, most workers report chemical shifts in **delta (δ) units,** or **parts per million (ppm),** of the main spectrometer frequency. On this scale, the resonance of the protons in TMS comes at exactly 0.00 ppm (by definition).

> **Note:** An old scale of chemical shift, no longer used, was called the tau (τ) scale. On this scale, the resonance position of TMS was defined to be 10.00. To convert δ values to τ values, merely subtract them from 10. You will find τ values in the older literature.

The NMR spectrometer actually scans from high δ values to low ones (as will be discussed in Section 3.6). Following is a typical chemical shift scale with the sequence of δ values which would be found on a typical NMR spectrum chart.

Direction of scan $\Rightarrow$

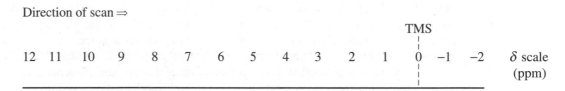

TMS

| 12 | 11 | 10 | 9 | 8 | 7 | 6 | 5 | 4 | 3 | 2 | 1 | 0 | −1 | −2 | δ scale (ppm) |

3.7 THE NUCLEAR MAGNETIC RESONANCE SPECTROMETER

A. The Continuous-Wave (CW) Instrument

Figure 3.11 schematically illustrates the basic elements of a classical 60-MHz NMR spectrometer. The sample is dissolved in a solvent containing no interfering protons (usually CCl_4 or $CDCl_3$), and a small amount of TMS is added to serve as an internal reference. The sample cell is a small cylindrical glass tube that is suspended in the gap between the faces of the pole pieces of the magnet. The sample is spun around its axis to ensure that all parts of the solution experience a relatively uniform magnetic field.

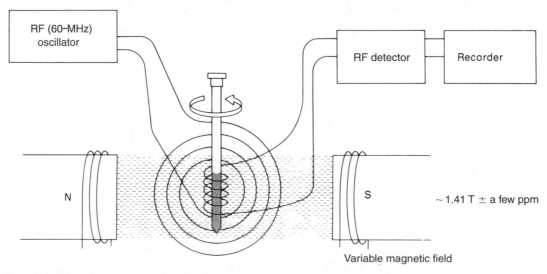

FIGURE 3.11 The basic elements of the classical nuclear magnetic resonance spectrometer.

Also in the magnet gap is a coil attached to a 60-MHz radiofrequency (RF) generator. This coil supplies the electromagnetic energy used to change the spin orientations of the protons. Perpendicular to the RF oscillator coil is a detector coil. When no absorption of energy is taking place, the detector coil picks up none of the energy given off by the RF oscillator coil. When the sample absorbs energy, however, the reorientation of the nuclear spins induces a radiofrequency signal in the plane of the detector coil, and the instrument responds by recording this as a **resonance signal,** or **peak.**

At a constant field strength, the distinct types of protons in a molecule precess at slightly different rates. Rather than changing the frequency of the RF oscillator to allow each of the protons in a molecule to come into resonance, the typical NMR spectrometer uses a constant-frequency RF signal and varies the magnetic field strength. As the magnetic field strength is increased, the precessional frequencies of all the protons increase. When the precessional frequency of a given type of proton reaches 60 MHz, it has resonance. The magnet that is varied is actually a two-part device. There is a main magnet, with a strength of about 1.41 Tesla, which is capped by electromagnet pole pieces. By varying the current through the pole pieces, the worker can increase the main field strength by as much as 20 parts per million (ppm). Changing the field in this way systematically brings all of the different types of protons in the sample into resonance.

As the field strength is increased linearly, a pen travels across a recording chart. A typical spectrum is recorded as in Figure 3.12. As the pen travels from left to right, the magnetic field is increasing. As each chemically distinct type of proton comes into resonance, it is recorded as a peak on the chart. The peak at $\delta = 0$ ppm is due to the internal reference compound TMS. Since highly shielded protons precess more slowly than relatively unshielded protons, it is necessary to increase the field to induce them to precess at 60 MHz. Hence, highly **shielded** protons appear to the right of this chart, and less shielded, or **deshielded,** protons appear to the left. The region of the chart to the

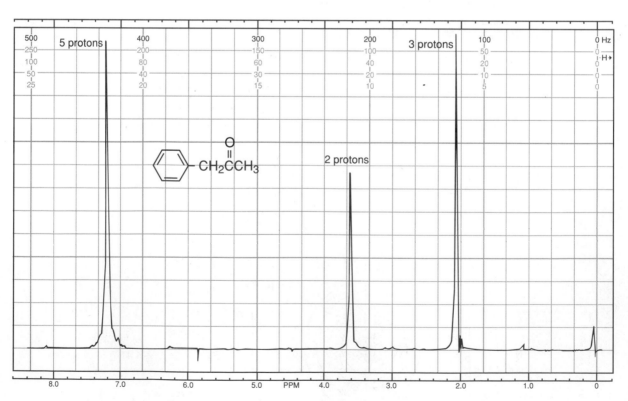

FIGURE 3.12 The 60-MHz ^{1}H nuclear magnetic resonance spectrum of phenylacetone (the absorption peak at the far right is caused by the added reference substance TMS).

left is sometimes said to be **downfield** (or at **low field**), and that to the right, **upfield** (or at **high field**). Varying the magnetic field as is done in the usual spectrometer is exactly equivalent to varying the RF frequency, and a change of 1 ppm in the magnetic field strength (increase) has the same effect as a 1-ppm change (decrease) in the RF frequency (see Eq. 3.6). Hence, changing the field strength instead of the RF frequency is only a matter of instrumental design. Instruments which vary the magnetic field in a continuous fashion, scanning from the downfield end to the upfield end of the spectrum, are called **continuous-wave (CW) instruments.** Because the chemical shifts of the peaks in this spectrum are calculated from *frequency* differences from TMS, this type of spectrum (Fig. 3.12) is said to be a **frequency-domain spectrum.**

A distinctive characteristic enables one to recognize a CW spectrum. Peaks generated by a CW instrument have **ringing,** a decreasing series of oscillations that occurs after the instrument has scanned through the peak (Fig. 3.13). Ringing occurs because the excited nuclei do not have time to relax back to their equilibrium state before the field, and pen, of the instrument have advanced to a new position. The excited nuclei have a relaxation rate that is slower than the rate of scan. As a result, they are still emitting an oscillating, rapidly decaying signal, which is recorded as ringing. Ringing is desirable in a CW instrument and is considered to indicate that the field homogeneity is well adjusted. Ringing is most noticeable when a peak is a sharp singlet (a single, isolated peak).

B. The Pulsed Fourier Transform (FT) Instrument

The CW type of NMR spectrometer, which was described in Section 3.6A, operates by exciting the nuclei of the isotope under observation one type at a time. In the case of ^{1}H nuclei, each distinct type of proton (phenyl, vinyl, methyl, and so on) is excited individually, and its resonance peak is observed and recorded independently of all the others. As we scan, we look at first one type of hydrogen and then another, scanning until all of the types have come into resonance.

An alternative approach, common to modern, sophisticated instruments, is to use a powerful but short burst of energy, called a **pulse,** that excites all of the magnetic nuclei in the molecule simultaneously. In an organic molecule, for instance, all of the ^{1}H nuclei are induced to undergo resonance at the same time. An instrument with a 2.1-Tesla magnetic field uses a short (1- to 10-μsec) burst of 90-MHz energy to accomplish this. The source is turned on and off very quickly, generating a pulse similar to that shown in Figure 3.14a. According to a variation of the Heisenberg Uncertainty Principle, even though the frequency of the oscillator generating this pulse is set to 90 MHz, if the duration of the pulse is very

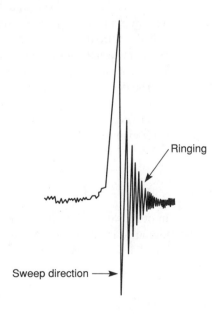

Ringing

Sweep direction ⟶

F I G U R E   3 . 1 3 A CW peak that shows ringing.

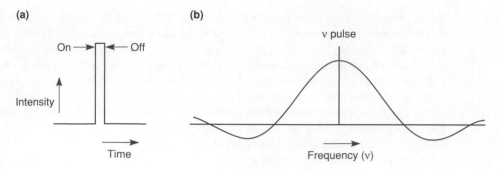

FIGURE 3.14 A short pulse. (a) The original pulse; (b) the frequency content of the same pulse.

short, the frequency content of the pulse is uncertain because the oscillator was not on long enough to establish a solid fundamental frequency. Therefore, the pulse actually contains a *range of frequencies* centered about the fundamental, as shown in Figure 3.14b. This range of frequencies is great enough to excite all of the distinct types of hydrogens in the molecule at once with this single burst of energy.

When the pulse is discontinued, the excited nuclei begin to lose their excitation energy and return to their original spin state, or **relax.** As each excited nucleus relaxes, it emits electromagnetic radiation. Since the molecule contains many different nuclei, many different frequencies of electromagnetic radiation are emitted simultaneously. This emission is called a **free-induction decay (FID)** signal (Fig. 3.15). Notice that the intensity of the FID decays with time as all of the nuclei eventually lose their excitation. The FID is a superimposed combination of all the frequencies emitted and can be quite complex. We usually extract the individual frequencies due to different nuclei by using a computer and a mathematical method called a Fourier transform (FT) analysis, which is described later in this section.

If we look at a very simple molecule such as acetone, we can avoid the inherent complexities of the Fourier transform and gain a clearer understanding of the method. Figure 3.16a shows the FID for the hydrogens in acetone. This FID was determined in an instrument with a 7.05-Tesla magnet operating at 300 MHz.

Since acetone has only one type of hydrogen (all six hydrogens are equivalent), the FID curve is composed of a single sinusoidal wave. The signal decays exponentially with time as the nuclei relax and their signal diminishes. Since the horizontal axis on this signal is time, the FID is sometimes called a **time-domain signal.** If the signal did not decay in intensity, it would appear as a sine (or cosine) wave of constant intensity, as shown in Figure 3.16b. One can calculate the frequency of this wave from its measured wavelength, λ (difference between the maxima).

The determined frequency is not the exact frequency emitted by the methyl hydrogens. Due to the design of the instrument, the basic frequency of the pulse is not the same as the frequency of the acetone resonance. The observed FID is actually an interference signal between the radiofrequency source (300 MHz in this case) and the frequency emitted by the excited nucleus, where the wavelength is given by

$$\lambda = \frac{1}{\nu_{acetone} - \nu_{pulse}}$$

Equation 3.9

In other words, this signal represents the difference in the two frequencies. Since the frequency of the pulse is known, we could readily determine the exact frequency. However, we do not need to know it since we are interested in the chemical shift of those protons, which is given by

$$\delta'_{acetone} = \frac{\nu_{acetone} - \nu_{pulse}}{\nu_{pulse}}$$

Equation 3.10

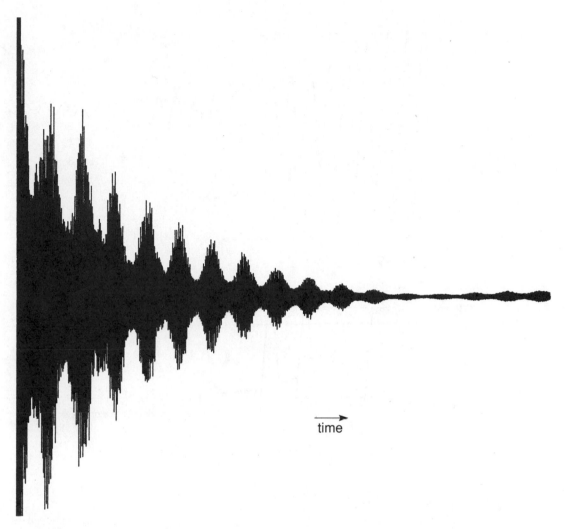

FIGURE 3.15 The ^1H free-induction decay (FID) signal of ethyl phenylacetate (300 MHz).

which can be reduced to the unit analysis

$$ppm = \frac{(Hz)}{MHz}$$

showing that $\delta_{acetone}$ is the chemical shift of the protons of acetone from the position of the pulse, not from TMS. If we know δ_{TMS}, the position of TMS from the pulse, the actual chemical shift of this peak can be calculated by the adjustment

$$\delta_{actual} = (\delta'_{acetone} - \delta'_{TMS})$$

Equation 3.11

We can now plot this peak as a chemical shift on a standard NMR spectrum chart (Fig. 3.16c). The peak for acetone appears at about 2.1 ppm. We have converted the time-domain signal to a **frequency-domain signal,** which is the standard format for a spectrum obtained by a CW instrument.

Now consider the ^1H FID from ethyl phenylacetate (Fig. 3.15). This complex molecule has many types of hydrogens, and the FID is the superimposition of many *different* frequencies, each of which could have a *different* decay rate! A mathematical method called a **Fourier transform,** however,

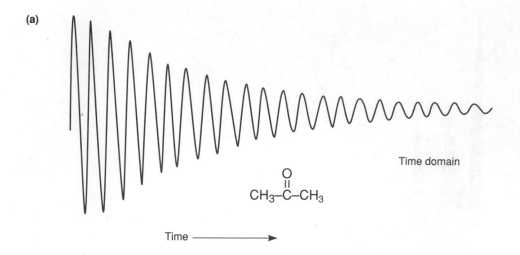

(a)

Time domain

$$CH_3-\overset{\overset{\textstyle O}{\|}}{C}-CH_3$$

Time ⟶

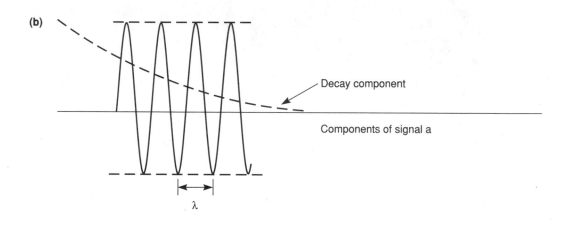

(b)

Decay component

Components of signal a

λ

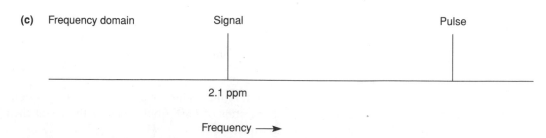

(c) Frequency domain   Signal   Pulse

2.1 ppm

Frequency ⟶

FIGURE 3.16 (a) An FID curve for the hydrogens in acetone (time domain); (b) the appearance of the FID when the decay is removed; (c) the frequency of this sine wave plotted on a frequency chart (frequency domain).

will separate each of the individual components of this signal and convert them to frequencies. The Fourier transform breaks the FID into its separate sine or cosine wave components. This procedure is too complex to be carried out by eye or by hand; it requires a computer. Most modern pulsed FT-NMR spectrometers have computers built into them that not only can work up the data by this method but can control all of the settings of the instrument.

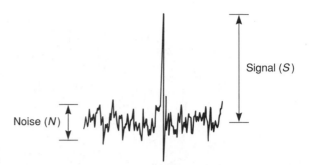

FIGURE 3.17 The signal-to-noise ratio.

The pulsed FT method described here has several advantages over the CW method. It is more sensitive, and it can measure weaker signals. Five to 10 minutes are required to scan and record a CW spectrum; a pulsed experiment is much faster, and a measurement of an FID can be performed in a few seconds. With a computer and fast measurement, it is possible to repeat and average the measurement of the FID signal. This is a real advantage when the sample is small, in which case the FID is weak in intensity and has a great amount of noise associated with it. **Noise** is random electronic signals that are usually visible as fluctuations of the baseline in the signal (Fig. 3.17). Since noise is random, it normally cancels out of the spectrum after many iterations of the spectrum are added together. Using this procedure, one can show that the signal-to-noise ratio improves as a function of the square root of the number of scans, n:

$$\frac{S}{N} = f\sqrt{n}$$

Pulsed FT-NMR is therefore especially suitable for the examination of nuclei that are not very abundant in nature, nuclei that are not strongly magnetic, or very dilute samples.

The most modern NMR spectrometers use supercooled magnets, which can have field strengths as high as 14 Tesla and operate at 600 MHz. A superconducting magnet is made of special alloys and must be cooled to liquid helium temperatures. The magnet is usually surrounded by a Dewar flask (an insulated chamber) containing liquid helium; in turn, this chamber is surrounded by another one containing liquid nitrogen. Instruments operating at frequencies above 100 MHz have superconducting magnets. NMR spectrometers with frequencies of 90 MHz, 200 MHz, 300 MHz, and 400 MHz are now common in chemistry; instruments with frequencies up to 800 MHz are used for special research projects.

3.8 CHEMICAL EQUIVALENCE—A BRIEF OVERVIEW

All of the protons found in chemically identical environments within a molecule are **chemically equivalent,** and they often exhibit the same chemical shift. Thus, all the protons in tetramethylsilane (TMS), or all the protons in benzene, cyclopentane, or acetone—which are molecules that have protons which are equivalent by symmetry considerations—have resonance at a single value of δ (but a different value from that of each of the other molecules in the same group). Each such compound gives rise to a single absorption peak in its NMR spectrum. The protons are said to be chemically equivalent. On the other hand, a molecule that has sets of protons that are chemically distinct from one another may give rise to a different absorption peak from each set, in which case the sets of protons are chemically nonequivalent. The following examples should help to clarify these relationships.

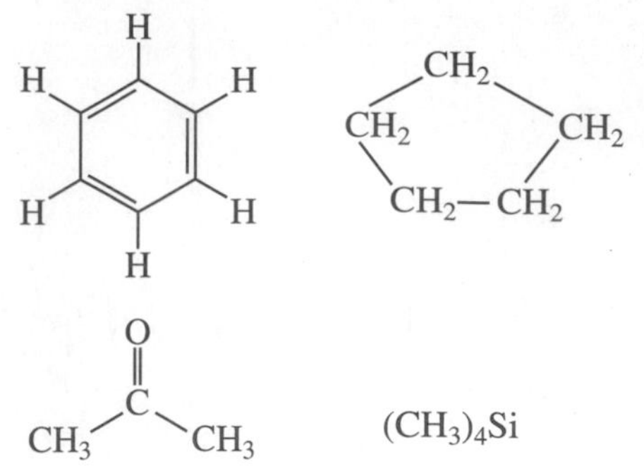

Molecules giving rise to one NMR absorption peak—all protons chemically equivalent

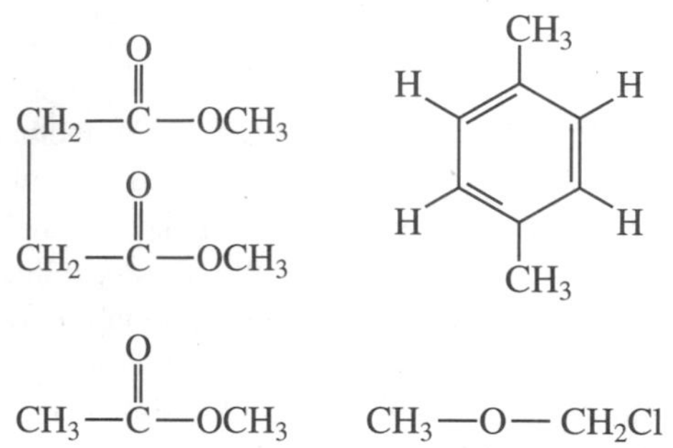

Molecules giving rise to two NMR absorption peaks—two different sets of chemically equivalent protons

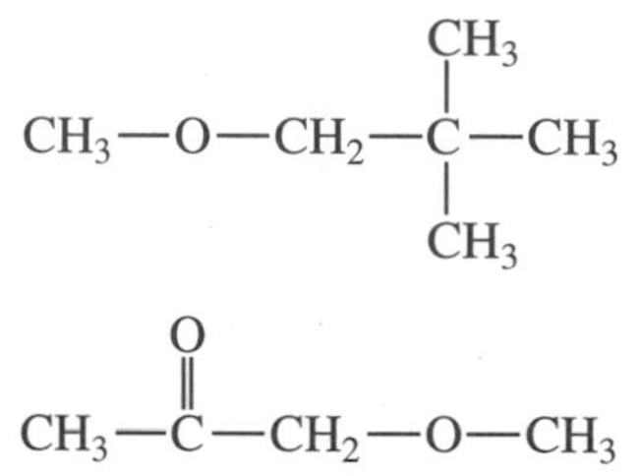

Molecules giving rise to three NMR absorption peaks—three different sets of chemically equivalent protons

You can see that an NMR spectrum furnishes a valuable type of information on the basis of the number of different peaks observed; that is, the number of peaks corresponds to the number of chemically distinct types of protons in the molecule. Often, protons that are chemically equivalent are also **magnetically equivalent.** Note, however, that *in some instances, protons that are chemically equivalent are not magnetically equivalent.* We will explore this circumstance in Chapter 5, which examines chemical and magnetic equivalence in more detail.

◼ 3.9 INTEGRALS AND INTEGRATION

The NMR spectrum not only distinguishes how many different types of protons a molecule has, but also reveals how many of each type are contained within the molecule. In the NMR spectrum, the area under each peak is proportional to the number of hydrogens generating that peak. Hence, in phenylacetone (see Fig. 3.12), the area ratio of the three peaks is 5:2:3, the same as the ratio of the numbers of the three types of hydrogens. The NMR spectrometer has the capability to electronically **integrate** the area under each peak. It does this by tracing over each peak a vertically rising line, called the **integral,** which rises in height by an amount proportional to the area under the peak. Figure 3.18 is an NMR spectrum of benzyl acetate, showing each of the peaks integrated in this way.

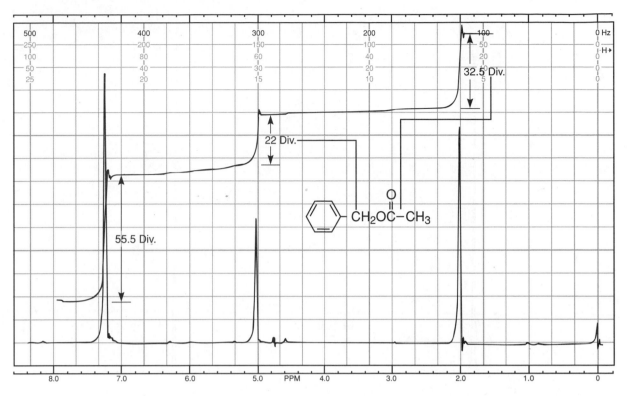

FIGURE 3.18 Determination of the integral ratios for benzyl acetate.

Note that the height of the integral line does not give the absolute number of hydrogens. It gives the relative number of each type of hydrogen. For a given integral to be of any use, there must be a second integral to which it may be referred. Benzyl acetate provides a good example of this. The first integral rises for 55.5 divisions on the chart paper; the second, 22.0 divisions; and the third, 32.5 divisions. These numbers are relative. One can find ratios of the types of protons by dividing each of the larger numbers by the smallest number:

$$\frac{55.5 \text{ div}}{22.0 \text{ div}} = 2.52 \qquad \frac{22.0 \text{ div}}{22.0 \text{ div}} = 1.00 \qquad \frac{32.5 \text{ div}}{22.0 \text{ div}} = 1.48$$

Thus, the number ratio of the protons of all the types is 2.52:1.00:1.48. If we assume that the peak at 5.1 ppm is really due to two hydrogens, and if we assume that the integrals are slightly (as much as 10%) in error, then we arrive at the true ratio by multiplying each figure by 2 and rounding off to 5:2:3. Clearly, the peak at 7.3 ppm, which integrates for five protons, arises from the resonance of the aromatic ring protons, whereas that at 2.0 ppm, which integrates for three protons, is due to the methyl protons. The two-proton resonance at 5.1 ppm arises from the benzyl protons. Notice that the integrals give the simplest ratio, but not necessarily the true ratio, of numbers of protons of each type.

In addition to the rising integral line, modern instruments usually give digitized numerical values for the integrals. Like the heights of the integral lines, these digitized integral values are not absolute but relative, and they should be treated as explained in the preceding paragraph. These digital values are also not exact; like the integral lines, they have the potential for a small degree of error (up to 10%). Figure 3.19 is an example of an integrated spectrum determined on a 300-MHz pulsed FT-NMR instrument. The digitized values appear under the peaks.

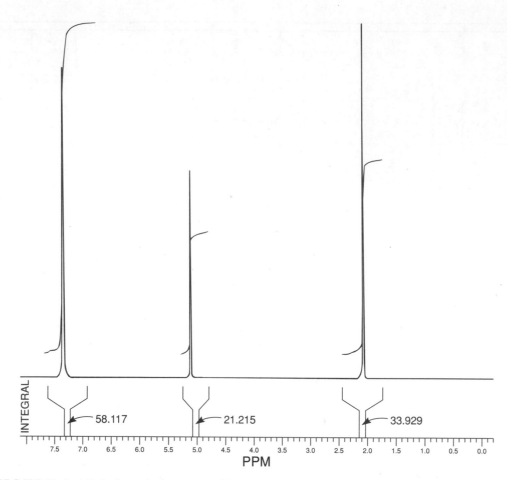

FIGURE 3.19 An integrated spectrum of benzyl acetate determined on a 300-MHz FT-NMR.

3.10 CHEMICAL ENVIRONMENT AND CHEMICAL SHIFT

If the resonance frequencies of all protons in a molecule were the same, NMR would be of little use to the organic chemist. Not only do different types of protons have different chemical shifts, but each also has a characteristic value of chemical shift. Every type of proton has only a limited range of δ values over which it gives resonance. Hence, the numerical value (in δ units or ppm) of the chemical shift for a proton gives a clue as to the type of proton originating the signal, just as an infrared frequency gives a clue as to the type of bond or functional group.

For instance, notice that the aromatic protons of both phenylacetone (Fig. 3.12) and benzyl acetate (Fig. 3.18) have resonance near 7.3 ppm and that both of the methyl groups attached directly to a carbonyl have resonance at about 2.1 ppm. Aromatic protons characteristically have resonance near 7 to 8 ppm, whereas acetyl groups (methyl groups of this type) have their resonance near 2 ppm. These values of chemical shift are diagnostic. Notice also how the resonance of the benzyl ($-CH_2-$) protons comes at a higher value of chemical shift (5.1 ppm) in benzyl acetate than in phenylacetone (3.6 ppm). Being attached to the electronegative element oxygen, these protons are more deshielded (see Section 3.11) than those in phenylacetone. A trained chemist would readily recognize the probable presence of the oxygen from the value of chemical shift shown by these protons.

It is important to learn the ranges of chemical shifts over which the most common types of protons have resonance. Figure 3.20 is a correlation chart that contains the most essential and fre-

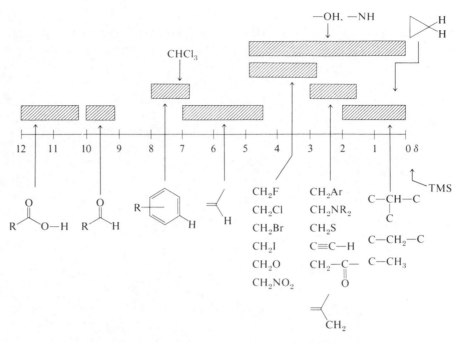

FIGURE 3.20 A simplified correlation chart for proton chemical shift values.

quently encountered types of protons. Table 3.4 lists the chemical shift ranges for selected types of protons. For the beginner it is often difficult to memorize a large body of numbers relating to chemical shifts and proton types. One actually need do this only crudely. It is more important to "get a feel" for the regions and the types of protons than to know a string of actual numbers. To do this, study Figure 3.20 carefully. Table 3.4 and Appendix 2 give more detailed listings of chemical shifts.

▢ 3.11 LOCAL DIAMAGNETIC SHIELDING

A. Electronegativity Effects

The trend of chemical shifts that is easiest to explain is that involving electronegative elements substituted on the same carbon to which the protons of interest are attached. The chemical shift simply increases as the electronegativity of the attached element increases. Table 3.5 illustrates this relationship for several compounds of the type CH_3X.

Multiple substituents have a stronger effect than a single substituent. The influence of the substituent drops off rapidly with distance, an electronegative element having little effect on protons that are more than three carbons distant. Table 3.6 illustrates these effects for the underlined protons.

Section 3.6 briefly discussed the origin of the electronegativity effect. Electronegative substituents attached to a carbon atom, because of their electron-withdrawing effects, reduce the valence electron density around the protons attached to that carbon. These electrons, it will be recalled, shield the proton from the applied magnetic field. Figure 3.10 illustrates this effect, called local diamagnetic shielding. Electronegative substituents on carbon reduce the local diamagnetic shielding in the vicinity of the attached protons because they reduce the electron density around those protons. Substituents which have this type of effect are said to deshield the proton. The greater the electronegativity of the substituent, the more it deshields protons, and hence the greater is the chemical shift of those protons.

TABLE 3.4
APPROXIMATE CHEMICAL SHIFT RANGES (PPM) FOR SELECTED TYPES OF PROTONS[a]

$R-CH_3$		0.7 – 1.3
$R-CH_2-R$		1.2 – 1.4
R_3CH		1.4 – 1.7

$R-\overset{\mid}{C}=\overset{\mid}{C}-\overset{\mid}{C}-H$ — 1.6 – 2.6

$R-\overset{O}{\overset{\|}{C}}-\overset{\mid}{\underset{\mid}{C}}-H,\ H-\overset{O}{\overset{\|}{C}}-\overset{\mid}{\underset{\mid}{C}}-H$ — 2.1 – 2.4

$RO-\overset{O}{\overset{\|}{C}}-\overset{\mid}{\underset{\mid}{C}}-H,\ HO-\overset{O}{\overset{\|}{C}}-\overset{\mid}{\underset{\mid}{C}}-H$ — 2.1 – 2.5

$N\equiv C-\overset{\mid}{\underset{\mid}{C}}-H$ — 2.1 – 3.0

(phenyl)$-\overset{\mid}{\underset{\mid}{C}}-H$ — 2.3 – 2.7

$R-C\equiv C-H$ — 1.7 – 2.7

$R-S-H$	var	1.0 – 4.0[b]
$R-N-H$	var	0.5 – 4.0[b]
$R-O-H$	var	0.5 – 5.0[b]
(phenyl)$-O-H$	var	4.0 – 7.0[b]
(phenyl)$-N-H$	var	3.0 – 5.0[b]
$R-\overset{O}{\overset{\|}{C}}-N-H$	var	5.0 – 9.0[b]

$R-N-\overset{\mid}{\underset{\mid}{C}}-H$ — 2.2 – 2.9

$R-S-\overset{\mid}{\underset{\mid}{C}}-H$ — 2.0 – 3.0

$I-\overset{\mid}{\underset{\mid}{C}}-H$ — 2.0 – 4.0

$Br-\overset{\mid}{\underset{\mid}{C}}-H$ — 2.7 – 4.1

$Cl-\overset{\mid}{\underset{\mid}{C}}-H$ — 3.1 – 4.1

$R-\overset{O}{\underset{O}{\overset{\|}{\underset{\|}{S}}}}-O-\overset{\mid}{\underset{\mid}{C}}-H$ — ca. 3.0

$RO-\overset{\mid}{\underset{\mid}{C}}-H,\ HO-\overset{\mid}{\underset{\mid}{C}}-H$ — 3.2 – 3.8

$R-\overset{O}{\overset{\|}{C}}-O-\overset{\mid}{\underset{\mid}{C}}-H$ — 3.5 – 4.8

$O_2N-\overset{\mid}{\underset{\mid}{C}}-H$ — 4.1 – 4.3

$F-\overset{\mid}{\underset{\mid}{C}}-H$ — 4.2 – 4.8

$R-\overset{\mid}{C}=\overset{\mid}{C}-H$ — 4.5 – 6.5

(phenyl)$-H$ — 6.5 – 8.0

$R-\overset{O}{\overset{\|}{C}}-H$ — 9.0 – 10.0

$R-\overset{O}{\overset{\|}{C}}-OH$ — 11.0 – 12.0

[a]For those hydrogens shown as $-\overset{\mid}{C}-H$, if that hydrogen is part of a methyl group (CH_3) the shift is generally at the low end of the range given, if the hydrogen is in a methylene group ($-CH_2-$) the shift is intermediate, and if the hydrogen is in a methine group ($-CH-$), the shift is typically at the high end of the range given.

[b] The chemical shift of these groups is variable, depending not only on the chemical environment in the molecule, but also on concentration, temperature, and solvent.

TABLE 3.5
DEPENDENCE OF THE CHEMICAL SHIFT OF CH$_3$X ON THE ELEMENT X

Compound CH$_3$X	CH$_3$F	CH$_3$OH	CH$_3$Cl	CH$_3$Br	CH$_3$I	CH$_4$	(CH$_3$)$_4$Si
Element X	F	O	Cl	Br	I	H	Si
Electronegativity of X	4.0	3.5	3.1	2.8	2.5	2.1	1.8
Chemical shift δ	4.26	3.40	3.05	2.68	2.16	0.23	0

TABLE 3.6
SUBSTITUTION EFFECTS

C<u>H</u>Cl$_3$	C<u>H</u>$_2$Cl$_2$	C<u>H</u>$_3$Cl	—C<u>H</u>$_2$Br	—C<u>H</u>$_2$—CH$_2$Br	—C<u>H</u>$_2$—CH$_2$CH$_2$Br
7.27	5.30	3.05	3.30	1.69	1.25

B.   Hybridization Effects

The second important set of trends is that due to differences in the hybridization of the atom to which hydrogen is attached.

sp^3 Hydrogens

Referring to Figure 3.20 and Table 3.4, notice that all hydrogens attached to purely *sp^3* carbon atoms (C—CH$_3$, C—CH$_2$—C, C—CH—C, cycloalkanes) have resonance in the limited range from
$\qquad\qquad\qquad\qquad\qquad\qquad\quad$ |
$\qquad\qquad\qquad\qquad\qquad\qquad\quad$ C
0 to 2 ppm, provided that no electronegative elements or π-bonded groups are nearby. At the extreme right of this range are TMS (0 ppm) and hydrogens attached to carbons in highly strained rings (0–1 ppm)—as occurs, for example, with cyclopropyl hydrogens. Most methyl groups occur near 1 ppm if they are attached to other *sp^3* carbons. Methylene-group hydrogens (attached to *sp^3* carbons) appear at greater chemical shifts (near 1.2 to 1.4 ppm) than do methyl-group hydrogens. Tertiary methine hydrogens occur at higher chemical shift than secondary hydrogens, which, in turn, have a greater chemical shift than do primary or methyl hydrogens. The following diagram illustrates these relationships.

Aliphatic region

$$
\begin{array}{ccc}
\text{C} & \text{H} & \text{H} \\
| & | & | \\
\text{C}-\text{C}-\text{H} & \text{C}-\text{C}-\text{H} & \text{C}-\text{C}-\text{H} \\
| & | & | \\
\text{C} & \text{C} & \text{H}
\end{array}
$$

$\qquad$ 3° $\quad > \quad$ 2° $\quad > \quad$ 1° $\quad > \quad$ Strained ring

$\qquad$ 2 $\qquad\qquad\qquad\qquad$ 1 $\qquad\qquad\qquad\qquad$ 0 δ

Of course, hydrogens on an *sp^3* carbon that is attached to a heteroatom (—O—CH$_2$—, and so on) or to an unsaturated carbon (—C=C—CH$_2$—) do not fall in this region but have greater chemical shifts.

sp² Hydrogens

Simple vinyl hydrogens ($-C=C-H$) have resonance in the range from 4.5 to 7 ppm. In an sp^2-$1s$ C$-$H bond, the carbon atom has more s character (33% s), which effectively renders it "more electronegative" than an sp^3 carbon (25% s). Remember that s orbitals hold electrons closer to the nucleus than do the carbon p orbitals. If the sp^2 carbon atom holds its electrons more tightly, this results in less shielding for the H nucleus than in an sp^3-$1s$ bond. Thus, vinyl hydrogens have a greater chemical shift (5 to 6 ppm) than aliphatic hydrogens on sp^3 carbons (1 to 4 ppm). Aromatic hydrogens appear in a range farther downfield (7 to 8 ppm). The downfield positions of vinyl and aromatic resonances are, however, greater than one would expect based on these hybridization differences. Another effect, called **anisotropy,** is responsible for the largest part of these shifts (and will be discussed in Section 3.12). Aldehyde protons (also attached to sp^2 carbon) appear even farther downfield (9 to 10 ppm) than aromatic protons, since the inductive effect of the electronegative oxygen atom further decreases the electron density on the attached proton. Aldehyde protons, like aromatic and alkene protons, exhibit an anomalously large chemical shift due to anisotropy (Section 3.12).

An aldehyde

sp Hydrogens

Acetylenic hydrogens (C$-$H, sp-$1s$) appear anomalously at 2 to 3 ppm owing to anisotropy (to be discussed in Section 3.12). On the basis of hybridization alone, as already discussed, one would expect the acetylenic proton to have a chemical shift greater than that of the vinyl proton. An sp carbon should behave as if it were more electronegative than an sp^2 carbon. This is the opposite of what is actually observed.

C.   Acidic and Exchangeable Protons; Hydrogen Bonding

Acidic Hydrogens

Some of the least shielded protons are those attached to carboxylic acids. These protons have their resonances at 10 to 12 ppm.

Both resonance and the electronegativity effect of oxygen withdraw electrons from the acid proton.

Hydrogen Bonding and Exchangeable Hydrogens

Protons that can exhibit hydrogen bonding (e.g., hydroxyl or amino protons) exhibit extremely variable absorption positions over a wide range. They are usually found attached to a heteroatom. Table 3.7 lists the ranges over which some of these types of protons are found. The more hydrogen bonding that takes place, the more deshielded a proton becomes. The amount of hydrogen bonding is

TABLE 3.7
TYPICAL RANGES FOR PROTONS WITH VARIABLE CHEMICAL SHIFT

Acids	RCOOH	10.5–12.0 ppm
Phenols	ArOH	4.0–7.0
Alcohols	ROH	0.5–5.0
Amines	RNH_2	0.5–5.0
Amides	$RCONH_2$	5.0–8.0
Enols	CH=CH—OH	>15

often a function of concentration and temperature. The more concentrated the solution, the more molecules can come into contact with each other and hydrogen bond. At high dilution (no H bonding), hydroxyl protons absorb near 0.5–1.0 ppm; in concentrated solution their absorption is closer to 4–5 ppm. Protons on other heteroatoms show similar tendencies.

Free (dilute solution)   Hydrogen-bonded (concentrated solution)

Hydrogens which can exchange either with the solvent medium or with one another also tend to be variable in their absorption positions. The following equations illustrate possible situations.

$$R—O—H_a + R'—O—H_b \rightleftharpoons R—O—H_b + R'—O—H_a$$

$$R—O—H + H{:}SOLV \rightleftharpoons R—\overset{+}{\underset{H}{O}}—H + {:}SOLV^-$$

$$R—O—H + {:}SOLV \rightleftharpoons H{:}SOLV^+ + R—O{:}^-$$

Chapter 6 will discuss all of these situations in more detail.

3.12 MAGNETIC ANISOTROPY

Figure 3.20 clearly shows that there are some types of protons whose chemical shifts are not easily explained by simple considerations of the electronegativity of the attached groups. For instance, consider the protons of benzene and other aromatic systems. Aryl protons generally have a chemical shift as large as that of the proton of chloroform! Alkenes, alkynes, and aldehydes also have protons whose resonance values are not in line with the expected magnitudes of any electron-withdrawing or hybridization effects. In each of these cases, the anomalous shift is due to the presence of an unsaturated system (one with π electrons) in the vicinity of the proton in question.

Take benzene, for example. When it is placed in a magnetic field, the π electrons in the aromatic ring system are induced to circulate around the ring. This circulation is called a **ring current.** The moving electrons generate a magnetic field much like that generated in a loop of wire through which a current is induced to flow. The magnetic field covers a spatial volume large enough that it influences the shielding of the benzene hydrogens. Figure 3.21 illustrates this phenomenon.

The benzene hydrogens are said to be deshielded by the diamagnetic anisotropy of the ring. In electromagnetic terminology, an isotropic field is one of either uniform density or spherically symmetric distribution; an anisotropic field is not isotropic; that is, it is nonuniform. An applied magnetic field is anisotropic in the vicinity of a benzene molecule because the labile electrons in the ring interact with the applied field. This creates a nonhomogeneity in the immediate vicinity of the molecule. Thus, a proton attached to a benzene ring is influenced by three magnetic fields: the strong magnetic field applied by the electromagnets of the NMR spectrometer and two weaker fields, one due to the usual shielding by the valence electrons around the proton, and the other due to the anisotropy generated by the ring-system π electrons. It is the anisotropic effect that gives the benzene protons a chemical shift that is greater than expected. These protons just happen to lie in a deshielding region of the anisotropic field. If a proton were placed in the center of the ring rather than on its periphery, it would be found to be shielded, since the field lines there would have the opposite direction from those at the periphery.

All groups in a molecule that have π electrons generate secondary anisotropic fields. In acetylene, the magnetic field generated by induced circulation of the π electrons has a geometry such that the acetylenic hydrogens are shielded (Fig. 3.22). Hence, acetylenic hydrogens have resonance at

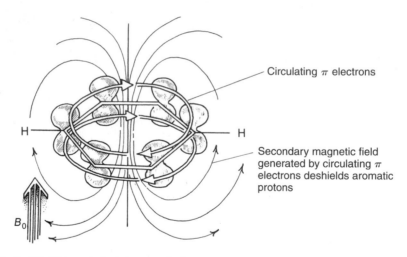

Circulating π electrons

Secondary magnetic field generated by circulating π electrons deshields aromatic protons

B_0

FIGURE 3.21 Diamagnetic anisotropy in benzene.

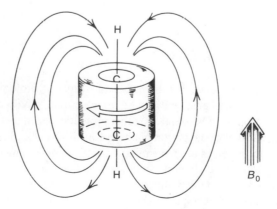

B_0

FIGURE 3.22 Diamagnetic anisotropy in acetylene.

higher field than expected. The shielding and deshielding regions due to the various π-electron functional groups have characteristic shapes and directions, and Figure 3.23 illustrates these for a number of groups. Protons falling within the conical areas are shielded, and those falling outside the conical areas are deshielded. The magnitude of the anisotropic field diminishes with distance, and beyond a certain distance there is essentially no anisotropic effect. Figure 3.24 shows the effects of anisotropy in several actual molecules.

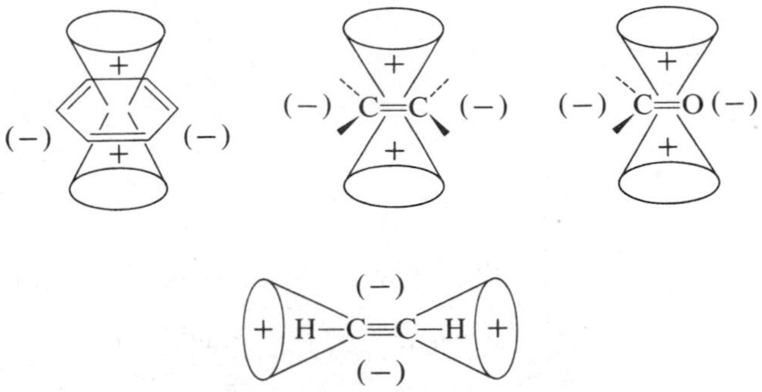

FIGURE 3.23 Anisotropy caused by the presence of π electrons in some common multiple-bond systems.

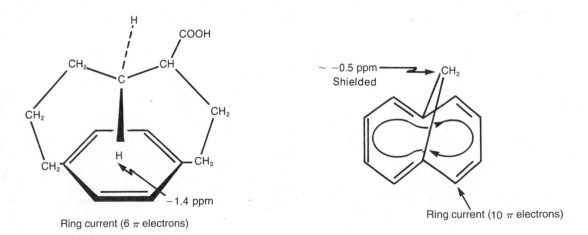

FIGURE 3.24 The effects of anisotropy in some actual molecules.

3.13 SPIN–SPIN SPLITTING ($n + 1$) RULE

We have discussed the manner in which the chemical shift and the integral (peak area) can give information about the number and types of hydrogens contained in a molecule. A third type of information to be found in the NMR spectrum is that derived from the spin–spin splitting phenomenon. Even in simple molecules, one finds that each type of proton rarely gives a single resonance peak. For instance, in 1,1,2-trichloroethane there are two chemically distinct types of hydrogens:

$$Cl-\underset{\underset{Cl}{|}}{\overset{\overset{\textcircled{H}}{|}}{C}}-\textcircled{CH_2}-Cl$$

On the basis of the information given thus far, one would predict two resonance peaks in the NMR spectrum of 1,1,2-trichloroethane, with an area ratio (integral ratio) of 2:1. In reality, the high-resolution NMR spectrum of this compound has five peaks: a group of three peaks (called a **triplet**) at 5.77 ppm and a group of two peaks (called a **doublet**) at 3.95 ppm. Figure 3.25 shows this spectrum. The methine (CH) resonance (5.77 ppm) is said to be split into a triplet, and the methylene resonance (3.95 ppm) is split into a doublet. The area under the three triplet peaks is 1, relative to an area of 2 under the two doublet peaks.

This phenomenon, called **spin–spin splitting,** can be explained empirically by the so-called $n + 1$ **Rule.** Each type of proton "senses" the number of equivalent protons (n) on the carbon atom(s) next to the one to which it is bonded, and its resonance peak is split into ($n + 1$) components.

Examine the case at hand, 1,1,2-trichloroethane, utilizing the $n + 1$ Rule. First the lone methine hydrogen is situated next to a carbon bearing two methylene protons. According to the rule, it has

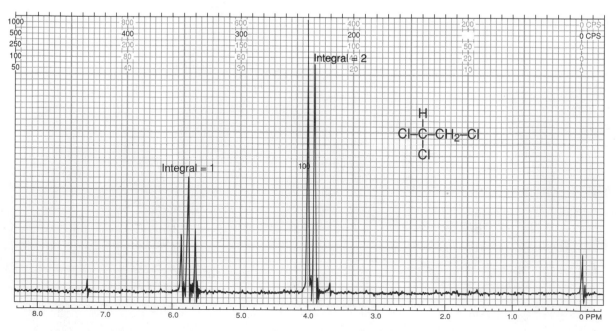

FIGURE 3.25 The ^{1}H NMR spectrum of 1,1,2-trichloroethane (60 MHz).

two equivalent neighbors ($n = 2$) and is split into $n + 1 = 3$ peaks (a triplet). The methylene protons are situated next to a carbon bearing only one methine hydrogen. According to the rule, these protons have one neighbor ($n = 1$) and are split into $n + 1 = 2$ peaks (a doublet).

Two neighbors give a triplet
($n + 1 = 3$) (area = 1)

One neighbor gives a doublet
($n + 1 = 2$) (area = 2)

Equivalent protons behave as a group

Before proceeding to explain the origin of this effect, let us examine two simpler cases predicted by the $n + 1$ Rule. Figure 3.26 is the spectrum of ethyl iodide (CH_3CH_2I). Notice that the methylene protons are split into a quartet (four peaks) and the methyl group is split into a triplet (three peaks). This is explained as follows:

Three equivalent neighbors give a quartet
($n + 1 = 4$) (area = 2)

Two equivalent neighbors give a triplet
($n + 1 = 3$) (area = 3)

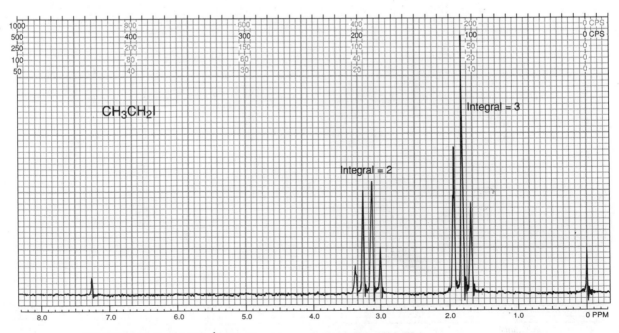

FIGURE 3.26 The ^{1}H NMR spectrum of ethyl iodide (60 MHz).

Finally, consider 2-nitropropane, which has the spectrum given in Figure 3.27.

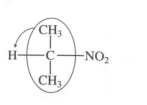 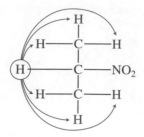

One neighbor gives a doublet
($n + 1 = 2$) (area = 6)

Six equivalent neighbors give a septet
($n + 1 = 7$) (area = 1)

Notice that in the case of 2-nitropropane there are two adjacent carbons that bear hydrogens (two carbons, each with three hydrogens) and that all six hydrogens as a group split the methine hydrogen into a **septet.**

Also notice that the chemical shifts of the various groups of protons make sense according to the discussions in Sections 3.10 and 3.11. Thus, in 1,1,2-trichloroethane, the methine hydrogen (on a carbon bearing two Cl atoms) has a larger chemical shift than the methylene protons (on a carbon bearing only one Cl atom). In ethyl iodide, the hydrogens on the carbon bearing iodine have a larger chemical shift than those of the methyl group. In 2-nitropropane, the methine proton (on the carbon bearing the nitro group) has a larger chemical shift than the hydrogens of the two methyl groups.

Finally, note that the spin–spin splitting gives a new type of structural information. It reveals how many hydrogens are adjacent to each type of hydrogen that is giving an absorption peak or, as in these cases, an absorption multiplet. For reference, some commonly encountered spin–spin splitting patterns are collected in Table 3.8.

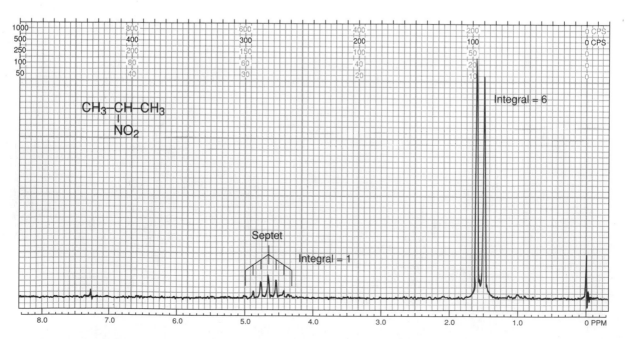

FIGURE 3.27 The ^{1}H NMR spectrum of 2-nitropropane (60 MHz).

TABLE 3.8
SOME COMMONLY OBSERVED SPLITTING PATTERNS

⅃⅃	$X-\overset{\mid}{C}H-\overset{\mid}{C}H-Y$ (X ≠ Y)	⅃⅃
⅃⅃	$-CH_2-\overset{\mid}{\underset{\mid}{C}}H$	⅃⅃⅃
⅃⅃⅃	$X-CH_2-CH_2-Y$ (X ≠ Y)	⅃⅃⅃
⅃⅃	$CH_3-\overset{\mid}{\underset{\mid}{C}}H$	⅃⅃⅃⅃
⅃⅃⅃	CH_3-CH_2-	⅃⅃⅃⅃
⅃⅃	$\begin{cases} CH_3 \\ \!\!\!\!\searrow \!\! CH- \\ CH_3 \end{cases}$	⅃⅃⅃⅃⅃⅃⅃

3.14 THE ORIGIN OF SPIN–SPIN SPLITTING

Spin–spin splitting arises because hydrogens on adjacent carbon atoms can "sense" one another. The hydrogen on carbon A can sense the spin direction of the hydrogen on carbon B. In some molecules of the solution, the hydrogen on carbon B has spin $+\frac{1}{2}$ (X-type molecules); in other molecules of the solution, the hydrogen on carbon B has spin $-\frac{1}{2}$ (Y-type molecules). Figure 3.28 illustrates these two types of molecules.

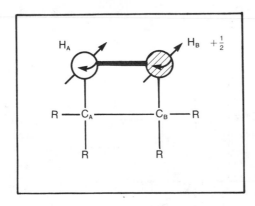

X-type molecule

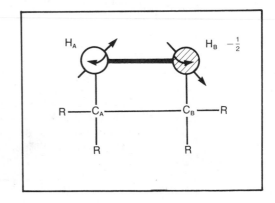

Y-type molecule

FIGURE 3.28 Two different molecules in a solution with differing spin relationships between protons H_A and H_B.

The chemical shift of proton A is influenced by the direction of the spin in proton B. Proton A is said to be **coupled** to proton B. Its magnetic environment is affected by whether proton B has a $+\frac{1}{2}$ or a $-\frac{1}{2}$ spin state. Thus, proton A absorbs at a slightly different chemical shift value in type X molecules than in type Y molecules. In fact, in X-type molecules, proton A is slightly deshielded because the field of proton B is aligned with the applied field and its magnetic moment adds to the applied field. In Y-type molecules, proton A is slightly shielded with respect to what its chemical shift would be in the absence of coupling. In this latter case, the field of proton B diminishes the effect of the applied field on proton A.

Since in a given solution there are approximately equal numbers of X- and Y-type molecules at any given time, two absorptions of nearly equal intensity are observed for proton A. The resonance of proton A is said to have been split by proton B, and the general phenomenon is called **spin–spin splitting.** Figure 3.29 summarizes the spin–spin splitting situation for proton A.

Of course, proton A also "splits" proton B, since proton A can adopt two spin states as well. The final spectrum for this situation consists of two doublets:

$$
\begin{array}{cc}
\mathrm{H_A} & \mathrm{H_B} \\
| & | \\
-\mathrm{C} - \mathrm{C}- \\
| & |
\end{array}
$$

Two doublets will be observed in any situation of this type except one in which protons A and B are identical by symmetry, as in the case of the first of the following molecules.

$$
\mathrm{Br - \underset{\underset{Cl}{|}}{\overset{\overset{H_A}{|}}{C}} - \underset{\underset{Cl}{|}}{\overset{\overset{H_B}{|}}{C}} - Br}
\qquad
\mathrm{Cl - \underset{\underset{Cl}{|}}{\overset{\overset{H_A}{|}}{C}} - \underset{\underset{OCH_3}{|}}{\overset{\overset{H_B}{|}}{C}} - OCH_3}
$$

The first molecule would give only a single NMR peak, since protons A and B have the same chemical shift value and are, in fact, identical. The second molecule would probably exhibit the two-doublet spectrum, since protons A and B are not identical and would surely have different chemical shifts.

Note that except in unusual cases, coupling (spin–spin splitting) occurs only between hydrogens on adjacent carbons. Hydrogens on nonadjacent carbon atoms generally do not couple strongly enough to produce observable splitting, although there are some important exceptions to this generalization, which Chapter 5 will discuss.

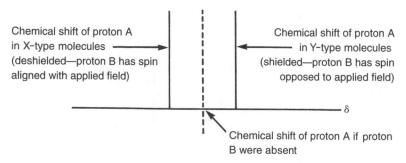

FIGURE 3.29 The origin of spin–spin splitting in proton A's NMR spectrum.

■ 3.15 THE ETHYL GROUP (CH₃CH₂—)

Now consider ethyl iodide, which has the spectrum shown in Figures 3.26 and 3.30. The methyl protons give rise to a triplet centered at 1.83 ppm, and the methylene protons give a quartet centered at 3.20 ppm. This pattern and the relative intensities of the component peaks can be explained with the use of the model for the two-proton case outlined in Section 3.13. First look at the methylene protons and their pattern, which is a quartet. The methylene protons are split by the methyl protons, and to understand the splitting pattern, you must examine the various possible spin arrangements of the protons for the methyl group, which are shown in Figure 3.31.

Some of the eight possible spin arrangements are identical to each other, since one methyl proton is indistinguishable from another and since there is free rotation in a methyl group. Taking this into consideration, there are only four different types of arrangements. There are, however, three possible ways to obtain the arrangements with net spins of $+\frac{1}{2}$ and $-\frac{1}{2}$. Hence, these arrangements are three times more probable statistically than are the $+\frac{3}{2}$ and $-\frac{3}{2}$ spin arrangements. Thus, one notes in the splitting pattern of the methylene protons that the center two peaks are more intense than the outer ones. In fact, the intensity ratios are 1:3:3:1. Each of these different spin arrangements of the methyl protons (except the sets of degenerate ones, which are effectively identical) gives the methylene protons in that molecule a different chemical shift value. Each of the spins in the $+\frac{3}{2}$ arrangement tends to deshield the methylene proton with respect to its position in the absence of coupling. The $+\frac{1}{2}$ arrangement also deshields the methylene proton, but only slightly, since the two opposite spins cancel each other's effects. The $-\frac{1}{2}$ arrangement shields the methylene proton slightly, whereas the $-\frac{3}{2}$ arrangement shields the methylene proton more strongly.

Keep in mind that there are actually four different "types" of molecules in the solution, each type having a different methyl spin arrangement. Each spin arrangement causes the methylene protons in that molecule to have a chemical shift different from those in a molecule with another methyl spin arrangement (except, of course, when the spin arrangements are indistinguishable, or degenerate). Molecules having the $+\frac{1}{2}$ and $-\frac{1}{2}$ spin arrangements are three times more numerous in solution than those with the $+\frac{3}{2}$ and $-\frac{3}{2}$ spin arrangements.

Figure 3.32 provides a similar analysis of the methyl splitting pattern, showing the four possible spin arrangements of the methylene protons. Examination of this figure makes it easy to explain the origin of the triplet for the methyl group and the intensity ratios of 1:2:1.

FIGURE 3.30 The ethyl splitting pattern.   δ :   3.20   1.83   CH₃CH₂I

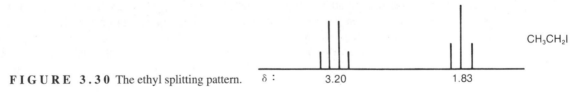

FIGURE 3.31 The splitting pattern of methylene protons due to the presence of an adjacent methyl group.

Net spin: $+\frac{3}{2}$ $+\frac{1}{2}$ $-\frac{1}{2}$ $-\frac{3}{2}$

Possible spin arrangements of the methyl protons

↑ = spin $+\frac{1}{2}$

↓ = spin $-\frac{1}{2}$

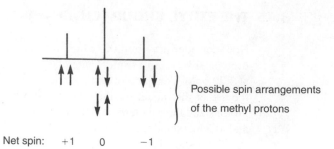

Possible spin arrangements
of the methyl protons

FIGURE 3.32 The splitting pattern of methyl protons due to the presence of an adjacent methylene group. Net spin: +1 0 −1

Now one can see the origin of the ethyl pattern and the explanation of its intensity ratios. The occurrence of spin–spin splitting is very important for the organic chemist, as it gives additional structural information about molecules. Namely, it reveals the number of nearest proton neighbors each type of proton has. From the chemical shift one can determine what type of proton is being split, and from the integral (the area under the peaks) one can determine the relative numbers of the types of hydrogen. This is a great amount of structural information, and it is invaluable to the chemist attempting to identify a particular compound.

3.16 PASCAL'S TRIANGLE

We can easily verify that the intensity ratios of multiplets derived from the $n + 1$ Rule follow the entries in the mathematical mnemonic device called **Pascal's triangle** (Fig. 3.33). Each entry in the triangle is the sum of the two entries above it and to its immediate left and right. Notice that the intensities of the outer peaks of a multiplet such as a septet are so small compared to the inner peaks that they are often obscured in the baseline of the spectrum. Figure 3.27 is an example of this phenomenon.

3.17 THE COUPLING CONSTANT

Section 3.15 discussed the splitting pattern of the ethyl group and the intensity ratios of the multiplet components but did not address the quantitative amount by which the peaks were split. The distance between the peaks in a simple multiplet is called the **coupling constant,** J. The coupling constant is a measure of how strongly a nucleus is affected by the spin states of its neighbor. The spacing between the multiplet peaks is measured on the same scale as the chemical shift, and the coupling constant is always expressed in Hertz (Hz). In ethyl iodide, for instance, the coupling constant J is 7.5 Hz. To see how this value was determined, consult Figures 3.26 and 3.34.

The spectrum in Figure 3.26 was determined at 60 MHz; thus, each ppm of chemical shift (δ unit) represents 60 Hz. Inasmuch as there are 12 grid lines per ppm, each grid line represents (60 Hz)/12 =

Singlet							1						
Doublet						1		1					
Triplet					1		2		1				
Quartet				1		3		3		1			
Quintet			1		4		6		4		1		
Sextet		1		5		10		10		5		1	
Septet	1		6		15		20		15		6		1

FIGURE 3.33 Pascal's triangle.

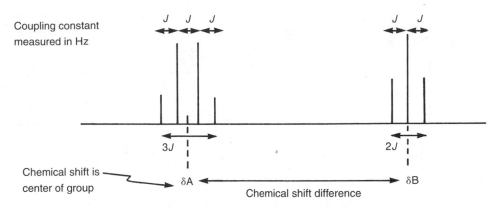

FIGURE 3.34 The definition of the coupling constants in the ethyl splitting pattern.

5 Hz. Notice the top of the spectrum. It is calibrated in cycles per second (cps), which are the same as Hertz, and since there are 20 chart divisions per 100 cps, one division equals (100 cps)/20 = 5 cps = 5 Hz. Now examine the multiplets. The spacing between the component peaks is approximately 1.5 chart divisions, so

$$J = 1.5 \text{ div} \times \frac{5 \text{ Hz}}{1 \text{ div}} = 7.5 \text{ Hz}$$

That is, the coupling constant between the methyl and methylene protons is 7.5 Hz. When the protons interact, the magnitude (in ethyl iodide) is always of this same value, 7.5 Hz. The amount of coupling is *constant*, and hence J can be called a coupling constant.

The constant nature of the coupling constant can be observed when the NMR spectrum of ethyl iodide is determined at both 60 MHz and 100 MHz. A comparison of the two spectra indicates that the 100-MHz spectrum is greatly expanded over the 60-MHz spectrum. The chemical shift in Hertz for the CH_3 and CH_2 protons is much larger in the 100-MHz spectrum, although the chemical shifts in δ units (ppm) for these protons remain identical to those in the 60-MHz spectrum. Despite the expansion of the spectrum determined at the higher spectrometer frequency, careful examination of the spectra indicates that the coupling constant between the CH_3 and CH_2 protons is 7.5 Hz in both spectra! The spacings of the lines of the triplet and the lines of the quartet do not expand when the spectrum of ethyl iodide is determined at 100 MHz. The extent of coupling between these two sets of protons remains constant irrespective of the spectrometer frequency at which the spectrum was determined (Fig. 3.35).

For the interaction of most aliphatic protons in acyclic systems, the magnitudes of coupling constants are always near 7.5 Hz. Compare, for example, 1,1,2-trichloroethane (Fig. 3.25), for which J = 6 Hz, and 2-nitropropane (Fig. 3.27), for which $J = 7$ Hz. These coupling constants are typical for the interaction of two hydrogens on adjacent sp^3-hybridized carbon atoms. Different types of protons have different magnitudes of J. For instance, the *cis* and *trans* protons substituted on a double bond commonly have values of approximately J_{cis} = 10 Hz and J_{trans} = 17 Hz. In ordinary compounds, coupling constants may range anywhere from 0 to 18 Hz.

The magnitude of J often provides structural clues. For instance, one can usually distinguish between a *cis* olefin and a *trans* olefin on the basis of the observed coupling constants for the vinyl protons. Table 3.9 gives the approximate values of some representative coupling constants. A more extensive list of coupling constants appears in Appendix 5.

Before closing this section, we should take note of an axiom: *the coupling constants of the groups of protons that split one another must be identical.* This axiom is extremely useful in interpreting a spectrum that may have several multiplets, each with a different coupling constant.

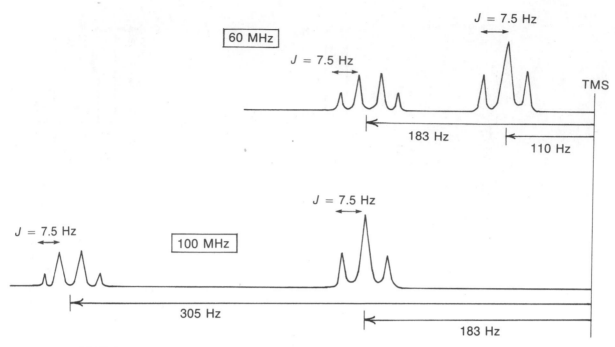

F I G U R E   3 . 3 5 An illustration of the relationship between the chemical shift and the coupling constant.

T A B L E   3 . 9
SOME REPRESENTATIVE COUPLING CONSTANTS AND THEIR APPROXIMATE VALUES (HZ)

H H | | C—C 6 to 8	*ortho* 6 to 10	a,a 8 to 14 a,e 0 to 7 e,e 0 to 5
H ⟍ ⟋ H 11 to 18	8 to 11	*cis* 6 to 12 *trans* 4 to 8
H H 6 to 15		*cis* 2 to 5 *trans* 1 to 3
H CH 4 to 10		5 to 7

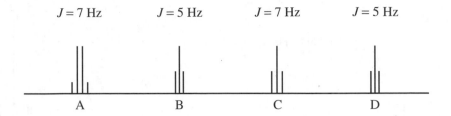

Take, for example, the preceding spectrum, which shows three triplets and one quartet. Which triplet is associated with the quartet? It is, of course, the one that has the same J values as are found in the quartet. The protons in each group interact to the same extent. In this example, with the J values given, clearly quartet A ($J = 7$ Hz) is associated with triplet C ($J = 7$ Hz) and not with triplet B or D ($J = 5$ Hz). It is also clear that triplets B and D are related to each other in the interaction scheme.

Multiplet skewing is another effect which can sometimes be used to link interacting multiplets. There is a tendency for the outermost lines of a multiplet to have nonequivalent heights. For instance, in a triplet, line 3 may be slightly taller than line 1, causing the multiplet to "lean." When this happens, the taller peak is usually in the direction of the proton or group of protons causing the splitting. This second group of protons leans toward the first one in the same fashion. If arrows are drawn on both multiplets in the directions of their respective skewing, these arrows will point at each other. See Figures 3.25 and 3.26 for examples.

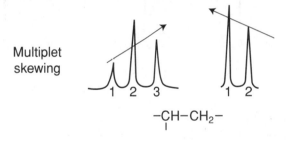

3.18 A COMPARISON OF NMR SPECTRA AT LOW AND HIGH FIELD STRENGTHS

Section 3.17 showed that, for a given proton, the frequency shift (in Hertz) from TMS is larger when the spectrum is determined at a higher field; however, all coupling constants remain the same as they were at low field (see Fig. 3.35). Even though the shifts in Hertz increase, the chemical shifts (in ppm) of a given proton at low field and high field are the same, because we divide by a different operating frequency in each case to determine the chemical shift (Eq. 3.8). If we compare the spectra of a compound determined at both low field and high field, however, the gross appearances of the spectra will differ because, although the coupling constant has the same magnitude in Hertz regardless of operating frequency, the number of Hertz per ppm unit changes. At 60 MHz, for instance, a ppm unit equals 60 Hz, whereas at 300 MHz a ppm unit equals 300 Hz. The coupling constant does not change, but it becomes a smaller fraction of a ppm unit!

When we plot the two spectra on paper to the same parts-per-million scale (same spacing in length for each ppm), the splittings in the high-field spectrum appear compressed, as in Figure 3.36, which shows the 60-MHz and 300-MHz spectra of 1-nitropropane. The coupling has not changed in size; it has simply become a smaller fraction of a ppm unit. At higher field it becomes necessary to use an expanded parts-per-million scale (more space per ppm) to observe the splittings. The 300-MHz multiplets are identical to those observed at 60 MHz. This can be seen in Figure 3.36B, which

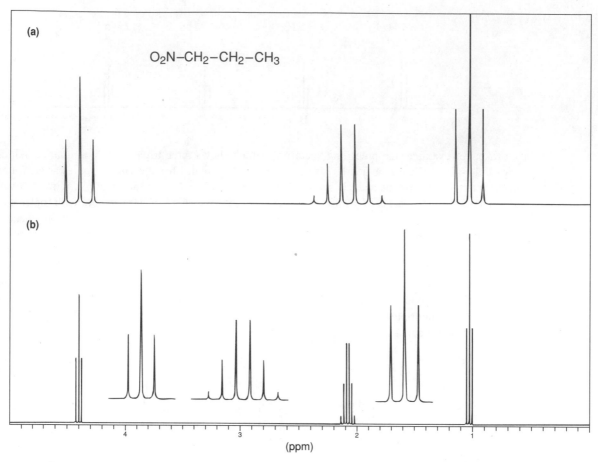

$O_2N-CH_2-CH_2-CH_3$

(ppm)

FIGURE 3.36 The NMR spectrum of 1-nitropropane. (a) Spectrum determined at 60 MHz; (b) spectrum determined at 300 MHz (with expansions).

shows expansions of the multiplets in the 300-MHz spectrum.

With 300-MHz spectra, therefore, it is frequently necessary to show expansions if one wishes to see the details of the multiplets. In some of the examples in this chapter, we have used 60-MHz spectra—not because we are old-fashioned but because these spectra show the multiplets more clearly without the need for expansions.

In most cases, the expanded multiplets from a high-field instrument are identical to those observed with a low-field instrument. However, there are also cases in which complex multiplets become simplified when higher field is used to determine the spectrum. This simplification occurs because the multiplets are moved farther apart, and a type of interaction called second-order interaction is reduced or even completely removed. Chapter 5 will discuss second-order interactions.

3.19 SURVEY OF TYPICAL ¹H NMR ABSORPTIONS BY TYPE OF COMPOUND

In this section we will review the typical NMR absorptions that may be expected for compounds in each of the most common classes of organic compounds. These guidelines can be consulted whenever you are trying to establish the class of an unknown compound. Coupling behaviors commonly observed in these compounds are also included in the tables. This coupling information was not covered in this chapter, but it is discussed in Chapters 5 and 6. It is included here so that it will be useful if you wish to use this survey later on.

A. Alkanes

Alkanes have three different types of hydrogens, methyl, methylene, and methine, each of which appears in its own region of the NMR spectrum.

SPECTRAL ANALYSIS BOX—Alkanes

CHEMICAL SHIFTS

R—CH$_3$	0.7–1.3 ppm	Methyl groups are often recognizable as a tall singlet, doublet, or triplet even when overlapping other CH absorptions.
R—CH$_2$—R	1.2–1.4 ppm	In long chains all of the methylene (CH$_2$) absorptions may be overlapped in an unresolvable group.
R$_3$CH	1.4–1.7 ppm	Note that methine hydrogens (CH) have a larger chemical shift than those in methylene or methyl groups.

COUPLING BEHAVIOR

—CH—CH—	$^3J \approx 7$–8 Hz	In hydrocarbon chains, adjacent hydrogens will generally couple, with the spin–spin splitting following the $n+1$ Rule.

In alkanes (aliphatic or saturated hydrocarbons) all of the **CH** hydrogen absortions are typically found from about 0.7 to 1.7 ppm. Hydrogens in methyl groups are the most highly shielded type of proton and are found at chemical shift values lower (0.7–1.3 ppm) then methylene (1.2–1.2 ppm) or methine hydrogens (1.4–1.7 ppm).

In long hydrocarbon chains, or in larger rings, all of the **CH** and **CH$_2$** absorptions may overlap in an unresolvable group. Methyl group peaks are usually separated from other types of hydrogens, being found at lower chemical shifts (higher field). However, even when methyl hydrogens are located within an unresolved cluster of peaks, the methyl peaks can often be recognized as tall singlets, doublets, or triplets clearly emerging from the absorptions of the other types of protons. Methine protons are usually separated from the other protons, being shifted further downfield.

Figure 3.37 shows the spectrum of the hydrocarbon octane. Note that the integral can be used to estimate the total number of hydrogens (the ratio of CH$_3$ to CH$_2$-type carbons) since all of the CH$_2$ hydrogens are in one group and the CH$_3$ hydrogens are in the other.

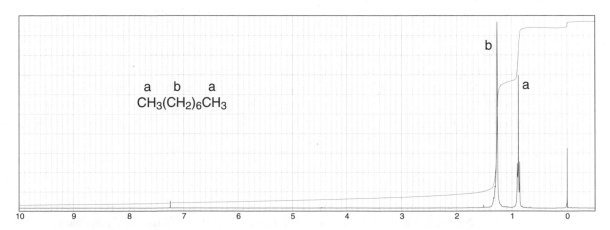

FIGURE 3.37 ^{1}H Spectrum of octane.

B.   Alkenes

Alkenes have two types of hydrogens: vinyl (those attached directly to the double bond) and allylic hydrogens (those attached to the *α carbon,* the carbon atom attached to the double bond). Each type has a characteristic chemical shift region.

SPECTRAL ANALYSIS BOX—Alkenes

CHEMICAL SHIFTS

C=C—H	4.5–6.5 ppm	Hydrogens attached to a double bond (vinyl hydrogens) are deshielded by the anisotropy of the adjacent double bond.
C=C—C—H	1.6–2.6 ppm	Hydrogens attached to a carbon adjacent to a double bond (allylic hydrogens) are also deshielded by the anisotropy of the double bond but, because the double bond is more distant, the effect is smaller.

COUPLING BEHAVIOR

H—C=C—H	$^3J_{trans} \approx 11$–18 Hz $^3J_{cis} \approx 6$–15 Hz	The splitting patterns of vinyl protons may be complicated by the fact that they may not be equivalent even when located on the same carbon of the double bond (Section 5.6)

—C=C—H
  |
  H
	$^2J \approx 0$–1 Hz	

—C=C—C—H
  |
  H
	$^4J \approx 0$–3 Hz	When allylic hydrogens are present in an alkene they may show long-range allylic coupling (Section 5.7) to hydrogens on the far double bond carbon as well as the usual splitting due to the hydrogen on the adjacent (nearest) carbon.

Two types of NMR absorptions are typically found in alkenes: **vinyl absorptions** due to protons directly attached to the double bond (4.5–6.5 ppm) and **allylic absorptions** due to protons located on a carbon atom adjacent to the double bond (1.6–2.6 ppm). Both types of hydrogens are deshielded due to the anisotropic field of the π electrons in the double bond. The effect is smaller for the allylic hydrogens because they are more distant from the double bond. A spectrum of 2-methyl-1-pentene is shown in Figure 3.38. Note the vinyl hydrogens at 4.7 ppm and the allylic methyl group at 1.7 ppm.

The splitting patterns of both vinyl and allylic hydrogens can be quite complex due to the fact that the hydrogens attached to a double bond are rarely equivalent and to the additional complication that allylic hydrogens can couple to all of the hydrogens on a double bond causing additional splittings. These situations are discussed in Chapter 5, Sections 5.6–5.7.

C.   Aromatic Compounds

Aromatic compounds have two characteristic types of hydrogens: aromatic ring hydrogens (benzene ring hydrogens) and benzylic hydrogens (those attached to an adjacent carbon atom).

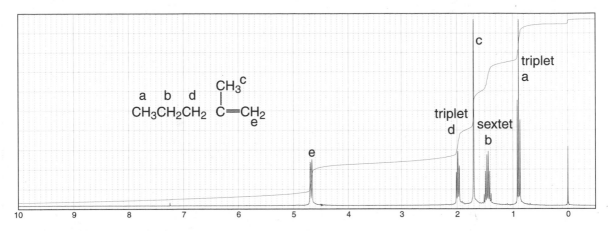

FIGURE 3.38 ^{1}H Spectrum of 2-methyl-1-pentene.

SPECTRAL ANALYSIS BOX—Aromatic Compounds

CHEMICAL SHIFTS

6.5–8.0 ppm   Hydrogens attached to an aromatic (benzenoid) ring have a large chemical shift, usually near 7.0 ppm. They are deshielded by the large anisotropic field generated by the electrons in the ring's π system.

2.3–2.7 ppm   Benzylic hydrogens are also deshielded by the anisotropic field of the ring, but they are more distant from the ring and the effect is smaller.

COUPLING BEHAVIOR

$^3J_{ortho} \approx 7{-}10$ Hz
$^4J_{meta} = 2{-}3$ Hz
$^5J_{para} = 0{-}1$ Hz

Splitting patterns for the protons on a benzene ring are discussed in Section 5.11. It is often possible to determine the positions of the substituents on the ring from these splitting patterns and the magnitudes of the coupling constants.

The hydrogens attached to aromatic rings are easily identified. They are found in a region of their own (6.5–8.0 ppm) in which few other types of hydrogens show absorption. Occasionally a highly deshielded vinyl hydrogen will have its absorption in this range, but this is not frequent. The hydrogens on an aromatic ring are more highly deshielded than those attached to double bonds due to the large anisotropic field that is generated by the circulation of the π electrons in the ring (ring current). See Section 3.12 for a review of this specialized behavior of aromatic rings.

The largest chemical shifts are found for ring hydrogens when electron-withdrawing groups such as $-NO_2$ are attached to the ring. These groups deshield the attached hydrogens by withdrawing electron density from the ring through resonance interaction. Conversely, electron-donating groups like methoxy ($-OCH_3$) increase the shielding of these hydrogens causing them to move upfield.

Nonequivalent hydrogens attached to a benzene ring will interact with one another to produce spin–spin splitting patterns. The amount of interaction between hydrogens on the ring is dependent

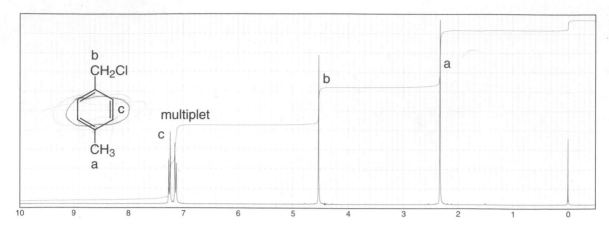

FIGURE 3.39 ^{1}H Spectrum of α-chloro-p-xylene.

on the number of intervening bonds, or the distance between them. *Ortho* hydrogens ($^3J \approx 7-10$ Hz) couple more strongly than *meta* hydrogens ($^4J \approx 2-3$ Hz), which in turn couple more strongly than *para* hydrogens ($^5J \approx 0-1$ Hz). It is frequently possible to determine the substitution pattern of the ring by the observed splitting patterns of the ring hydrogens (Section 5.11). One pattern that is easily recognized is that of a *para*-disubstituted benzene ring (Fig. 5.44). The spectrum of α-chloro-*p*-xylene is shown in Figure 3.39. The highly deshielded ring hydrogens appear at 7.2 ppm and clearly show a *para*-disubstitution pattern. The chemical shift of the methyl protons at 2.3 ppm shows a smaller deshielding effect. The large shift of the methylene hydrogens is due to the electronegativity of the attached chlorine.

D. Alkynes

Terminal alkynes (those with a triple bond at the end of a chain) will show an acetylenic hydrogen, as well as the α hydrogens found on carbon atoms next to the triple bond. The acetylenic hydrogen will be absent if the triple bond is in the middle of a chain.

SPECTRAL ANALYSIS BOX—Alkynes

CHEMICAL SHIFTS

C≡C−**H**	1.7–2.7 ppm	The terminal or acetylenic hydrogen has a chemical shift near 1.8 ppm due to anisotropic shielding by the adjacent π bonds.
C≡C−**CH**−	1.6–2.6 ppm	Protons on a carbon next to the triple bond are also affected by the π system.

COUPLING BEHAVIOR

H−C≡C−C−**H**	$^4J = 2-3$ Hz	"Allylic coupling" is often observed in alkynes, but it will often be quite small or even nonexistent ($J = 0$).

In terminal alkynes the acetylenic hydrogen (the one attached to the triple bond) has resonance near 1.9 ppm. It is shifted to a higher field than might be expected, or shielded, due to the

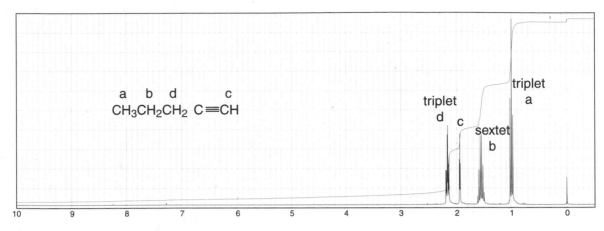

FIGURE 3.40 ^{1}H Spectrum of 1-pentyne.

anisotropic field of the π-bonded electrons. The hydrogens on the adjacent carbon are also shielded by this anisotropic field. A spectrum of 1-pentyne is shown in Figure 3.40. Note the acetylenic hydrogen at 1.95 ppm and the allylic hydrogens at 2.2 ppm. The acetylenic hydrogen is not a sharp singlet since it is showing a slight allylic coupling (Section 5.7) and is a finely split triplet.

E.   Alkyl Halides

In alkyl halides the α hydrogen (the one attached to the same carbon as the halogen) will be deshielded.

SPECTRAL ANALYSIS BOX—Alkyl Halides

CHEMICAL SHIFTS

$-$CH$-$I	2.0–4.0 ppm	The chemical shift of a hydrogen atom attached to the same carbon as a halide atom will increase (move further downfield).
$-$CH$-$Br	2.7–4.1 ppm	This deshielding effect is due to the electronegativity of the attached halogen atom. The extent of the shift is increased as
$-$CH$-$Cl	3.1–4.1 ppm	the electronegativity of the attached atom increases, with the largest shift found in compounds containing fluorine.
$-$CH$-$F	4.2–4.8 ppm	

COUPLING BEHAVIOR

$-$CH$-$F	$^2J \approx 50$ Hz	Compounds containing fluorine will show spin–spin splitting
$-$CH$-$CF$-$	$^3J \approx 20$ Hz	due to coupling between the fluorine and the hydrogens on either the same or the adjacent carbon atom. Fluorine has a spin of $\frac{1}{2}$. The other halogens (I, Cl, Br) do not show any coupling.

Hydrogens attached to the same carbon as a halogen are deshielded (local diamagnetic shielding) due to the electronegativity of the attached halogen (Section 3.11A). The amount of deshielding increases as the electronegativity of the halogen increases, and it is further increased when multiple halogens are present.

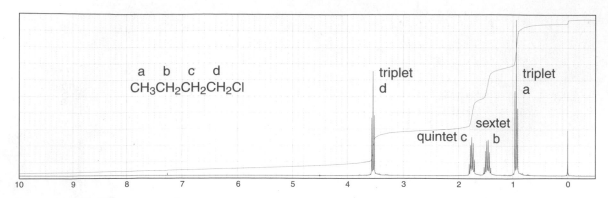

FIGURE 3.41 ¹H Spectrum of 1-chlorobutane.

Compounds containing fluorine will show coupling between the fluorine and the hydrogens both on the same carbon (−**CHF**) and those hydrogens on the adjacent carbon (**CH**−**CF**−). Since the spin of fluorine is $\frac{1}{2}$, the $n + 1$ Rule can be used to predict the multiplicities of the attached hydrogens. Other halogens do not cause spin–spin splitting of hydrogen peaks.

The spectrum of 1-chlorobutane is shown in Figure 3.41. Note the large downfield shift (deshielding) of the hydrogens on carbon 1 due to the attached chlorine.

F.   Alcohols

In alcohols, both the hydroxyl proton and the α hydrogens (those on the same carbon as the hydroxyl group) have characteristic chemical shifts.

SPECTRAL ANALYSIS BOX—Alcohols

CHEMICAL SHIFTS

C−OH	0.5–5.0 ppm	The chemical shift of the −OH hydrogen is variable, its position depending on concentration, solvent, and temperature. The peak may be broadened at its base by the same set of factors.
CH−O−H	3.2–3.8 ppm	Protons on the α carbon are deshielded by the electronegative oxygen atom and are shifted downfield in the spectrum.

COUPLING BEHAVIOR

CH−OH	*No coupling (usually)*, or $^3J = 5$ Hz	Because of the rapid chemical exchange of the −OH proton in many solutions, coupling is not usually observed between the −OH proton and those hydrogens attached to the α carbon.

The chemical shift of the −**OH** hydrogen is variable, its position depending on concentration, solvent, temperature, and presence of water or of acidic or basic impurities. This peak can be found anywhere in the range of 0.5–5.0 ppm. The variability of this absorption is dependant on the rates of −**OH** proton exchange and the amount of hydrogen bonding in the solution (Section 6.1).

The −**OH** hydrogen is usually not split by hydrogens on the adjacent carbon (−**CH**−**OH**) because rapid exchange decouples this interaction (Section 6.1).

$$-\text{CH}-\text{OH} + \text{HA} \rightleftharpoons -\text{CH}-\text{OH} + \text{HA} \qquad \textbf{no coupling if}$$
$$\textbf{exchange is rapid}$$

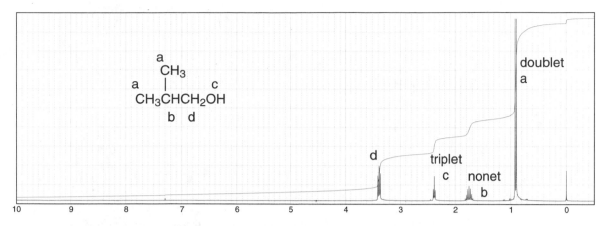

FIGURE 3.42 ^{1}H Spectrum of 2-methyl-1-propanol.

Exchange is promoted by increased temperature, small amounts of acid impurities, or the presence of water in the solution. In ultrapure alcohol samples −**CH**−**OH** coupling is observed. A freshly purified and distilled sample, or a previously unopened commercial bottle, may show this coupling.

On occasion one may use the rapid exchange of an alcohol as a method for identifying the −OH absorption. In this method, D_2O is placed in the NMR tube containing the alcohol solution. After sitting for a few minutes the −**OH** hydrogen is replaced by deuterium, causing it to disappear from the spectrum (or to have its intensity reduced).

$$-CH-OH + D_2O \rightleftharpoons -CH-OD + HOD \quad \textbf{deuterium exchange}$$

The hydrogen on the adjacent carbon (−**CH**−OH) appears in the range 3.2–3.8 ppm, being deshielded by the attached oxygen. If exchange is taking place this hydrogen will not show any coupling with the −**OH** hydrogen, but may show coupling to any hydrogens on the adjacent carbon located further along the carbon chain. If exchange is not occurring, the pattern of this hydrogen may be complicated by differently sized coupling constants for the −**CH**−**OH** and −**CH**−**CH**−**O**− couplings (Section 6.1).

A spectrum of 2-methyl-1-propanol is shown in Figure 3.42. Note the large downfield shift (3.4 ppm) of the hydrogens attached to the same carbon as the oxygen of the hydroxyl group. The hydroxyl group appears at 2.4 ppm, and in this sample it shows some coupling to the hydrogens on the adjacent carbon.

G.   Ethers

In ethers, the α hydrogens (those attached to the α *carbon*, which is the carbon atom attached to the oxygen) are highly deshielded.

SPECTRAL ANALYSIS BOX—Ethers

CHEMICAL SHIFTS

R−O−CH−   3.2–3.8 ppm   The hydrogens on the carbons attached to the oxygen are deshielded due to the electronegativity of the oxygen.

In ethers, the hydrogens on the carbon next to oxygen are deshielded due to the electronegativity of the attached oxygen, and they appear in the range 3.2–3.8 ppm. Methoxy groups are especially

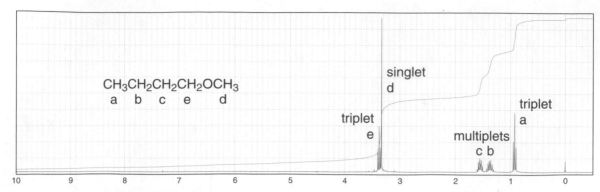

FIGURE 3.43 [1]H Spectrum of butyl methyl ether.

easy to identify as they appear as a tall singlet in this area. Ethoxy groups are also easy to identify, having both an upfield triplet and a distinct quartet in the region of 3.2–3.8 ppm. An exception is found in epoxides in which, due to ring strain, the deshielding is not as great and the hydrogens on the ring appear in the range 2.5–3.5 ppm.

The spectrum of butyl methyl ether is shown in Figure 3.43. The absorption of the methyl and methylene hydrogens next to the oxygen are both seen at about 3.4 ppm. The methoxy peak is un-split and stands out as a tall sharp singlet. The methylene hydrogens are split into a triplet by the hydrogens on the adjacent carbon of the chain.

H. Amines

Two characteristic types of hydrogens are found in amines: those attached to nitrogen (the hydrogens of the amino group) and those attached to the α carbon (the same carbon to which the amino group is attached).

SPECTRAL ANALYSIS BOX—Amines

CHEMICAL SHIFTS

R–N–H	0.5–4.0 ppm	Hydrogens attached to a nitrogen have a variable chemical shift depending on the temperature, acidity, amount of hydrogen bonding, and solvent.
–CH–N–	2.2–2.9 ppm	The α hydrogen is slightly deshielded due to the electronegativity of the attached nitrogen.
⬡–N–H	3.0–5.0 ppm	This hydrogen is deshielded due to the anisotropy of the ring and the resonance that removes electron density from nitrogen and changes its hybridization.

COUPLING BEHAVIOR

–N–H	$^{1}J \approx 50$ Hz	Direct coupling between a nitrogen and an attached hydrogen is not usually observed, but is quite large when it occurs. More commonly this coupling is obscured by quadrupole broadening by nitrogen or by proton exchange. See Sections 6.4 and 6.5.
–N–CH	$^{2}J \approx 0$ Hz	This coupling is usually not observed.
C–N–H | H	$^{3}J \approx 0$ Hz	Due to chemical exchange, this coupling is usually not observed.

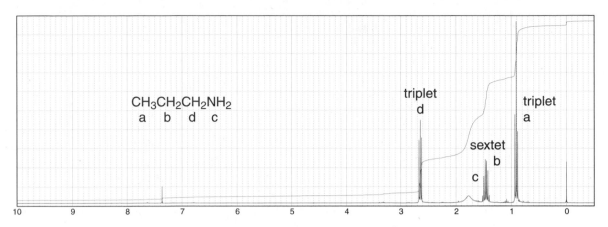

FIGURE 3.44 ^{1}H Spectrum of propylamine.

Location of the $-$NH absorptions is not a reliable method for the identification of amines. These peaks are extremely variable, appearing over a wide range of 0.5–4.0 ppm, and the range is extended in aromatic amines. The position of resonance is affected by temperature, acidity, amount of hydrogen bonding, and solvent. In addition to this variability in position, the $-$NH peaks are often very broad and weak without any distinct coupling to hydrogens on an adjacent carbon atom. This condition can be caused by chemical exchange of the $-$NH proton, or by a property of nitrogen atoms called quadrupole broadening. See Section 6.5. The amino hydrogens will exchange with D_2O, as already described for alcohols, causing the peak to disappear.

$$-N-H + D_2O \rightleftharpoons \ N-D + DOH$$

The $-$NH peaks are strongest in aromatic amines (anilines) in which resonance appears to strengthen the NH bond by changing the hybridization. Although nitrogen is a spin-active element (I = 1) coupling is usually not observed between either attached or adjacent hydrogen atoms, but it can appear in certain specific cases. Reliable prediction is difficult.

The hydrogens α to the amino group are slightly deshielded by the presence of the electronegative nitrogen atom, and they appear in the range 2.2–2.9 ppm. A spectrum of propylamine is shown in Figure 3.44. Notice the weak, broad NH absorptions at 1.8 ppm and that there appears to be a lack of coupling between the hydrogens on the nitrogen and those on the adjacent carbon atom.

I.   Nitriles

In nitriles, only the α hydrogens (those attached to the same carbon as the cyano group) have a characteristic chemical shift.

SPECTRAL ANALYSIS BOX—Nitriles

CHEMICAL SHIFTS

$-CH-C{\equiv}N$ 2.1–3.0 ppm   The α hydrogens are slightly deshielded by the cyano group.

Hydrogens on the adjacent carbon of a nitrile are slightly deshielded by the anisotropic field of the π-bonded electrons appearing in the range 2.1–3.0 ppm. A spectrum of valeronitrile is shown in Figure 3.45. The hydrogens next to the cyano group appear near 2.35 ppm.

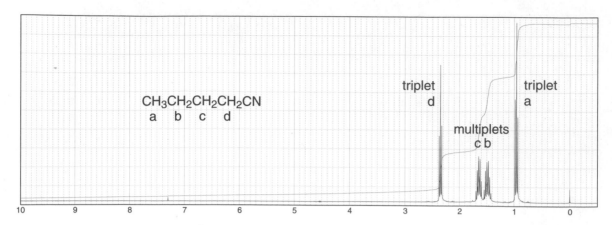

FIGURE 3.45 ^{1}H Spectrum of valeronitrile.

J.   Aldehydes

Two types of hydrogens are found in aldehydes: the aldehyde hydrogen and the α hydrogens (those hydrogens attached to the same carbon as the aldehyde group).

SPECTRAL ANALYSIS BOX—Aldehydes

CHEMICAL SHIFTS

R—CHO	9.0–10.0 ppm	The aldehyde hydrogen is shifted far downfield due to the anisotropy of the carbonyl group (C=O).
R—CH—CH=O	2.1–2.4 ppm	Hydrogens on the carbon adjacent to the C=O group are also deshielded due to the carbonyl group, but they are more distant and the effect is smaller.

COUPLING BEHAVIOR

—CH—CHO	$^3J \approx 1$–3 Hz	Coupling frequently occurs between the aldehyde hydrogen and hydrogens on the adjacent carbon, but J is small.

The chemical shift of the hydrogen in the aldehyde group (—CHO) is found in the range 9–10 ppm. This peak is very indicative of an aldehyde since no other type of hydrogen appears in this region. The aldehyde hydrogen (—CHO) couples weakly ($^3J \approx 3$ Hz) with any hydrogens substituted on the carbon atom on the other side of the carbonyl group. When the —CHO multiplet is cleanly resolved in the spectrum, it is possible to determine the number of hydrogens on the adjacent carbon by using the $n + 1$ Rule.

The hydrogen(s) on the carbon adjacent to the carbonyl group (—CH$_n$—C=O) appear in the range 2.1–2.4 ppm and may be split not only by the aldehyde hydrogen ($^3J \approx 3$ Hz) but also by any hydrogens on the next adjacent carbon in the alkyl chain ($^3J \approx 8$ Hz). All hydrogens on a carbon next to a carbonyl group give absorptions in the same area (2.1–2.4 ppm). Therefore, ketones, aldehydes, esters, amides, and carboxylic acids would all give rise to NMR absorptions in this area, and it is necessary to look for the aldehyde hydrogen to confirm the compound as an aldehyde. Infrared

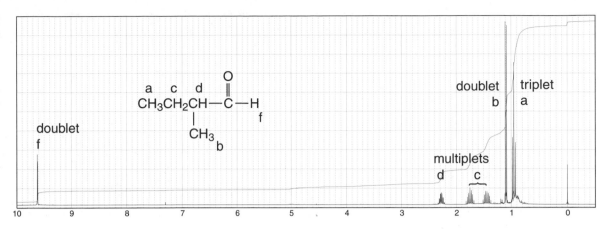

FIGURE 3.46 ¹H Spectrum of 2-methylbutyraldehyde.

spectroscopy would also be a great help in identifying the compound as an aldehyde—look for the —CHO absorptions and a carbonyl peak of the correct frequency.

The spectrum of 2-methylbutyraldehyde is shown in Figure 3.46. Notice the distinct aldehyde peak at 9.6 ppm. Note also that this peak is split into a finely split doublet by the hydrogen on the other side of the carbonyl group. The hydrogen next to the carbonyl group appears at 2.3 ppm.

K.   Ketones

Ketones have only one distinct type of hydrogen atom—those attached to the α carbon.

SPECTRAL ANALYSIS BOX—Ketones

CHEMICAL SHIFTS

R—CH—C=O   2.1–2.4 ppm   The α hydrogens in ketones are deshielded by the anisotropy
     | of the adjacent C=O group.
     R

In a ketone, the hydrogens on the carbon next to the carbonyl group appear in the range 2.1–2.4 ppm. If these hydrogens are part of a longer chain they will be split by any hydrogens on the adjacent carbon, which is further along the chain. Methyl ketones are quite easy to distinguish since they show a sharp three-proton singlet near 2.1 ppm. Be aware that all hydrogens on a carbon next to a carbonyl group give absorptions within the range of 2.1–2.4 ppm. Therefore, ketones, aldehydes, esters, amides, and carboxylic acids would all give rise to NMR absorptions in this same area. It is necessary to look for the absence of other absorptions (—CHO, —OH, —NH₂, —OCH₂R, etc.) to confirm the compound as a ketone. Infrared spectroscopy would also be of great assistance in differentiating these types of compounds. Absence of the aldehyde, hydroxyl, amino, or ether stretching absorptions would help to confirm the compound as a ketone. A spectrum of 3-methyl-2-pentanone is shown in Figure 3.47. Note the tall singlet at 2.1 ppm, which is quite characteristic of a methyl ketone.

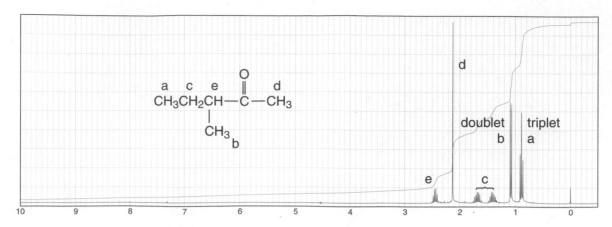

FIGURE 3.47 ^{1}H Spectrum of 3-methyl-2-pentanone.

L. Esters

Two distinct types of hydrogen are found in esters: those on the carbon atom attached to the oxygen atom in the *alcohol part* of the ester and those on the α carbon in the *acid part* of the ester (that is, those attached to the carbon next to the C=O group).

SPECTRAL ANALYSIS BOX—Esters

CHEMICAL SHIFTS

R—CH—COOR	2.1–2.5 ppm	The α hydrogens in esters are deshielded by the anisotropy of the adjacent (C=O) group.
R—COO—CH—	3.5–4.8 ppm	Hydrogens on the carbon attached to the oxygen are deshielded due to the electronegativity of oxygen.

All hydrogens on a carbon next to a carbonyl group give absorptions in the same area (2.1–2.4 ppm). The anisotropic field of the carbonyl group deshields these hydrogens. Therefore, ketones, aldehydes, esters, amides, and carboxylic acids would all give rise to NMR absorptions in this same area. The peak in the 3.4–4.8 ppm region is the key to identifying an ester. The large chemical shift of these hydrogens is due to the deshielding effect of the electronegative oxygen atom, which is attached to the same carbon. Either of the two types of hydrogens mentioned may be split into multiplets if they are part of a longer chain. Methyl esters are easy to recognize by the appearance of a strong singlet ($-OCH_3$) near 4.0 ppm. Ethyl esters will show a quartet ($-O-CH_2CH_3$) near 4.0 ppm and an associated triplet near 1.3 ppm ($-O-CH_2CH_3$).

A spectrum of isobutyl acetate is shown in Figure 3.48. Note the tall singlet of the acyl methyl group at 2.1 ppm, and the highly deshielded methylene group (3.8 ppm) attached to the oxygen and split into a doublet by its methine neighbor.

M. Carboxylic Acids

Carboxylic acids have the acid proton (the one attached to the $-COOH$ group) and the α hydrogens (those attached to the same carbon as the carboxyl group).

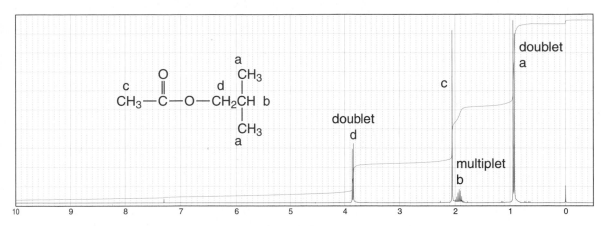

FIGURE 3.48 ^{1}H Spectrum of isobutyl acetate.

SPECTRAL ANALYSIS BOX—Carboxyllic Acids

CHEMICAL SHIFTS

R—COOH	11.0–12.0 ppm	This hydrogen is deshielded by the attached oxygen and it is highly acidic. This is a very characteristic peak for carboxylic acids.
—CH—COOH	2.1–2.5 ppm	Hydrogens adjacent to the carbonyl group are slightly deshielded.

In carboxylic acids the hydrogen of the carboxyl group (—COOH) has resonance in the range 11.0–12.0 ppm. With the exception of the special case of a hydrogen in an enolic OH group that has strong internal hydrogen bonding, no other type of hydrogen appears in this region. A peak in this region is a strong indication of a carboxyl group. Since the carboxyl hydrogen has no neighbors, it is usually unsplit; however, hydrogen bonding and exchange many cause the peak to become *broadened* (become very wide at the base of the peak) and show very low intensity. Under these conditions the integral should not be trusted to be accurate. If this peak is assigned as 1H, the integral may be low and cause an overestimation of the number of other types of hydrogens. As with alcohols, this hydrogen will exchange with water and D_2O. In D_2O proton exchange will convert the group to —COOD and the —COOH absorption near 12.0 ppm will disappear.

$$R—COOH + D_2O \rightleftharpoons R—COOD + DOH \qquad \textbf{exchange in } D_2O$$

Carboxylic acids are often insoluble in $CDCl_3$ and it is common practice to determine their spectra in D_2O to which a small amount of sodium metal is added. This basic solution (NaOD, D_2O) will remove the proton making a soluble sodium salt of the acid. However, when this is done the —COOH absorption will disappear from the spectrum.

$$R—COOH + NaOD \rightleftharpoons R—COO^-Na^+ + DOH$$
insoluble   **soluble**

Hydrogens on the carbon next to the carboxyl group appear in the same range that all hydrogens next to a carbonyl, regardless of functional group type, are found: 2.1–2.5 ppm.

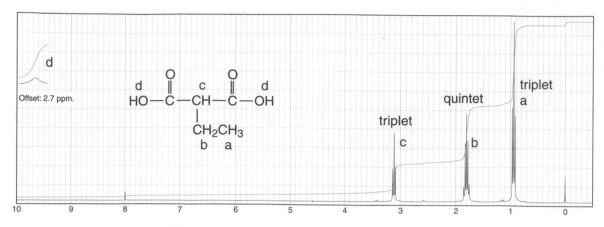

FIGURE 3.49 ^{1}H Spectrum of ethylmalonic acid.

   A spectrum of ethylmalonic acid is shown in Figure 3.49. The —COOH absorption is off the chart to the left. It has been *offset* (moved to the right) by 2.7 ppm. Since it appears at about 9.65 ppm, adding 2.7 ppm brings it to 12.35 ppm, a characteristic position for a carboxyl hydrogen. Notice also that this peak is not sharp—it has been broadened by hydrogen bonding and exchange. This too is characteristic of a carboxyl hydrogen. The hydrogen between the two carbonyl groups is deshielded by both groups and appears at a greater chemical shift (3.1 ppm) than is normal for a hydrogen next to a single C=O.

N.   Amides

SPECTRAL ANALYSIS BO—Amides

CHEMICAL SHIFTS

R(CO)—N—**H**	5.0–9.0 ppm	Hydrogens attached to an amide nitrogen are variable in chemical shift, the value being dependent on the temperature, concentration, and solvent.
—C**H**—CONH—	2.1–2.5 ppm	The α hydrogens in amides absorb in the same range as other acyl (next to C=O) hydrogens. They are slightly deshielded by the carbonyl group.
R(CO)—N—C**H**	2.2–2.9 ppm	Hydrogens on the carbon next to the nitrogen of an amide are slightly deshielded by the electronegativity of the attached nitrogen.

COUPLING BEHAVIOR

—N—**H**	$^1J \approx 50$ Hz	In cases in which this coupling is seen (rare) it is quite large, typically 50 Hz or more. In most cases either the quadrupole moment of the nitrogen atom or chemical exchange decouples this interaction.
—N—C**H**—	$^2J \approx 0$ Hz	Usually not seen for the same reasons stated above.

—N—C**H**—
   |
   H
	$^3J \approx 0$ Hz	Usually not seen for the same reasons stated above.

Amides have three distinct types of hydrogens: those attached to nitrogen, α hydrogens attached to the carbon atom on the carbonyl side of the amide group, and hydrogens attached to a carbon atom that is also attached to the nitrogen atom.

The —NH absorptions of an amide group are highly variable, depending not only on their environment in the molecule, but also on temperature and the solvent used. Because of resonance between the unshared pairs on nitrogen and the carbonyl group, rotation is restricted in most amides. Without rotational freedom, the two hydrogens attached to the nitrogen in an unsubstituted amide are not equivalent and *two different absorption peaks* will be observed, one for each hydrogen (Section 6.6). Nitrogen atoms also have a quadrupole moment (Section 6.5), its magnitude depending on the particular molecular environment. If the nitrogen atom has a large quadrupole moment, the attached hydrogens will show peak broadening (a widening of the peak at its base) and an overall reduction of its intensity.

Hydrogens adjacent to a carbonyl group (regardless of type) all absorb in the same region of the NMR spectrum: 2.1–2.5 ppm.

The spectrum of butyramide is shown in Figure 3.50. Notice the separate absorptions for the two —NH hydrogens (6.6 and 7.2 ppm). This occurs due to restricted rotation in this compound. The hydrogens next to the C=O group appear characteristically at 2.1 ppm.

O. Nitroalkanes

In nitroalkanes, α hydrogens, those hydrogen atoms that are attached to the same carbon atom to which the nitro group is attached, have a characteristically large chemical shift.

SPECTRAL ANALYSIS BOX—Nitroalkanes		
—CH—NO$_2$	4.1–4.3 ppm	Deshielded by the nitro group.

Hydrogens on a carbon next to a nitro group are highly deshielded and appear in the range 4.1–4.3 ppm. The electronegativity of the attached nitrogen and the positive formal charge assigned to that nitrogen clearly indicate the deshielding nature of this group.

A spectrum of 1-nitrobutane is shown in Figure 3.51. Note the large chemical shift (4.4 ppm) of the hydrogens on the carbon adjacent to the nitro group.

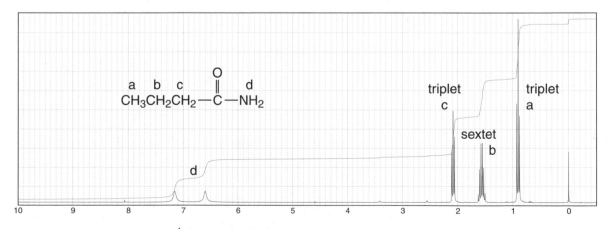

FIGURE 3.50 ^{1}H Spectrum of butyramide.

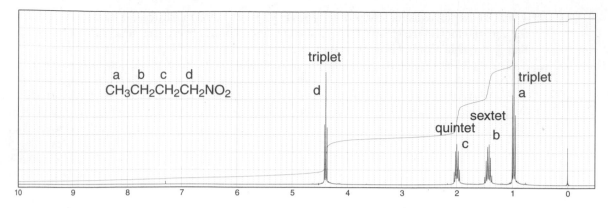

FIGURE 3.51 ^{1}H Spectrum of 1-nitrobutane.

PROBLEMS

*1. What are the allowed nuclear spin states for the following atoms?

    (a) ^{14}N (b) ^{13}C (c) ^{17}O (d) ^{19}F

*2. Calculate the chemical shift in parts per million (δ) for a proton that has resonance 128 Hz downfield from TMS on a spectrometer that operates at 60 MHz.

*3. A proton has resonance 90 Hz downfield from TMS when the field strength is 1.41 Tesla (14,100 Gauss) and the oscillator frequency is 60 MHz.

    (a) What will be its shift in Hertz if the field strength is increased to 2.82 Tesla and the oscillator frequency to 120 MHz?

    (b) What will be its chemical shift in parts per million (δ)?

*4. Acetonitrile (CH_3CN) has resonance at 1.97 ppm, whereas methyl chloride (CH_3Cl) has resonance at 3.05 ppm, even though the dipole moment of acetonitrile is 3.92 D and that of methyl chloride is only 1.85 D. The larger dipole moment for the cyano group suggests that the electronegativity of this group is greater than that of the chlorine atom. Explain why the methyl hydrogens on acetonitrile are actually more shielded than those in methyl chloride, in contrast with the results expected on the basis of electronegativity. (*Hint:* What kind of spatial pattern would you expect for the magnetic anisotropy of the cyano group, CN?)

*5. The position of the OH resonance of phenol varies with concentration in solution, as the following table shows. On the other hand, the hydroxyl proton of *ortho*-hydroxyacetophenone appears at 12.05 ppm and does not show any great shift upon dilution. Explain.

Concentration w/v in CCl$_4$	δ (ppm)
100%	7.45
20%	6.75
10%	6.45
5%	5.95
2%	4.88
1%	4.37

Phenol

o-Hydroxyacetophenone

12.05 ppm

*6. The chemical shifts of the methyl groups of three related molecules, pinane, α-pinene, and β-pinene, follow.

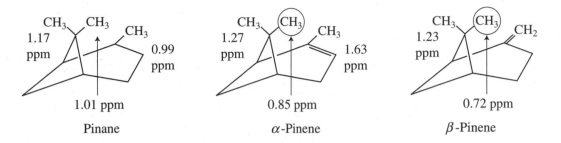

Pinane α-Pinene β-Pinene

Build models of these three compounds, and then explain why the two circled methyl groups have such small chemical shifts.

*7. In benzaldehyde, two of the ring protons have resonance at 7.87 ppm, and the other three have resonance in the range from 7.5 to 7.6 ppm. Explain.

*8. Make a three-dimensional drawing illustrating the magnetic anisotropy in 15,16-dihydro-15, 16-dimethylpyrene, and explain why the methyl groups have resonance at −4.2 ppm!

15,16-Dihydro-15,16-dimethylpyrene

*9. Work out the spin arrangements and splitting patterns for the following spin system.

$$\text{Cl}\!-\!\underset{\underset{\text{Cl}}{|}}{\overset{\overset{\text{H}_A}{|}}{\text{C}}}\!-\!\underset{\underset{\text{H}_B}{|}}{\overset{\overset{\text{H}_B}{|}}{\text{C}}}\!-\!\text{Br}$$

*10. Explain the patterns and intensities of the isopropyl group in isopropyl iodide.

$$\underset{\text{CH}_3}{\overset{\text{CH}_3}{\diagdown}}\!\!\text{CH}\!-\!\text{I}$$

*11. What spectrum would you expect for the following molecule?

$$\text{H}\!-\!\underset{\underset{\text{Cl}}{|}}{\overset{\overset{\text{Cl}}{|}}{\text{C}}}\!-\!\underset{\underset{\text{Cl}}{|}}{\overset{\overset{\text{H}}{|}}{\text{C}}}\!-\!\underset{\underset{\text{Cl}}{|}}{\overset{\overset{\text{Cl}}{|}}{\text{C}}}\!-\!\text{H}$$

*12. What arrangement of protons would give two triplets of equal area?

*13. Predict the appearance of the NMR spectrum of propyl bromide.

*14. The following compound, with the formula $C_4H_8O_2$, is an ester. Give its structure and assign the chemical shift values.

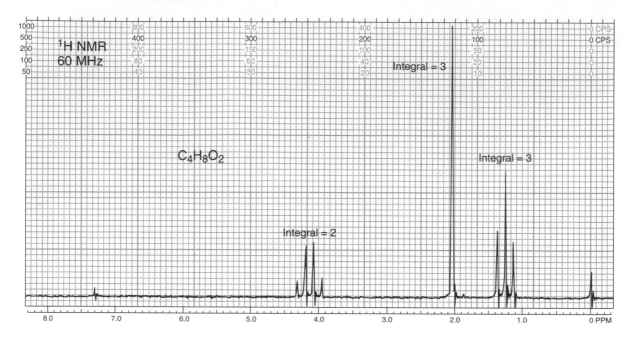

*15. The following compound is a monosubstituted aromatic hydrocarbon with the formula C_9H_{12}. Give its structure and assign the chemical shift values.

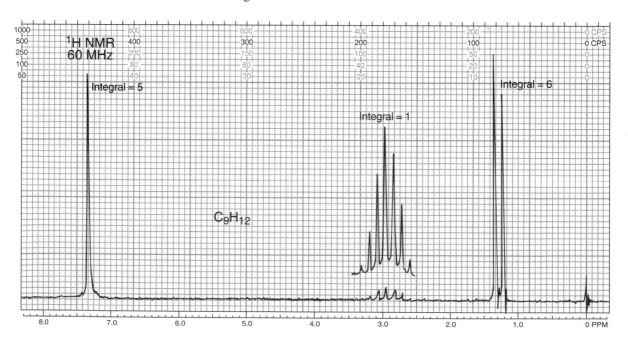

*16. The following compound is a carboxylic acid that contains a bromine atom: $C_4H_7O_2Br$. The peak at 10.97 ppm was moved onto the chart (which runs only from 0 to 8 ppm) for clarity. What is the structure of the compound?

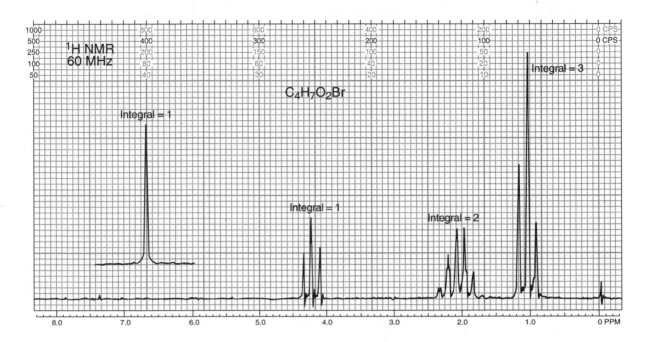

*17. The following compounds are isomeric esters derived from acetic acid, each with formula $C_5H_{10}O_2$. The peaks of the spectrum have been labeled to indicate the degrees of splitting. With the first spectrum as an example, use the integral curve traced on the spectrum to calculate the number of hydrogens represented in each multiplet (pp. 118–120). The multiplets appear both on the spectrum and in the first column of the following table. The second column is obtained by dividing through by the lowest number (1.7 div). The third column is obtained by multiplying by 2 and rounding off the values. Notice that the sum of the numbers in the third column equals the number of hydrogen atoms (10) present in the formula. Often one can inspect the spectrum and visually approximate the relative numbers of hydrogen atoms, thus avoiding the more mathematical approach demonstrated in the following table. Using either method, the second spectrum yields a ratio of 1:3:6. What are the structures of the two esters?

1.7 div	1.0	2 H
2.5 div	1.47	3 H
1.7 div	1.0	2 H
2.5 div	1.47	3 H

(a)

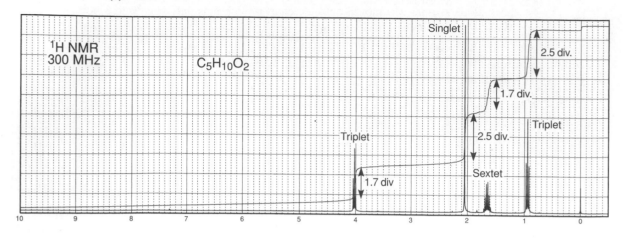

(b)

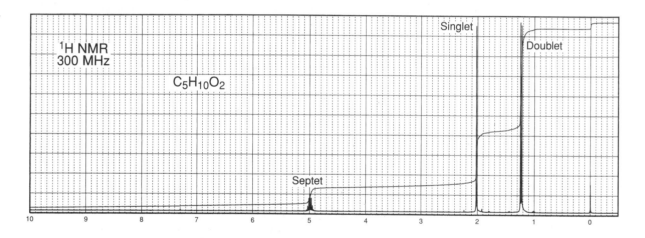

*18. The compound that gives the following NMR spectrum has the formula $C_3H_6Br_2$. Draw the structure.

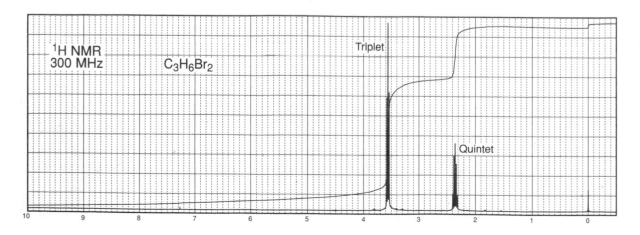

*19. Draw the structure of an ether with formula $C_5H_{12}O_2$ that fits the following NMR spectrum.

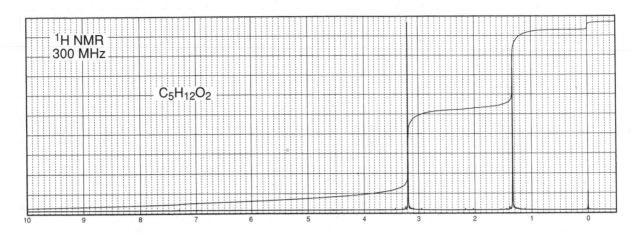

*20. Following are the NMR spectra of three isomeric esters with the formula $C_7H_{14}O_2$, all derived from propanoic acid. Provide a structure for each.

(a)

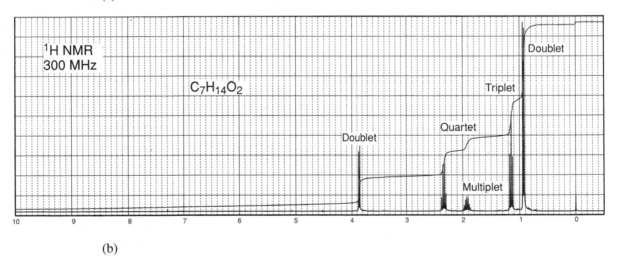

(b)

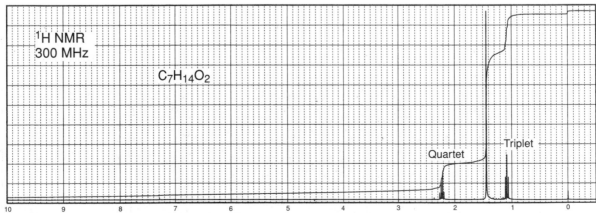

(c)

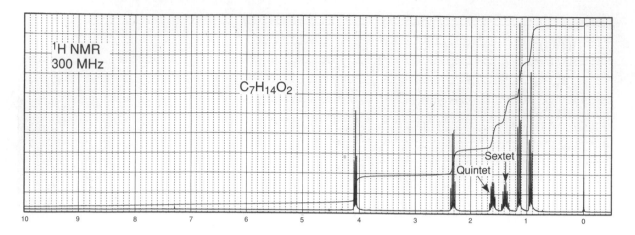

*21. The two isomeric carboxylic acids that give the following NMR spectra both have the formula $C_3H_5ClO_2$. Draw their structures.

(a)

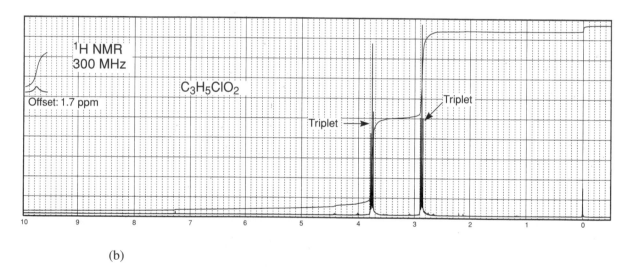

(b)

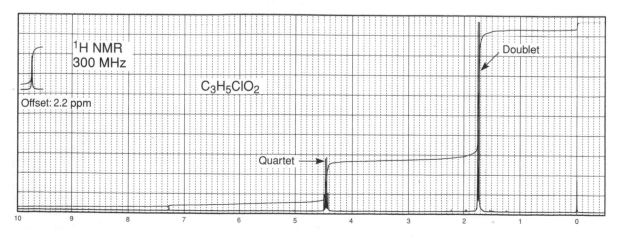

*22. The following NMR spectra are of monosubstituted aromatic hydrocarbon compounds with the formula $C_{10}H_{14}$. Make no attempt to interpret the aromatic proton area between 7.1 and 7.3 ppm except to determine the relative number of hydrogen atoms. Draw structures for these compounds.

(a)

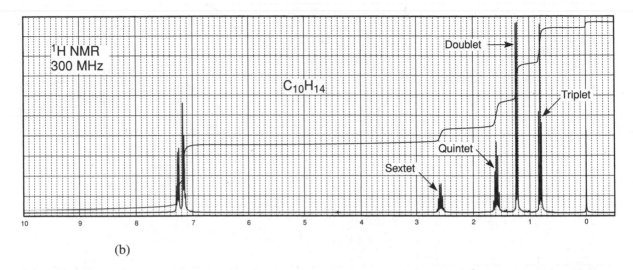

(b)

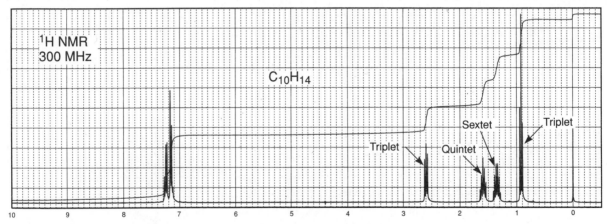

*23. The following compound, with formula $C_8H_{11}N$, shows a doublet at about 3350 cm^{-1} and bands in the range from 1600 to 1450 cm^{-1} in the infrared spectrum. Draw its structure.

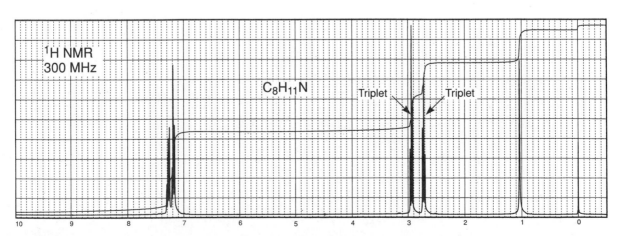

24. The following compounds are isomers with formula $C_{10}H_{12}O$. Their infrared spectra show strong bands near 1715 cm^{-1} and in the range from 1600 to 1450 cm^{-1}. Draw their structures.

(a)

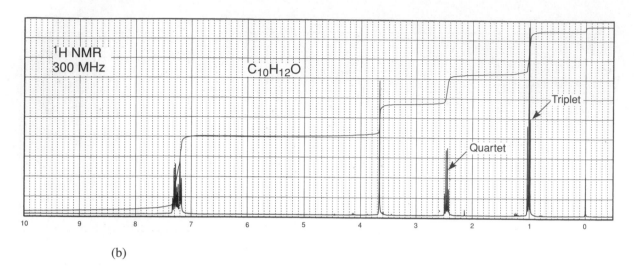

(b)

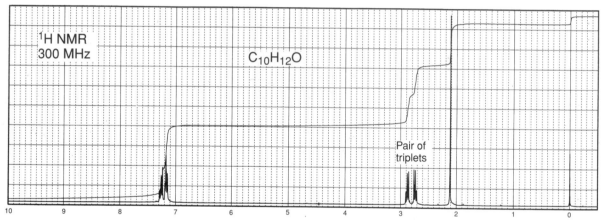

25. The following four NMR spectra are of isomeric monosubstituted aromatic esters with formula $C_{10}H_{12}O_2$. Make no attempt to interpret the aromatic proton areas between 7.1 and 7.4 ppm. Draw the structures of the compounds.

(a)

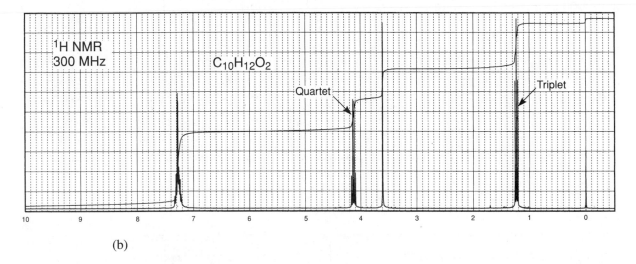

(b)

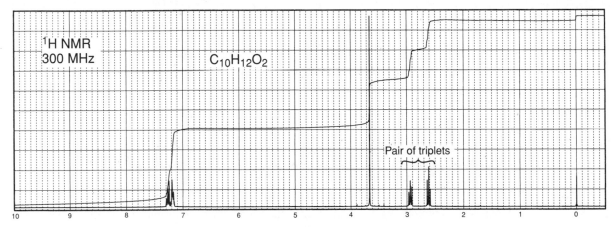

(c)

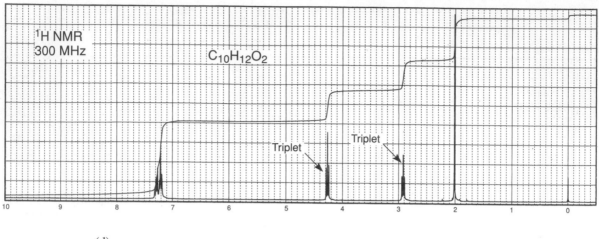

(d)

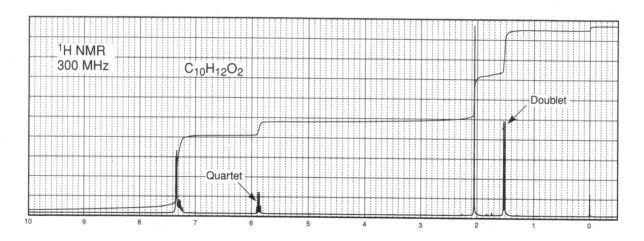

26. Along with the following NMR spectrum, this compound, with formula $C_5H_{10}O_2$, shows bands at 3450 cm^{-1} (broad) and 1713 cm^{-1} (strong) in the infrared spectrum. Draw its structure.

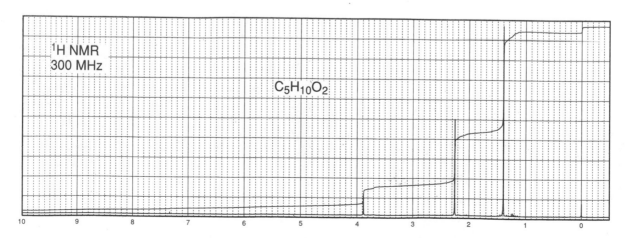

27. The following ester, with formula $C_5H_6O_2$, shows medium bands in the infrared spectrum at 3270 and 2118 cm^{-1}. Draw the structure of the compound.

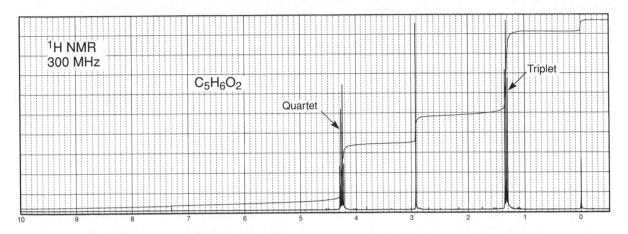

28. The following compound, with formula $C_7H_{12}O_4$, shows strong absorption at 1734 cm^{-1} and has several strong bands centering at about 1200 cm^{-1} in the infrared spectrum. Draw its structure.

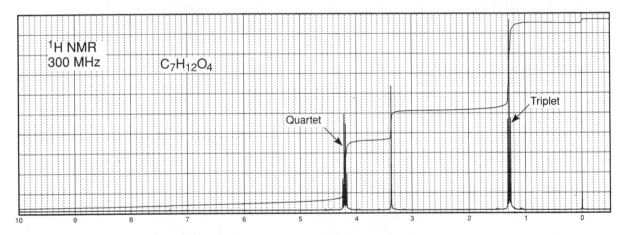

REFERENCES

Textbooks

Ault, A., and G. O. Dudek, *NMR—An Introduction to Nuclear Magnetic Resonance Spectroscopy,* Holden–Day, San Francisco, 1976.

Friebolin, H., *Basic One- and Two-Dimensional NMR Spectroscopy,* 2nd ed., VCH Publishers, New York, 1993.

Gunther, H., *NMR Spectroscopy,* 2nd ed., John Wiley and Sons, New York, 1995.

Jackman, L. M., and S. Sternhell, *Nuclear Magnetic Resonance Spectroscopy in Organic Chemistry,* 2d ed., Pergamon Press, New York, 1969.

Macomber, R. S., *NMR Spectroscopy: Essential Theory and Practice,* College Outline Series, Harcourt, Brace Jovanovich, New York, 1988.

Macomber, R. S., *A Complete Introduction to Modern NMR Spectroscopy,* John Wiley and Sons, New York, 1997.

Sanders, J. K. M., and B. K. Hunter, *Modern NMR Spectroscopy—A Guide for Chemists,* 2nd ed., Oxford University Press, Oxford, 1993.

Silverstein, R. M., and F. X. Webster, *Spectrometric Identification of Organic Compounds,* 6th ed., John Wiley and Sons, New York, 1998.

Williams, D. H., and I. Fleming, *Spectroscopic Methods in Organic Chemistry,* 4th ed., McGraw-Hill Book Co. Ltd., London–New York, 1987.

Yoder, C. H., and C. D. Schaeffer, *Introduction to Multinuclear NMR,* Benjamin-Cummings, Menlo Park, CA, 1987.

Compilations of Spectra

Ault, A., and M. R. Ault, *A Handy and Systematic Catalog of NMR Spectra, 60 MHz with Some 270 MHz,* University Science Books, Mill Valley, CA, 1980.

Pouchert, C. J., *The Aldrich Library of NMR Spectra, 60 MHz,* 2d ed., Aldrich Chemical Company, Milwaukee, WI, 1983.

Pouchert, C. J., and J. Behnke, *The Aldrich Library of ^{13}C and ^{1}H FT-NMR Spectra, 300 MHz,* Aldrich Chemical Company, Milwaukee, WI, 1993.

Pretsch, E., T. Clerc, J. Seibl, and W. Simon, *Tables of Spectral Data for Structure Determination of Organic Compounds,* 2d ed., Springer-Verlag, Berlin and New York, 1989. Translated from the German by K. Biemann.

Computer Programs that Teach Spectroscopy

Clough, F. W., "Introduction to Spectroscopy," Version 2.0 for MS-DOS and Macintosh, Trinity Software, 607 Tenney Mtn Highway, Suite 215, Plymouth, NH, 03264, www.trinitysoftware.com.

Pavia, D. L., "Spectral Interpretation," MS-DOS Version, Trinity Software, 607 Tenney Mtn Highway, Suite 215, Plymouth, NH, 03264, www.trinitysoftware.com.

Schatz, P. F., "Spectrabook I and II," MS-DOS Version, and "Spectradeck I and II," Macintosh Version, Falcon Software, One Hollis Street, Wellesley, MA, 02482, www.falconsoftware.com.

Web sites

http://www.aist.go.jp/RIODB/SDBS/menu-e.html
Integrated Spectral DataBase System for Organic Compounds, National Institute of Materials and Chemical Research, Tsukuba, Ibaraki 305-8565, Japan. This database includes infrared, mass spectra, and NMR data (proton and carbon-13) for a large number of compounds.

http://www.chem.ucla.edu/~webnmr
UCLA Department of Chemistry and Biochemistry in connection with Cambridge University Isotope Laboratories, maintains a website, WebSpecta, that provides NMR and IR spectroscopy problems for students to interpret. They provide links to other sites with problems for students to solve.

http://www.nd.edu/~smithgrp/structure/workbook.html
Combined structure problems provided by the Smith group at Notre Dame University.

NUCLEAR MAGNETIC RESONANCE SPECTROSCOPY

Part Two: Carbon-13 Spectra, Including Heteronuclear Coupling with Other Nuclei

The study of carbon nuclei through nuclear magnetic resonance (NMR) spectroscopy is an important technique for determining the structures of organic molecules. Using it together with proton NMR and infrared spectroscopy, organic chemists can often determine the complete structure of an unknown compound without "getting their hands dirty" in the laboratory! Modern Fourier transform–NMR (FT-NMR) instrumentation makes it possible to obtain routine carbon spectra easily.

Carbon spectra can be used to determine the number of nonequivalent carbons and to identify the types of carbon atoms (methyl, methylene, aromatic, carbonyl, and so on) that may be present in a compound. Thus, carbon NMR provides direct information about the carbon skeleton of a molecule. Some of the principles of proton NMR apply to the study of carbon NMR; however, structural determination is generally easier with carbon-13 NMR spectra than with proton NMR. Typically, both techniques are used together to determine the structure of an unknown compound.

4.1 THE CARBON-13 NUCLEUS

Carbon-12, the most abundant isotope of carbon, is NMR inactive since it has a spin of zero (see Section 3.1). Carbon-13, or ^{13}C, however, has odd mass and does have nuclear spin, with $I = \frac{1}{2}$. Unfortunately, the resonances of ^{13}C nuclei are more difficult to observe than those of protons (^{1}H). They are about 6000 times weaker than proton resonances, for two major reasons.

First, the natural abundance of carbon-13 is very low; only 1.08% of all carbon atoms in nature are ^{13}C atoms. If the total number of carbons in a molecule is low, it is very likely that a majority of the molecules in a sample will have no ^{13}C nuclei at all. In molecules containing a ^{13}C isotope, it is unlikely that a second atom in the same molecule will be a ^{13}C atom. Therefore, when we observe a ^{13}C spectrum, we are observing a spectrum built up from a collection of molecules, where each molecule supplies no more than a single ^{13}C resonance. No *single* molecule supplies a complete spectrum.

Second, since the magnetogyric ratio of a ^{13}C nucleus is smaller than that of hydrogen (Table 3.2), ^{13}C nuclei always have resonance at a frequency lower than protons. Recall that at lower frequencies, the population of excess nuclei is reduced (Table 3.3); this, in turn, reduces the sensitivity of NMR detection procedures.

Through the use of modern Fourier transform instrumentation (Section 3.7B), it is possible to obtain ^{13}C NMR spectra of organic compounds even though detection of carbon signals is difficult compared to detection of proton spectra. To compensate for the low natural abundance of carbon, a greater number of individual scans of the spectrum must be accumulated than is common for a proton spectrum.

For a given magnetic field strength, the resonance frequency of a ^{13}C nucleus is about one-fourth the frequency required to observe proton resonances (see Table 3.2). For example, in a 7.05-Tesla applied magnetic field, protons are observed at 300 MHz while ^{13}C nuclei are observed at about 75 MHz. With modern instrumentation, it is a simple matter to switch the transmitter frequency from the value required to observe proton resonances to the value required for ^{13}C resonances.

4.2 CARBON-13 CHEMICAL SHIFTS

A.   Correlation Charts

An important parameter derived from carbon-13 spectra is the chemical shift. The correlation chart in Figure 4.1 shows typical ^{13}C chemical shifts, listed in parts per million (ppm) from tetramethylsilane (TMS), where the carbons of the methyl groups of TMS (not the hydrogens) are used for reference. Approximate ^{13}C chemical shift ranges for selected types of carbon are also given in Table 4.1. Notice that the chemical shifts appear over a range (0 to 220 ppm) much larger than that observed for protons (0 to 12 ppm). Because of the very large range of values, nearly every nonequivalent carbon atom in an organic molecule gives rise to a peak with a different chemical shift. Peaks rarely overlap as they often do in proton NMR.

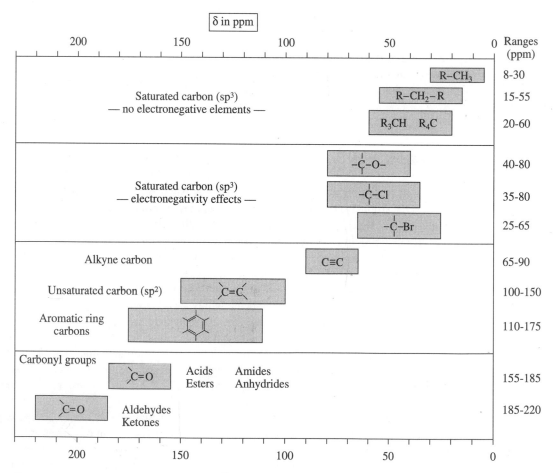

F I G U R E   4 . 1 A correlation chart for ^{13}C chemical shifts (chemical shifts are listed in parts per million from TMS).

TABLE 4.1
APPROXIMATE ^{13}C CHEMICAL SHIFT RANGES (ppm) FOR SELECTED TYPES OF CARBON

$R-CH_3$	8–30	$C{\equiv}C$	65–90
R_2CH_2	15–55	$C{=}C$	100–150
R_3CH	20–60	$C{\equiv}N$	110–140
$C-I$	0–40		110–175
$C-Br$	25–65		
$C-N$	30–65	$\underset{O}{R-\overset{\displaystyle O}{C}-OR}, R-\overset{\displaystyle O}{C}-OH$	155–185
$C-Cl$	35–80	$R-\overset{\displaystyle O}{C}-NH_2$	155–185
$C-O$	40–80	$R-\overset{\displaystyle O}{C}-R, R-\overset{\displaystyle O}{C}-H$	185–220

The correlation chart is divided into four sections. Saturated carbon atoms appear at highest field, nearest to TMS (8 to 60 ppm). The next section of the correlation chart demonstrates the effect of electronegative atoms (40 to 80 ppm). The third section of the chart includes alkene and aromatic ring carbon atoms (100 to 175 ppm). Finally, the fourth section of the chart contains carbonyl carbons, which appear at the lowest field values (155 to 220 ppm).

Electronegativity, hybridization, and anisotropy all affect ^{13}C chemical shifts in nearly the same fashion as they affect ^{1}H chemical shifts; however ^{13}C chemical shifts are about 20 times larger[1]. Electronegativity (Section 3.11A) produces the same deshielding effect in carbon NMR as in proton NMR—the electronegative element produces a large downfield shift. The shift is greater for a ^{13}C atom than for a proton since the electronegative atom is directly attached to the ^{13}C atom, and the effect occurs through only a single bond, C−X. With protons, the electronegative atoms are attached to carbon, not hydrogen; the effect occurs through two bonds, H−C−X, rather than one.

In ^{1}H NMR, the effect of an electronegative element on chemical shift diminishes with distance, but it is always in the same direction (deshielding and downfield). In ^{13}C NMR, an electronegative element also causes a downfield shift in the α and β carbons, but it usually leads to a small *upfield* shift for the γ carbon. This effect is clearly seen in the carbons of hexanol:

$$14.2 \quad 22.8 \quad 32.0 \quad 25.8 \quad 32.8 \quad 61.9 \text{ ppm}$$
$$CH_3{-}CH_2{-}CH_2{-}CH_2{-}CH_2{-}CH_2{-}OH$$
$$\omega \qquad \varepsilon \qquad \delta \qquad \gamma \qquad \beta \qquad \alpha$$

The shift for C3, the γ carbon, seems quite at odds with the expected effect of an electronegative substituent. This anomaly points up the need to consult detailed correlation tables for ^{13}C chemical shifts. Such tables appear in Appendix 7 and are discussed in the next section.

---

[1]This is sometimes called the *20x Rule.* See Macomber, R., "Proton-Carbon Chemical Shift Correlations," *Journal of Chemical Education, 68(a),* 284–5, 1991.

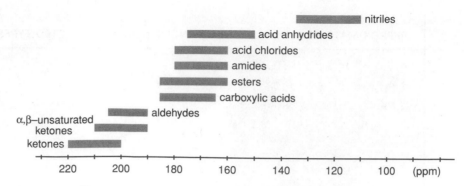

FIGURE 4.2 A ^{13}C correlation chart for carbonyl and nitrile functional groups.

Analogous with ^{1}H shifts, changes in hybridization (Section 3.11B) also produce larger shifts for the carbon-13 that is *directly involved* (no bonds) than they do for the hydrogens attached to that carbon (one bond). In ^{13}C NMR, the carbons of carbonyl groups have the largest chemical shifts, due both to sp^2 hybridization and to the fact that an electronegative oxygen is directly attached to the carbonyl carbon, deshielding it even further. Anisotropy (Section 3.12) is responsible for the large chemical shifts of the carbons in aromatic rings and alkenes.

Notice that the range of chemical shifts is larger for carbon atoms than for hydrogen atoms. Because the factors affecting carbon shifts operate either through one bond or directly on carbon, they are greater than those for hydrogen, which operate through more bonds. As a result, the entire range of chemical shifts becomes larger for ^{13}C (0 to 220 ppm) than for ^{1}H (0 to 12 ppm).

Many of the important functional groups of organic chemistry contain a carbonyl group. In determining the structure of a compound containing a carbonyl group, it is frequently helpful to have some idea of the type of carbonyl group in the unknown. Figure 4.2 illustrates the typical ranges of ^{13}C chemical shifts for some carbonyl-containing functional groups. Although there is some overlap in the ranges, ketones and aldehydes are easy to distinguish from the other types. Chemical shift data for carbonyl carbons are particularly powerful when combined with data from an infrared spectrum.

B.   Calculation of ^{13}C Chemical Shifts

Nuclear magnetic resonance spectroscopists have accumulated, organized, and tabulated a great deal of data for ^{13}C chemical shifts. It is possible to predict the chemical shift of almost any ^{13}C atom from these tables, starting with a base value for the molecular skeleton and then adding increments that correct the value for each substituent. Corrections for the substituents depend on both the type of substituent and its position relative to the carbon atom being considered. Corrections for rings are different from those for chains, and they frequently depend on stereochemistry.

Consider *m*-xylene (1,3-dimethylbenzene) as an example. Consulting the tables, you will find that the base value for the carbons in a benzene ring is 128.5 ppm. Next, look in the substituent tables that relate to benzene rings for the methyl substituent corrections (Table A8.7). These values are

	ipso	*ortho*	*meta*	*para*
CH$_3$:	8.9	0.7	−0.1	−2.9 ppm

The *ipso* carbon is the one to which the substituent is directly attached. The calculations for *m*-xylene start with the base value and add these increments as follows:

$$C1 = \text{base} + ipso + meta = 128.5 + 8.9 + (-0.1) = 137.3 \text{ ppm}$$

$$C2 = \text{base} + ortho + ortho = 128.5 + 0.7 + 0.7 \quad = 129.9 \text{ ppm}$$

$$C3 = C1$$

$$C4 = \text{base} + ortho + para = 128.5 + 0.7 + (-2.9) = 126.3 \text{ ppm}$$

$$C5 = \text{base} + meta + meta = 128.5 + 2(-0.1) \quad = 128.3 \text{ ppm}$$

$$C6 = C4$$

The observed values for C1, C2, C4, and C5 of *m*-xylene are 137.6, 130.0, 126.2, and 128.2 ppm, respectively, and the calculated values agree well with those actually measured.

Appendix 8 presents some ^{13}C chemical shift correlation tables with instructions. Complete ^{13}C chemical shift correlation tables are too numerous to include in this book. If you are interested, consult the textbooks by Levy, Macomber, Silverstein, and Friebolin, which are listed in the references at the end of this chapter. Even more convenient than tables are computer programs that calculate ^{13}C chemical shifts. In the more advanced programs, the operator need only sketch the molecule on the screen, using a mouse, and the program will calculate both the chemical shifts and the rough appearance of the spectrum. Some of these programs are also listed in the references.

4.3 PROTON-COUPLED ^{13}C SPECTRA—SPIN–SPIN SPLITTING OF CARBON-13 SIGNALS

Unless a molecule is artificially enriched by synthesis, the probability of finding two ^{13}C atoms in the same molecule is low. The probability of finding two ^{13}C atoms adjacent to each other in the same molecule is even lower. Therefore, we rarely observe **homonuclear** (carbon–carbon) spin–spin splitting patterns where the interaction occurs between two ^{13}C atoms. However, the spins of protons attached directly to ^{13}C atoms do interact with the spin of carbon and cause the carbon signal to be split according to the *n* + 1 Rule. This is **heteronuclear** (carbon–hydrogen) coupling involving two different types of atoms. With ^{13}C NMR, we generally examine splitting that arises from the protons *directly attached* to the carbon atom being studied. This is a one-bond coupling. Remember that in proton NMR, the most common splittings are *homonuclear* (hydrogen–hydrogen) and occur between protons attached to *adjacent* carbon atoms. In these cases, the interaction is a three-bond coupling, H—C—C—H.

Figure 4.3 illustrates the effect of protons directly attached to a ^{13}C atom. The *n* + 1 Rule predicts the degree of splitting in each case. The resonance of a ^{13}C atom with three attached protons, for instance, is split into a quartet (*n* + 1 = 3 + 1 = 4). The possible spin combinations for the three protons are the same as those illustrated in Figure 3.33, and each spin combination interacts with carbon to give a different peak of the multiplet. Since the hydrogens are directly attached to the carbon-13 (one-bond couplings), the coupling constants for this interaction are quite large, with *J* values of about 100 to 250 Hz. Compare the typical three-bond H—C—C—H couplings that are common in NMR spectra, which have *J* values of about 1 to 20 Hz.

It is important to note while examining Figure 4.3 that you are not "seeing" protons directly when looking at a ^{13}C spectrum (proton resonances occur at frequencies outside the range used to obtain ^{13}C spectra); you are observing only the effect of the protons on ^{13}C atoms. Also, remember that we cannot observe ^{12}C, because it is NMR inactive.

Spectra that show the spin–spin splitting, or coupling, between carbon-13 and the protons directly attached to it are called **proton-coupled spectra** or sometimes **nondecoupled spectra** (see the next section). Figure 4.4a is the proton-coupled ^{13}C NMR spectrum of ethyl phenylacetate. In this spectrum, the first quartet downfield from TMS (14.2 ppm) corresponds to the carbon of the methyl group. It is split into a quartet (*J* = 127 Hz) by the three attached hydrogen atoms (^{13}C—H,

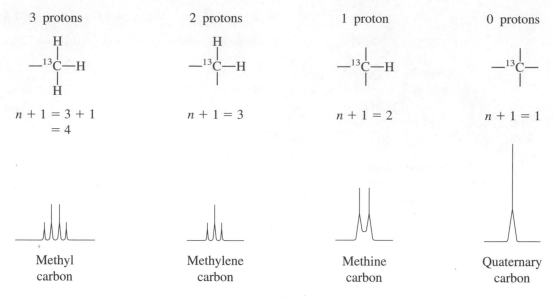

F I G U R E   4 . 3 The effect of attached protons on ^{13}C resonances.

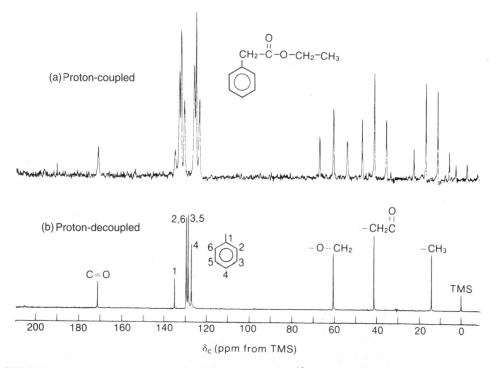

F I G U R E   4 . 4 Ethyl phenylacetate. (a) The proton-coupled ^{13}C spectrum (20 MHz). (b) The proton-decoupled ^{13}C spectrum (20 MHz). (From Moore, J. A., and D. L. Dalrymple, *Experimental Methods in Organic Chemistry,* W. B. Saunders, Philadelphia, 1976.)

one-bond couplings). In addition, although it cannot be seen on the scale of this spectrum (an expansion must be used), each of the quartet lines is split into a closely spaced triplet (J = ca. 1 Hz). This additional fine splitting is caused by the two protons on the adjacent −CH$_2$− group. These are two-bond couplings (H—C—^{13}C) of a type that occurs commonly in ^{13}C spectra, with coupling constants that are generally quite small ($J = 0$–2 Hz) for systems with carbon atoms in an aliphatic chain. Because of their small size, these couplings are frequently ignored in the routine analysis of spectra, with greater attention being given to the larger one-bond splittings seen in the quartet itself.

There are two —CH$_2$— groups in ethyl phenylacetate. The one corresponding to the ethyl —CH$_2$— group is found farther downfield (60.6 ppm), as this carbon is deshielded by the attached oxygen. It is a triplet because of the two attached hydrogens (one-bond couplings). Again, although it is not seen in this unexpanded spectrum, the three hydrogens on the adjacent methyl group finely split each of the triplet peaks into a quartet. The benzyl —CH$_2$— carbon is the intermediate triplet (41.4 ppm). Farthest downfield is the carbonyl-group carbon (171.1 ppm). On the scale of this presentation it is a singlet (no directly attached hydrogens), but because of the adjacent benzyl —CH$_2$— group, it is actually split finely into a triplet. The aromatic-ring carbons also appear in the spectrum, and they have resonances in the range from 127 to 136 ppm. Section 4.12 will discuss aromatic-ring ^{13}C resonances.

Proton-coupled spectra for large molecules are often difficult to interpret. The multiplets from different carbons commonly overlap because the ^{13}C—H coupling constants are frequently larger than the chemical shift differences of the carbons in the spectrum. Sometimes, even simple molecules such as ethyl phenylacetate (Fig. 4.4a) are difficult to interpret. Proton decoupling, which is discussed in the next section, avoids this problem.

4.4 PROTON-DECOUPLED ^{13}C SPECTRA

By far the great majority of ^{13}C NMR spectra are obtained as **proton-decoupled spectra.** The decoupling technique obliterates all interactions between protons and ^{13}C nuclei; therefore, only **singlets** are observed in a decoupled ^{13}C NMR spectrum. Although this technique simplifies the spectrum and avoids overlapping multiplets, it has the disadvantage that the information on attached hydrogens is lost.

Proton **decoupling** is accomplished in the process of determining a ^{13}C NMR spectrum by simultaneously irradiating all of the protons in the molecule with a broad spectrum of frequencies in the proper range. Most modern NMR spectrometers provide a second, tunable radiofrequency generator, the **decoupler,** for this purpose. Irradiation causes the protons to become saturated, and they undergo rapid upward and downward transitions, among all their possible spin states. These rapid transitions decouple any spin–spin interactions between the hydrogens and the ^{13}C nuclei being observed. In effect, all spin interactions are averaged to zero by the rapid changes. The carbon nucleus "senses" only one average spin state for the attached hydrogens rather than two or more distinct spin states.

Figure 4.4b is a proton-decoupled spectrum of ethyl phenylacetate. The proton-coupled spectrum (Fig. 4.4a) was discussed in Section 4.3. It is interesting to compare the two spectra to see how the proton decoupling technique simplifies the spectrum. Every chemically and magnetically distinct carbon gives only a single peak. Notice, however, that the two *ortho* ring carbons (carbons 2 and 6) and the two *meta* ring carbons (carbons 3 and 5) are equivalent by symmetry and that each gives only a single peak.

Figure 4.5 is a second example of a proton-decoupled spectrum. Notice that the spectrum shows three peaks, corresponding to the exact number of carbon atoms in 1-propanol. If there are no equivalent carbon atoms in a molecule, a ^{13}C peak will be observed for *each* carbon. Notice also that the assignments given in Figure 4.5 are consistent with the values in the chemical shift table (Fig. 4.1). The carbon atom closest to the electronegative oxygen is farthest downfield, and the methyl carbon is at highest field.

The three-peak pattern centered at $\delta = 77$ ppm is due to the solvent CDCl$_3$. This pattern results from the coupling of a deuterium (^{2}H) nucleus to the ^{13}C nucleus (see Section 4.13). Often the CDCl$_3$ pattern is used as an internal reference, in place of TMS.

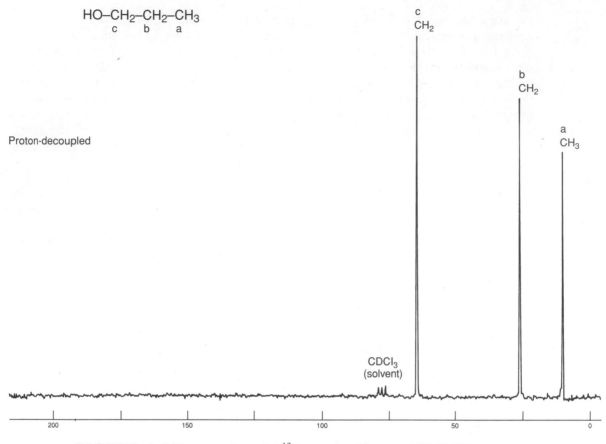

FIGURE 4.5 The proton-decoupled ^{13}C spectrum of 1-propanol (22.5 MHz).

▣ 4.5 NUCLEAR OVERHAUSER ENHANCEMENT (NOE)

When we obtain a proton-decoupled ^{13}C spectrum, the intensities of many of the carbon resonances increase significantly above those observed in a proton-coupled experiment. Carbon atoms with hydrogen atoms directly attached are enhanced the most, and the enhancement increases (but not always linearly) as more hydrogens are attached. This effect is known as the nuclear Overhauser effect, and the degree of increase in the signal is called the **nuclear Overhauser enhancement (NOE).** The NOE effect is *heteronuclear* in this case, operating between two dissimilar atoms (carbon and hydrogen). Both atoms exhibit spins and are NMR active. The nuclear Overhauser effect is general, showing up when one of two different types of atoms is irradiated while the NMR spectrum of the other type is determined. If the absorption intensities of the observed (i.e., nonirradiated) atom change, enhancement has occurred. The effect can be either positive or negative, depending on which atom types are involved. In the case of carbon-13 interacting with hydrogen-1, the effect is positive; irradiating the hydrogens increases the intensities of the carbon signals. The maximum enhancement that can be observed is given by the relationship

$$\text{NOE}_{max} = \frac{1}{2}\left(\frac{\gamma_{irr}}{\gamma_{obs}}\right)$$

Equation 4.1

where γ_{irr} is the magnetogyric ratio of the nucleus being irradiated and γ_{obs} is that of the nucleus being observed. Remember that NOE_{max} is the *enhancement* of the signal, and it must be added to the original signal strength:

$$\text{total predicted intensity (maximum)} = 1 + NOE_{max} \qquad \text{Equation 4.2}$$

For a proton-decoupled ^{13}C spectrum, we would calculate, using the values in Table 3.2,

$$NOE_{max} = \frac{1}{2}\left(\frac{267.5}{67.28}\right) = 1.988 \qquad \text{Equation 4.3}$$

indicating that the ^{13}C signals can be enhanced up to 200% by irradiation of the hydrogens. This value, however, is a theoretical maximum, and most actual cases exhibit less than ideal enhancement.

The heteronuclear NOE effect actually operates in both directions; either atom can be irradiated. If one were to irradiate carbon-13 while determining the NMR spectrum of the hydrogens—the reverse of the usual procedure—the hydrogen signals would increase by a very small amount. However, because there are few ^{13}C atoms in a given molecule, the result would not be very dramatic. In contrast, NOE is a definite *bonus* received in the determination of proton-decoupled ^{13}C spectra. The hydrogens are numerous, and carbon-13, with its low abundance, generally produces weak signals. Because NOE increases the intensity of the carbon signals, it substantially increases the sensitivity (signal-to-noise ratio) in the ^{13}C spectrum.

Signal enhancement due to NOE is an example of **cross-polarization,** in which a polarization of the spin states in one type of nucleus causes a polarization of the spin states in another nucleus. Cross-polarization will be explained in Section 4.6. In the current example (proton-decoupled ^{13}C spectra), when the hydrogens in the molecule are irradiated, they become saturated and attain a distribution of spins very different from their equilibrium (Boltzmann) state. There are more spins than normal in the *excited* state. Due to the interaction of spin dipoles, the spins of the carbon nuclei "sense" the spin imbalance of the hydrogen nuclei and begin to adjust themselves to a new equilibrium state that has more spins in the *lower* state. This increase of population in the lower spin state of carbon increases the intensity of the NMR signal.

In a proton-decoupled ^{13}C spectrum, the total NOE for a given carbon increases as the number of nearby hydrogens increases. Thus, we usually find that the intensities of the signals in a ^{13}C spectrum (assuming a single carbon of each type) assume the order

$$CH_3 > CH_2 > CH > C$$

Although the hydrogens producing the NOE effect influence carbon atoms more distant than the ones to which they are attached, their effectiveness drops off rapidly with distance. The interaction of the spin–spin dipoles operates through space, not through bonds, and its magnitude decreases as a function of the inverse of r^3, where r is the radial distance from the hydrogen of origin.

$$C \xrightarrow{r} H \qquad NOE = f\left(\frac{1}{r^3}\right)$$

Thus, nuclei must be rather close together in the molecule in order to exhibit the NOE effect. The effect is greatest for hydrogens that are directly attached to carbon.

In advanced work, NOEs are sometimes used to verify peak assignments. Irradiation of a selected hydrogen or group of hydrogens leads to a greater enhancement in the signal of the closer of the two carbon atoms being considered. In dimethylformamide, for instance, the two methyl groups are nonequivalent, showing two peaks at 31.1 and 36.2 ppm, because free rotation is restricted about the C—N bond due to resonance interaction between the lone pair on nitrogen and the π bond of the carbonyl group.

anti, 31.1 ppm

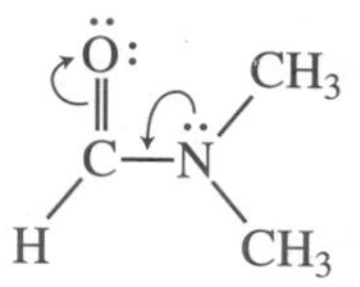

syn, 36.2 ppm

Dimethylformamide

Irradiation of the aldehyde hydrogen leads to a larger NOE for the carbon of the *syn* methyl group (36.2 ppm) than for that of the *anti* methyl group (31.1 ppm), allowing the peaks to be assigned. The *syn* methyl group is closer to the aldehyde hydrogen.

It is possible to retain the benefits of NOE even when determining a proton-coupled ^{13}C NMR spectrum that shows the attached hydrogen multiplets. The favorable perturbation of spin-state populations builds up slowly during irradiation of the hydrogens by the decoupler, and it persists for some time after the decoupler is turned off. In contrast, decoupling is available only while the decoupler is in operation and stops immediately when the decoupler is switched off. One can build up the NOE effect by irradiating with the decoupler during a period before the pulse and then turning off the decoupler during the pulse and free-induction decay (FID) collection periods. This technique gives an **NOE-enhanced proton-coupled spectrum,** with the advantage that peak intensities have been increased due to the NOE effect. See Section 10.1 for details.

4.6 CROSS-POLARIZATION: ORIGIN OF THE NUCLEAR OVERHAUSER EFFECT

To see how cross-polarization operates to give nuclear Overhauser enhancement, consider the energy diagram shown in Figure 4.6. Consider a two-spin system between atoms ^{1}H and ^{13}C. These two atoms may be spin coupled, but the following explanation is easier to follow if we simply ignore any spin–spin splitting. The following explanation is applied to the case of ^{13}C NMR spectroscopy, although the explanation is equally applicable to other possible combinations of atoms. Figure 4.6 shows four separate energy levels (N_1, N_2, N_3, and N_4), each with a different combination of spins of atoms ^{1}H and ^{13}C. The spins of the atoms are shown at each energy level.

The selection rules, derived from quantum mechanics, require that the only allowed transitions involve a change of only one spin at a time (these are called **single-quantum transitions**). The allowed transitions, proton excitations (labeled ^{1}H) and carbon excitations (labeled ^{13}C), are shown. Notice that both proton transitions and both carbon transitions have the same energy (remember that we are ignoring splitting due to *J* interactions).

Because the four spin states have different energies, they also have different *populations*. Because the spin states N_3 and N_2 have very similar energies, we can assume that their populations are approximately equal. Now use the symbol *B* to represent the equilibrium populations of these two spin states. The population of spin state N_1, however, will be higher (by an amount δ), and the population of spin state N_4 will be reduced (also by an amount δ). The intensities of the NMR lines will be proportional to the difference in populations between the energy levels where transitions are occurring. If we compare the populations of each energy level, we can see that the intensities of the two carbon lines (*X*) will be equal.

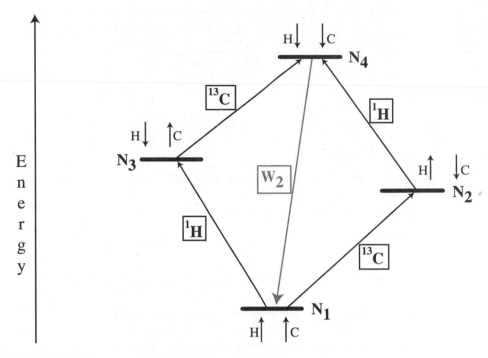

FIGURE 4.6 Spin energy level diagram for an **AX** System.

Level	Equilibrium Populations
N_1	$B + \delta$
N_2	B
N_3	B
N_4	$B - \delta$

Assuming that the populations of the ^{13}C energy levels are at equilibrium, the carbon signals will have intensities:

^{13}C Energy Levels at Equilibrium

$$N_3 - N_4 = B - B + \delta = \delta$$
$$N_1 - N_2 = B + \delta - B = \delta$$

Consider now what happens when we irradiate the proton transitions during the broad-band decoupling procedure. The irradiation of the protons causes the proton transitions to become **saturated**. In other words, the probability of an upward and a downward transition for these nuclei (the proton transitions shown in Fig. 4.6) now becomes *equal*. The population of level N_4 becomes equal to the population of level N_2, and the population of level N_3 is now equal to the population of level N_1. The populations of the spin states can now be represented by the following expressions.

PROTON DECOUPLED	
Level	**Populations**
N_1	$B + \frac{1}{2}\delta$
N_2	$B - \frac{1}{2}\delta$
N_3	$B + \frac{1}{2}\delta$
N_4	$B - \frac{1}{2}\delta$

Using these expressions, the intensities of the carbon lines can be represented:

^{13}C Energy Levels with Broad-Band Decoupling

$$N_3 - N_4 = B + \frac{1}{2}\delta - B + \frac{1}{2}\delta = \delta$$

$$N_1 - N_2 = B + \frac{1}{2}\delta - B + \frac{1}{2}\delta = \delta$$

So far, there has been no change in the intensity of the carbon transition.

At this point we need to consider that there is another process operating in this system. When the populations of the spin states have been disturbed from their equilibrium values, as in this case by irradiation of the proton signal, **relaxation processes** will tend to restore the populations to their equilibrium values. Unlike excitation of a spin from a lower to a higher spin state, relaxation process are not subject to the same quantum mechanical selection rules. Relaxation involving changes of both spins simultaneously (called **double-quantum transitions**) are allowed; in fact they are relatively important in magnitude. The relaxation pathway labeled W_2 in Fig. 4.6 tends to restore equilibrium populations by relaxing spins from state N_4 to N_1. We shall represent the number of spins that are relaxed by this pathway by the symbol d. The populations of the spin states thus become as follows:

Level	**Populations**
N_1	$B + \frac{1}{2}\delta + d$
N_2	$B - \frac{1}{2}\delta$
N_3	$B + \frac{1}{2}\delta$
N_4	$B - \frac{1}{2}\delta - d$

The intensities of the carbon lines can now be represented:

^{13}C Energy Levels with Broad-Band Decoupling and with Relaxation

$$N_3 - N_4 = B + \frac{1}{2}\delta - B + \frac{1}{2}\delta + d = \delta + d$$

$$N_1 - N_2 = B + \frac{1}{2}\delta + d - B + \frac{1}{2}\delta = \delta + d$$

Thus the intensity of each of the carbon lines has been enhanced by an amount d because of this relaxation.

The theoretical maximum value of d is 2.988 (see Equations 4.2 and 4.3). The amount of nuclear Overhauser enhancement that may be observed, however, is often less than this amount. The preceding discussion has ignored possible relaxation from state N_3 to N_2. This relaxation pathway would involve no net change in the total number of spins (a **zero-quantum transition**). This relaxation would tend to *decrease* the nuclear Overhauser enhancement. With relatively small molecules, this second relaxation pathway is much less important than W_2; therefore, we generally see a substantial enhancement.

4.7 PROBLEMS WITH INTEGRATION IN ^{13}C Spectra

Avoid attaching too much significance to peak sizes and integrals in proton-decoupled ^{13}C spectra. In fact, carbon spectra are usually not integrated in the same routine fashion as is accepted for proton spectra. Integral information derived from ^{13}C spectra is usually not reliable unless special techniques are used to ensure its validity. It is true that a peak derived from two carbons is larger than one derived from a single carbon. However, as we saw in Section 4.5, if decoupling is used, the intensity of a carbon peak is NOE enhanced by any hydrogens that are either attached to that carbon or found close by. Nuclear Overhauser enhancement is not the same for every carbon. Recall that as a very rough approximation (with some exceptions), a CH_3 peak generally has a greater intensity than a CH_2 peak, which in turn has a greater intensity than a CH peak, and quaternary carbons, those without any attached hydrogens, are normally the weakest peaks in the spectrum.

A second problem arises in the measurement of integrals in ^{13}C FT-NMR. Figure 4.7 shows the typical pulse sequence for an FT-NMR experiment. Repetitive pulse sequences are spaced at intervals of about 1 to 3 seconds. Following the pulse, the time allotted to collect the data (the FID) is called the **acquisition time.** A short **delay** usually follows the acquisition of data. When hydrogen spectra are determined, it is common for the FID to have decayed to zero before the end of the acquisition time. Most hydrogen atoms relax back to their original Boltzmann condition very quickly—within less than a second. For ^{13}C atoms, however, the time required for relaxation is quite variable, depending on the molecular environment of the particular atom (see Section 4.8). Some ^{13}C atoms relax very quickly (in seconds), but others require quite long periods (minutes) compared to hydrogen. If carbon atoms with long relaxation times are present in a molecule, collection of the FID signal may have already ceased before all of the ^{13}C atoms have relaxed. The result of this discrepancy is that some atoms have strong signals, as their contribution to the FID is complete, while others, those that have not relaxed completely, have weaker signals. When this happens, the resulting peak areas do not integrate to give the correct numbers of carbons.

It is possible to extend the data collection period (and the delay period) to allow all of the carbons in a molecule to relax; however, this is usually done only in special cases. Since repetitive scans are used in ^{13}C spectra, the increased acquisition time means that it would simply take too long to measure a complete spectrum with a reasonable signal-to-noise ratio.

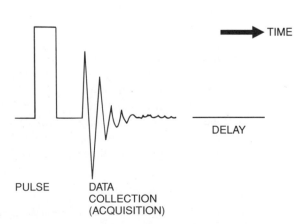

F I G U R E 4 . 7 A typical FT-NMR pulse sequence.

4.8 MOLECULAR RELAXATION PROCESSES

In the absence of an applied field, there is a nearly 50/50 distribution of the two spin states for a nucleus of spin $= \frac{1}{2}$. A short time after a magnetic field is applied, a slight excess of nuclei builds up in the lower-energy (aligned) spin state due to thermal equilibration. We call the relative numbers of nuclei in the upper and lower states the **Boltzmann equilibrium.** In Section 3.5 we used the Boltzmann equations to calculate the expected number of excess nuclei for NMR spectrometers that operate at various frequencies (Table 3.3). We rely on these excess nuclei to generate NMR signals. When we pulse the system at the resonance frequency, we disturb the Boltzmann equilibrium (alter the spin population ratios). Excess nuclei are excited to the upper spin state and, as they **relax,** or return to the lower spin state and equilibrium, they generate the FID signal, which is processed to give the spectrum.

If all of the excess nuclei absorb energy, **saturation,** a condition in which the populations of both spin states are once again equal, is reached and the population of the upper spin state cannot be increased further. This limitation exists because further irradiation induces just as many downward transitions as there are upward transitions when the populations of both states are equal. Net signals are observed only when the populations are unequal. If irradiation is stopped, either at or before saturation, the excited excess nuclei relax and the Boltzmann equilibrium is reestablished.

The methods by which excited nuclei return to their ground state and by which the Boltzmann equilibrium is reestablished are called **relaxation processes.** In NMR systems there are two principal types of relaxation processes: spin–lattice relaxation and spin–spin relaxation. Each occurs as a first-order rate process and is characterized by a **relaxation time,** which governs the rate of decay.

Spin–lattice, or *longitudinal,* **relaxation processes** are those that occur in the direction of the field. The spins lose their energy by transferring it to the surroundings—the *lattice*—as thermal energy. Ultimately the lost energy heats the surroundings. The **spin–lattice relaxation time, T_1,** governs the rate of this process. The inverse of the spin–lattice relaxation time, $1/T_1$, is the rate constant for the decay process.

Several processes, both within the molecule (intramolecular) and between molecules (intermolecular), contribute to spin–lattice relaxation. The principal contributor is dipole–dipole interaction. The spin of an excited nucleus interacts with the spins of other magnetic nuclei that are in the same molecule or in nearby molecules. These interactions can induce nuclear spin transitions and exchanges. Eventually the system relaxes back to the Boltzmann equilibrium. This mechanism is especially effective if there are hydrogen atoms nearby. For carbon nuclei, relaxation is fastest if hydrogen atoms are directly bonded, as in CH, CH_2, and CH_3 groups. Spin–lattice relaxation is also most effective in larger molecules, which tumble (rotate) slowly, and it is very inefficient in small molecules, which tumble faster.

Spin–spin, or *transverse*, **relaxation processes** are those that occur in a plane perpendicular to the direction of the field—the same plane in which the signal is detected. Spin–spin relaxation does not change the energy of the spin system. It is often described as an entropy process. When nuclei are induced to change their spin by the absorption of radiation, they all end up precessing in phase after resonance. This is called **phase coherence.** The nuclei lose the phase coherence by exchanging spins. The phases of the precessing spins randomize (increase entropy). This process occurs only between nuclei of the same type—those that are studied in the NMR experiment. The **spin–spin relaxation time, T_2,** governs the rate of this process.

Our interest is in spin–lattice relaxation times, T_1 (rather than spin–spin relaxation times), as they relate to the intensity of NMR signals and have other implications relevant to structure determination. T_1 relaxation times are relatively easy to measure by the **inversion recovery method.**[2] Spin–spin relaxation times, T_2, are more difficult to measure and do not provide useful structural in-

---

[2]Consult the references listed at the end of the chapter for the details of this method.

formation. Spin–spin relaxation (phase randomization) always occurs more quickly than the rate at which spin–lattice relaxation returns the system to Boltzmann equilibrium ($T_2 \leq T_1$). However, for nuclei with spin $= \frac{1}{2}$ and a solvent of low viscosity, T_1 and T_2 are usually very similar.

Spin–lattice relaxation times, T_1 values, are not of much use in proton NMR since protons have very short relaxation times. However, T_1 values are quite important to ^{13}C NMR spectra because they are much longer for carbon nuclei and can dramatically influence signal intensities. One can always expect quaternary carbons (including most carbonyl carbons) to have long relaxation times because they have no attached hydrogens. A common instance of long relaxation times is the carbons in an aromatic ring with a substituent group different from hydrogen. The ^{13}C T_1 values for isooctane (2,2,4-trimethylpentane) and toluene follow.

C	T_1
1, 6, 7	9.3 sec
2	68
3	13
4	23
5, 8	9.8

2,2,4-Trimethylpentane

C	T_1	NOE
α	16 sec	0.61
1	89	0.56
2	24	1.6
3	24	1.7
4	17	1.6

Toluene

Notice that in isooctane the quaternary carbon 2, which has no attached hydrogens, has the longest relaxation time (68 sec). Carbon 4, which has one hydrogen, has the next longest (23 sec) and is followed by carbon 3, which has two hydrogens (13 sec). The methyl groups (carbons 1, 5, 6, 7, and 8) have the shortest relaxation times in this molecule. The NOE factors for toluene have been listed along with the T_1 values. As expected, the *ipso* carbon 1, which has no hydrogens, has the longest relaxation time and the smallest NOE. In the ^{13}C NMR of toluene, the *ipso* carbon has the lowest intensity.

Remember also that T_1 values are greater when a molecule is small and tumbles rapidly in the solvent. The carbons in cyclopropane have a T_1 of 37 sec. Cyclohexane has a smaller value, 20 sec. In a larger molecule such as the steroid cholesterol, all of the carbons except those that are quaternary would be expected to have T_1 values less than 1 to 2 sec. The quaternary carbons would have T_1 values of about 4 to 6 sec due to the lack of attached hydrogens. In solid polymers, such as polystyrene, the T_1 values for the various carbons are around 10^{-2} sec.

To interpret ^{13}C NMR spectra, you should know what effects of NOE and spin–lattice relaxation to expect. We cannot fully cover the subject here, and there are many additional factors besides those that we have discussed. If you are interested, consult more advanced textbooks, such as the ones listed in the references.

The example of 2,3-dimethylbenzofuran will close this section. In this molecule, the quaternary (*ipso*) carbons have relaxation times that exceed 1 minute. As discussed in Section 4.7, to obtain a decent spectrum of this compound, it would be necessary to extend the data acquisition and delay periods so as to determine the entire spectrum of the molecule and see the carbons with high T_1 values.

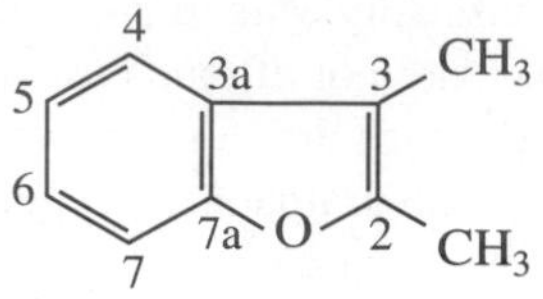

C	T_1	NOE
2	83 sec	1.4
3	92	1.6
3a	114	1.5
7a	117	1.3
Others	<10	1.6–2

2,3-Dimethylbenzofuran

4.9 OFF-RESONANCE DECOUPLING

The decoupling technique that is used to obtain typical proton-decoupled spectra has the advantage that all peaks become singlets. For carbon atoms bearing attached hydrogens, an added benefit is that peak intensities increase, owing to the nuclear Overhauser effect, and signal-to-noise ratios improve. Unfortunately, much useful information is also lost when carbon spectra are decoupled. We no longer have information about the number of hydrogens that are attached to a particular carbon atom.

In many cases it would be helpful to have the information about the attached hydrogens that a proton-coupled spectrum provides, but frequently the spectrum becomes too complex, with overlapping multiplets that are difficult to resolve or assign correctly. A compromise technique called **off-resonance decoupling** can often provide multiplet information while keeping the spectrum relatively simple in appearance.

In an off-resonance-decoupled ^{13}C spectrum, the coupling between each carbon atom and each hydrogen attached directly to it is observed. The $n + 1$ Rule can be used to determine whether a given carbon atom has three, two, one, or no hydrogens attached. However, when off-resonance decoupling is used, the *apparent magnitude* of the coupling constants is reduced, and overlap of the resulting multiplets is a less frequent problem. The off-resonance-decoupled spectrum retains the couplings between the carbon atom and directly attached protons (the one-bond couplings) but effectively removes the couplings between the carbon and more remote protons.

In this technique, the frequency of a second radiofrequency transmitter (the **decoupler**) is set either upfield or downfield from the usual sweep width of a normal proton spectrum (i.e., *off resonance*). In contrast, the frequency of the decoupler is set to *coincide exactly* with the range of proton resonances in a true decoupling experiment. Furthermore, in off-resonance decoupling, the power of the decoupling oscillator is held *low* to avoid complete decoupling.

Off-resonance decoupling can be a great help in assigning spectral peaks. The off-resonance-decoupled spectrum is usually obtained separately, along with the proton-decoupled spectrum. Figure 4.8 shows the off-resonance-decoupled spectrum of 1-propanol, in which the methyl carbon atom is split into a quartet and each of the methylene carbons appears as a triplet. Notice that the observed multiplet patterns are consistent with the $n + 1$ Rule and with the patterns shown in Figure 4.3. If TMS had been added, its methyl carbons would have appeared as a quartet centered at $\delta = 0$ ppm.

4.10 A QUICK DIP INTO DEPT

Despite its utility, off-resonance decoupling is now considered an old-fashioned technique. It has been replaced by more modern methods, the most important of which is **D**istortionless **E**nhancement by **P**olarization **T**ransfer, better know as **DEPT**. The DEPT technique requires an FT-pulsed spectrometer. It is more complicated than off-resonance decoupling and it requires a computer, but it gives the same information more reliably and more clearly. Chapter 10 will discuss the DEPT

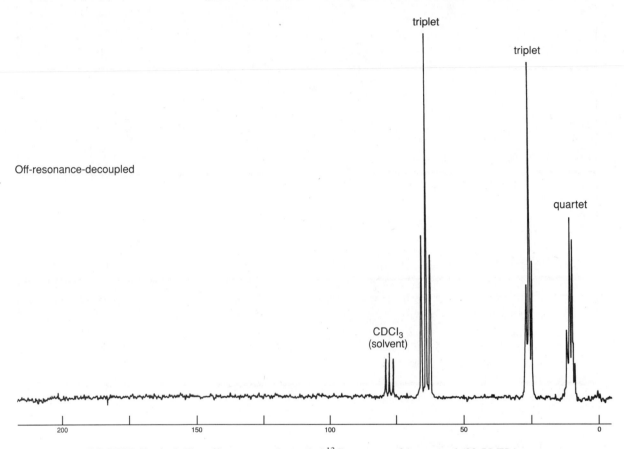

FIGURE 4.8 The off-resonance-decoupled ^{13}C spectrum of 1-propanol (22.5 MHz).

method in detail; only a brief introduction to the method and the results it provides will be provided here.

In the DEPT technique, the sample is irradiated with a complex sequence of pulses in both the ^{13}C and ^1H channels. The result of these pulse sequences[3] is that the ^{13}C signals for the carbon atoms in the molecule will exhibit different **phases,** depending on the number of hydrogens attached to each carbon. Each type of carbon will behave slightly differently, depending on the **duration** of the complex pulses. These differences can be detected, and spectra produced in each experiment can be plotted.

One common method of presenting the results of a DEPT experiment is to plot four different **subspectra**. Each subspectrum provides different information. A sample DEPT plot for **isopentyl acetate** is shown in Figure 4.9.

$$\overset{5}{CH_3}-\overset{\overset{\displaystyle O}{\|}}{\underset{6}{C}}-O-\overset{4}{CH_2}-\overset{3}{CH_2}-\overset{2}{\underset{\underset{1}{CH_3}}{CH}}-\overset{1}{CH_3}$$

---

[3]Pulse sequences were introduced in Section 4.7.

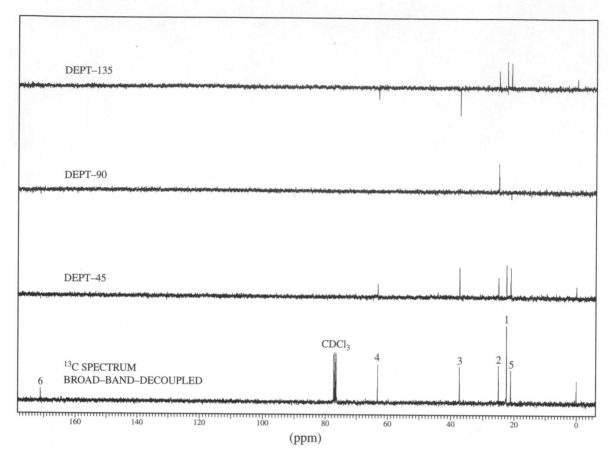

FIGURE 4.9 DEPT spectra of isopentyl acetate.

The lowest trace in the figure is the usual broad-band-decoupled ^{13}C spectrum. The second trace from the bottom is the result of a pulse sequence (called a **DEPT-45**) in which the only signals detected are those that arise from protonated carbons. You will notice that the carbonyl carbon (labeled **6**), at 172 ppm, is not seen. The solvent peaks arising from $CDCl_3$ (77 ppm) are also not seen. Deuterium (D or ^2H) behaves differently from ^1H, and as a result the carbon of $CDCl_3$ behaves as if it were not protonated. The third trace is the result of a slightly different pulse sequence (called a **DEPT-90**). In this trace, only those carbons that bear a single hydrogen are seen. Only the carbon at position **2** (25 ppm) is observed.

The uppermost trace is more complicated than the previous subspectra. The pulse sequence that gives rise to this subspectrum is called **DEPT-135**. Here all carbons that have an attached proton provide a signal, but the **phase** of the signal will be different, depending on whether the number of attached hydrogens is an odd or an even number. Signals arising from CH or CH_3 groups will give positive peaks, while signals arising from CH_2 groups will form negative (inverse) peaks. When we examine the upper trace in Figure 4.9, we can identify all of the carbon peaks in the spectrum of isopentyl acetate. The positive peaks at 21 and 22 ppm must represent CH_3 groups, as those peaks are not represented in the DEPT-90 subspectrum. When we look at the original ^{13}C spectrum, we see that the peak at 21 ppm is not as strong as the peak at 22 ppm. We conclude, therefore, that the peak at 21 ppm must come from the CH_3 carbon at position **5**, while the stronger peak at 22 ppm comes from the pair of equivalent CH_3 carbons at position **1**. We have already determined that the positive peak at 25 ppm is due to the CH carbon at position **2**, as it appears in both the DEPT-135 and the DEPT-90 subspectra. The inverse peak at 37 ppm is due to a CH_2 group, and we can identify it as coming from the carbon at position **3**. The inverse peak at 53 ppm is clearly caused by the CH_2

carbon at position **4**, deshielded by the attached oxygen atom. Finally, the downfield peak at 172 ppm has already been labeled as arising from the carbonyl carbon at **6**. This peak appears only in the original ^{13}C spectrum; therefore, it must not have any attached hydrogens.

Through the mathematical manipulation of the results of each of the different DEPT pulse sequences, it is also possible to present the results as a series of subspectra in which only the CH carbons appear in one trace, only the CH_2 carbons appear in the second trace, and only the CH_3 carbons appear in the third trace. Another common means of displaying DEPT results is to show only the result of the DEPT-135 experiment. The spectroscopist generally can interpret the results of this spectrum by applying knowledge of likely chemical shift differences to distinguish between CH and CH_3 carbons.

The results of DEPT experiments may be used from time to time in this textbook to help you solve assigned exercises. In an effort to save space, most often only the results of the DEPT experiment, rather than the complete spectrum, will be provided.

4.11 SOME SAMPLE SPECTRA—EQUIVALENT CARBONS

Equivalent ^{13}C atoms appear at the same chemical shift value. Figure 4.10 shows the proton-decoupled carbon spectrum for 2,2-dimethylbutane. The three methyl groups at the left side of the molecule are equivalent by symmetry.

$$CH_3-\overset{\overset{\displaystyle CH_3}{|}}{\underset{\underset{\displaystyle CH_3}{|}}{C}}-CH_2-CH_3$$

Although this compound has a total of six carbons, there are only four peaks in the ^{13}C NMR spectrum. The ^{13}C atoms that are equivalent appear at the same chemical shift. The single methyl carbon, **a**, appears at highest field (9 ppm), while the three equivalent methyl carbons, **b,** appear at 29 ppm. The quaternary carbon **c** gives rise to the small peak at 30 ppm, and the methylene carbon **d** appears at 37 ppm. The relative sizes of the peaks are related, in part, to the number of each type of carbon atom present in the molecule. For example, notice in Figure 4.10 that the peak at 29 ppm (**b**) is much larger than the others. This peak is generated by three carbons. The quaternary carbon at 30

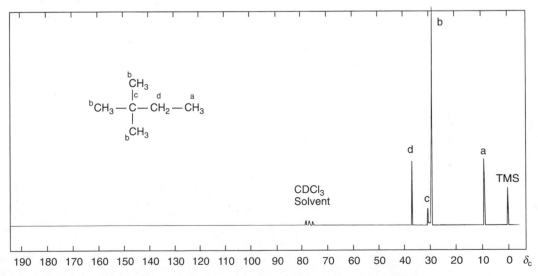

FIGURE 4.10 The proton-decoupled ^{13}C NMR spectrum of 2,2-dimethylbutane.

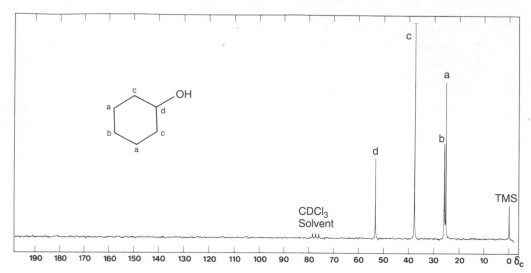

F I G U R E   4 . 1 1 The proton-decoupled ^{13}C NMR spectrum of cyclohexanol.

ppm (**c**) is very weak. Since no hydrogens are attached to this carbon, there is very little NOE enhancement. Without attached hydrogen atoms, relaxation times are also longer than for other carbon
atoms. Quaternary carbons, those with no hydrogens attached, frequently appear as weak peaks in
proton-decoupled ^{13}C NMR spectra (see Sections 4.5 and 4.7).

Figure 4.11 is a proton-decoupled ^{13}C spectrum of cyclohexanol. This compound has a plane of
symmetry passing through its hydroxyl group, and *it shows only four carbon resonances*. Carbons **a**
and **c** are doubled due to symmetry and give rise to larger peaks than carbons **b** and **d.** Carbon **d,**
bearing the hydroxyl group, is deshielded by oxygen and has its peak at 70.0 ppm. Notice that this
peak has the lowest intensity of all of the peaks. Its intensity is lower than that of carbon **b** in part
because the carbon **d** peak receives the least amount of NOE; there is only one hydrogen attached to
the hydroxyl carbon, whereas each of the other carbons has two hydrogens.

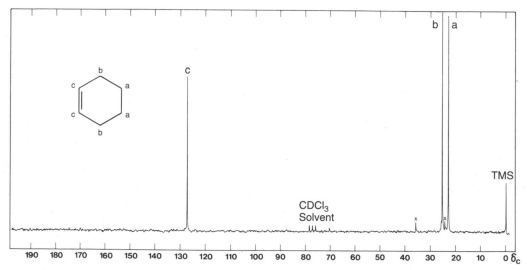

F I G U R E   4 . 1 2 The proton-decoupled ^{13}C NMR spectrum of cyclohexene.

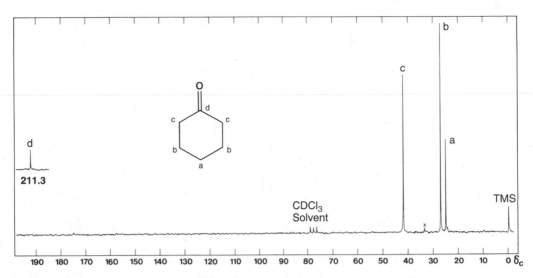

FIGURE 4.13 The proton-decoupled ^{13}C spectrum of cyclohexanone.

A carbon attached to a double bond is deshielded due to its sp^2 hybridization and some diamagnetic anisotropy. This effect can be seen in the ^{13}C NMR spectrum of cyclohexene (Fig. 4.12). Cyclohexene has a plane of symmetry that runs perpendicular to the double bond. As a result, we observe only three absorption peaks. There are two of each type of carbon. Each of the double-bond carbons **c** has only one hydrogen, whereas each of the remaining carbons has two. As a result of a reduced NOE, the double-bond carbons have a lower-intensity peak in the spectrum.

In Figure 4.13, the spectrum of cyclohexanone, the carbonyl carbon has the lowest intensity. This is due not only to reduced NOE (no hydrogen attached) but also to the long relaxation time of the carbonyl carbon. As you have already seen, quaternary carbons tend to have long relaxation times. Notice also that Figure 4.1 predicts the large chemical shift for this carbonyl carbon.

4.12 COMPOUNDS WITH AROMATIC RINGS

Compounds with carbon–carbon double bonds or aromatic rings give rise to chemical shifts in the range from 100 to 175 ppm. Since relatively few other peaks appear in this range, a great deal of useful information is available when peaks appear here.

A **monosubstituted** benzene ring shows *four* peaks in the aromatic carbon area of a proton-decoupled ^{13}C spectrum, since the *ortho* and *meta* carbons are doubled by symmetry. Often the carbon with no protons attached, the *ipso* carbon, has a very weak peak due to a long relaxation time and a weak NOE. In addition, there are two larger peaks for the doubled *ortho* and *meta* carbons and a medium-sized peak for the *para* carbon. In many cases, it is not important to be able to assign all of the peaks precisely. In the example of toluene, shown in Figure 4.14, notice that carbons **c** and **d** are not easy to assign by inspection of the spectrum. However, use of chemical shift correlation tables (see Section 4.2B and Appendix 8) would enable us to assign these signals.

Toluene

Difficult to assign without using chemical shift correlation tables

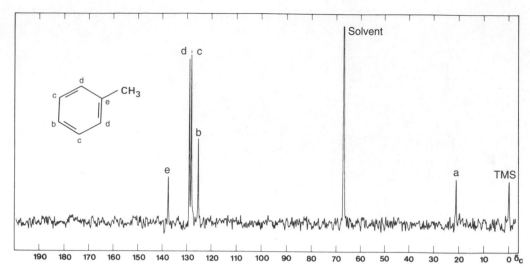

F I G U R E   4 . 1 4 The proton-decoupled ^{13}C NMR spectrum of toluene.

In an off-resonance-decoupled or proton-coupled ^{13}C spectrum, a monosubstituted benzene ring shows three doublets and one singlet. The singlet arises from the *ipso* carbon, which has no attached hydrogen. Each of the other carbons in the ring (*ortho, meta,* and *para*) has one attached hydrogen and yields a doublet.

Figure 4.4b is the proton-decoupled spectrum of ethyl phenylacetate, with the assignments noted next to the peaks. Notice that the aromatic ring region shows four peaks between 125 and 135 ppm, consistent with a monosubstituted ring. There is one peak for the methyl carbon (13 ppm) and two peaks for the methylene carbons. One of the methylene carbons is directly attached to an electronegative oxygen atom and appears at 61 ppm, while the other is more shielded (41 ppm). The carbonyl carbon (an ester) has resonance at 171 ppm. All of the carbon chemical shifts agree with the values in the correlation chart (Fig. 4.1).

Depending on the mode of substitution, a symmetrically **disubstituted** benzene ring can show two, three, or four peaks in the proton-decoupled ^{13}C spectrum. The following drawings illustrate this for the isomers of dichlorobenzene.

Three unique carbon atoms   Four unique carbon atoms   Two unique carbon atoms

Figure 4.15 shows the spectra of all three dichlorobenzenes, each of which has the number of peaks consistent with the analysis just given. You can see that ^{13}C NMR spectroscopy is very useful in the identification of isomers.

Most other polysubstitution patterns on a benzene ring yield six different peaks in the proton-decoupled ^{13}C NMR spectrum, one for each carbon. However, when identical substituents are present, watch carefully for planes of symmetry that may reduce the number of peaks.

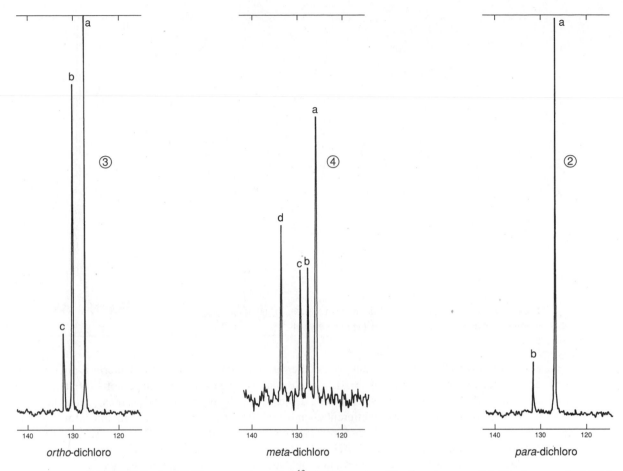

FIGURE 4.15 The proton-decoupled ^{13}C NMR spectra of the three isomers of dichlorobenzene (25 MHz).

4.13 CARBON-13 NMR SOLVENTS—HETERONUCLEAR COUPLING OF CARBON TO DEUTERIUM

Most FT-NMR spectrometers require the use of deuterated solvents because the instruments use the deuterium resonance signal as a "lock signal," or reference signal, to keep the magnet and the electronics adjusted correctly. Deuterium is the ^2H isotope of hydrogen and can easily substitute for it in organic compounds. Deuterated solvents present few difficulties in hydrogen spectra, as the deuterium nuclei are largely invisible when a proton spectrum is determined. Deuterium has resonance at a different frequency from hydrogen. In ^{13}C NMR, however, these solvents are frequently seen as part of the spectrum, as they all have carbon atoms. In this section we explain the spectra of some of the common solvents and, in the process, examine heteronuclear coupling of carbon and deuterium. Figure 4.16 shows the ^{13}C NMR peaks due to the solvents chloroform-d and dimethylsulfoxide-d_6.

Chloroform-d, $CDCl_3$, is the compound most commonly used as a solvent for ^{13}C NMR. It is also called deuteriochloroform or deuterated chloroform. Its use gives rise to a three-peak multiplet in the spectrum, with the center peak having a chemical shift of about 77 ppm. Figure 4.16 shows an example. Notice that this "triplet" is different from the triplets in a hydrogen spectrum (from two neighbors) or in a proton-coupled ^{13}C spectrum (from two attached hydrogens); the intensities are different. In this triplet, all three peaks have approximately the same intensity (1:1:1), whereas the other types of triplets have intensities that follow the entries in Pascal's triangle, with ratios of 1:2:1.

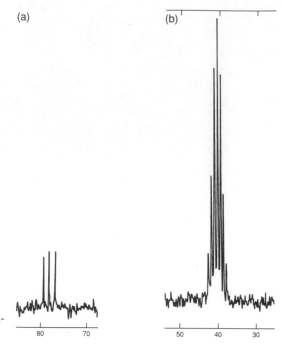

FIGURE 4.16 The ^{13}C NMR peaks of two common solvents. (a) Chloroform-d. (b) Dimethylsulfoxide-d$_6$.

In contrast with hydrogen (spin $= \frac{1}{2}$), deuterium has spin $= 1$. A single deuterium nucleus can adopt three different spins ($2I + 1 = 3$), where the spins have quantum numbers of -1, 0, and $+1$. In a solution of CDCl$_3$, molecules can have a deuterium with any one of these spins, and as they are equally probable, we see three different chemical shifts for the carbon atom in chloroform-d. The ^{13}C$-$D one-bond coupling constant for this interaction is about 45 Hz. At 75 MHz these three peaks are about 0.6 ppm apart (45 Hz/75 MHz $= 0.60$ ppm).

Because deuterium is not a spin $= \frac{1}{2}$ nucleus, the $n + 1$ Rule does not correctly predict the multiplicity of the carbon resonance. The $n + 1$ Rule works only for spin $= \frac{1}{2}$ nuclei and is a specialized case of a more general prediction formula:

$$\text{multiplicity} = 2nI + 1 \qquad \qquad \textbf{Equation 4.4}$$

where n is the number of nuclei and I is the spin of that type of nucleus. If we use this formula, the correct multiplicity of the carbon peak with *one deuterium* attached is predicted by

$$2 \cdot 1 \cdot 1 + 1 = 3$$

If there are *three hydrogens*, the formula correctly predicts a quartet for the proton-coupled carbon peak:

$$2 \cdot 3 \cdot \frac{1}{2} + 1 = 4$$

Dimethylsulfoxide-d$_6$, CD$_3-$SO$-$CD$_3$, is frequently used as a solvent for carboxylic acids and other compounds that are difficult to dissolve in CDCl$_3$. Equation 4.4 predicts a septet for the multiplicity of the carbon with three deuterium atoms attached:

$$2 \cdot 3 \cdot 1 + 1 = 7$$

This is exactly the pattern observed in Figure 4.16. The pattern has a chemical shift of 39.5 ppm, and the coupling constant is about 40 Hz.

n	$2nI+1$ Lines	Relative Intensities											
0	1	1											
1	3	1	1	1									
2	5	1	2	3	2	1							
3	7	1	3	6	7	6	3	1					
4	9	1	4	10	16	19	16	10	4	1			
5	11	1	5	15	30	45	51	45	30	15	5	1	
6	13	1	6	21	50	90	126	141	126	90	50	21	6	1

FIGURE 4.17 An intensity triangle for deuterium multiplets (n = number of deuterium atoms).

Because deuterium has spin = 1 instead of spin = $\frac{1}{2}$ like hydrogen, the Pascal triangle (Fig. 3.33 in Section 3.16) does not correctly predict the intensities in this seven-line pattern. Instead, a different intensity triangle must be used for splittings caused by deuterium atoms. Figure 4.17 is this intensity triangle, and Figure 4.18 is an analysis of the intensities for three-line and five-line multiplets. In the latter figure, an upward arrow represents spin = 1, a downward arrow represents spin = −1, and a large dot represents spin = 0. Analysis of the seven-line multiplet is left for the reader to complete.

Acetone-d₆, $CD_3-CO-CD_3$, shows the same ^{13}C septet splitting pattern as dimethylsulfoxide-d₆, but the multiplet is centered at 29.8 ppm with the carbonyl peak at 206 ppm. The carbonyl carbon is a singlet; three-bond coupling does not appear.

Acetone-d₅ frequently appears as an impurity in spectra determined in acetone-d₆. It leads to interesting results in both the hydrogen and the carbon-13 spectra. Although this chapter is predominantly about carbon-13 spectra, we will examine both cases.

Hydrogen Spectrum

In proton (^1H) NMR spectra, a commonly encountered multiplet arises from a small amount of acetone-d₅ impurity in acetone-d₆ solvent. Figure 4.19 shows the multiplet, which is generated by the hydrogen in the $-CHD_2$ group of the $CD_3-CO-CHD_2$ molecule. Equation 4.4 correctly predicts that there should be a quintet in the proton spectrum of acetone-d₅:

$$2 \cdot 2 \cdot 1 + 1 = 5$$

and this is observed.

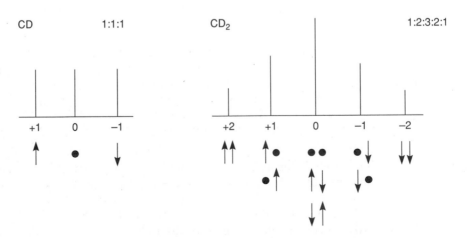

FIGURE 4.18 An intensity analysis of three- and five-line deuterium multiplets.

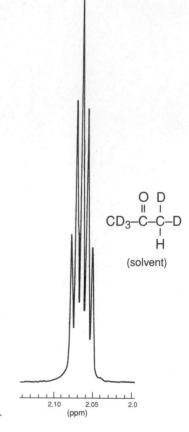

FIGURE 4.19 The 300-MHz ^{1}H spectrum of acetone-d$_5$ (CD$_3$—CO—CHD$_2$).

Carbon Spectrum

The proton-coupled ^{13}C spectrum of the —CHD$_2$ group is more complicated as both hydrogen (spin = $\frac{1}{2}$) and deuterium (spin = 1) interact with carbon. In this case we use the following formula, which is extended from Equation 4.4.

$$\text{total multiplicity} = \Pi_i (2n_i I_i + 1) \qquad \text{\textbf{Equation 4.5}}$$

$$\text{Condition: } I \geq \frac{1}{2}$$

The large Π_i indicates a product of terms for each different type of atom i that couples to the atom being observed. These atoms must have spin $\geq \frac{1}{2}$; atoms of spin = 0 do not cause splitting. In the present case (—CHD$_2$) there are two terms, one for hydrogen and one for deuterium.

$$\text{total multiplicity} = (2 \cdot 1 \cdot \frac{1}{2} + 1)(2 \cdot 2 \cdot 1 + 1) = 10$$

The ^{13}C—H and ^{13}C—D coupling constants would most likely be different, resulting in 10 lines that would not all be equally spaced. In addition, acetone has a second "methyl" group on the other side of the carbonyl group. The —CD$_3$ group (seven peaks) would overlap the 10 peaks from —CHD$_2$ and make a pattern that would be quite difficult to decipher! The ^{1}H and ^{13}C chemical shifts for common NMR solvents are provided in Appendix 10.

4.14 HETERONUCLEAR COUPLING OF CARBON TO FLUORINE-19

Heteronuclear $^{13}C-^{19}F$ coupling is observed when there are fluorine atoms in an organic compound. Figures 4.20 and 4.21 are two spectra that exhibit this effect.

The spectrum of $CFBr_3$ shows a doublet (two peaks) at 43.6 and 48.5 ppm. The $^{13}C-^{19}F$ coupling constant in this example is quite large, at about 368 Hz (4.9 ppm × 75 MHz = 368 Hz). The second example shows both one-bond and two-bond couplings. The large quartet centered at about 124 ppm is due to the $-CF_3$ carbon. The one-bond C$-$F coupling constant in this case is 276 Hz (using the center two peaks of 122.49 and 126.17 ppm: 3.68 ppm × 75 MHz = 276 Hz). The CH_2 carbon also has its resonance split by the three fluorines (two-bond coupling). This second quartet has a smaller coupling constant because the interaction operates over more bonds (using the center peaks of 61.01 and 61.47 ppm: 0.46 ppm × 75 MHz = 35 Hz).

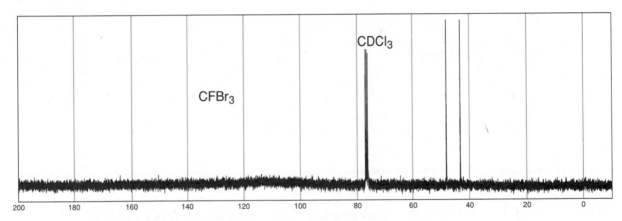

FIGURE 4.20 The ^{13}C proton-decoupled spectrum of $CFBr_3$ (75 MHz).

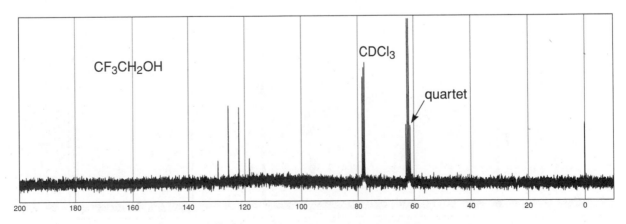

FIGURE 4.21 The ^{13}C proton-decoupled spectrum of CF_3CH_2OH (75 MHz).

■ 4.15 HETERONUCLEAR COUPLING OF CARBON TO PHOSPHORUS-31

The two spectra in Figures 4.22 and 4.23 demonstrate coupling between ^{13}C and ^{31}P. In the first compound, shown in Figure 4.22, the carbons of the methyl groups are split by phosphorus (spin $= \frac{1}{2}$) into a doublet with $J = 56$ Hz ($12.16 - 11.41 = 0.75$ ppm; 0.75 ppm $\times 75$ MHz $= 56$ Hz). This is a one-bond coupling. The second compound shows both one-bond and two-bond coupling. The one-bond coupling occurs between the phosphorus and the carbon of the methyl group, $P-CH_3$, and $J = 143$ Hz ($10.83 - 8.92 = 1.91$ ppm; 1.91 ppm $\times 75$ MHz $= 143$ Hz). The second coupling is a two-bond coupling, $P-O-CH_3$, and has a magnitude of about 7 Hz ($52.23 - 52.14 = 0.09$ ppm; 0.09 ppm $\times 75$ MHz $= 6.75$ Hz).

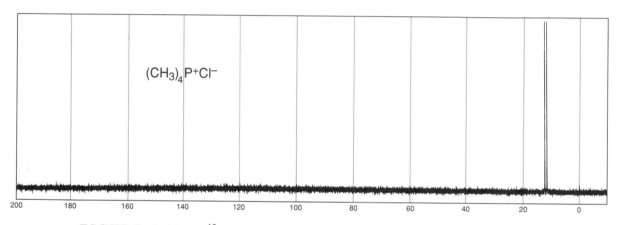

F I G U R E   4 . 2 2 The ^{13}C proton-decoupled spectrum of tetramethylphosphonium chloride, $(CH_3)_4P^+Cl^-$ (75 MHz).

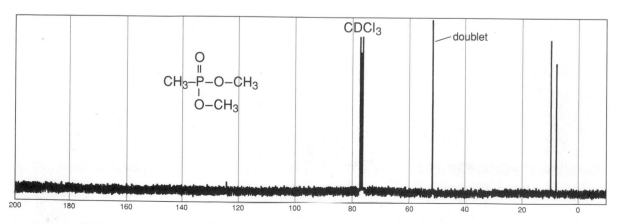

F I G U R E   4 . 2 3 The ^{13}C proton-decoupled spectrum of $CH_3PO(OCH_3)_2$ (75 MHz).

PROBLEMS

*1. A compound with the formula $C_3H_6O_2$ gives the following proton-decoupled and off-resonance-decoupled spectra. Determine the structure of the compound.

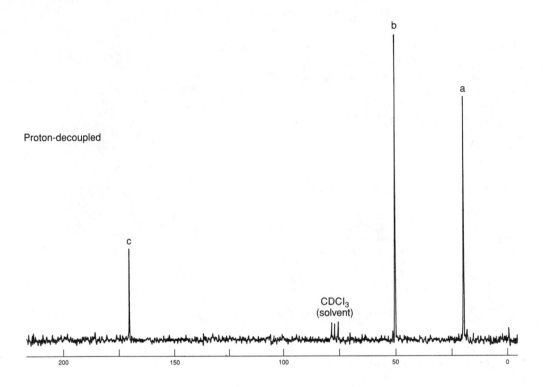

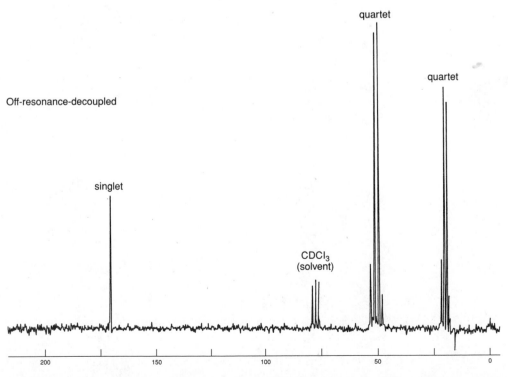

*2. Predict the number of peaks that you would expect in the proton-decoupled ^{13}C spectrum of each of the following compounds. Problems 2a and 2b are provided as examples. Dots are used to show the nonequivalent carbon atoms in these two examples.

(a)

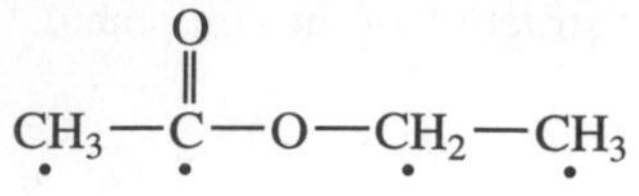

$$CH_3-\overset{\overset{O}{\|}}{C}-O-CH_2-CH_3$$ Four peaks

(b)

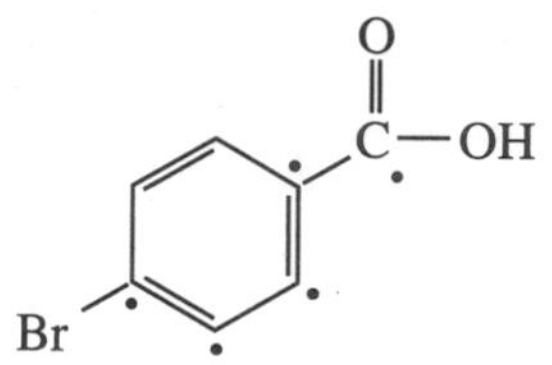

Five peaks

(c)

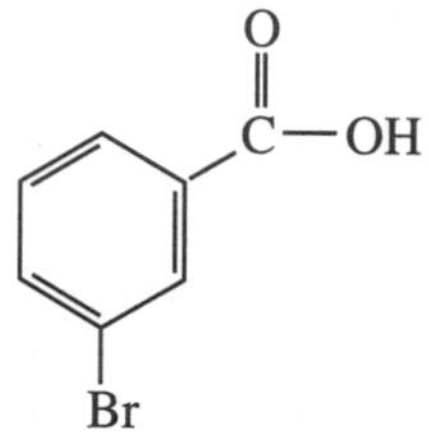

(d)

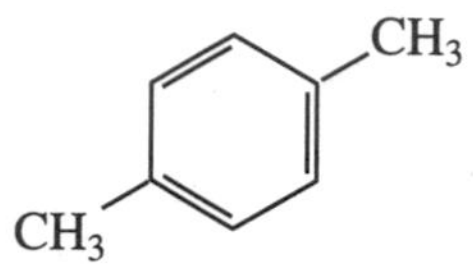

(e)

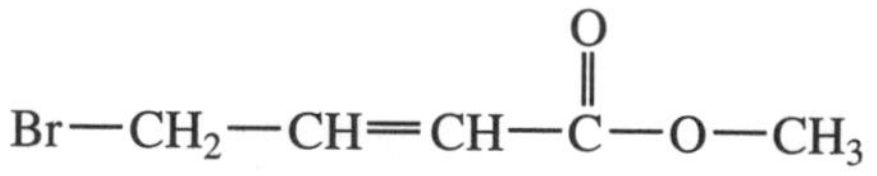

$$Br-CH_2-CH=CH-\overset{\overset{O}{\|}}{C}-O-CH_3$$

(f)

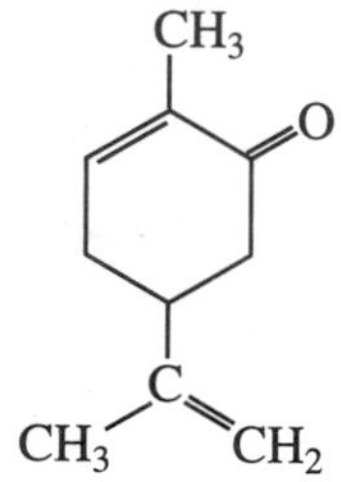

(g)

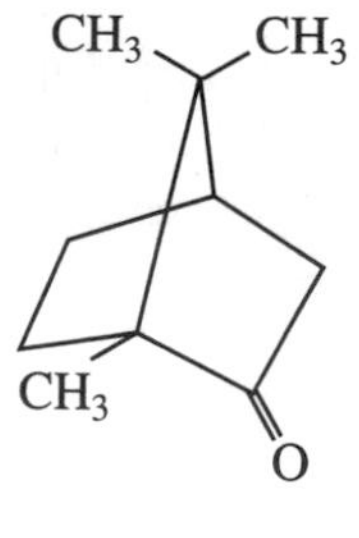

(h)

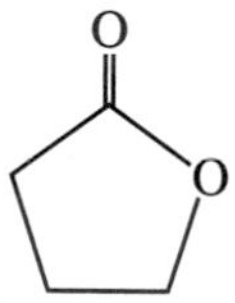

(i)

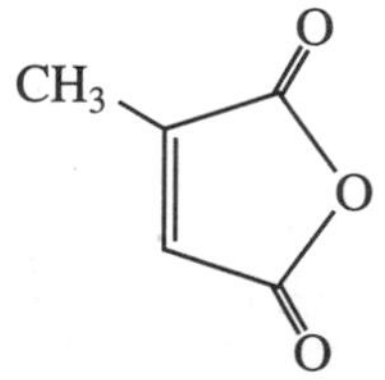

(j)

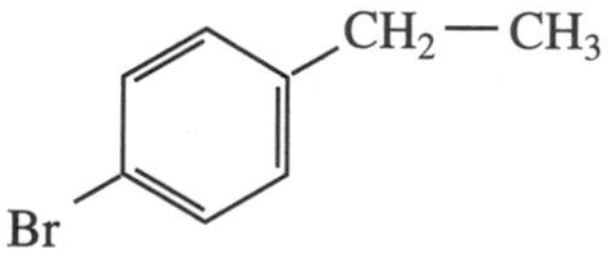

(k)

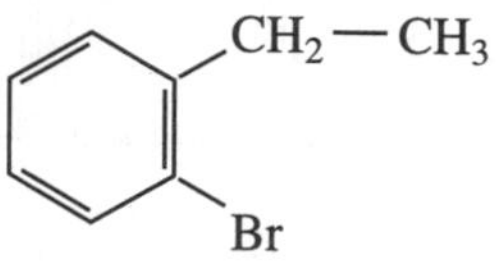

*3. Following are proton-decoupled ^{13}C spectra for three isomeric alcohols with the formula $C_4H_{10}O$. A DEPT or an off-resonance analysis yields the multiplicities shown; **s** = singlet, **d** = doublet, **t** = triplet, and **q** = quartet. Identify the alcohol responsible for each spectrum, and assign each peak to an appropriate carbon atom or atoms.

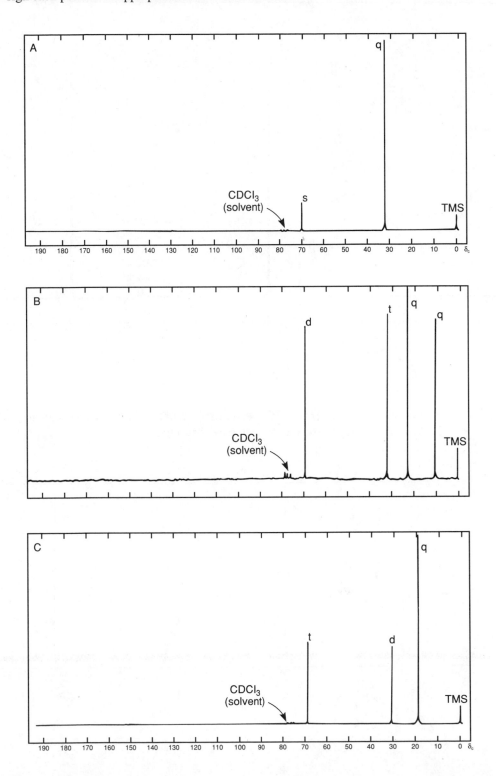

*4. The following spectrum is of an ester with formula $C_5H_8O_2$. Multiplicities are indicated. Draw the structure of the compound, and assign each peak.

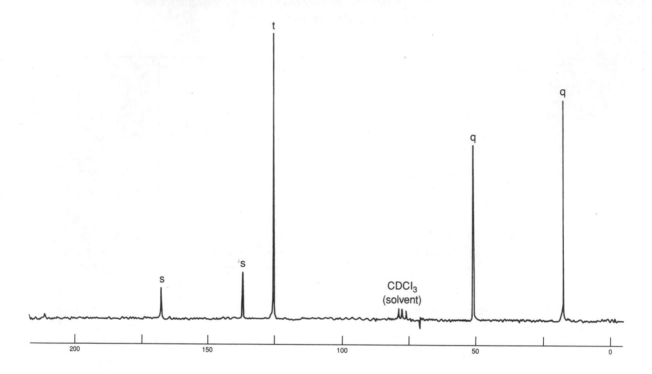

*5. Following are the ^1H and ^{13}C spectra for each of four isomeric bromoalkanes with formula C_4H_9Br. Assign a structure to each pair of spectra.

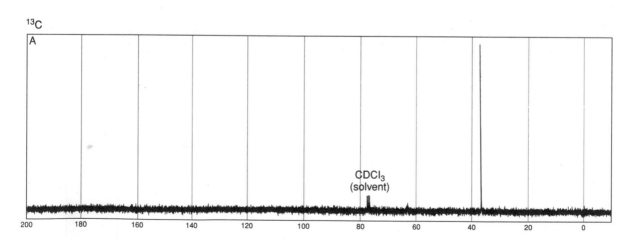

¹H

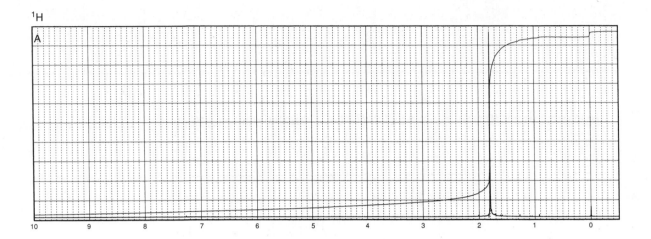

¹³C

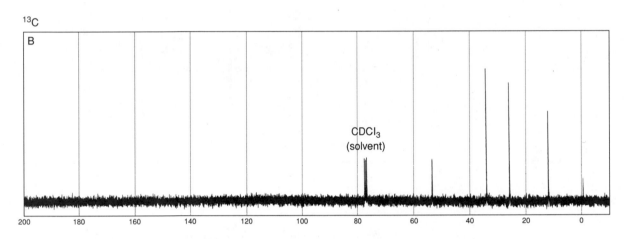

¹H

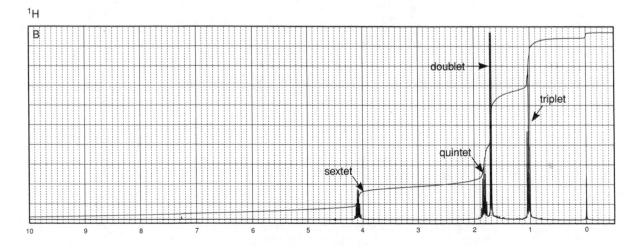

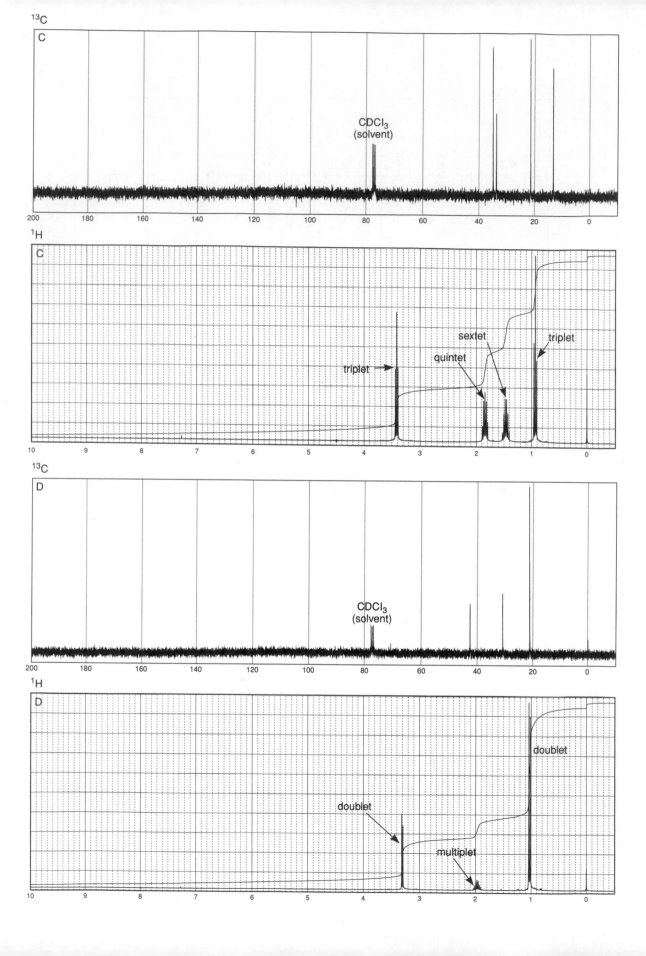

CDCl$_3$
(solvent)

200 180 160 140 120 100 80 60 40 20 0

sextet

quintet

triplet

triplet

10 9 8 7 6 5 4 3 2 1 0

CDCl$_3$
(solvent)

200 180 160 140 120 100 80 60 40 20 0

doublet

doublet

multiplet

10 9 8 7 6 5 4 3 2 1 0

*6. Following are the ^1H and ^{13}C spectra for each of three isomeric ketones with formula $C_7H_{14}O$. Assign a structure to each pair of spectra.

^{13}C

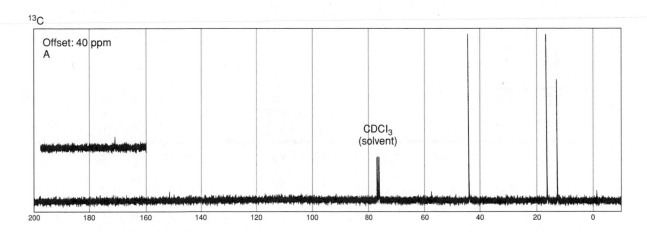

^1H

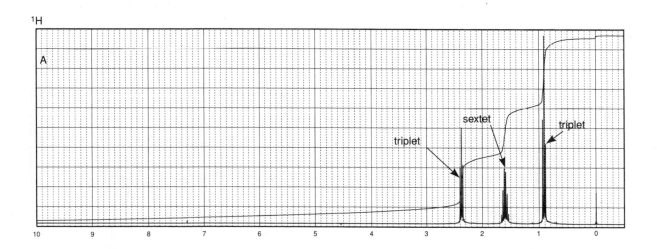

¹³C

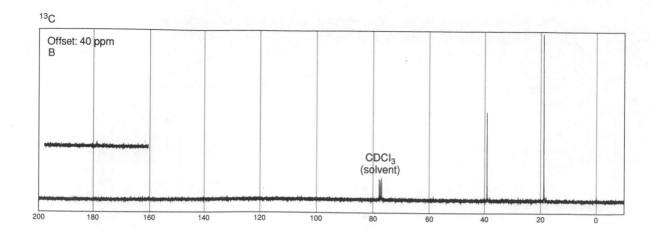

Offset: 40 ppm
B

CDCl₃
(solvent)

¹H

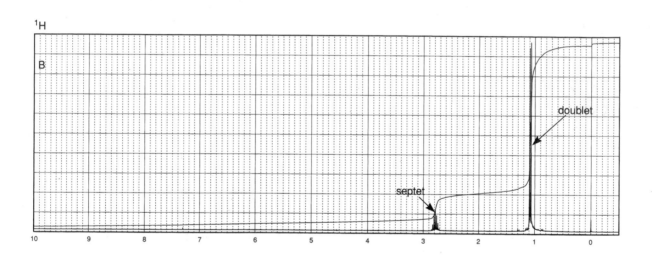

B

doublet

septet

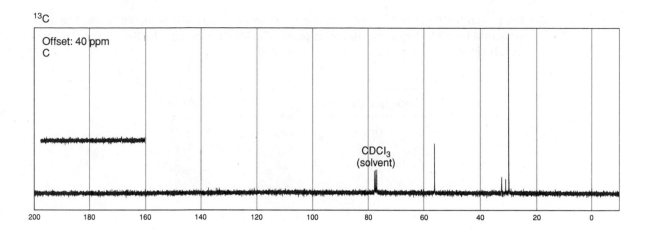

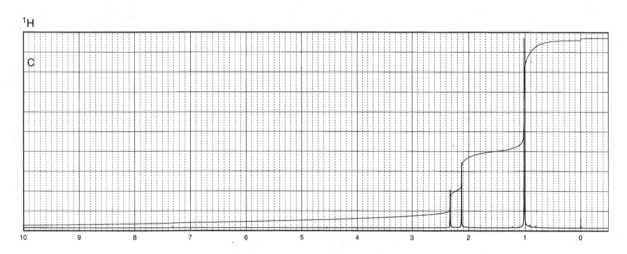

7. The proton NMR spectrum for a compound with formula C_8H_{18} shows only one peak at 0.86 ppm. The carbon-13 NMR spectrum has two peaks, a large one at 26 ppm and a small one at 35 ppm. Draw the structure of this compound.

8. The proton NMR spectrum for a compound with formula $C_5H_{12}O_2$ is shown below. The normal carbon-13 NMR spectrum has three peaks. The DEPT-135 and DEPT-90 spectral results are tabulated. Draw the structure of this compound.

Normal Carbon	DEPT-135	DEPT-90
15 ppm	Positive	No peak
63	Negative	No peak
95	Negative	No peak

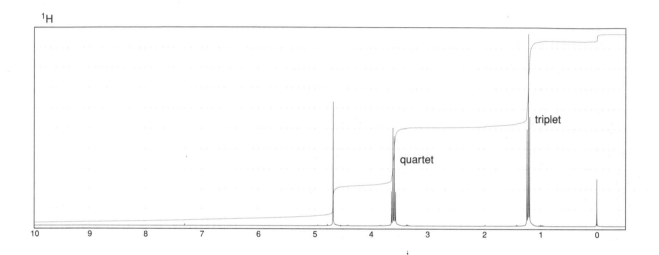

9. The proton NMR spectrum for a compound with formula $C_5H_{10}O$ is shown below. The normal carbon-13 NMR spectrum has three peaks. The DEPT-135 and DEPT-90 spectral results are tabulated. Draw the structure of this compound.

Normal Carbon	DEPT-135	DEPT-90
26 ppm	Positive	No peak
36	No peak	No peak
84	Negative	No peak

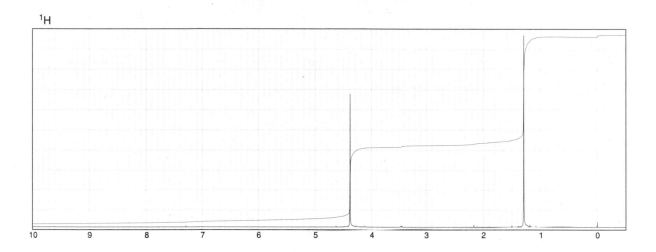

10. The proton NMR spectrum for a compound with formula $C_5H_{10}O_3$ is shown below. The normal carbon-13 NMR spectrum has four peaks. The infrared spectrum has a strong band at 1728 cm^{-1}. The DEPT-135 and DEPT-90 spectral results are tabulated. Draw the structure of this compound.

Normal Carbon	DEPT-135	DEPT-90
25 ppm	Positive	No peak
55	Positive	No peak
104	Positive	Positive
204	No peak	No peak

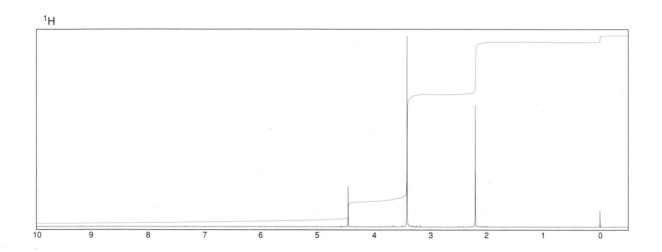

11. The proton NMR spectrum for a compound with formula C_9H_8O is shown below. The normal carbon-13 NMR spectrum has five peaks. The infrared spectrum has a strong band at 1746 cm^{-1}. The DEPT-135 and DEPT-90 spectral results are tabulated. Draw the structure of this compound.

Normal Carbon	DEPT-135	DEPT-90
44 ppm	Negative	No peak
125	Positive	Positive
127	Positive	Positive
138	No peak	No peak
215	No peak	No peak

¹H

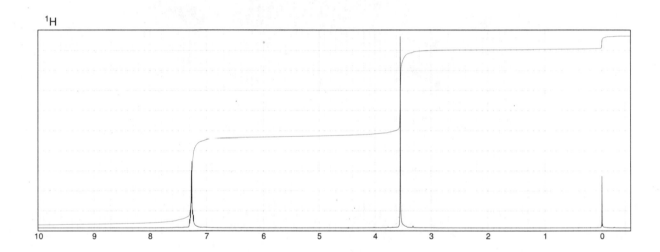

12. The proton NMR spectrum for a compound with formula $C_{10}H_{12}O_2$ is shown below. The infrared spectrum has a strong band at 1711 cm^{-1}. The normal carbon-13 NMR spectral results are tabulated along with the DEPT-135 and DEPT-90 information. Draw the structure of this compound.

Normal Carbon	DEPT-135	DEPT-90
29 ppm	Positive	No peak
50	Negative	No peak
55	Positive	No peak
114	Positive	Positive
126	No peak	No peak
130	Positive	Positive
159	No peak	No peak
207	No peak	No peak

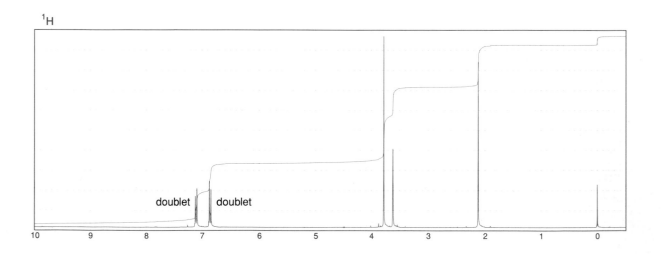

¹H

doublet   doublet

10   9   8   7   6   5   4   3   2   1   0

13. The proton NMR spectrum of a compound with formula $C_7H_{12}O_2$ is shown. The infrared spectrum displays a strong band at 1738 cm^{-1} and a weak band at 1689 cm^{-1}. The normal carbon-13 and the DEPT experimental results are tabulated. Draw the structure of this compound.

Normal Carbon	DEPT-135	DEPT-90
18 ppm	Positive	No peak
21	Positive	No peak
26	Positive	No peak
61	Negative	No peak
119	Positive	Positive
139	No peak	No peak
171	No peak	No peak

^{1}H

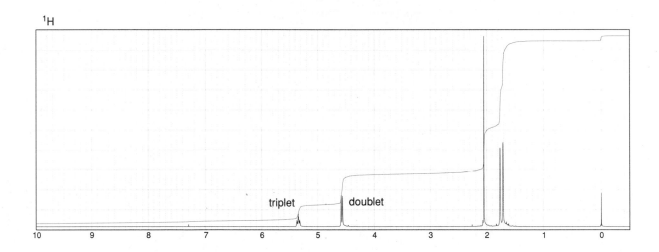

14. The proton NMR spectrum of a compound with formula $C_7H_{12}O_3$ is shown. The coupling constant for the triplet at 1.25 ppm is of the same magnitude as the one for the quartet at 4.15 ppm. The pair of distorted triplets at 2.56 and 2.75 ppm are coupled to each other. The infrared spectrum displays strong bands at 1720 and 1738 cm^{-1}. The normal carbon-13 and the DEPT experimental results are tabulated. Draw the structure of this compound.

Normal Carbon	DEPT-135	DEPT-90
14 ppm	Positive	No peak
28	Negative	No peak
30	Positive	No peak
38	Negative	No peak
61	Negative	No peak
173	No peak	No peak
207	No peak	No peak

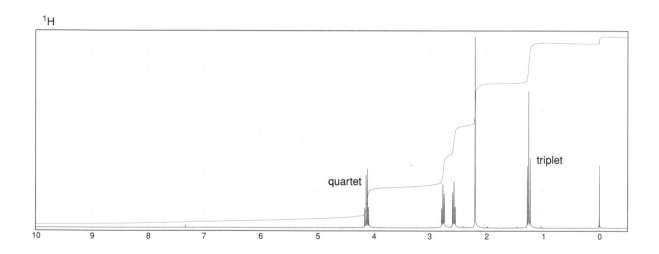

15. The proton NMR spectrum is shown for a compound with formula $C_6H_8Cl_2O_2$. The two chlorine atoms are attached to the same carbon atom. The infrared spectrum displays a strong band 1739 cm^{-1}. The normal carbon-13 and the DEPT experimental results are tabulated. Draw the structure of this compound.

Normal Carbon	DEPT-135	DEPT-90
18 ppm	Positive	No peak
31	Negative	No peak
35	No peak	No peak
53	Positive	No peak
63	No peak	No peak
170	No peak	No peak

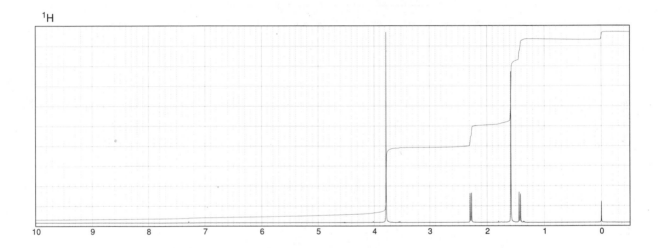

16. The proton NMR spectrum is shown for a compound with formula $C_5H_9NO_4$. The infrared spectrum displays strong bands at 1750 and 1562 cm^{-1} and a medium intensity band at 1320 cm^{-1}. The normal carbon-13 and the DEPT experimental results are tabulated. Draw the structure of this compound.

Normal Carbon	DEPT-135	DEPT-90
14 ppm	Positive	No peak
16	Positive	No peak
63	Negative	No peak
83	Positive	Positive
165	No peak	No peak

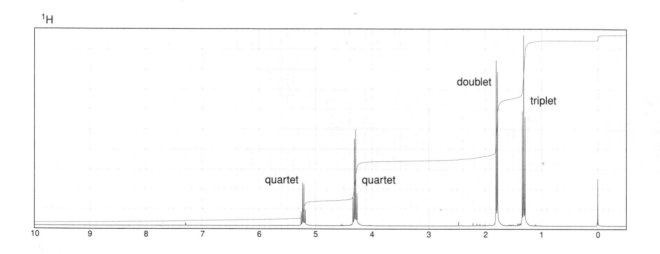

*17. The following alcohol undergoes elimination in the presence of concentrated sulfuric acid, but the product shown is not its chief product. Instead, another isomeric six-carbon alkene forms. This product shows a large peak at 20.4 ppm and a smaller one at 123.4 ppm in its proton-decoupled ^{13}C NMR spectrum. Draw the structure of the product, and interpret the spectrum. Outline a mechanism for the formation of the product that possesses this spectrum.

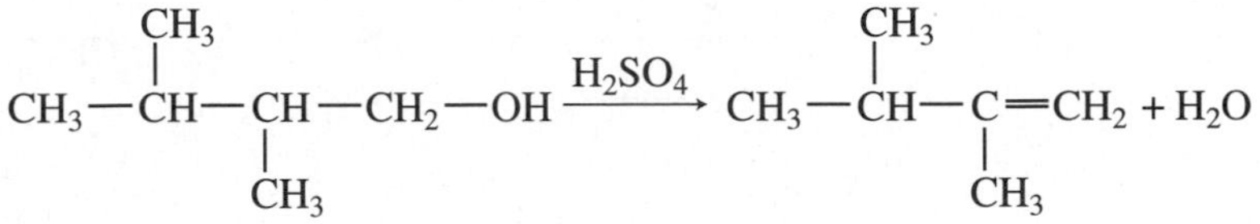

$$\underset{\underset{\underset{CH_3}{|}}{\overset{\overset{CH_3}{|}}{CH_3-CH-CH-CH_2-OH}}}{} \xrightarrow{\text{H}_2\text{SO}_4} \underset{\underset{\underset{CH_3}{|}}{CH_3-CH-C=CH_2}}{\overset{\overset{CH_3}{|}}{}} + H_2O$$

*18. Predict the appearances of the proton-decoupled ^{13}C spectra for the following compounds.

(a)

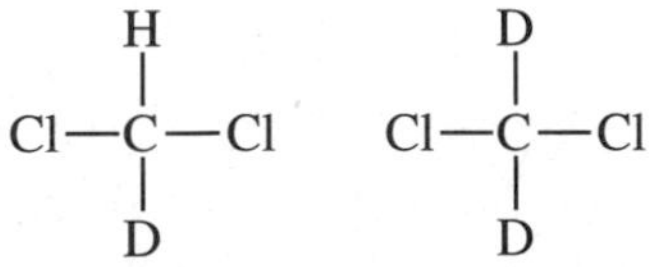

$$\underset{\underset{D}{|}}{\overset{\overset{H}{|}}{Cl-C-Cl}} \qquad \underset{\underset{D}{|}}{\overset{\overset{D}{|}}{Cl-C-Cl}}$$

$I = 1$

$J_{CD} \cong 20\text{–}30$ Hz (one bond)

(b)

$$\underset{\underset{H}{|}}{\overset{\overset{H}{|}}{F-C-H}} \qquad \underset{\underset{F}{|}}{\overset{\overset{F}{|}}{F-C-H}} \qquad \underset{\underset{F}{|}}{\overset{\overset{H}{|}}{F-C-CH_2-Cl}} \qquad \underset{\underset{F}{|}}{\overset{\overset{F}{|}}{F-C-CH_2-Cl}}$$

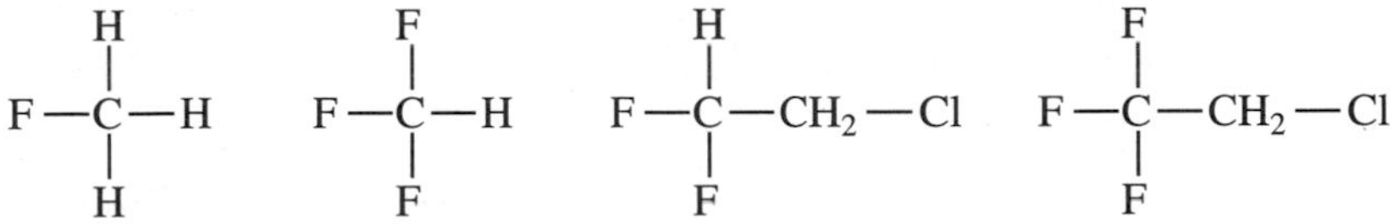

$I = \frac{1}{2}$

$J_{CF} > 180$ Hz (one bond)

$J_{CF} \cong 40$ Hz (two bonds)

*19. Figure 4.14 (page 188) is the ^{13}C NMR spectrum of toluene. We indicated in Section 4.12 that it was difficult to assign the **c** and **d** carbons to peaks in this spectrum. Using Table 7 in Appendix 8, calculate the expected chemical shifts of all the carbons in toluene, and assign all of the peaks.

*20. Using the tables in Appendix 8, calculate the expected carbon-13 chemical shifts for the indicated carbon atoms in the following compounds.

(a) 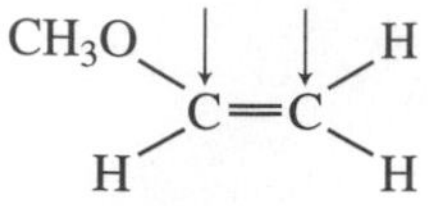 CH_3O \ | \ | \ H
$\quad\quad$ C=C
$\quad$ H $\quad\quad$ H

(b)
$\quad$ OH

(c) H $\quad\quad\quad$ CH_3
$\quad\quad$ C=C
$\quad$ CH_3CH_2 ↑ $\quad$ ↑ H

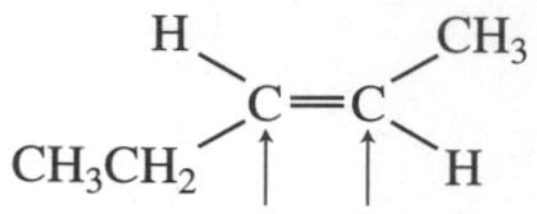

(d) 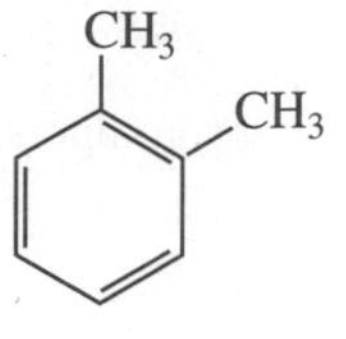 CH_3
$\quad\quad$ CH_3

$\quad\quad$ CH_3
$\quad\quad\quad\quad$ CH_3

$\quad\quad$ CH_3
$\quad\quad\quad\quad$ CH_3 **All**

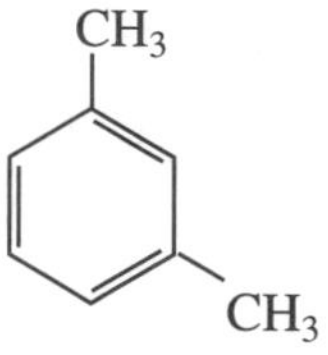

All $\quad\quad\quad$ **All**

(e) 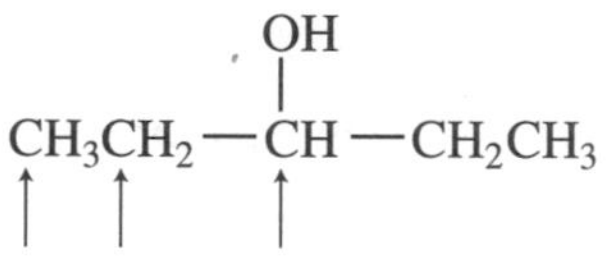 $\quad\quad$ OH
$\quad\quad\quad$ |
CH_3CH_2—CH—CH_2CH_3
$\quad$ ↑ $\quad$ ↑ $\quad$ ↑

(f) 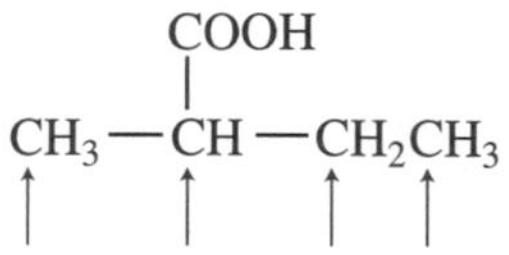 $\quad$ COOH
$\quad\quad\quad$ |
CH_3—CH—CH_2CH_3
$\quad$ ↑ $\quad$ ↑ $\quad$ ↑ $\quad$ ↑

(g) 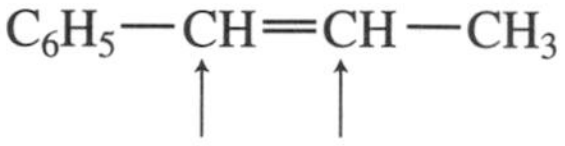 C_6H_5—CH=CH—CH_3
$\quad\quad\quad\quad$ ↑ $\quad$ ↑

(h) 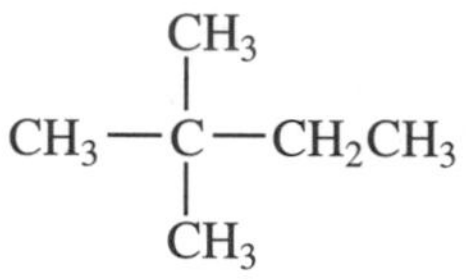 $\quad\quad$ CH_3
$\quad\quad\quad$ |
CH_3—C—CH_2CH_3
$\quad\quad\quad$ |
$\quad\quad$ CH_3 $\quad\quad$ **All**

(i) 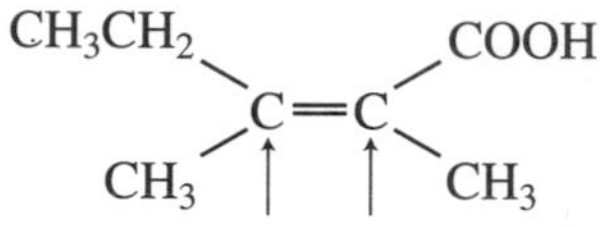 CH_3CH_2 $\quad\quad$ COOH
$\quad\quad\quad\quad$ C=C
$\quad\quad$ CH_3 ↑ $\quad$ ↑ CH_3

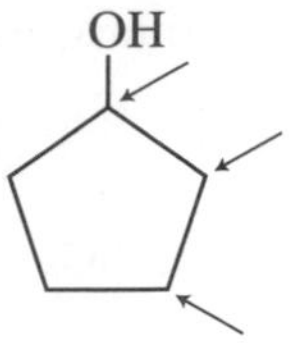

(j) 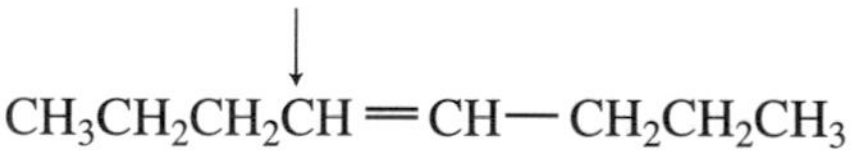 $\quad\quad\quad\quad$ |
$CH_3CH_2CH_2CH$=CH—$CH_2CH_2CH_3$

(k) 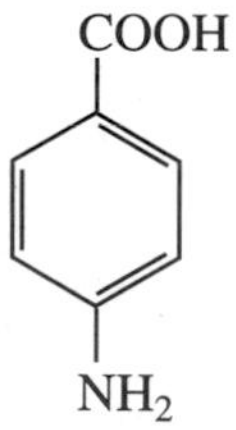 $\quad$ COOH

$\quad\quad$ NH_2 $\quad$ **Ring carbons**

(l) 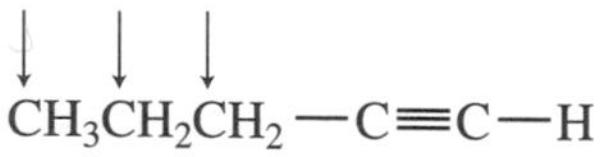 $\quad$ ↓ $\quad$ ↓ $\quad$ ↓
$\quad\quad$ $CH_3CH_2CH_2$—C≡C—H

(m) 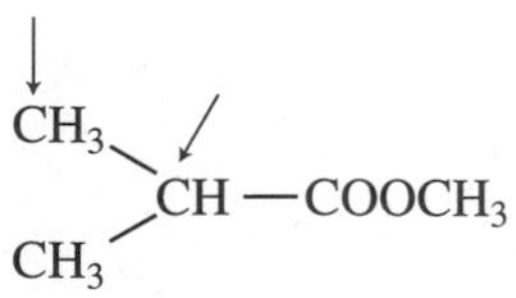 $\quad\quad$ ↓
$\quad$ CH_3 \ /
$\quad\quad\quad$ CH—$COOCH_3$
$\quad$ CH_3

(n)

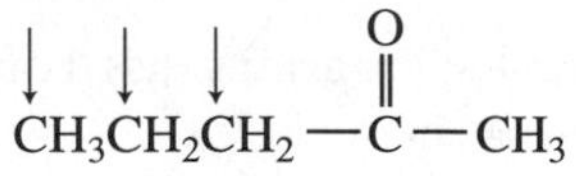

$$CH_3CH_2CH_2 - \overset{\overset{\displaystyle O}{\parallel}}{C} - CH_3$$

(o)

Br

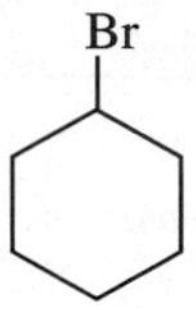

All

(p)

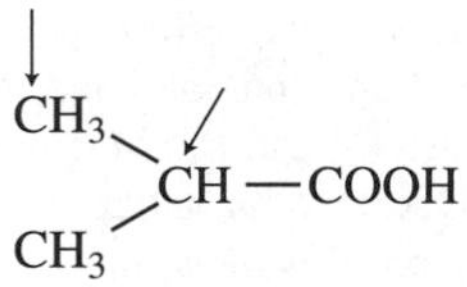

$$\underset{CH_3}{\overset{\displaystyle CH_3}{\diagdown}} CH - COOH$$

(q)

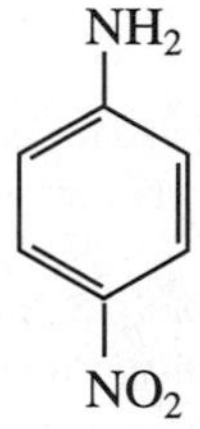

NH$_2$

NO$_2$ **All**

NH$_2$

NO$_2$

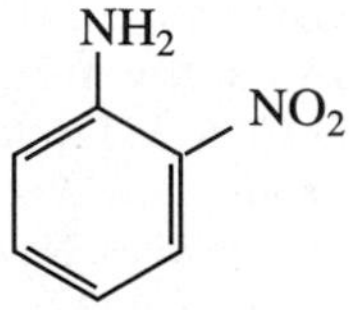

All

(r)

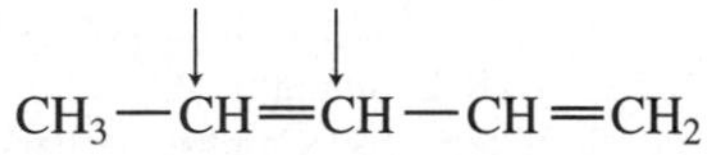

$$CH_3 - CH = CH - CH = CH_2$$

(s)

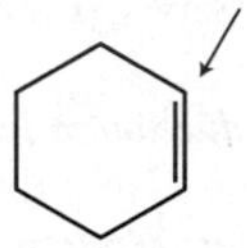

(t)

$$CH_3 - \underset{\underset{\displaystyle}{\overset{\displaystyle CH_3}{\mid}}}{CH} - CH = CH - CH_3$$

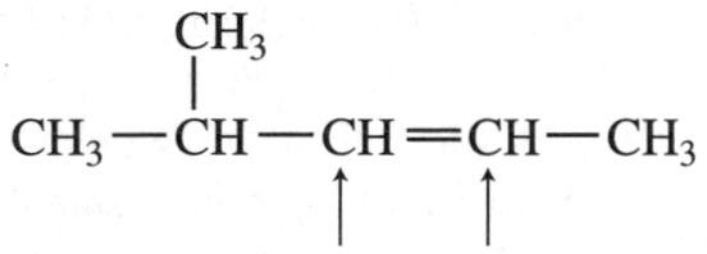

REFERENCES

Textbooks

Friebolin, H., *Basic One- and Two-Dimensional NMR Spectroscopy*, 2nd ed., VCH Publishers, New York, 1993.

Gunther, H., *NMR Spectroscopy*, 2nd ed., John Wiley and Sons, New York, 1995.

Levy, G. C., *Topics in Carbon-13 Spectroscopy*, John Wiley and Sons, New York, 1984.

Levy, G. C., R. L. Lichter, and G. L. Nelson, *Carbon-13 Nuclear Magnetic Resonance Spectroscopy*, 2d ed., John Wiley and Sons, New York, 1980.

Levy, G. C., and G. L. Nelson, *Carbon-13 Nuclear Magnetic Resonance for Organic Chemists*, John Wiley and Sons, New York, 1979.

Macomber, R. S., *NMR Spectroscopy—Essential Theory and Practice*, College Outline Series, Harcourt, Brace Jovanovich, New York, 1988.

Macomber, R. S., *A Complete Introduction to Modern NMR Spectroscopy*, John Wiley and Sons, New York, 1997.

Sanders, J. K. M., and B. K. Hunter, *Modern NMR Spectroscopy—A Guide for Chemists*, 2d ed., Oxford University Press, Oxford, England, 1993.

Silverstein, R. M., and F. X. Webster, *Spectrometric Identification of Organic Compounds*, 6th ed., John Wiley and Sons, New York, 1998.

Yoder, C. H., and C. D. Schaeffer, *Introduction to Multinuclear NMR*, Benjamin–Cummings, Menlo Park, CA, 1987.

Compilations of Spectra

Ault, A., and M. R. Ault, *A Handy and Systematic Catalog of NMR Spectra*, 60 MHz with some 270 MHz, University Science Books, Mill Valley, CA, 1980.

Fuchs, P. L., *Carbon-13 NMR Based Organic Spectral Problems*, 25 MHz, John Wiley and Sons, New York, 1979.

Johnson, L. F., and W. C. Jankowski, *Carbon-13 NMR Spectra: A Collection of Assigned, Coded, and Indexed Spectra*, 25 MHz, Wiley–Interscience, New York, 1972.

Pouchert, C. J., and J. Behnke, *The Aldrich Library of ^{13}C and ^{1}H FT-NMR Spectra,* 75 and 300 MHz, Aldrich Chemical Company, Milwaukee, WI, 1993.

Pretsch, E., T. Clerc, J. Seibl, and W. Simon, *Tables of Spectral Data for Structure Determination of Organic Compounds*, 2nd ed., Springer-Verlag, Berlin and New York, 1989. Translated from the German by K. Biemann.

Computer Programs that Teach Carbon-13 NMR Spectroscopy

Clough, F. W., "Introduction to Spectroscopy," Version 2.0 for MS-DOS and Macintosh, Trinity Software, 607 Tenney Mtn. Highway, Suite 215, Plymouth, NH, 03264, www.trinitysoftware.com.

Schatz, P. F., "Spectrabook I and II," MS-DOS Version, and "Spectradeck I and II," Macintosh Version, Falcon Software, One Hollis Street, Wellesley, MA, 02482, www.falconsoftware.com.

Computer Estimation of Carbon-13 Chemical Shifts

"C-13 NMR Estimate," IBM PC/Windows, Software for Science, 2525 N. Elston Ave., Chicago, IL 60647.

"^{13}C NMR Estimation," CS ChemDraw Ultra, Cambridge SoftCorp., 100 Cambridge Park Drive, Cambridge, MA 02140.

"Carbon 13 NMR Shift Prediction Module" requires ChemWindow (IBM PC) or ChemIntosh (Macintosh), SoftShell International, Ltd., 715 Horizon Drive, Grand Junction, CO 81506.

"HyperNMR," IBM PC/Windows, Hypercube, Inc., 419 Phillip Street, Waterloo, Ontario, Canada N2L 3X2.

"TurboNMR," Silicon Graphics Computers, Biosym Technologies, Inc., 4 Century Drive, Parsippany, NJ 07054.

Web sites

http://www.aist.go.jp/RIODB/SDBS/menu-e.html
Integrated Spectral DataBase System for Organic Compounds, National Institute of Materials and Chemical Research, Tsukuba, Ibaraki 305-8565, Japan. This database includes infrared, mass spectra, and NMR data (proton and carbon-13) for a number of compounds.

http://www.chem.ucla.edu/~webnmr
UCLA Department of Chemistry and Biochemistry, in connection with Cambridge University Isotope Laboratories, maintains a website, WebSpecta, that provides NMR and IR spectroscopy problems for students to interpret. They provide links to other sites with problems for students to solve.

http://www.nd.edu/~smithgrp/structure/workbook.html
Combined structure problems provided by the Smith group at Notre Dame University.

NUCLEAR MAGNETIC RESONANCE SPECTROSCOPY
Part Three: Spin–Spin Coupling

Chapters 3 and 4 covered only the most essential elements of NMR theory. Now we will consider applications of the basic concepts to more complicated situations. In this chapter (Chapter 5) the emphasis will be on the origin and behavior of coupling constants, and on advanced cases of spin–spin coupling, such as those leading to second-order spectra. Enantiotopic and diastereotopic systems will also be covered in this chapter. In Chapter 6, the NMR spectra of molecules that have internal hydrogen bonding or exchangeable protons (alcohols, acids, amines, and amides) will be discussed. The NMR spectra of equilibrating systems, such as tautomers, will be included. The uses of chemical shift reagents (both chiral and achiral) will also be introduced in Chapter 6.

5.1 COUPLING CONSTANTS: SYMBOLS

Chapter 3, Sections 3.17 and 3.18, introduced coupling constants. For simple multiplets, the coupling constant, J, is easily measured by determining the spacing (in Hertz) between the individual peaks of the multiplet. This coupling constant is independent of field strength, and it has the same value regardless of the field strength or operating frequency of the NMR spectrometer. J is a *constant*.

Coupling between two nuclei of the same type is called **homonuclear coupling.** Chapter 3 examined the three-bond couplings between hydrogens on adjacent carbon atoms, which gave multiplets governed by the $n + 1$ Rule. They are examples of homonuclear coupling. Coupling between two different types of nuclei is called **heteronuclear coupling.** The couplings between carbon-13 and attached hydrogens are one-bond heteronuclear couplings.

The magnitude of the coupling constant depends on the number of bonds intervening between the two atoms or groups of atoms that interact. Various factors influence the strength of interaction between two nuclei, but in general, one-bond couplings are larger than two-bond couplings, which in turn are larger than three-bond couplings, and so on. Consequently, the symbols used to represent coupling are often extended to include additional information about the type of atoms involved and the number of bonds through which the coupling constant operates.

We frequently add a superscript to the symbol J to indicate the number of bonds through which the interaction occurs. If the identity of the two nuclei involved is not obvious, we add this information in parentheses. Thus, the symbol

$$^1J\,(^{13}\text{C}-^1\text{H}) = 156 \text{ Hz}$$

indicates a one-bond coupling between carbon-13 and a hydrogen atom (C−H) with a value of 156 Hz. The symbol

$$^3J\,(^1\text{H}-^1\text{H}) = 8 \text{ Hz}$$

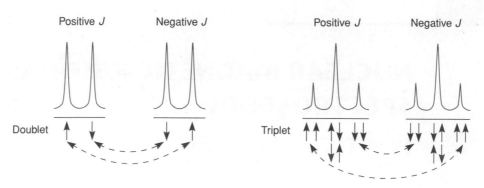

FIGURE 5.1   The dependence of multiplet assignments on the sign of *J*, the coupling constant.

indicates a three-bond coupling between two hydrogen atoms, as in H—C—C—H. Subscripts may also be used to give additional information. $J_{1,3}$, for instance, might indicate a coupling between atoms 1 and 3 in a structure. J_{CH} or J_{HH} would clearly indicate the types of atoms involved. The different coupling constants in a molecule might be designated simply as J_1, J_2, J_3, and so forth. Expect to see many variants in the usage of *J* symbols.

Although it makes no difference to the gross appearance of a spectrum, some coupling constants are positive and others are negative. With a negative *J*, the meanings of the individual lines in a multiplet are reversed—the upfield and downfield peaks exchange places—as shown in Figure 5.1. In the measurement of a coupling constant from a spectrum, it is impossible to tell whether the constant is positive or negative. Therefore, a measured value should always be regarded as the *absolute value* of *J*.

5.2 COUPLING CONSTANTS: THE MECHANISM OF COUPLING

A physical picture of spin–spin coupling, the way in which the spin of one nucleus influences that of another, is not easy to develop. Several theoretical models are available. The best theories we have are based on the Dirac vector model. This model has limitations, but its predictions are substantially correct. According to the Dirac model, the electrons in the intervening bonds between the two nuclei transfer spin information from one nucleus to another by means of interaction between the nuclear and electronic spins. An electron near the nucleus is assumed to have the lowest energy of interaction with the nucleus when the spin of the electron (small arrow) has its spin direction opposed to (or "paired" with) that of the nucleus (heavy arrow).

Spins of nucleus and electron paired,   Spins of nucleus and electron parallel
  or opposed (lower energy)   (higher energy)

This picture makes it easy to understand why the size of the coupling constant diminishes as the number of intervening bonds increases. As we will see, it also explains why some coupling constants are negative while others are positive. Theory shows that couplings involving an odd number of intervening bonds (^1J, ^3J . . .) are expected to be positive, while those involving an even number of intervening bonds (^2J, ^4J . . .) are expected to be negative.

A. One-Bond Couplings (^1J)

A one-bond coupling occurs when a single bond links two spin-active nuclei. The bonding electrons in a single bond are assumed to avoid each other such that when one electron is near nucleus A, the other is near nucleus B. According to the Pauli Principle, pairs of electrons in the same orbital have opposed spins; therefore, the Dirac model predicts that the most stable condition in a bond is where both nuclei have opposed spins. Following is an illustration of a ^{13}C—H bond, where the nucleus of the ^{13}C atom (heavy black arrow) has a spin opposite to that of the hydrogen nucleus (heavy white arrow).

$$^{13}\text{C—H}$$

The alignments shown would be typical for a ^{13}C—H bond or for any other type of bond where both nuclei have spin (for instance, ^{1}H—^{1}H or ^{31}P—H). Notice that in this arrangement the two nuclei prefer to have opposite spins. When two spin-active nuclei prefer an opposed alignment (have opposite spins), the coupling constant, J, is usually positive. If the nuclei are parallel or aligned (have the same spin), J is usually negative. Thus, most one-bond couplings have positive J values. Keep in mind, however, that there are some prominent exceptions, such as ^{13}C—^{19}F, where the coupling constants are negative (see Table 5.1).

It is not unusual for coupling constants to depend on the hybridization of the atoms involved. ^1J values for ^{13}C—H coupling constants vary with the amount of s character in the carbon hybrid, according to the following relationship:

$$^1J_{CH} = (500 \text{ Hz})\left(\frac{1}{n+1}\right) \text{ for hybridization type } sp^n \qquad \textbf{Equation 5.1}$$

Notice the specific values given for the ^{13}C—H couplings of ethane, ethene, and ethyne in Table 5.1.

Using the Dirac nuclear–electronic spin model, we can also develop an explanation for the origin of the spin–spin splitting multiplets that are the results of coupling. As a simple example, consider a ^{13}C—^{1}H bond. Recall that a ^{13}C atom that has one hydrogen attached appears as a doublet (two peaks) in a proton-coupled ^{13}C NMR spectrum (Section 4.3 and Fig. 4.3, page 171).There are two lines (peaks) in the ^{13}C NMR spectrum because the hydrogen nucleus can have two spins ($+\frac{1}{2}$ or $-\frac{1}{2}$), leading to two different energy transitions for the ^{13}C nucleus. Figure 5.2 illustrates these two situations.

TABLE 5.1
SOME ONE-BOND COUPLING CONSTANTS (^1J)

^{13}C—^{1}H	110–270 Hz
	sp^3 115–125 Hz (ethane = 125 Hz)
	sp^2 150–170 Hz (ethene = 156 Hz)
	sp 240–270 Hz (ethyne = 249 Hz)
^{13}C—^{19}F	−165 to −370 Hz
^{13}C—^{31}P	48–56 Hz
^{13}C—D	20–30 Hz
^{31}P—^{1}H	190–700 Hz

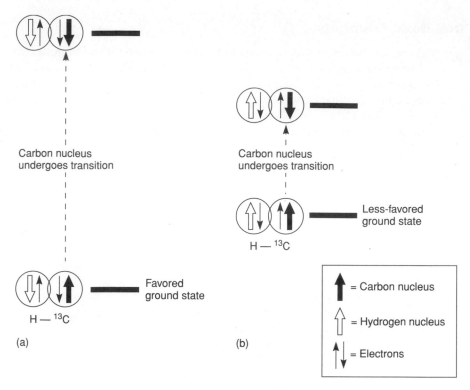

FIGURE 5.2   The two different energy transitions for a ^{13}C nucleus in a C—H bond. (a) The favored ground state (all spins paired); (b) the less-favored ground state (impossible to pair all spins).

At the bottom of part a in Figure 5.2 is the favored ground state for the ^{13}C—^{1}H bond. In this arrangement, the carbon nucleus is in its lowest energy state (spin $= +\frac{1}{2}$), and all of the spins, both nuclear and electronic, are paired, resulting in the lowest energy for the system. The spin of the nucleus of the hydrogen atom is opposed to the spin of carbon. A higher energy results for the system if the spin of the hydrogen is reversed. This less-favored ground state is shown at the bottom of part b of Figure 5.2.

Now assume that the carbon nucleus undergoes transition and inverts its spin. The excited state that results from the less-favored ground state (seen at the top of part b) turns out to have a lower energy than the one resulting from the favored ground state (top of part a) because all of its nuclear and electronic spins are paired. Thus, we see two different transitions for the ^{13}C nucleus (spin $= +\frac{1}{2}$), depending on the spin of the attached hydrogen. As a result, in a proton-coupled NMR spectrum a doublet is observed for a methine carbon (^{13}C—^{1}H).

B.   Two-Bond Couplings (^2J)

Two-bond couplings are quite common in NMR spectra. They are sometimes called **geminal coupling,** because the two nuclei that interact are attached to the same central atom (Latin *gemini* = "twins"). Two-bond coupling constants are abbreviated ^2J. They occur in carbon compounds whenever two or more spin-active atoms are attached to the same carbon atom. Table 5.2 lists some two-bond coupling constants that involve carbon as the central atom. Two-bond coupling constants are typically, although not always, smaller in magnitude than one-bond couplings (Table 5.2). Notice that the most common type of two-bond coupling, HCH, is frequently (but not always) negative.

TABLE 5.2
SOME TWO-BOND COUPLING CONSTANTS (^2J)

$\overset{H}{\underset{H}{>C<}}$ −9 to −15 Hz	$\overset{H}{\underset{^{19}F}{>C<}}$ ~50 Hz[a]
$\overset{H}{\underset{H}{=C<}}$ 0 to 2 Hz	$\overset{H}{\underset{^{13}C}{>C<}}$ ~5 Hz[a]
$\overset{H}{\underset{D}{>C<}}$ ~2 Hz[a]	$\overset{H}{\underset{^{31}P}{>C<}}$ 7 – 14 Hz
$\overset{^{19}F}{\underset{^{19}F}{>C<}}$ ~160 Hz[a]	

[a]*Absolute values.*

The mechanistic picture for geminal coupling (^2J) invokes nuclear–electronic spin coupling as a means of transmitting spin information from one nucleus to the other. It is consistent with the Dirac model that we discussed at the beginning of Section 5.2 and in Section 5.2A. Figure 5.3 shows this mechanism. In this case another atom (without spin) intervenes between two interacting orbitals. When this happens, theory predicts that the interacting electrons, and hence the nuclei, prefer to have parallel spins, resulting in a negative coupling constant. The preferred alignment is shown on the left side of Figure 5.3.

The amount of geminal coupling depends on the HCH angle, α. The graph in Figure 5.4 shows this dependence, where the amount of *electronic* interaction between the two C−H orbitals determines the magnitude of the coupling constant ^2J. In general, ^2J geminal coupling constants increase as the angle α decreases. As the angle α decreases, the two orbitals shown in Figure 5.3 move closer, and the electron spin correlations become greater. Note, however, that the graph in Figure 5.4 is very approximate, showing only the general trend; actual values vary quite widely.

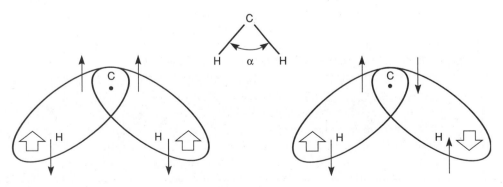

FIGURE 5.3 The mechanism of geminal coupling.

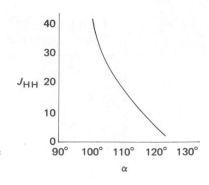

FIGURE 5.4 The dependence of the magnitude of $^2J_{HCH}$, the geminal coupling constant, on the HCH bond angle α.

Following are some systems that show geminal coupling, along with their approximate HCH bond angles. Notice that the coupling constants become smaller, as predicted, when the HCH angle becomes larger.

$$\alpha \cong 109° \qquad\qquad \alpha \cong 118° \qquad\qquad \alpha \cong 120°$$
$$^2J_{HH} \cong 12\text{–}18 \text{ Hz} \qquad ^2J_{HH} \cong 5 \text{ Hz} \qquad ^2J_{HH} \cong 0\text{–}3 \text{ Hz}$$

Table 5.3 shows a larger range of variation, with approximate values for a selected series of cyclic compounds and alkenes. Notice that as ring size decreases the absolute value of the coupling constant ^2J also decreases. Compare, for instance, cyclohexane, where ^2J is −13, and cyclopropane, where ^2J is −4. As the angle CCC in the ring becomes smaller (as *p* character increases), the complementary HCH angle grows larger (*s* character increases) and consequently the geminal coupling constant decreases. Note that hybridization is important and that the sign of the coupling constant for alkenes changes to positive except where they have an electronegative element attached.

In many cases, no geminal HCH coupling (no spin–spin splitting) is observed, either because the geminal protons are equivalent by symmetry or because free rotation renders them equivalent (see Section 5.3).

TABLE 5.3
VARIATIONS IN $^2J_{HH}$ WITH HYBRIDIZATION AND RING SIZE

+2	−2	−4	−9	−11	−13	−9 to −15 Hz

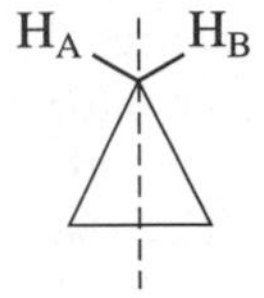

Plane of symmetry—
no splitting

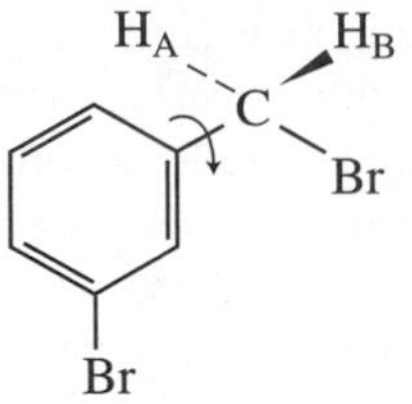

Free rotation—
no splitting

You have already seen, in our discussions of the $n + 1$ Rule, that in a hydrocarbon chain the protons attached to the same carbon may be treated as a group and do not split one another. How, then, can it be said that coupling exists in such cases if no spin–spin splitting is observed? The answer comes from two sources. First is the fact that geminal coupling is often observed in cyclic compounds that are conformationally rigid—for instance, bicyclic compounds such as bicyclo[2.1.1]hexane.

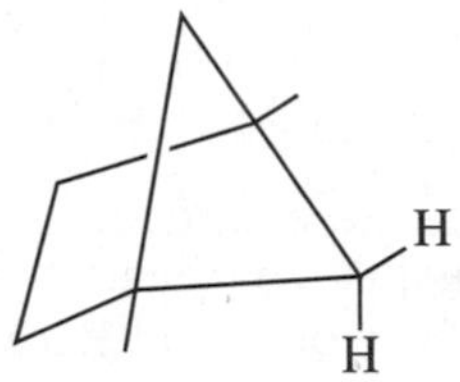

$$^2J = -5 \text{ Hz}$$

The second source of information about geminal coupling is deuterium substitution experiments. If one of the hydrogens in a compound that shows no spin–spin splitting is replaced by a deuterium, geminal splitting with deuterium ($I = 1$) is observed. Since deuterium and hydrogen are electronically the same atom (they differ by a neutron), it can be assumed that if there is interaction for HCD there is also interaction for HCH. In fact, the HCH and HCD coupling constants are related ($J_{HH} = 6.55 \times J_{HD}$)[1]. Section 5.3 will show that coupling can exist and still not lead to observable spin–spin splitting.

C. Three-Bond Couplings (^3J)

In a typical hydrocarbon, the spin of a hydrogen nucleus in one C—H bond is coupled to the spins of those hydrogens in adjacent C—H bonds.

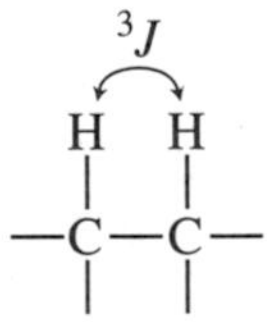

Three-bond vicinal coupling

These H—C—C—H couplings are sometimes called **vicinal couplings** because the hydrogens are on neighboring carbon atoms (Latin *vicinus* = "neighbor"). Vicinal couplings are three-bond couplings and have a coupling constant designated as ^3J. In Sections 3.13 through 3.17 you saw that these couplings produce spin–spin splitting patterns that follow the $n + 1$ Rule in simple aliphatic hydrocarbon chains.

---

[1]The multiplier 6.55 is the ratio of H and D magnetic moments: $\mu_H/\mu_D = 6.55$

Once again, nuclear and electronic spin interactions carry the spin information from one hydrogen to its neighbor. Since the σ C—C bond is nearly orthogonal (perpendicular) to the σ C—H bonds, there is no overlap between the orbitals, and the electrons cannot interact strongly through the sigma bond system. According to theory, they transfer the nuclear spin information via the small amount of parallel orbital *overlap* that exists between adjacent C—H bond orbitals. The spin interaction between the electrons in the two adjacent C—H bonds is the major factor determining the size of the coupling constant.

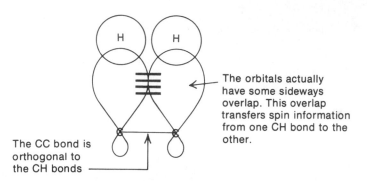

The orbitals actually have some sideways overlap. This overlap transfers spin information from one CH bond to the other.

The CC bond is orthogonal to the CH bonds

Figure 5.5 illustrates the two possible arrangements of nuclear and electronic spins for two coupled protons that are on adjacent carbon atoms. Recall that the carbon nuclei (^{12}C) have zero spin. The drawing on the left side of the figure, where the spins of the hydrogen nuclei are paired and where the spins of the electrons that are interacting through orbital overlap are also paired, is expected to represent the lowest energy and have the favored interactions. Because the interacting nuclei are spin paired in the favored arrangement, three-bond H—C—C—H couplings are expected to be positive. In fact, most three-bond couplings, regardless of atom types, are found to be positive.

That our current picture of three-bond vicinal coupling is substantially correct can be seen best in the effect of the dihedral angle between adjacent C—H bonds on the magnitude of the spin interaction. Recall that two nonequivalent adjacent protons give rise to a pair of doublets, each proton splitting the other.

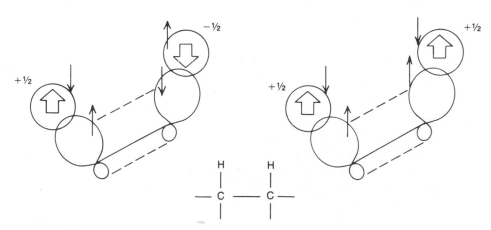

FIGURE 5.5   The method of transferring spin information between two adjacent C—H bonds.

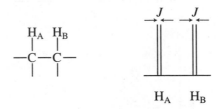

The parameter $^3J_{HH}$, the vicinal coupling constant, measures the magnitude of the splitting. This coupling constant measures the separation in Hertz between the multiplet peaks. The actual magnitude of the coupling constant between two adjacent C—H bonds can be shown to depend directly on the dihedral angle α between these two bonds. Figure 5.6 defines the dihedral angle α as a perspective drawing and a Newman diagram.

The magnitude of the splitting between H_A and H_B is greatest when $\alpha = 0°$ or $180°$ and is smallest when $\alpha = 90°$. The side–side overlap of the two C—H bond orbitals is at a maximum at $0°$, where the C—H bond orbitals are parallel, and at a minimum at $90°$, where they are perpendicular. At $\alpha = 180°$, overlap with the back lobes of the orbitals occurs.

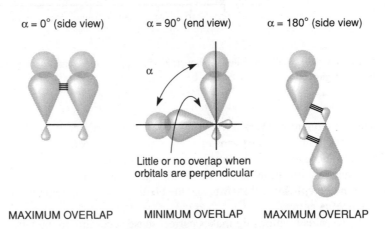

Martin Karplus was the first to study the variation of the coupling constant $^3J_{HH}$ with angle α. He was the first worker to develop an equation that gave a good fit to the experimental data shown in the graph in Figure 5.7. The Karplus equation takes the form

$$^3J_{HH} = A + B \cos \alpha + C \cos 2\alpha$$

$$A = 7 \qquad B = -1 \qquad C = 5$$

Equation 5.2

FIGURE 5.6 The definition of a dihedral angle α.

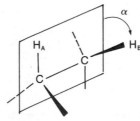

Side View

End View

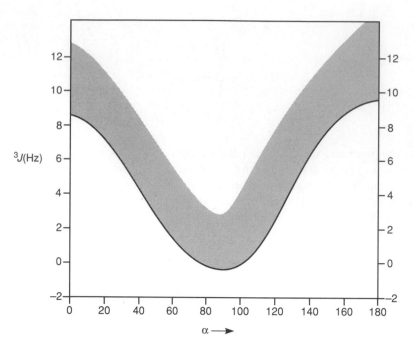

FIGURE 5.7 The Karplus relationship—the approximate variation of the coupling constant ^3J with the dihedral angle α.

Many subsequent workers have modified this equation—particularly its set of constants, A, B, and C—and several different forms of it are found in the literature. The constants shown are now accepted as those that give the best predictions. Note, however, that actual experimental data exhibit a wide range of variation, as shown by the shaded area of the curve in Figure 5.7. Equation 5.2 is called the **Karplus relationship,** and the predictive curve it generates (Fig. 5.7) is sometimes called the **Karplus curve.**

The Karplus relationship makes perfect sense according to the Dirac model. When the two adjacent C–H σ bonds are orthogonal ($\alpha = 90°$, perpendicular), there should be minimal orbital overlap, with little or no spin interaction between the electrons in these orbitals. As a result, nuclear spin information is not transmitted, and $^3J_{HH} \cong 0$. Conversely, when these two bonds are parallel ($\alpha = 0°$) or antiparallel ($\alpha = 180°$), the coupling constant should have its greatest magnitude ($^3J_{HH}$ = max).

The variation of $^3J_{HH}$ indicated by the shaded area in Figure 5.7 is a result of factors other than the dihedral angle, α. These factors (Fig. 5.8) include the bond length R_{CC}, the valence angles θ_1 and θ_2, and the electronegativity of any substituents X attached to the carbon atoms.

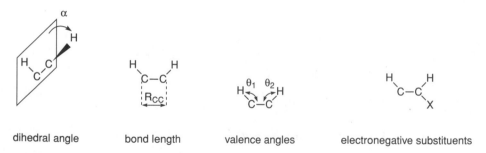

dihedral angle   bond length   valence angles   electronegative substituents

FIGURE 5.8 Factors influencing the magnitude of $^3J_{HH}$.

In any hydrocarbon, the magnitude of interaction between any two adjacent C—H bonds is always close to the values given in Figure 5.7. Cyclohexane derivatives that are conformationally rigid are the best illustrative examples of this principle. In the following molecule, the ring is "locked" in the indicated conformation because the bulky *tert*-butyl group must be placed equatorially. The coupling constant between two axial hydrogens, J_{aa}, is normally 10 to 14 Hz ($\alpha = 180°$), whereas the magnitude of interaction between an axial hydrogen and an equatorial hydrogen, J_{ae}, is generally 4 to 5 Hz ($\alpha = 60°$). A diequatorial interaction also has $J_{ee} = 4$ to 5 Hz ($\alpha = 60°$).

a,a

$J = 10 - 14$ Hz
$\alpha = 180°$

e,e

$J = 4 - 5$ Hz
$\alpha = 60°$

a,e

$J = 4 - 5$ Hz
$\alpha = 60°$

Cyclopropane derivatives provide another conformationally rigid example. Notice that J_{cis} ($\alpha = 0°$) is a larger coupling constant than J_{trans} ($\alpha = 120°$).

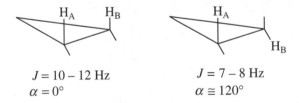

$J = 10 - 12$ Hz
$\alpha = 0°$

$J = 7 - 8$ Hz
$\alpha \cong 120°$

Table 5.4 lists some representative three-bond coupling constants. Notice that in the alkenes the *trans* coupling constant is always larger than the *cis* coupling constant. In Table 5.5 an interesting variation is seen with ring size in cyclic alkenes. Wider HCH valence angles in the smaller ring sizes cause smaller coupling constants ($^3J_{HH}$).

D. Long-Range Couplings ($^4J-^nJ$)

Longer-range couplings, those that involve more than three bonds, are common in systems with allylic hydrogens (Section 5.7) or aromatic ring hydrogens (Section 5.11) and in compounds that are rigid bicyclic systems (Section 5.13).

TABLE 5.4
SOME THREE-BOND COUPLING CONSTANTS ($^3J_{XY}$)

H–C–C–H	6–8 Hz	H–C≡C–H	*cis*	6–15 Hz
			trans	11–18 Hz
^{13}C–C–C–H	5 Hz	H–C=C–^{19}F	*cis*	18 Hz
			trans	40 Hz
^{19}F–C–C–H	5–20 Hz	^{19}F–C=C–^{19}F	*cis*	30–40 Hz
			trans	–120 Hz
^{19}F–C–C–^{19}F	–3 to –20			
^{31}P–C–C–H	13 Hz			
^{31}P–O–C–H	5–15 Hz			

TABLE 5.5
VARIATION OF $^3J_{HH}$ WITH VALENCE ANGLES IN CYCLIC ALKENES (Hz)

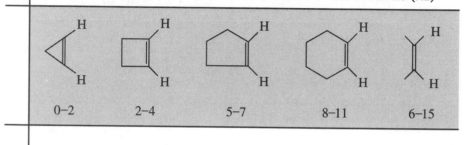

0–2	2–4	5–7	8–11	6–15

5.3 MAGNETIC EQUIVALENCE

In Chapter 3, Section 3.8, we discussed the idea of chemical equivalence. If two or more nuclei are equivalent by symmetry, they are said to be **chemically equivalent.** A plane of symmetry or an axis of symmetry renders nuclei chemically equivalent.

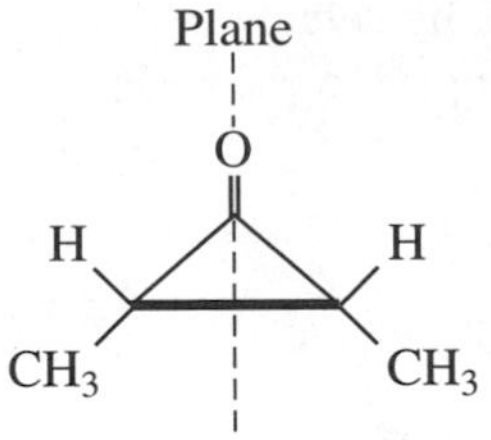

In molecule **1,** acetone, a plane of symmetry renders the two methyl groups chemically equivalent. The two methyl carbon atoms yield a single peak in the ^{13}C NMR spectrum. In addition, free rotation of the methyl group around the CC bond ensures that all six hydrogen atoms are equivalent and have resonance at the same frequency, producing a singlet in the ^1H NMR spectrum. In molecule **2,** 1,2-dichloroethane, there is also a plane of symmetry, rendering the two methylene (CH_2) groups equivalent. Even though the hydrogens on these two carbon atoms are close enough for vicinal (three-bond) coupling, ^3J, all four hydrogens appear as a single peak in the ^1H NMR spectrum, and no spin–spin splitting is seen. In molecules **3** and **4** there is a twofold axis of symmetry that renders the carbons and hydrogens chemically equivalent. Because of symmetry, the adjacent *trans* vinyl hydrogens in **3** do not show spin–spin splitting, and they appear as a singlet (both hydrogens having the same resonance frequency). The two hydrogens in molecule **4,** *trans*-2,3-dimethylcyclopropanone (axis of symmetry), and the two hydrogens in molecule **5,** *cis*-2,3-dimethylcyclopropanone (plane of symmetry), are also chemically equivalent.

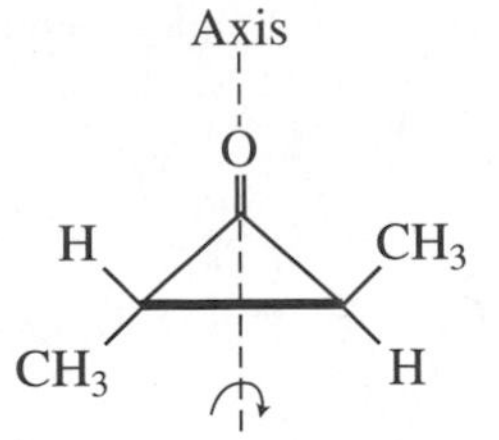

In most cases, chemically equivalent nuclei have the same resonance frequency (chemical shift), do not split each other, and give a single NMR signal. When this happens, the nuclei are said to be **magnetically equivalent** as well as chemically equivalent. However, it is possible for nuclei to be chemically equivalent and still not be magnetically equivalent. As we will show, magnetic equivalence has requirements that are stricter than those for chemical equivalence. For a group of nuclei to be magnetically equivalent, their magnetic environments, including *all coupling interactions*, must be of identical type.

Magnetic equivalence has two strict requirements:

1. Magnetically equivalent nuclei must be **isochronous;** that is, they must have identical chemical shifts.

2. Magnetically equivalent nuclei must have equal coupling (same *J* values) to all other nuclei in the molecule.

A corollary that follows from magnetic equivalence is:

Magnetically equivalent nuclei, even if they are close enough to be coupled, do not split one another, and they give only one signal (for both nuclei) in the NMR spectrum.

This corollary does not imply that no coupling occurs between magnetically equivalent nuclei, it means only that no observable spin–spin splitting results from the coupling.

Some simple examples will help you understand these requirements. In chloromethane, all of the hydrogens of the methyl group are chemically and magnetically equivalent because of the threefold axis of symmetry (coincident with the C—Cl bond axis) and three planes of symmetry (each containing one hydrogen and the C—Cl bond) in this molecule. In addition, the methyl group rotates freely about the C—Cl axis. Taken alone, this rotation would ensure that all three hydrogens experience the same *average* magnetic environment. The three hydrogens in chloromethane give a single signal in the NMR and are isochronous. Because there are no adjacent hydrogens in this one-carbon compound, by default all three hydrogens are equally coupled to all adjacent nuclei (a null set) and equally coupled to each other.

$$3\text{-Pentanone} \qquad CH_3CH_2 \overset{\overset{\displaystyle O}{\overset{\displaystyle \|}{}}}{-C} -CH_2CH_3$$

When a molecule has a plane of symmetry that divides it into equivalent halves, the observed spectrum is that of "half" of the molecule. In 3-pentanone, a plane of symmetry renders the two ethyl groups equivalent. The spectrum shows only one quartet (CH_2 with three neighbors) and one triplet (CH_3 with two neighbors). Both of the ethyl groups give the same spectrum. Because of this plane of symmetry, the two methyl groups are chemically equivalent and the two methylene groups are chemically equivalent. The coupling of any of the hydrogens in the methyl group to any of the hydrogens in the methylene group (^3J) is also equivalent (due to free rotation), and the coupling is the same on one side of the molecule as on the other. Each type of hydrogen is chemically equivalent.

para-Disubstituted benzene   1, 1-Difluoroethene

Now consider a *para*-disubstituted benzene ring, where the *para* substituents X and Y are not the same. This molecule has a plane of symmetry that renders the hydrogens on opposite sides of the ring chemically equivalent. You might expect the ^1H spectrum to be that of one-half of the molecule—two doublets. It is not, however, since the corresponding hydrogens in this molecule are not *magnetically equivalent*. Let us label the chemically equivalent hydrogens H_a and $H_{a'}$ (and H_b and $H_{b'}$). We would expect both H_a and $H_{a'}$ or H_b and $H_{b'}$ to have the same chemical shift (be isochronous), but their coupling constants to the other nuclei are not the same. H_a, for instance, would not have the same coupling constant to H_b (three bonds, ^3J) as $H_{a'}$ has to H_b (five bonds, ^5J). Because H_a and $H_{a'}$ do not have the same coupling constant to H_b, they cannot be magnetically equivalent even though they are chemically equivalent (see Section 5.11B for more information). This analysis also applies to $H_{a'}$, H_b, and $H_{b'}$, none of which has equivalent couplings to the other hydrogens.

Why is this subtle difference between the two kinds of equivalence important? Often protons that are chemically equivalent are also magnetically equivalent; however, when chemically equivalent protons *are not* magnetically equivalent, there are usually consequences in the spectrum. Nuclei that are magnetically equivalent will give "first-order spectra" that can be analyzed using the $n + 1$ Rule or a simple "tree diagram" (Section 5.4). Nuclei that are not magnetically equivalent will give "second-order spectra," where unexpected peaks may appear in multiplets (Sections 5.10 and 5.11B). Because coupling constants do not change, an increase in magnetic field *will not* simplify these second-order spectra.

A simpler case than benzene, which has chemical equivalence (due to symmetry) but not magnetic equivalence, is 1,1-difluoroethene. Both hydrogens couple to the fluorines (^{19}F, spin $= \frac{1}{2}$); however, the two hydrogens are not magnetically equivalent, because H_a and H_b do not couple to F_a with the same coupling constant ($^3J_{HF}$). One of these couplings is *cis* ($^3J_{cis}$) and the other is *trans* ($^3J_{trans}$). In Table 5.4 it was shown that *cis* and *trans* coupling constants in alkenes were different in magnitude, with $^3J_{trans}$ having the larger value. Because these hydrogens have different coupling constants to the same atom, they cannot be magnetically equivalent. A similar argument applies to the two fluorine atoms, which also are not magnetically equivalent.

1-Chloropropane $CH_3 - CH_2 - CH_2 - Cl$

                             c   **b**   **a**

If conformation is locked (no rotation)

Now consider 1-chloropropane. The hydrogens within a group (those on C1, C2, and C3) are isochronous, but each group is on a different carbon and, as a result, each group of hydrogens has a different chemical shift. The hydrogens in each group experience identical average magnetic environments, mainly because of free rotation, and are magnetically equivalent. Furthermore, also because of rotation, the hydrogens in each group are equally coupled to the hydrogens in the other groups. If we consider the two hydrogens on C2, H_b and $H_{b'}$, and pick any other hydrogen on either C1 or C3, both H_b and $H_{b'}$ will have the same coupling constant to the designated hydrogen. Without free rotation (see the preceding illustration) there would be no magnetic equivalence. Because of the fixed dissimilar dihedral angles, J_{ab} and $J_{ab'}$ would not be the same. Free rotation can be slowed or stopped by lowering the temperature.

5.4 NONEQUIVALENCE WITHIN A GROUP—THE USE OF TREE DIAGRAMS WHEN THE *n* + 1 RULE FAILS

When the protons attached to a single carbon are chemically equivalent (have the same chemical shift), the *n* + 1 Rule successfully predicts the splitting patterns. In contrast, when the protons attached to a single carbon are chemically nonequivalent (have different chemical shifts), the *n* + 1 Rule no longer applies. We shall examine two cases, one in which the *n* + 1 Rule applies (1,1,2-trichloroethane) and one in which it fails (styrene oxide).

Chapter 3, Section 3.13 and Figure 3.25 (page 128), addressed the spectrum of 1,1,2-trichloroethane. This molecule has a three-proton system, configured as $-CH_2-CH-$, with the two methylene (CH_2) protons having the same chemical shift. The group of protons on the methylene carbon is rendered equivalent due to free rotation around the C—C bond. Because of this rotation, and due to a lack of asymmetry in this molecule, the methylene protons each experience the same averaged environment—they have the same chemical shift (chemical equivalence), and they do not split each other. In addition, the rotation ensures that they both have the same averaged coupling constant, *J*, to the methine (CH) hydrogen. As a result, they behave as a group, and geminal coupling between them does not lead to any splitting. The *n* + 1 Rule correctly predicts a doublet for the CH_2 protons (one neighbor) and a triplet for the CH proton (two neighbors). Figure 5.9a illustrates the parameters for this molecule.

Figure 5.10, the ^1H spectrum of styrene oxide, shows how chemical nonequivalence complicates the spectrum. The ring blocks rotation, causing protons H_A and H_B to have different chemical shift values; they are chemically and magnetically nonequivalent. Hydrogen H_A is on the same side of the ring as the phenyl group; hydrogen H_B is on the opposite side of the ring. These hydrogens have slightly different chemical shift values, $H_A = 2.75$ ppm and $H_B = 3.09$ ppm, and they show geminal splitting with respect to each other. The third proton, H_C, has yet another value of chemical shift ($H_C = 3.81$ ppm) and is coupled differently to H_A (which is *trans*) than to H_b (which is *cis*). Because H_A and H_B are nonequivalent, and because H_C is coupled differently to H_A than to H_B ($^3J_{AC} \neq {}^3J_{BC}$), the *n* + 1 Rule fails and the

(a) FREE ROTATION
THE $n + 1$ RULE APPLIES

(b) LOCKED CONFORMATION
A TREE DIAGRAM IS REQUIRED

$\delta_A = \delta_B$
$J_{AC} = J_{BC}$
$J_{AB} = 0$

$\delta_A \neq \delta_B$
$J_{AC} \neq J_{BC}$
$J_{AB} \neq 0$

FIGURE 5.9 Two cases of coupling.

spectrum of styrene oxide becomes more complicated. To explain the spectrum, one must examine each hydrogen individually and take into account its coupling with every other hydrogen. Each coupling is independent of the others. Figure 5.9b shows the parameters for this situation.

An analysis of the splitting pattern in styrene oxide is carried out splitting-by-splitting with **graphical analyses,** or **tree diagrams** (Fig. 5.11). Begin with an examination of hydrogen H_C. First, the two possible spins of H_B split H_C ($^3J_{BC}$) into a doublet; second, H_A splits each of the doublet peaks ($^3J_{AC}$) into another doublet. The resulting pattern of two doublets is sometimes called a

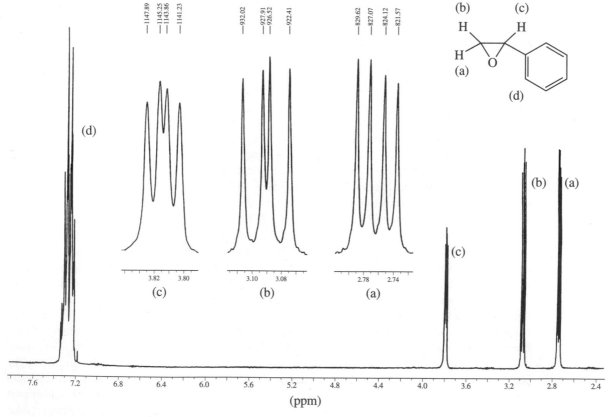

FIGURE 5.10 The ^{1}H NMR spectrum of styrene oxide.

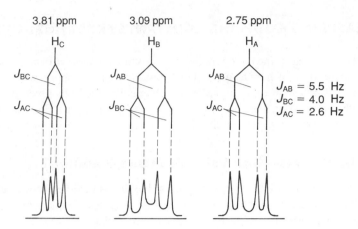

FIGURE 5.11 An analysis of the splitting pattern in styrene oxide.

doublet of doublets. You may also look at the same splitting from H_A first and from H_B second. It is customary to show the *largest* splitting first, but it is not necessary to follow this convention in order to obtain the correct result. If the actual coupling constants are known, it is very convenient to perform this analysis (*to scale*) on graph paper with 1-mm squares.

Note that $^3J_{BC}$ (*cis*) is larger than $^3J_{AC}$ (*trans*). This is typical for small ring compounds where there is more interaction between protons that are *cis* to each other than between protons that are *trans* to each other (see Section 5.2C). Thus, we see that H_C gives rise to *four* peaks (a doublet of doublets) centered at 3.81 ppm. Similarly, H_A and H_B each gives four peaks (a doublet of doublets) at 2.75 ppm and 3.09 ppm, respectively. Figure 5.11 also shows these splittings. Notice that the magnetically nonequivalent protons H_A and H_B give rise to geminal splitting ($^2J_{AB}$) that is quite significant.

As you see, the splitting situation becomes quite complicated for molecules that contain non-equivalent groups of hydrogens. In fact, you may ask, how can one be sure that the graphic analysis just given is the correct one? First, this analysis explains the entire pattern; second, it is internally consistent. Notice that the coupling constants have the same magnitude wherever they are used. Thus, in the analysis, $^3J_{BC}$ (*cis*) is given the same magnitude when it is used in splitting H_C as when it is used in splitting H_B. Similarly, $^3J_{AC}$ (*trans*) has the same magnitude in splitting H_C as in splitting H_A. The coupling constant $^2J_{AB}$ (geminal) has the same magnitude for H_A as for H_B. If this kind of consistency were not apparent in the analysis, the splitting analysis would have been incorrect. To complete the analysis, note that the NMR peak at 7.28 ppm is due to the protons of the phenyl ring. It integrates for five protons, while the other three four-peak multiplets integrate for one proton each.

We must sound one note of caution at this point. In many molecules the splitting situation becomes so complicated that it is virtually impossible for the beginning student to derive it. There are also situations involving apparently simple molecules for which a graphical analysis of the type we have just completed does not suffice. Section 5.10 will describe a few of these cases.

We have now discussed three situations in which the $n + 1$ Rule fails: (1) when the coupling involves nuclei other than hydrogen that do not have spin $\frac{1}{2}$ (e.g., deuterium, Section 4.13), (2) when there is nonequivalence in a set of protons attached to the same carbon, and (3) when (sometimes unexpectedly) the chemical shift difference between two sets of protons is small compared to the coupling constant linking them. Section 5.10 will discuss this situation.

5.5 MEASURING COUPLING CONSTANTS FROM FIRST-ORDER SPECTRA

When you determine the coupling constants of a compound from an actual spectrum there is always some question of how to go about the task. In this section we will attempt to give some guidelines that will help you to approach this problem. The methods given here apply to first-order spectra; methods for second-order spectra are given in Section 5.10. Fortunately, most spectra determined at 300 MHz or greater will be first-order spectra.

A.   Simple Multiplets—One Value of *J* (One Coupling)

For simple multiplets, where only one value of *J* is involved (one coupling), there is little difficulty in measuring the coupling constant. In this case it is a simple matter of determining the spacing (in Hertz) between the successive peaks in the multiplet. This was discussed in Chapter 3, Section 3.17. Also discussed in that section was the method of converting differences in parts per million (ppm) to Hertz (Hz). Use of the relationship

$$1 \text{ ppm (in Hertz)} = \text{Spectrometer Frequency in Hertz} \div 1{,}000{,}000$$

gives the simple correspondence values given in Table 5.6, which shows that if the spectrometer frequency is *n* **MHz**, one ppm of it will be *n* **Hz**. This relationship allows an easy determination of the coupling constant linking two peaks when their chemical shifts are known only in ppm: just find the chemical shift difference in ppm and multiply by the Hertz equivalent.

For most modern FT-NMR instruments it is not necessary to make the conversion from ppm to Hertz since the instrument will generally give the peak locations in both Hertz and ppm. Figure 5.12 is an example of the printed output from a modern 300-MHz FT-NMR. In this septet, the chemical shift values of the peaks (ppm) are obtained from the scale printed at the bottom of the spectrum, and the values of the peaks in Hertz are printed vertically above each peak. To obtain the coupling constant it is necessary only to subtract the Hertz values for the successive peaks. In doing this, however, you will note that not all of the differences are identical. In this case they are 6.889, 6.858, 6.852, 6.895, 6.871, and 6.820 Hz, starting from the left. There are two reasons for the inconsistencies. First, these values are given to more places than the appropriate number of significant figures would warrant. When these values are rounded off to two or three significant figures they will agree more closely. In most cases, two significant figures is appropriate. Second, the values given for the peaks are not always exact due to the way the data points are taken when determining the spectrum. Data points are taken at equally spaced intervals. If the apex of a peak does not correspond exactly to a point at which a data point is recorded, it will be in error by a small amount.

TABLE 5.6
THE HERTZ EQUIVALENT OF A ppm UNIT AT VARIOUS SPECTROMETER OPERATING FREQUENCIES

Spectrometer Frequency	Hertz Equivalent of 1 ppm
60 MHz	60 Hz
100 MHz	100 Hz
300 MHz	300 Hz
500 MHz	500 Hz

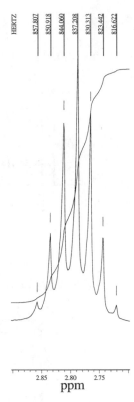

F I G U R E   5 . 1 2   A septet determined at 300 MHz showing both ppm and Hz values.

When several conflicting *J* values are determined for a multiplet it is usually appropriate to round them off to two significant figures, or to take an average of the similar values and round that average to two significant figures (or three for larger-value coupling constants). For most purposes, it is sufficient if all the measured *J* values agree to within 0.2–0.5 Hz. In the septet shown in Figure 5.12, the average of all the differences is 6.864 Hz, and an appropriate value for the coupling constant would be 6.9 Hz.

B.   More Complex Multiplets—More Than One Value of *J*

When measuring coupling constants in a system with more than one coupling you will often notice that none of the multiplet peaks is located at the appropriate chemical shift values to directly determine a value for an intermediate *J* value. This is illustrated in Figure 5.13 where a doublet of doublets is illustrated. In this case, none of the peaks is located at the chemical shift values that would result from the first coupling J_1. To a beginning student, it may be tempting to average the chemical shift values for peaks 1 and 2, and for peaks 3 and 4, and then take the difference (dotted lines). This is not necessary. With a little thought you will see that the distance between peaks 1 and 3, and also the distance between peaks 2 and 4 (solid arrows), can yield the desired value with much less work. This type of situation will occur whenever there is an even number of subpeaks (doublets, quartets, etc.) in the separated multiplets. In these systems you should look for an appropriately spaced pair of offset subpeaks that will yield the value you need. It will usually be necessary to construct a splitting diagram (tree) in order to decide which of the peaks are the appropriate ones.

When the separated multiplets have an odd number of subpeaks, one of the subpeaks will inevitably fall directly on the desired chemical shift value without a need to look for appropriate offset peaks. Figure 5.14 shows a doublet of triplets. Note that peaks 2 and 5 are ideally located for a determination of J_1.

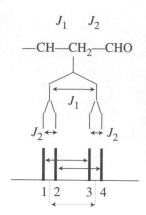

J_1 J_2

—CH—CH$_2$—CHO

Doublet of Doublets (dd)

To obtain J_1 measure the difference between lines 1 and 3, or 2 and 4, in Hz.*

*Do not try to find the centers of the doublets!

J_2 is the spacing between lines 1 and 2, or 3 and 4

FIGURE 5.13 Determining coupling constants for a doublet of doublets (dd).

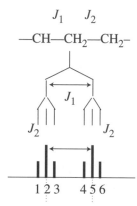

J_1 J_2

—CH—CH$_2$—CH$_2$—

Doublet of Triplets (dt)

To obtain J_1 measure the difference between the most intense lines (2 and 5) in Hz

J_2 is the spacing between lines 1 and 2, or 2 and 3, or those in the other triplet.

FIGURE 5.14 Determining coupling constants for a doublet of triplets (dt).

Figure 5.15 shows a pattern that may be called a doublet of doublets of doublets. After constructing a tree diagram, it is relatively easy to select appropriate peaks to use for the determination of the three coupling constants (solid arrows).

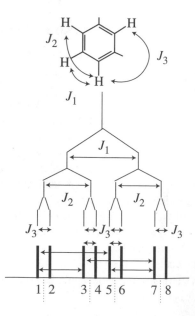

A number of approaches are possible. Generally you should choose an appropriately spaced pair of peaks based on their intensities and sharpness.

FIGURE 5.15 Determining coupling constants for a doublet of doublets of doublets (ddd).

5.6 ALKENES

The NMR chemical shifts of protons attached to double bonds are much larger than those of protons attached to sp^3 carbon atoms. This arises, in part, from the change in hybridization. Also important is the deshielding due to the diamagnetic anisotropy generated by the π electrons of the double bond. We can estimate the chemical shifts of protons attached to double bonds (vinyl protons) by using Table 2 in Appendix 6. The protons on double bonds also differ in that they are rarely magnetically equivalent, and they often give rise to splitting patterns that cannot be explained by the $n + 1$ Rule. In the typical alkene, as follows, the chemical shifts of the three protons H_A, H_B, and H_C rarely coincide.

$$\delta_{H_A} \neq \delta_{H_B} \neq \delta_{H_C}$$
$$^3J_{AB} \neq {}^3J_{AC} \neq {}^3J_{BC}$$

Furthermore, the extent of coupling between the protons is usually quite different, and three distinct types of spin interaction are observed:

cis
$^3J \cong 6 - 15$ Hz

trans
$^3J \cong 11 - 18$ Hz

Terminal methylene (geminal)
$^2J \cong 0 - 5$ Hz

Protons substituted *trans* on a double bond couple most strongly, with a typical value for ^3J of about 16 Hz. The *cis* coupling constant is commonly half this value, being about 8 Hz. Coupling between terminal methylene protons (geminal coupling) is much smaller yet. As an example of the ^{1}H NMR spectrum of a simple alkene, Figure 5.16 shows the spectrum of *trans*-cinnamic acid.

The phenyl protons appear as a large group between 7.4 and 7.6 ppm, and the acid proton is a singlet that appears off-scale at 13.21 ppm. The two vinyl protons H_A and H_C split each other into two doublets, one centered at 7.83 ppm to the left of the phenyl resonances and the other at 6.46 ppm to the right of the phenyl resonances. The proton H_C, attached to the carbon bearing the phenyl ring, is assigned the larger chemical shift, as it should be in a deshielding area of the anisotropic field generated by the π electrons of the aromatic ring. The coupling constant $^3J_{AC}$ can be determined quite easily from the 300-MHz spectrum shown in Figure 5.16. The splitting is about 16 Hz—a common value for *trans* proton–proton coupling across a double bond. The *cis* isomer would exhibit a smaller splitting.

A molecule that has a symmetry element (a plane or axis of symmetry) passing through the C=C double bond does not show any *cis* or *trans* splitting, since the vinyl protons are chemically and magnetically equivalent. An example of each type can be seen in *cis*- and *trans*-stilbene, respectively. In each compound, the vinyl protons H_A and H_B give rise to only a single *unsplit* resonance peak.

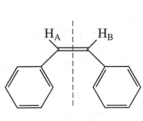

Plane of symmetry
cis-Stilbene

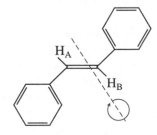

Twofold axis of symmetry
trans-Stilbene

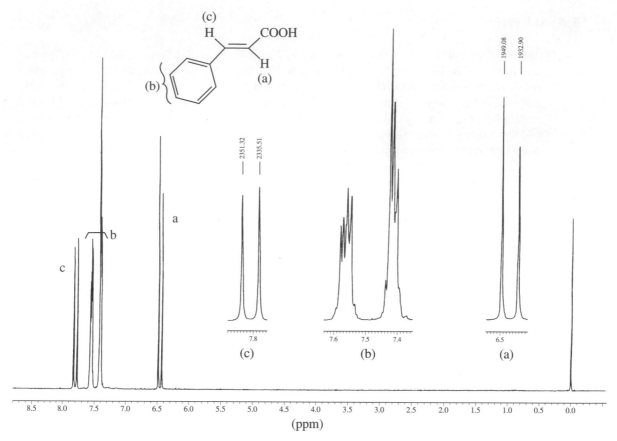

FIGURE 5.16   The NMR spectrum of *trans*-cinnamic acid.

Vinyl acetate gives an NMR spectrum typical of a compound with three hydrogens substituted on a double bond. Each proton has a chemical shift and a coupling constant different from those of each of the other protons. This spectrum, shown in Figure 5.17, is not unlike that of styrene oxide (Fig. 5.10). Each hydrogen is split into a doublet of doublets (four peaks). Figure 5.18 is a graphical analysis of the vinyl portion. Notice that $^3J_{BC}$ (*trans*) is larger than $^3J_{AC}$ (*cis*) and that $^2J_{AB}$ (geminal) is very small—the usual situation for vinyl compounds.

Again, as in the last section, we close with a note of caution. Cases exist in which the splitting patterns of apparently simple alkenes cannot be explained either by this simple graphical approach or by use of the $n + 1$ Rule.

5.7 MECHANISMS OF COUPLING IN ALKENES; ALLYLIC COUPLING

The mechanism of *cis* and *trans* coupling in alkenes is no different from that of any other three-bond vicinal coupling, and that of the terminal methylene protons is just a case of two-bond geminal coupling. All three types have been discussed already and are illustrated in Figure 5.19.

To obtain an explanation of the relative magnitudes of the ^3J coupling constants, notice that the two C—H bonds are parallel in *trans* coupling, while they are tilted away from each other in *cis* coupling. Also note that the H—C—H angle for geminal coupling is nearly 120°, a virtual minimum in the graph of Figure 5.4. In addition to these three types of coupling, alkenes often show small couplings between protons substituted on carbons α to the double bond and those on the opposite end of the double bond:

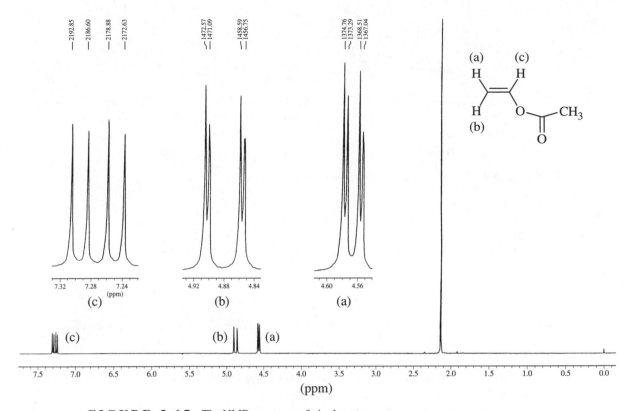

FIGURE 5.17 The NMR spectrum of vinyl acetate.

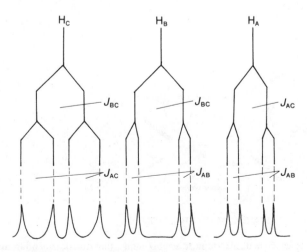

FIGURE 5.18 A graphical analysis of the splittings in vinyl acetate.

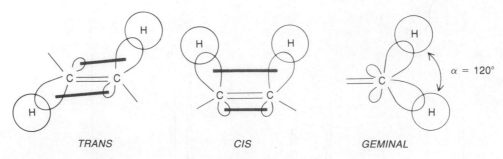

FIGURE 5.19   The types of coupling present in alkenes.

$^4J \sim 0 - 3$ Hz $^4J \sim 0 - 3$ Hz

This four-bond coupling (^4J) is called **allylic coupling.** The π electrons of the double bond apparently help to transmit the spin information from one nucleus to the other, as shown in Figure 5.20.

When all nuclei are coplanar, there is no interaction of the allylic C—H bond orbital with the π system, and $^4J = 0$ Hz. However, when the allylic C—H bond is perpendicular to the C=C plane (as is the π bond), the interaction assumes the maximum value, $^4J = 3$ Hz. Figure 5.20 illustrates these geometries.

Allylic splitting is observed in compounds such as the following:

1   **2**   **3**

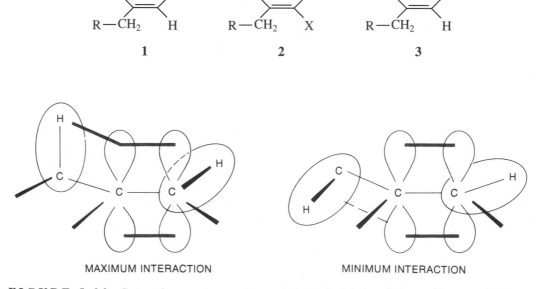

MAXIMUM INTERACTION   MINIMUM INTERACTION

FIGURE 5.20   Geometric arrangements that maximize and minimize allylic coupling.

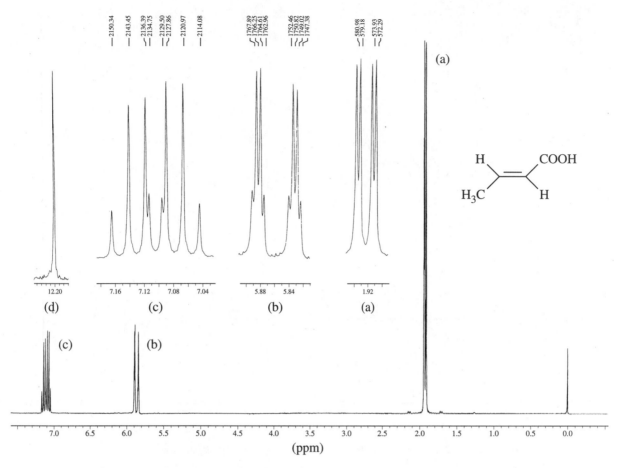

FIGURE 5.21 The NMR spectrum of crotonic acid.

Figure 5.21 is a spectrum of crotonic acid. See if you can assign the peaks and explain the couplings (draw a tree diagram) in this compound. The acid peak is not shown on the main trace, but is shown in the expansions at 12.2 ppm. Also remember that $^3J_{trans}$ is quite large in an alkene while the allylic couplings will be small. The multiplets may be described as a doublet of doublets (1.92 ppm), a doublet of quartets (5.86 ppm), and a doublet of quartets (7.10 ppm) with the peaks of the two quartets overlapping.

In many molecules, ^4J is small or vanishingly small for the allylic coupling of type $H_A-C-C=C-H_B$, and often *no spin–spin splitting* is observed. It is difficult to generalize, but this ($^4J \approx 0$) is particularly true for molecules in which the allylic carbon bearing H_A has free rotation. Rigid systems of the correct geometry (see Fig. 5.20) are the systems in which allylic coupling is most common.

5.8 MEASURING COUPLING CONSTANTS—ANALYSIS OF AN ALLYLIC SYSTEM

In this section we will work through the analysis of the 300-MHz FT-NMR spectrum of 4-allyloxyanisole. The complete spectrum is shown in Figure 5.22.

4–allyloxyanisole

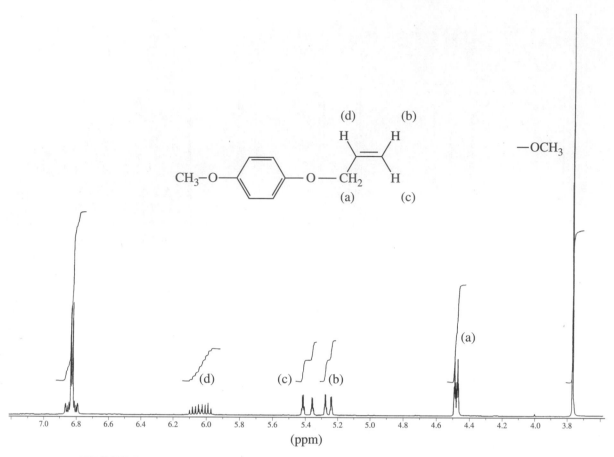

FIGURE 5.22   The 300-MHz FT-NMR spectrum of 4-allyloxyanisole.

The hydrogens of the allylic system are labeled a–d. Also shown are the methoxy group hydrogens (three-proton singlet at 3.78 ppm) and the *para*-disubstituted benzene ring resonances (multiplet at 6.84 ppm). The origin of the *para*-disubstitution pattern will be discussed later in Section 5.11B. The main concern here will be to explain the allylic splitting patterns and to extract the various coupling constants.

The exact assignment of the multiplets in the allylic group depends not only on their chemical shift values, but also on the splitting patterns observed. Some initial analysis must be done before any assignments can be definitely made.

Initial Analysis

The allylic OCH$_2$ group (4.48 ppm) is labeled **a** on the spectrum and is the easiest multiplet to identify since it shows an integral of 2H. It is also in the chemical shift range expected for a group of protons on a carbon atom that is attached to an oxygen atom. It has a larger chemical shift than the upfield methoxy group (3.77 ppm) because it is attached to the carbon–carbon double bond as well as the oxygen atom.

The hydrogen attached to the same carbon of the double bond as the OCH$_2$ group will be expected to have the broadest and most complicated pattern and is labeled **d** on the spectrum. This pattern should be spread out because the first splitting it will experience is a large splitting $^3J_{cd}$ from the *trans*-hydrogen **c**, followed by another large coupling $^3J_{bd}$ from the *cis*-hydrogen **b**. The adjacent OCH$_2$ group will yield additional (and smaller) splitting into triplets $^4J_{ad}$. Finally, this entire pattern integrates for only 1H.

Assigning the two terminal vinyl hydrogens relies on the difference in the magnitude of a *cis*- and a *trans*-coupling constant. Proton **c** will have a *wider* pattern than proton **b** because it will have a *trans* coupling $^3J_{cd}$ to proton **d**, while proton **b** will experience a smaller $^3J_{bd}$ *cis* coupling. Therefore, the multiplet with wider spacing is assigned to proton **c** and the narrower multiplet is assigned to proton **b**. Notice also that each of these multiplets integrates for 1H.

The preliminary assignments just given are tentative, and they must pass the test of a full tree analysis with coupling constants. This will require expansion of all the multiplets so that the exact value (in Hertz) of each subpeak can be measured. Within reasonable error limits, all coupling constants must agree in magnitude wherever they appear.

Tree-Based Analysis and Determination of Coupling Constants

The best way to start the analysis of a complicated system is to start with the **simplest** of the splitting patterns. In this case, we will start with the OCH_2 protons in multiplet **a**. The expansion of this multiplet is shown in Figure 5.23a. It appears to be a doublet of triplets (dt). However, examination of the molecular structure (see Figure 5.22) would lead us to believe that this multiplet should be a doublet of doublets of doublets (ddd), the OCH_2 group being split first by proton **d** ($^3J_{ad}$), then by proton **b** ($^4J_{ab}$) and then by proton **c** ($^4J_{ac}$), each of which is a single proton. A doublet of triplets could result only if protons **b** and **c** behave as a group of two, yielding a triplet ($n + 1$ Rule). This can happen only if they both have the same degree of coupling to the OCH_2 protons, that is, if $^4J_{ab}$ and $^4J_{ac}$ are *accidentally* equivalent ($^4J_{ab} = {}^4J_{ac}$). We can find out if this is the case by extracting the coupling constants and constructing a tree diagram. Figure 5.23b gives the positions of the peaks in the multiplet. By taking appropriate differences (see Section 5.5), we can extract two coupling constants with magnitudes of 1.47 Hz and 5.15 Hz. The larger value is in the correct range for a vicinal coupling ($^4J_{ad}$) and the smaller value must be identical for both the *cis* and *trans* allylic couplings ($^4J_{ab}$ and $^4J_{ac}$). This would lead to the tree diagram shown in Figure 5.23c. Notice that when the two smaller couplings are equivalent (or nearly equivalent) the central lines in the final doublet coincide, or overlap, and effectively give triplets instead of pairs of doublets. We will begin by assuming that this is correct. If we are in error, there will be a problem in trying to make the rest of the patterns consistent with these values.

Next consider proton **b**. The expansion of this multiplet (Figure 5.24a) shows it to be an apparent doublet of quartets. The largest coupling should be the *cis* coupling $^3J_{bd}$, which should yield a doublet. The geminal coupling $^2J_{bc}$ should produce another pair of doublets (dd) and the allylic geminal coupling $^4J_{ab}$ should produce triplets (two interacting H). The expected final pattern would be a doublet of doublet of triplets (ddt) with six peaks in each arm of the splitting pattern. Since only four peaks are observed, there must be overlap such as was discussed for H_a. Figure 5.24c shows that could happen if $^2J_{bc}$ and $^4J_{ab}$ are both small and have nearly the same magnitude. In fact, the two Js appear to be accidentally the same (or similar), and this is expected (see the typical geminal and allylic values on pp. 237 and 240). Figure 5.24b also shows that only two different J values can be extracted from the

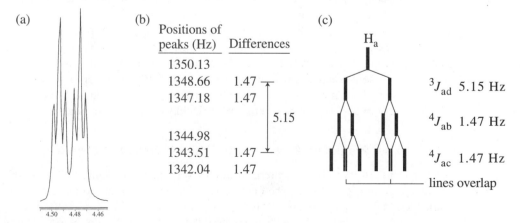

F I G U R E   5 . 2 3 Allyloxyanisole. (a) Expansion of proton H_a. (b) Peak positions (Hz) and selected differences. (c) A tree diagram showing the origin of the splitting pattern.

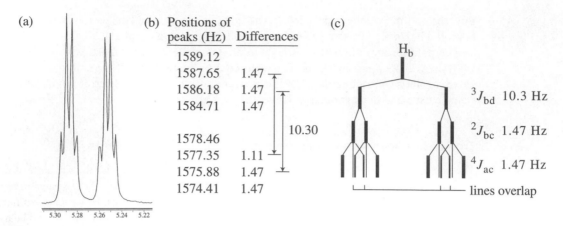

FIGURE 5.24 Allyloxyanisole. (a) Expansion of proton H_b. (b) Peak positions (Hz) and selected differences. (c) A tree diagram showing the origin of the splitting pattern.

peaks (1.47 and 10.30 ppm). Examine the tree diagram in Figure 5.24c to see the final solution, a doublet of doublet of triplets (ddt) pattern, which appears to be a doublet of quartets due to the overlaps.

Proton **c** is also expected to be a doublet of doublet of triplets (ddt), but shows a doublet of quartets for reasons similar to those explained for proton **b**. Examination of Figure 5.25 should explain how this occurs. Notice that the first coupling is larger than for proton **b**. The first coupling for proton **c** is $^3J_{cd}$-*trans*, which is larger than $^3J_{bd}$-*cis*, which applies to proton **b**.

At this point we have extracted all six of the coupling constants for the system:

$$^3J_{cd}\text{-}trans = 17.3 \text{ Hz}$$

$$^3J_{bd}\text{-}cis = 10.3 \text{ Hz}$$

$$^3J_{ad}\text{-}vic = 5.5 \text{ Hz}$$

$$^2J_{bc}\text{-}gem = 1.47 \text{ Hz}$$

$$^4J_{ab}\text{-}allyl = 1.47 \text{ Hz}$$

$$^4J_{ac}\text{-}allyl = 1.47 \text{ Hz}$$

Proton **d** has not been analyzed, but we will do this *by prediction* in the next paragraph. Note that three of the coupling constants (all of which are expected to be small ones) are equivalent, or nearly equivalent. This is either coincidence or could have to do with an inability of the NMR instrument to resolve them more clearly. These coincidences probably result from the fact that the computer did not have enough data points to distinguish the exact apex of each peak. Note that this has clearly happened in Figure 5.24b where one of the differences is 1.11 Hz instead of the expected 1.47 Hz. There may be some second-order interaction (Section 5.10), but it appears that all of the protons are sufficiently well separated form one another that this seems unlikely. More advanced techniques would be required to determine whether the coincidences are real or not.

Proton d—A Prediction Based on the *J* Values Already Determined

An expansion of the splitting pattern for proton **d** is shown in Figure 5.26a and the peak values in Hz are given in Figure 5.26b. The observed pattern will be predicted using the *J* values just determined as a way of checking our results. If we have extracted the constants correctly, we should be able to correctly predict the splitting pattern. This is done in Figure 5.26c, where the tree is constructed to scale using the *J* values already determined. The predicted pattern is a doublet of doublet of triplets (ddt), which should have six peaks on each arm of the multiplet. However, due to over-

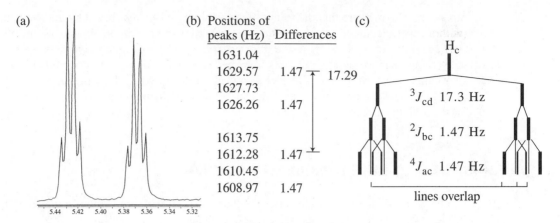

FIGURE 5.25 Allyloxyanisole. (a) Expansion of proton H_c. (b) Peak positions (Hz) and selected differences. (c) A tree diagram showing the origin of the splitting pattern.

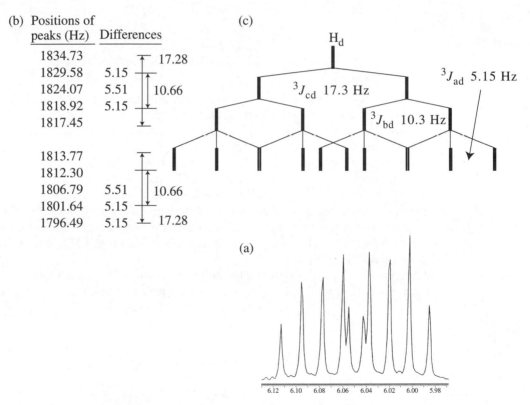

FIGURE 5.26 Allyloxyanisole. (a) Expansion of proton H_d. (b) Peak positions (Hz) and selected differences. (c) A tree diagram showing the origin of the splitting pattern.

laps, we see what appears to be two overlapping quintets. This agrees perfectly with the observed spectrum, thereby validating our analysis.

The Method

Notice that we started with the *simplest* pattern, determined its tree, and extracted the relevant coupling constants. Then we moved to the next most complicated pattern, doing essentially the same procedure, making sure that the values of any coupling constants shared by the two patterns agreed (within experimental error). If they do not agree, something is in error, and you must back up and start again. With the analysis of the third pattern, all of the coupling constants were obtained. Finally, rather than extracting constants from the last pattern, the pattern was predicted using the constants already determined. It is always a good idea to use prediction on the final pattern as a method of validation. If the predicted pattern matches the experimentally determined pattern, then it is almost certainly correct.

■ 5.9 IS THE $n + 1$ RULE EVER REALLY OBEYED?

In a linear chain, the $n + 1$ Rule is strictly obeyed only if the vicinal interproton coupling constants (^3J) are the same for every successive pair of carbons.

$$\text{(A)} \quad \text{(B)} \quad \text{(C)} \quad \text{(D)}$$

$$-CH_2-CH_2-CH_2-CH_2-$$

$$J_{AB} = J_{BC} = J_{CD} \text{, and so on}$$

Condition under which the $n + 1$ Rule is strictly obeyed

Consider a three-carbon chain as an example. The protons on carbons A and C split those on carbon B. If there is a total of four protons on carbons A and C, the $n + 1$ Rule predicts a quintet. This occurs only if $^3J_{AB} = {}^3J_{BC}$. Figure 5.27 represents the situation graphically.

First, the protons on carbon A split those on carbon B ($^3J_{AB}$), yielding a triplet (intensities 1:2:1). The protons on carbon C then split each component of the triplet ($^3J_{BC}$) into another triplet (1:2:1). At this stage, many of the lines from the second splittings *overlap* because they have the same spacings as the lines from the first splitting. Because of this coincidence, only five lines are observed. But we can easily confirm that they arise in the fashion indicated by adding the intensities of the splittings to predict the intensities of the final five-line pattern (see Fig. 5.27). These intensities agree with those predicted by the use of Pascal's triangle (Section 3.16). Thus, the $n + 1$ Rule depends on a special condition.

In many molecules J_{AB} is only slightly different from J_{BC}. This leads to peak broadening in the multiplet, since the lines do not quite overlap. (Broadening occurs because the peak separation in

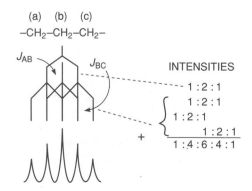

FIGURE 5.27 Construction of a quintet for a methylene group with four neighbors all coupled to the same extent.

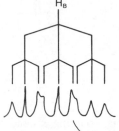

F I G U R E   5 . 2 8 Loss of the simple quintet when $^3J_{AB} \neq {}^3J_{BC}$.

Broadening due to partial overlap

Hertz is too small in magnitude for the NMR instrument to be able to distinguish the separate peak components.) Nonequivalence of $^3J_{AB}$ and $^3J_{BC}$, of course, happens most often when the chemical shifts of the protons on carbons A, B, and C are all quite different.

Sometimes the perturbation of the quintet is only slight, and then either a shoulder is seen on the side of a peak or a dip is obvious in the middle of a peak. At other times, when there is a large difference between $^3J_{AB}$ and $^3J_{BC}$, distinct peaks, more than five in number, can be seen. Deviations of this type are most common in a chain of the type X—$CH_2CH_2CH_2$—Y, where X and Y are widely different in character. Figure 5.28 illustrates the origin of some of these deviations.

Chains of any length can exhibit this phenomenon, whether or not they consist solely of methylene groups. For instance, the spectrum of the protons in the second methylene group of propylbenzene is simulated as follows.

The splitting pattern gives a crude sextet, but the second line has a shoulder on the left, and the fourth line shows an unresolved splitting. The other peaks are somewhat broadened.

5.10 SECOND-ORDER SPECTRA—STRONG COUPLING

A. First-Order and Second-Order Spectra

Spectra that can be interpreted by using the $n + 1$ Rule or a simple graphical analysis (splitting trees) are said to be **first-order spectra.** In certain cases, however, neither graphical analysis nor the $n + 1$ Rule suffices to explain the splitting patterns, intensities, and numbers of peaks observed. In these latter cases a mathematical analysis must be carried out, usually by computer, to explain the spectrum. Spectra which require such advanced analysis are said to be **second-order spectra.**

Second-order spectra are most commonly observed in situations where the difference in chemical shift between two groups of protons is similar in magnitude (in Hertz) to the coupling constant, J (also in Hertz), which links them. That is, second-order spectra are observed for couplings between nuclei that have *nearly equivalent chemical shifts* but are not exactly identical. In contrast, if two sets of nuclei are separated by a large chemical shift difference, they show first-order coupling.

Strong coupling, second-order spectra
($\Delta v/J$ small)

Weak coupling, first-order spectra
($\Delta v/J$ large)

Another way of expressing this generalization is by means of the ratio $\Delta v/J$, where Δv is the chemical shift difference and J is the coupling constant that links the two groups. Both values are expressed in Hertz, and their absolute values are used for the calculation. When $\Delta v/J$ is large (>10), the splitting pattern typically approximates first-order splitting. However, when the chemical shifts of the two groups of nuclei move closer together and $\Delta v/J$ approaches unity, we begin to see second-order changes in the splitting pattern. When $\Delta v/J$ is large and we see first-order splitting, the system is said to be **weakly coupled;** if $\Delta v/J$ is small and we see second-order coupling, the system is said to be **strongly coupled.**

B. Spin System Notation

NMR spectroscopists have developed a convenient shorthand notation to designate the type of spin system. Each chemically different type of proton is given a letter: A, B, C, and so forth. If a group has two or more protons of one type, they are distinguished by subscripts, as in A_2 or B_3. Protons of similar chemical shift values are assigned letters that are close to one another in the alphabet, such as A, B, and C. Protons of widely different chemical shift are assigned letters far apart in the alphabet: X, Y, Z versus A, B, C. The two-proton system $-CH_A-CH_X-$, where H_A and H_X are widely separated, and that exhibits first-order splitting, is called an AX system. A system in which the two protons have similar chemical shifts, and which exhibits second-order splitting, is called an AB system. When the two protons have identical chemical shifts, are magnetically equivalent, and give rise to a singlet, the system is designated A_2. Two protons that have the same chemical shift but are not magnetically equivalent are designated as AA′. If three protons are involved and they all have very different chemical shifts, a letter from the middle of the alphabet is used, usually M, as in AMX. In contrast, ABC would be used for the strongly coupled situation where all three protons have similar chemical shifts. We will use designations similar to these throughout this section.

C. The A_2, AB, and AX Spin Systems

Start by examining the system with two protons, H_A and H_B, on adjacent carbon atoms. Using the $n + 1$ Rule, we expect to see two doublets with components of equal intensity in the ${}^1H$ NMR spectrum. In actuality, we see two doublets of equal intensity in this situation only if the difference in chemical shift (Δv) between protons A and B is large compared to the magnitude of the coupling constant (${}^3J_{AB}$) that links them. Figure 5.29 illustrates this case.

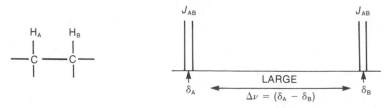

FIGURE 5.29 The case in which first-order coupling ($n + 1$ Rule) applies (Δv large).

Figure 5.30 shows how the splitting pattern for the two-proton system H_AH_B changes as the chemical shifts of H_A and H_B come closer together and the ratio $\Delta v/J$ becomes smaller. The figure is drawn to scale, with $^3J_{AB} = 7$ Hz. When $\delta H_A = \delta H_B$ (that is, when the protons H_A and H_B have the same chemical shift), then $\Delta v/J = 0/J = 0$ and no splitting is observed; both protons give rise to a single absorption peak. Between one extreme, where there is no splitting due to chemical shift equivalence ($\Delta v/J = 0$), and the other, the simple first-order spectrum ($\Delta v/J = 15$) that follows the $n + 1$ Rule, subtle and continuous changes in the splitting pattern take place. Most obvious is the decrease in intensity of the outer peaks of the doublets, with a corresponding increase in the intensity of the inner peaks. Other changes that are not as obvious also occur. Mathematical analysis by theoreticians has shown that although the chemical shifts of H_A and H_B in the simple first-order spectrum correspond to the center point of each doublet, a more complex situation holds in the second-order cases: the chemical shifts of H_A and H_B are closer to the inner peaks than to the outer peaks. The actual positions of δ_A and δ_B must be calculated. The difference in chemical shift must be determined from the line positions (in Hertz) of the individual peak components of the group, using the equation

$$(\delta_A - \delta_B) = \sqrt{(\delta_1 - \delta_4)(\delta_2 - \delta_3)}$$

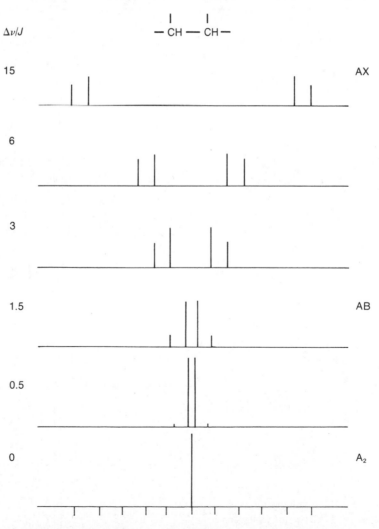

FIGURE 5.30 The splitting patterns of a two-proton system H_AH_B for various ratios of $\Delta v/J$.

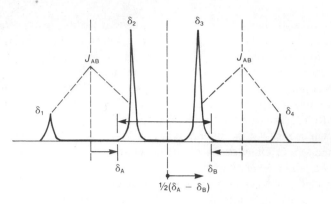

F I G U R E   5 . 3 1   The relationships among the chemical shifts, line positions, and coupling constant in a two-proton spectrum that exhibits second-order splitting effects.

where δ_1 is the position (in Hertz downfield from TMS) of the first line of the group and δ_2, δ_3, and δ_4 are the second, third, and fourth lines, respectively (Fig. 5.31). The chemical shifts of H_A and H_B are then displaced $\frac{1}{2}(\delta_A - \delta_B)$ to each side of the center of the group, as shown in Figure 5.31.

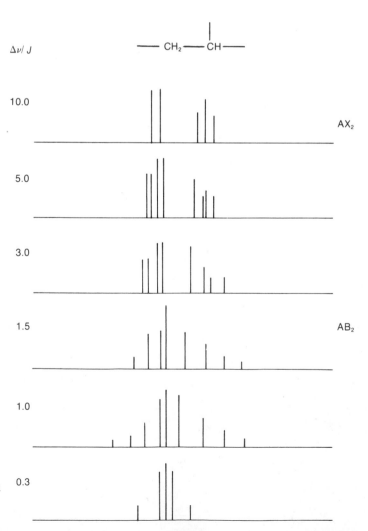

F I G U R E   5 . 3 2   The splitting patterns of a three-proton system $-CH-CH_2-$ for various ratios of $\Delta\nu/J$.

D.   The AB$_2$ ... AX$_2$ and A$_2$B$_2$ ... A$_2$X$_2$ Spin Systems

To provide some idea of the magnitude of second-order variations from simple behavior, Figures 5.32 and 5.33 illustrate the calculated ^{1}H spectra of two additional systems ($-CH-CH_2-$ and $-CH_2-CH_2-$). The first-order spectra appear at the top ($\Delta\nu/J > 10$), while increasing amounts of second-order complexity are encountered as we move toward the bottom ($\Delta\nu/J$ approaches zero).

The two systems shown in Figures 5.32 and 5.33 are, then, AB$_2$ ($\Delta\nu/J < 10$) and AX$_2$ ($\Delta\nu/J > 10$) in one case, and A$_2$B$_2$ ($\Delta\nu/J < 10$) and A$_2$X$_2$ ($\Delta\nu/J > 10$) in the other. We will leave discussion of these types of spin systems to more advanced texts, such as those in the reference list at the end of this chapter. Remember that situations occasionally arise that cannot possibly be interpreted with only a limited knowledge.

Figures 5.34 through 5.37 (pages 252–253) show actual 60-MHz ^{1}H NMR spectra of some molecules of the A$_2$B$_2$ type. It is convenient to examine these spectra and compare them with the expected patterns in Figure 5.33, which were calculated from theory by use of a computer.

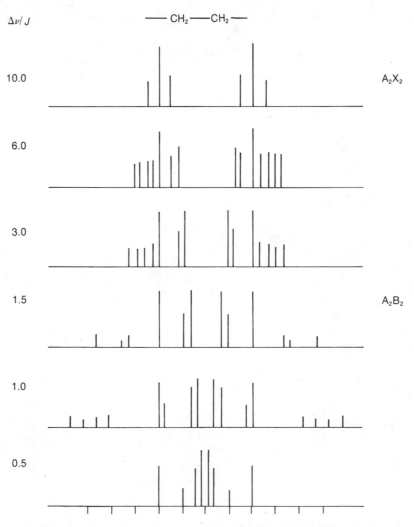

FIGURE 5.33 The splitting patterns of a four-proton system $-CH_2-CH_2-$ for various ratios of $\Delta\nu/J$.

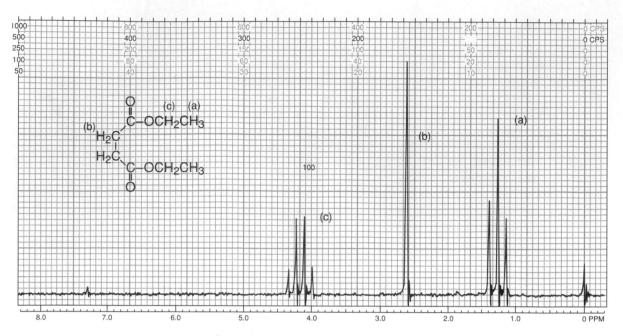

FIGURE 5.34   The ^{1}H NMR spectrum of diethyl succinate (60 MHz).

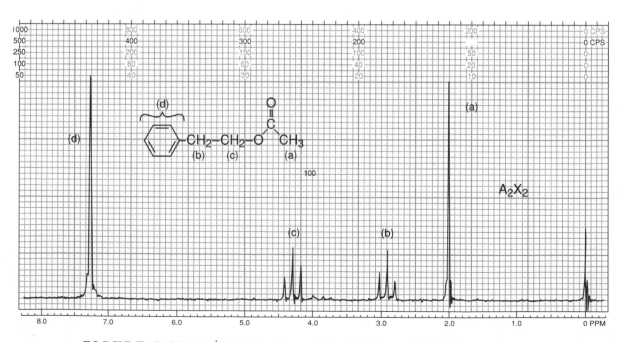

FIGURE 5.35   The ^{1}H NMR spectrum of phenylethyl acetate (60 MHz).

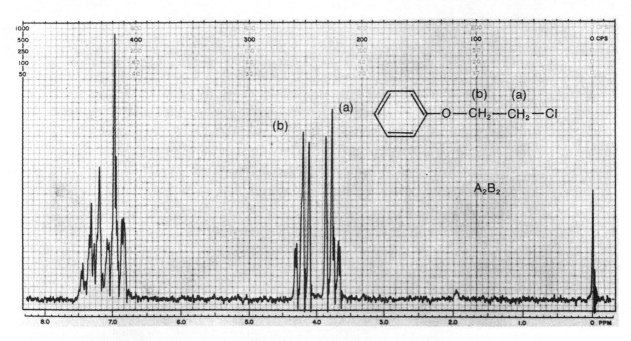

FIGURE 5.36 The ¹H NMR spectrum of β-chlorophenetole (60 MHz).

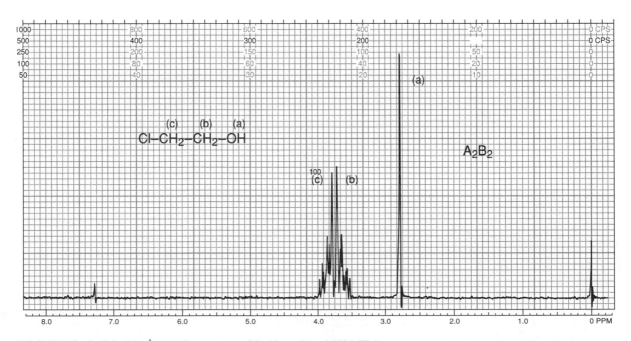

FIGURE 5.37 The ¹H NMR spectrum of 2-chloroethanol (60 MHz).

E.   Simulation of Spectra

We will not consider all the possible second-order spin systems in this text. Splitting patterns can often be more complicated than expected, especially when the chemical shifts of the interacting groups of protons are very similar. In many cases, only a trained NMR spectroscopist using a computer can interpret spectra of this type. Today there are many computer programs, for both PC and UNIX work stations, that can simulate the appearances of NMR spectra (at any frequency) if the operator provides a chemical shift and a coupling constant for each of the peaks in the interacting spin system. In addition, there are programs that will attempt to match a calculated spectrum to an actual NMR spectrum. In these programs, the user initially guesses at the parameters (chemical shifts and coupling constants), and the program varies these parameters until it finds the best fit. Some of these programs are included in the reference list at the end of this chapter.

F.   The Absence of Second-Order Effects at Higher Field

Many of the second-order effects we have discussed are absent from spectra determined on a high-field instrument, where proton frequencies are 300 MHz or greater. In Sections 3.17 and 3.18 you saw that the chemical shift increases when a spectrum is determined at higher field but that the coupling constants do not change in magnitude (see Fig. 3.35). In other words, Δv (the chemical shift difference in Hertz) increases, but J (the coupling constant) does not. This causes the $\Delta v/J$ ratio to increase, and second-order effects begin to disappear. At high field, many spectra are first order and are therefore easier to interpret than spectra determined at lower field strengths.

As an example, Figure 5.37 is the 60-MHz ^1H spectrum of 2-chloroethanol. This is an A_2B_2 spectrum showing substantial second-order interaction. In Figure 5.38, which shows the ^1H spectrum taken at 300 MHz, the formerly complicated and second-order patterns have almost reverted to

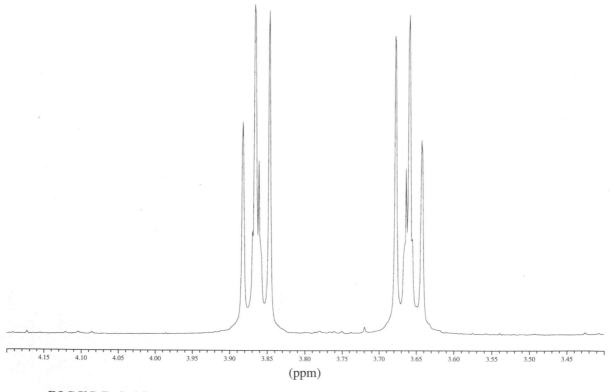

F I G U R E   5 . 3 8   A nearly first-order A_2X_2 pattern in the proton spectrum of 2-chloroethanol at 300 MHz.

two triplets just as the $n + 1$ Rule would predict. At 500 MHz (not shown) a complete transition to the predicted A_2X_2 pattern would be expected.

G. Deceptively Simple Spectra

It is not always obvious when a spectrum has become completely first order. Consider the $A_2B_2 \ldots A_2X_2$ splitting patterns shown in Figure 5.33. At which value of $\Delta v/J$ does this spectrum become truly first order? Someplace between $\Delta v/J = 6$ and $\Delta v/J = 10$ the spectrum seems to become A_2X_2. The number of observed lines decreases from 14 lines (peaks) to only 6 lines. However, if spectra are simulated, changing $\Delta v/J$ slowly from 6 to 10, we find that the change is not abrupt but gradual. Some of the lines disappear by decreasing in intensity, and some merge together, increasing their intensities. It is possible for weak lines to be lost in the noise of the baseline or for merging lines to approach so closely that the spectrometer cannot resolve them any longer, in which case the spectrum would appear to be first order, but in fact it would not be.

Also notice in Figure 5.30 that the AB spectra with $\Delta v/J$ equal to 3, 6, and 15 all appear roughly first order, but the doublets observed in the range $\Delta v/J = 3$ to 6 have chemical shifts that do not correspond to the center of the doublet (see Fig. 5.31). Unless the worker recognizes the possibility of second-order effects and does a *mathematical* extraction of the chemical shifts, the chemical shift values will be in error. Spectra that appear to be first order, but actually are not, are called **deceptively simple spectra.** The pattern appears to the casual observer to be first order and capable of being explained by the $n + 1$ Rule. However, there may be second-order lines that are either too weak or too closely spaced to observe, and there may be other subtle changes.

Is it important to determine whether a system is deceptively simple? In many cases the system is so close to first order that it doesn't matter. However, there is always the possibility that if we assume the spectrum is first order and measure the chemical shifts and coupling constants, we will get incorrect values. Only a complete mathematical analysis tells the truth. For the organic chemist trying to identify an unknown compound, it rarely matters whether the system is deceptively simple. However, if you are trying to use the chemical shift values or coupling constants to prove a troublesome structural point, take the time to be more careful. Unless they are obvious cases, we will treat deceptively simple spectra as though they follow the $n + 1$ Rule or as though they can be analyzed by simple tree diagrams. In doing your own work, always realize that there is a considerable margin for error.

H. Some Other Patterns

In two earlier sections, we introduced the spectra of some other commonly encountered types of systems. The NMR spectra of both styrene oxide (Fig. 5.10) and vinyl acetate (Fig. 5.17) contain examples of AMX systems, where M, being in the middle of the alphabet, indicates a chemical shift intermediate between A and X. That is, all three chemical shifts (A, M, and X) are separated widely. *trans*-Cinnamic acid (Fig. 5.16) is an example of an AX system.

▪ 5.11 AROMATIC COMPOUNDS—SUBSTITUTED BENZENE RINGS

Phenyl rings are so common in organic compounds that it is important to know a few facts about NMR absorptions in compounds that contain them. In general, the ring protons of a benzenoid system have resonance near 7.3 ppm; however, electron-withdrawing ring substituents (e.g., nitro, cyano, carboxyl, or carbonyl) move the resonance of these protons downfield, and electron-donating ring substituents (e.g., methoxy or amino) move the resonance of these protons upfield. Table 5.7 shows these trends for a series of symmetrically *p*-disubstituted benzene compounds. The *p*-disubstituted compounds were chosen because their two planes of symmetry render all of the hydrogens equivalent. Each compound gives only one aromatic peak (a singlet) in the proton

TABLE 5.7
^{1}H CHEMICAL SHIFTS IN *p*-DISUBSTITUTED BENZENE COMPOUNDS

Substituent X	δ (ppm)	
—OCH$_3$	6.80	Electron donating
—OH	6.60	
—NH$_2$	6.36	
—CH$_3$	7.05	
—H	7.32	
—COOH	8.20	Electron withdrawing
—NO$_2$	8.48	

NMR spectrum. Later you will see that some positions are affected more strongly than others in systems with substitution patterns different from this one. Table 6.3 in Appendix 6 enables us to make rough estimates of some of these chemical shifts.

In the sections that follow, we will attempt to cover some of the most important types of benzene ring substitution. In many cases it will be necessary to examine sample spectra taken at both 60 MHz and 300 MHz. Many benzenoid rings show second-order splittings at 60 MHz but are essentially first order at 300 MHz.

A.   Monosubstituted Rings

Alkylbenzenes

In monosubstituted benzenes in which the substituent is neither a strongly electron-withdrawing nor a strongly electron-donating group, all the ring protons give rise to what appears to be a *single resonance* when the spectrum is determined at 60 MHz. This is a particularly common occurrence in alkyl-substituted benzenes. Although the protons *ortho, meta,* and *para* to the substituent are not chemically equivalent, they generally give rise to a single unresolved absorption peak. A possible explanation is that the chemical shift differences, which should be small in any event, are somehow eliminated by the presence of the ring current, which tends to equalize them. All of the protons are nearly equivalent under these conditions. The NMR spectra of the aromatic portions of alkylbenzene compounds are good examples of this type of circumstance. Figure 5.39a is the 60-MHz ^{1}H spectrum of ethylbenzene.

The 300-MHz spectrum of ethylbenzene, shown in Figure 5.39b, presents quite a different picture. With the increased frequency shifts at 300 MHz (see Fig. 3.37), the nearly equivalent (at 60 MHz) protons are neatly separated into two groups. The *ortho* and *para* protons appear upfield from the *meta* protons. The splitting pattern is clearly second order.

Electron-Donating Groups

When electron-donating groups are attached to the ring, the ring protons are not equivalent, even at 60 MHz. A highly activating substituent such as methoxy clearly increases the electron density at the *ortho* and *para* positions of the ring (by resonance) and helps to give these protons greater shielding than those in the *meta* positions and, thus, a substantially different chemical shift.

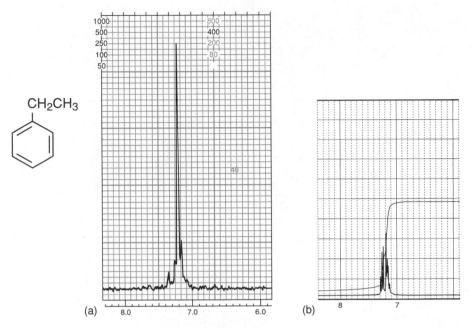

FIGURE 5.39 The aromatic-ring portions of the ^{1}H NMR spectra of ethylbenzene at (a) 60 MHz and (b) 300 MHz.

At 60 MHz, this chemical shift difference results in a complicated second-order splitting pattern for anisole (methoxybenzene), but the protons do fall clearly into two groups, the *ortho/para* protons and the *meta* protons. The 60-MHz NMR spectrum of the aromatic portion of anisole (Fig. 5.40) has a complex multiplet for the *o,p* protons (integrating for three protons) that is upfield from the *meta* protons (integrating for two protons), with a clear distinction (gap) between the two types. Aniline (aminobenzene) provides a similar spectrum, also with a 3:2 split, owing to the electron-releasing effect of the amino group.

The 300-MHz spectrum of anisole shows the same separation between the *ortho/para* hydrogens (upfield) and the *meta* hydrogens (downfield). However, because the actual shift, Δv (in Hertz), between the two types of hydrogens is greater, there is less second-order interaction, and the lines in the pattern are sharper at 300 MHz. In fact, it might be tempting to try to interpret the observed pattern as if it were first order, but remember that the protons on opposite sides of the ring are not magnetically equivalent even though there is a plane of symmetry (see Section 5.3). Anisole is an AA'BB'C spin system.

Anisotropy—Electron-Withdrawing Groups

A carbonyl or a nitro group would be expected to show (aside from anisotropy effects) a reverse effect, since these groups are electron withdrawing. One would expect that the group would act to

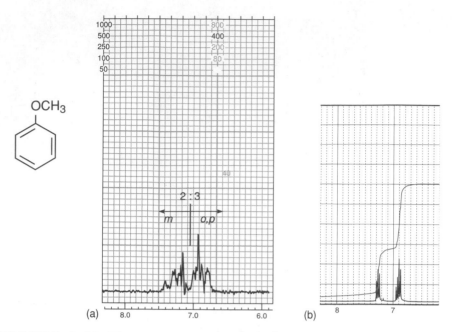

FIGURE 5.40   The aromatic-ring portions of the ^{1}H NMR spectra of anisole at (a) 60 MHz and (b) 300 MHz.

decrease the electron density around the *ortho* and *para* positions, thus deshielding the *ortho* and *para* hydrogens and providing a pattern exactly the reverse of the one shown for anisole (3:2 ratio, downfield:upfield). Convince yourself of this by drawing resonance structures. Nevertheless, the actual NMR spectra of nitrobenzene and benzaldehyde do not have the appearances that would be predicted on the basis of resonance structures. Instead, the *ortho* protons are much more deshielded than the *meta* and *para* protons, due to the magnetic anisotropy of the π bonds in these groups.

Anisotropy is observed when a substituent group bonds a carbonyl group directly to the benzene ring (Fig. 5.41). Once again, the ring protons fall into two groups, with the *ortho* protons downfield from the *meta/para* protons. Benzaldehyde (Fig. 5.42) and acetophenone both show this effect in their NMR spectra. A similar effect is sometimes observed when a carbon–carbon double bond is attached to the ring. The 300-MHz spectrum of benzaldehyde (Fig. 5.42b) is a nearly first-order spectrum (probably a deceptively simple AA′BB′C spectrum) and shows a doublet (H_C, 2 H), a triplet (H_B, 1 H), and a triplet (H_A, 2 H). It can be analyzed by the $n + 1$ Rule, assuming that protons farther than three bonds away do not interact.

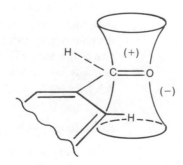

FIGURE 5.41   Anisotropic deshielding of the *ortho* protons of benzaldehyde.

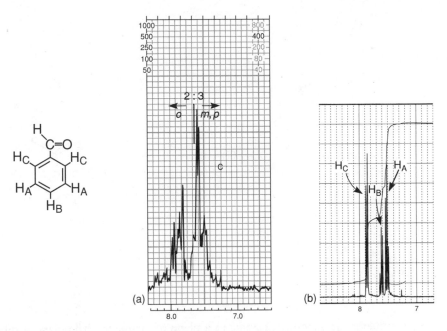

FIGURE 5.42 The aromatic-ring portions of the ^{1}H NMR spectra of benzaldehyde at (a) 60 MHz and (b) 300 MHz.

B. *para*-Disubstituted Rings

Of the possible substitution patterns of a benzene ring, only a few are easily recognized. One of these is the *para*-disubstituted benzene ring. Examine anethole (Fig. 5.43a) as a first example.

Because this compound has a plane of symmetry (passing through the methoxy and propenyl groups), the protons H_a and $H_{a'}$ (both *ortho* to OCH_3) would be expected to have the same chemical shift. Similarly, the protons H_b and $H_{b'}$ should have the same chemical shift. This is found to be the case. You might think that both sides of the ring should then have identical splitting patterns. With this assumption, one is tempted to look at each side of the ring separately, expecting a pattern in which proton H_b splits proton H_a into a doublet, and proton H_a splits proton H_b into a second doublet. Examination of the NMR spectrum of anethole (Fig. 5.44a) shows (crudely) just such a four-line pattern for the ring protons. In fact, a *para*-disubstituted ring is easily recognized by this four-line pattern. However, the four lines do not correspond to a simple first-order splitting pattern. That is because the two protons H_a and $H_{a'}$, although chemically equivalent (have the same chemical shift), are not magnetically equivalent. Protons H_a and $H_{a'}$ interact with each other and have finite coupling constant $J_{aa'}$. Similarly, H_b and $H_{b'}$ interact with each other and have coupling constant $J_{bb'}$. More important, H_a does not interact equally with H_b (*ortho* to H_a) and with $H_{b'}$ (*para* to H_a); that is, $J_{HaHb} \neq J_{HaHb'}$. If H_b and $H_{b'}$ are coupled differently to H_a, they cannot be magnetically equivalent. Turning the argument around, neither can H_a and $H_{a'}$ be magnetically equivalent, because they are coupled differently to H_b and also to $H_{b'}$. This fact suggests that the situation is more complicated than it might at first appear. A closer look at the pattern in Figure 5.44a shows that this is indeed the case. With an expansion of the parts-per-million scale, this pattern actually resembles four triplets, as shown in Figure 5.45. The pattern is an AA′BB′ spectrum.

We will leave the analysis of this second-order pattern to more advanced texts. Note, however, that a crude four-line spectrum is characteristic of a *para*-disubstituted ring. It is also characteristic of an *ortho*-disubstituted ring of the type shown in Figure 5.43b, where the two *ortho*

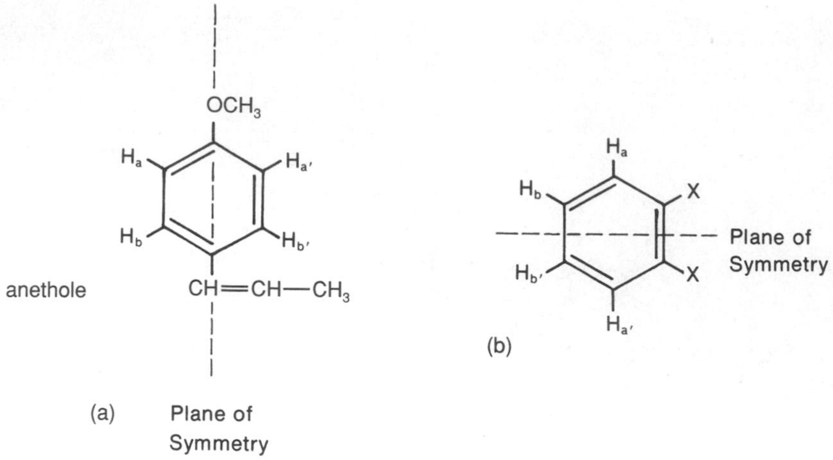

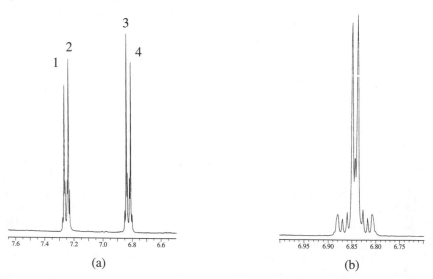

FIGURE 5.43   The planes of symmetry present in (a) a *para*-disubstituted benzene ring (anethole) and (b) a symmetric *ortho*-disubstituted benzene ring.

substituents are identical, leading to a plane of symmetry. The *ortho*-disubstituted case is relatively uncommon.

As the chemical shifts of H_a and H_b approach each other, the *p*-disubstituted pattern becomes similar to that of 4-allyloxyanisole (Fig. 5.44b). The inner peaks move closer together, and the outer ones become smaller or even disappear. Ultimately, when H_a and H_b approach each other closely enough in chemical shift, the outer peaks disappear and the two inner peaks merge into a *singlet*; *p*-xylene, for instance, gives a singlet at 7.05 ppm (Table 5.5). Hence, a single aromatic resonance integrating for four protons could easily represent a *para*-disubstituted ring, but the substituents would obviously be either identical or very similar. Note also that a monosubstituted ring with alkyl groups frequently gives an apparent singlet at 60 MHz, but this singlet integrates for five protons instead of four.

FIGURE 5.44   The aromatic-ring portions of the 300 MHz ^{1}H-NMR spectra of (a) anethole and (b) 4-allyloxyanisole.

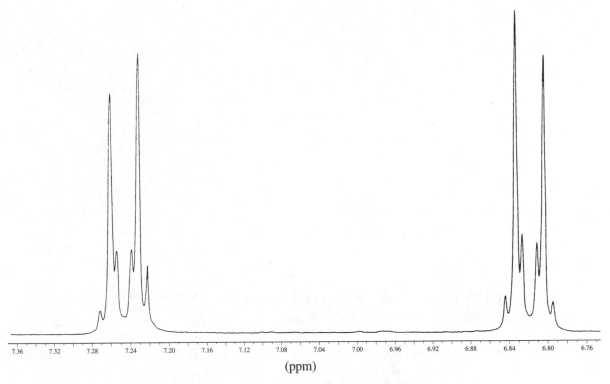

FIGURE 5.45   The expanded *para*-disubstituted benzene AA'BB' pattern.

C.   Other Substitution

Other modes of ring substitution can often lead to splitting patterns more complicated than those of the aforementioned cases. In aromatic rings, coupling usually extends beyond the adjacent carbon atoms. In fact, *ortho, meta,* and *para* protons can all interact, although the last interaction (*para*) is not usually observed. Following are typical *J* values for these interactions.

ortho	*meta*	*para*
$^3J = 7 - 10$ Hz	$^4J = 1 - 3$ Hz	$^5J = 0 - 1$ Hz

The trisubstituted compound 2,4-dinitroanisole shows all of the types of interactions mentioned. Figure 5.46 shows the aromatic-ring portion of the ^{1}H NMR spectrum of 2,4-dinitroanisole, and Figure 5.47 is its analysis. In this example, as is typical, the coupling between the *para* protons is essentially zero. Also notice the effects of the nitro groups on the chemical shifts of the adjacent protons. Proton H_D, which lies between two nitro groups, has the largest chemical shift (8.72 ppm). Proton H_C, which is affected only by the anisotropy of a single nitro group, is not shifted as far downfield.

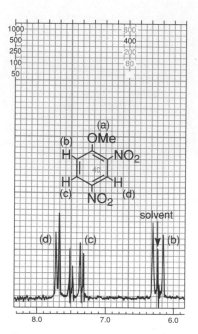

FIGURE 5.46 The 60-MHz ^{1}H NMR spectrum of the aromatic-ring portion of 2,4-dinitroanisole.

Figure 5.48 gives the 300-MHz ^{1}H spectra of the aromatic-ring portions of 2-, 3-, and 4-nitroaniline (the *ortho*, *meta*, and *para* isomers). The characteristic pattern of a *para*-disubstituted ring makes it easy to recognize 4-nitroaniline. Here the protons on opposite sides of the ring are not magnetically equivalent and the observed splitting is a second order. In contrast, the splitting patterns for 2- and 3-nitroaniline are simpler and at 300 MHz a first-order analysis will suffice to explain the spectra. As an exercise, see if you can analyze these patterns, assigning the multiplets to specific protons on the ring. Use the indicated multiplicities (s, d, t) and expected chemical shifts to help your assignments. You may ignore any *meta* or *para* interactions, remembering that ^{4}J and ^{5}J couplings will be too small in magnitude to be observed on the scale that these figures are presented.

In Figures 5.49 and 5.50 the expanded ring-proton spectra of 2-nitrophenol and 3-nitrobenzoic acid are presented. The resonances of the protons in the functional groups are not shown. In these spectra the position of each subpeak is given in Hertz. For these spectra it should be possible not only to assign peaks to specific hydrogens, but also to derive tree diagrams with discrete coupling constants for each interaction (see Problem 1 at the end of this chapter).

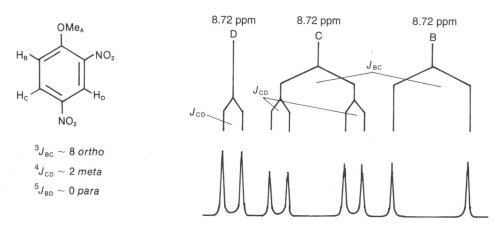

FIGURE 5.47 An analysis of the splitting pattern in the ^{1}H NMR spectrum of 2,4-dinitroanisole.

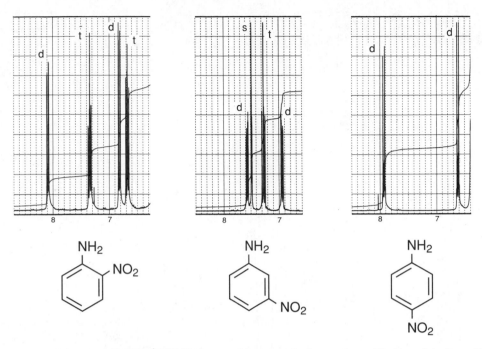

FIGURE 5.48 The 300-MHz ^{1}H NMR spectra of the aromatic-ring portions of 2-, 3-, and 4-nitroaniline.

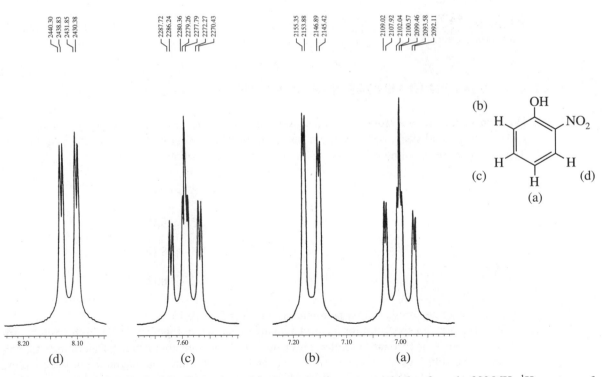

FIGURE 5.49 Expansions of the aromatic-ring-proton multiplets from the 300 MHz ^{1}H spectrum of 2-nitrophenol. The accompanying hydroxyl absorption (OH) is not shown.

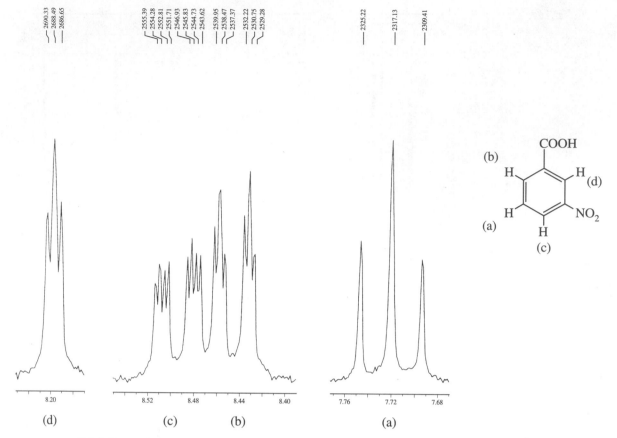

FIGURE 5.50   Expansions of the aromatic-ring-proton multiplets from the 300 MHz ^{1}H spectrum of 3-nitrobenzoic acid. The accompanying carboxylic acid group resonance (COOH) is not shown.

5.12 COUPLING IN HETEROAROMATIC SYSTEMS

Heteroaromatic systems (furans, pyrroles, pyridines, thiophenes, etc.) show couplings analogous to those in benzenoid systems. In furan, for instance, couplings occur between all of the ring protons. Typical values of coupling constants for furanoid rings are as follows.

$$^3J_{\alpha\beta} = 1.6 - 2.0 \text{ Hz}$$

$$^4J_{\alpha\beta'} = 0.3 - 0.8 \text{ Hz}$$

$$^4J_{\alpha\alpha'} = 1.3 - 1.8 \text{ Hz}$$

$$^3J_{\beta\beta'} = 3.2 - 3.8 \text{ Hz}$$

Values for other heterocycles may be found in Appendix 5, p. A-15.

The structure and spectrum for furfuryl alcohol are shown in Figure 5.51. Only the ring hydrogens are shown—the resonances of the side chain ($-$CH$_2$OH) are not included. Determine a tree diagram for the splittings shown in this molecule, and also determine the magnitude of the coupling constants (see Problem 1 at the end of this chapter). Notice that proton H$_a$ not only shows coupling to the other two ring hydrogens (H$_b$ and H$_c$), but appears to have small unresolved *cis*-allylic interaction with the methylene (CH$_2$) group as well.

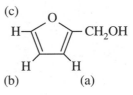

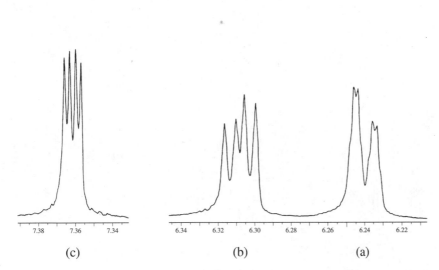

FIGURE 5.51 Expansions of the ring-proton resonances from the 300-MHz ^{1}H spectrum of furfuryl alcohol. The resonances of the side chain (-CH$_2$OH) are not shown.

Figure 5.52 shows the ring-proton resonances of 2-picoline (2-methylpyridine)—the part of the spectrum originating from the methyl group is not included. Determine a tree diagram that explains the observed splittings and extract the values of the coupling constants (see Problem 1 at the end of this chapter). Typical values of coupling constants for a pyridine ring will be found in Appendix 5, p. A-15. Do the values you have determined agree with those in the table? Notice that the peaks originating from proton H$_d$ are quite broad, suggesting that some long-range splitting interactions may not be resolved. There may be some coupling of this hydrogen to the adjacent nitrogen (I = 1) or there may be a quadrupole-broadening effect operating (Section 6.5).

▪ 5.13 LONG-RANGE COUPLING

Proton–proton coupling is normally observed between protons on *adjacent* atoms (if they are not equivalent) and is only sometimes observed between protons on the *same* atom (if they are magnetically nonequivalent). These are called **vicinal** (^3J) and **geminal** (^2J) **couplings,** respectively. Only under special circumstances does coupling occur between protons that lie farther apart than in these two types of coupling. In addition, these more distant couplings, which are made possible through overlapping orbitals, usually have a stereochemical requirement. Couplings that occur over greater distances than those of the geminal (^2J) or vicinal (^3J) type are conveniently collected under the heading **long-range coupling.** In fact, in earlier sections we discussed two examples of long-range coupling: allylic coupling of the type H—C=C—CH (^4J, Section 5.6) and coupling of *meta* and *para* hydrogens in an aromatic ring (^4J and ^5J, Section 5.11C).

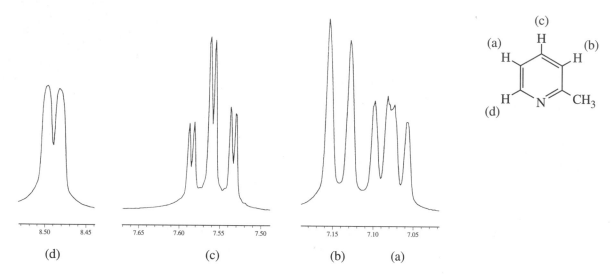

FIGURE 5.52 Expansions of the ring-proton resonances from the 300-MHz ^{1}H spectrum of 2-picoline (2-methylpyridine). The methyl group is not shown.

In allylic coupling in alkenes (see Section 5.7) we saw that the magnitude of ^4J depends upon the extent of overlap of the carbon–hydrogen σ bond with the π bond. A similar type of interaction occurs in acetylenes. In addition, in some alkenes coupling can occur between the C—H σ bonds on either side of the double bond. This type of coupling is generally very small or even nonexistent in most molecules, but it sometimes appears in NMR spectra. It is called **homoallylic coupling,** and it occurs over five bonds (^5J). Homoallylic coupling is naturally weaker than allylic coupling because it occurs over a greater distance. Figure 5.53 illustrates allylic coupling in acetylenes (^4J), while Figure 5.54 shows homoallylic coupling (^5J).

The long-range coupling that occurs in aromatic rings must also transmit the spin information through the π system of the ring. Figure 5.55 illustrates this requirement for the ^4J coupling between *meta* protons.

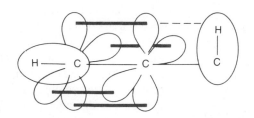

FIGURE 5.53 Allylic coupling in acetylenes (^4J).

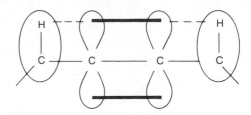

FIGURE 5.54 Homoallylic coupling (^5J).

Similar types of coupling exist in all of the aromatic heterocycles (furans, pyrroles, thiophenes, pyridines, and so on). In furan (see p. 264), for instance, couplings occur between all of the ring protons. In these compounds, the spin information is also transmitted through the π system of the ring. The tables of coupling constants in Appendix 5 list the magnitudes of the various types of proton–proton couplings for all the major heterocyclic systems. Note that couplings of the allylic (^4J) and homoallylic (^5J) types seldom have magnitudes of more than 2 to 3 Hz. However, couplings of the $^3J_{\alpha\beta}$ or $^3J_{\beta\beta'}$ type, either in furan or in other heterocycles, are vicinal couplings and may have values up to 8 or 9 Hz. In practice, the values are generally lower than this maximum.

Long-range couplings in compounds without π systems are less common but do occur in special cases. One case of long-range coupling in saturated systems occurs through an arrangement of atoms in the form of a W (^4J), with hydrogens occupying the end positions. Two possible types of orbital overlap have been suggested to explain this type of coupling, but we do not currently know which mechanism for transferring the spin information is correct. Figure 5.56 illustrates the two possible mechanisms.

The magnitude of ^4J for W coupling is usually small except in highly strained ring systems. Apparently these rigid strained arrangements somehow have a favorable geometry for the overlaps involved.

$J_{AB} \sim 1$ Hz $J_{AB} \sim 3$ Hz $J_{AB} \sim 7$ Hz

Increasing strain

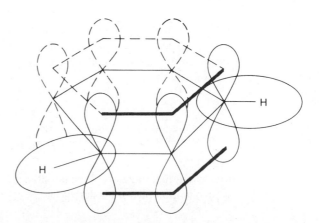

FIGURE 5.55 Coupling between the *meta* protons in an aromatic ring (^4J).

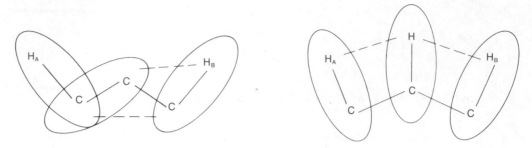

FIGURE 5.56 Two possible overlap mechanisms explaining the transfer of spin information between H_A and H_B when a molecule is arranged in a (^4J) W pattern.

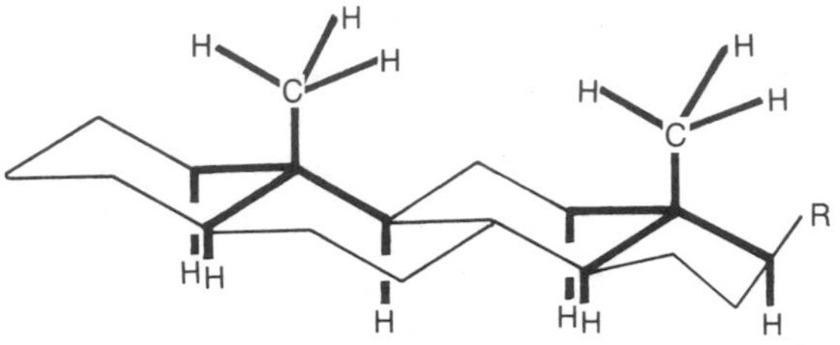

FIGURE 5.57 A steroid ring skeleton showing several possible couplings of the W type (^4J).

In other systems, the magnitude of ^4J is often less than 1 Hz. The common NMR instrument cannot distinguish between such closely spaced peak components—it may not resolve them because the electronics are not sufficiently sensitive. Peaks that have spacings less than the resolving capabilities of the spectrometer are usually broadened. That is, two peaks very close to each other are seen as a single "fat," or broad, peak. Their spacing is said to be beyond the limits of resolution of the spectrometer. Most W couplings are of this type; they give rise to peak broadening, rather than a discernible splitting, for the protons that are coupled. Small allylic couplings $(^4J < 1$ Hz) also give risk to peak broadening.

Angular methyl groups in steroids, or those at the ring junctions in a decalin system, often exhibit peak broadening due to W coupling with several hydrogens of the ring. Figure 5.57 illustrates several such interactions. Because these systems are relatively unstrained, ^4J_W is usually quite small. Hence, one sees a broadening of the methyl resonance peak rather than any splitting.

5.14 HOMOTOPIC, ENANTIOTOPIC, AND DIASTEREOTOPIC SYSTEMS

A situation that arises frequently in the analysis of NMR spectra is the need to determine whether two groups attached to the same carbon (geminal groups) are equivalent or nonequivalent. Methylene groups (geminal protons) and isopropyl groups (geminal methyl groups) are frequently the subjects of interest.

Methylene group: $\quad$ Geminal dimethyl group:

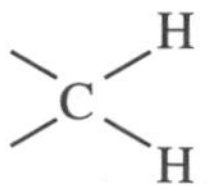

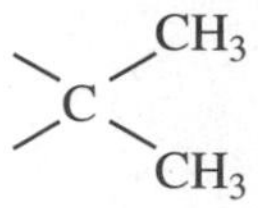

It turns out that there are three possible relationships for such geminal groups: They can be homotopic, enantiotopic, or diastereotopic.

Homotopic groups are always equivalent, and in the absence of couplings from another group of nuclei, they are isochronous and give a single NMR absorption. The simplest way to recognize homotopic groups is by means of a substitution test. In this test, first one member of the group is substituted for a different group, then the other is substituted in the same fashion. The results of the substitution are examined to see the relationship of the resulting new structures. If the new structures are *identical*, the two original groups are homotopic. Figure 5.58a shows the substitution procedure for a molecule with two homotopic methylene hydrogens. In this molecule, the structures resulting from the replacement of first H_A and then H_B are identical. Notice that for this homotopic molecule, the substituents X are the same. The starting compound is completely symmetric because it has both a plane and a twofold axis of symmetry.

Enantiotopic groups only appear to be equivalent, but they are typically isochronous and give a single NMR absorption except when they are placed in a chiral environment or acted upon by a chiral reagent. Enantiotopic groups can also be recognized by the substitution test. Figure 5.58b shows the substitution procedure for a molecule with two enantiotopic methylene hydrogens. In this molecule, the resultant structures from the replacement of first H_A and then H_B are *enantiomers*. Replacement of the two hydrogens gives different results. Although these two hydrogens appear to be equivalent and are isochronous in a typical NMR spectrum, they are not equivalent on replacement, each hydrogen giving a different stereoisomer. Notice that the structure of this enantiotopic molecule is not chiral but that substituents X and Y are different groups. There is a plane of symmetry, but the twofold axis is lost.

Enantiotopic groups are sometimes called **prochiral** groups. When one or the other of these groups is replaced by a different one, a *chiral* molecule results. The original molecule is not chiral, but it can produce a chiral molecule. The reaction of prochiral molecules with a chiral reagent, such

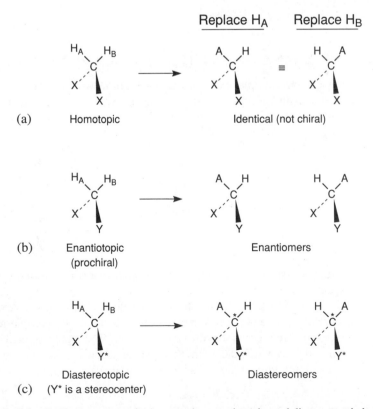

FIGURE 5.58 Replacement tests for homotopic, enantiotopic, and diastereotopic hydrogens.

as an enzyme in a biological system, produces a chiral result. If these molecules are placed in a chiral environment, the two groups are no longer equivalent. We will examine two chiral environments, chiral solvents and chiral shift reagents in Chapter 6 (Section 6.9).

Diastereotopic groups are not equivalent and are not isochronous; they have different chemical shifts in the NMR spectrum. When the diastereotopic groups are hydrogens, they frequently split each other with a geminal coupling constant ^2J. Figure 5.58c shows the substitution procedure for a molecule with two diastereotopic hydrogens. In this molecule, the replacement of first H_A and then H_B yields a pair of *diastereomers*. Diastereomers are produced when substituent Y* already has an adjacent stereocenter. Section 5.15 will cover this situation in detail.

5.15 SPECTRA OF DIASTEREOTOPIC SYSTEMS

In this section we examine some molecules that have diastereotopic groups (discussed in Section 5.14). Diastereotopic groups are not equivalent, and two different NMR signals are observed. The most common instance of diastereotopic groups is when two similar groups, G and G', are substituents on a carbon *adjacent to a stereocenter*. If first group G and then group G' are replaced by another group, a pair of diastereomers forms (see Fig. 5.58c).[2]

A. Diastereotopic Methyl Groups: (*S*)-(+)-3-Methyl-2-Butanol

As a first chemical example, examine the spectra of (*S*)-(+)-3-methyl-2-butanol. This molecule has diastereotopic methyl groups on carbon 3. Figures 5.59 and 5.60 show the ^{13}C and ^1H proton NMR spectra, respectively.

First examine the ^{13}C spectrum in Figure 5.59. If this compound were not diastereotopic, we would expect only two different peaks for the methyl carbons, as there are only two different types of methyl groups. However, the spectrum shows three methyl peaks. A very closely spaced pair of peaks occurs at 17.90 and 18.16 ppm, representing the diastereotopic methyl groups, and a third peak at 19.99 ppm represents the third methyl group. There are two peaks for the geminal dimethyl groups! Carbon 3, to which the methyl groups are attached, is seen at about 35 ppm, and carbon 2, which has the deshielding hydroxyl attached, appears near 72 ppm.

---

[2]Note that groups farther down the chain may also be diastereotopic, but the effect may be small or unobserved due to the increased distance. Also, it is not essential that the stereocenter be a carbon atom.

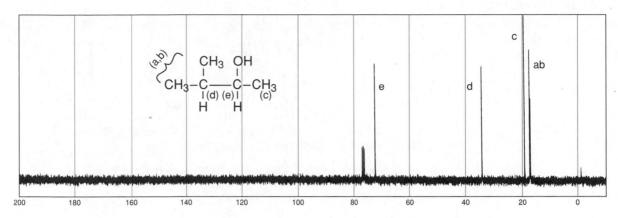

FIGURE 5.59   The ^{13}C NMR spectrum of (S)-(+)-3methyl-2-butanol (75 MHz).

The two methyl groups have slightly different chemical shifts because the adjacent carbon atom is a stereocenter. The two methyl groups are always nonequivalent in this molecule, even in the presence of free rotation. You can confirm this fact by examining the various fixed, staggered rotational conformations, using Newman projections. There are no planes of symmetry in any of these conformations; neither of the methyl groups is ever enantiomeric. Not even rotational averaging can render them equivalent.

The ^{1}H proton NMR spectrum (Fig. 5.60) is a bit more complicated, but, just as the two diastereotopic carbons have different chemical shifts, so do the hydrogens that are attached to them. The hydrogen atom that is attached to the same carbon as the two methyl groups splits each methyl

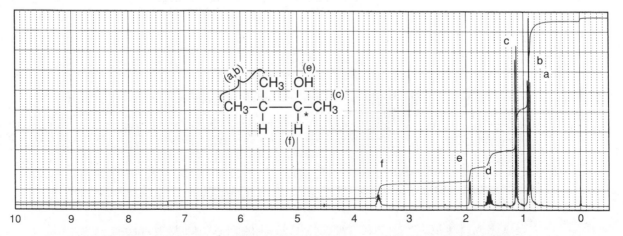

FIGURE 5.60   The ^{1}H NMR spectrum of (S)-(+)-3-methyl-2-butanol (300 MHz).

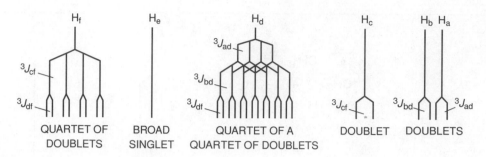

FIGURE 5.61 An analysis of the splitting patterns in the ^{1}H NMR spectrum of (*S*)-(+)-3-methyl-2-butanol.

group into a doublet. By coincidence, the two doublets overlap and appear as a triplet. Figure 5.61 is an analysis of the splittings in the spectrum.

Although not explicitly shown, both (*R*)-(−)-3-methyl-2-butanol and *racemic* 3-methyl-2-butanol produce spectra identical to the spectrum of the *S* stereoisomer. The diastereomeric effect is not limited to a solution of one pure enantiomer, free of the other.

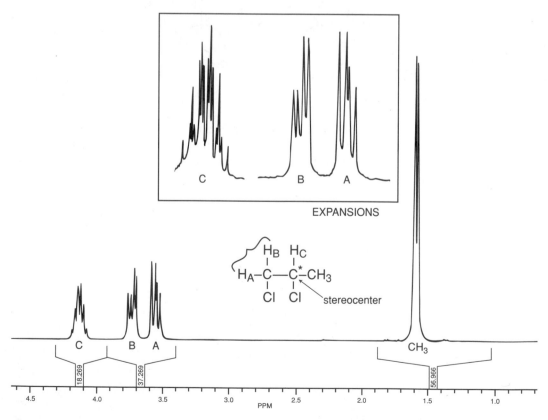

FIGURE 5.62 The ^{1}H NMR spectrum of 1,2-dichloropropane (300 MHz).

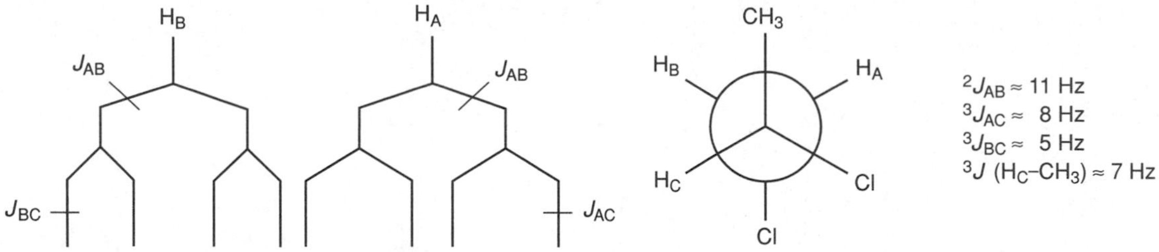

FIGURE 5.63   An analysis of the splitting of diastereotopic protons in 1,2-dichloropropane.

B.   Diasteretopic Hydrogens: 1,2-Dichloropropane

As with diastereotopic methyl groups, a pair of hydrogens located on a carbon atom adjacent to a stereocenter is expected to be diastereotopic. In many compounds of the type expected to have diastereotopic hydrogens, the difference between the chemical shifts of the diastereotopic hydrogens, H_A and H_B, is so small that neither this difference nor any coupling between H_A and H_B is easily detectable. In this case, the two protons act as a single group. In other compounds, however, the chemical shifts of H_A and H_B are quite different, and they split each other ($^2J_{AB}$) into doublets. If there is an adjacent proton, H_C, large differences between $^3J_{AC}$ and $^3J_{BC}$ are seen as well. The spectrum of 1,2-dichloropropane shows all of these effects.

Figure 5.62 is the ^1H NMR spectrum of 1,2-dichloropropane, with expansions of the downfield peaks to make the splitting patterns clear. Figure 5.63 is an analysis of the diastereotopic protons H_A and H_B. Notice that the coupling constant $^2J_{AB}$ = 11 Hz, which is a very characteristic value for geminal coupling in noncyclic linear aliphatic systems (Section 5.2B). The coupling constant $^3J_{AC}$ (α = 180°) is larger than $^3J_{BC}$ (α = 60°), which is in accordance with the predictions of the Karplus relationship (Section 5.2C). The third hydrogen, H_C, is coupled not only to H_A and H_B but to the methyl group, with ^3J (HC−CH$_3$) = 7 Hz. Because quartets from the methyl and doublets from H_A and H_B complicate the splitting of H_C, a splitting tree is not shown for this proton.

An interesting case of diastereotopic hydrogens is found in citric acid, shown in Figure 5.64. Citric acid is a symmetric molecule without any obvious stereocenter. Yet, even though there is no adjacent stereocenter, the methylene protons H_a and H_b are diastereotopic, and they not only have different chemical shifts, but they also show splitting interaction. The nonequivalence appears to be a result of restricted rotation in citric acid. As the structure in Figure 5.64 shows, the protons H_a and H_b are on different sides of the molecule. Proton H_a is on the *same side* of the molecule as the hydroxyl group on the central carbon, while proton H_b is in a different environment on the *opposite side* of the molecule from the hydroxyl group on the central carbon—they would be expected to have different chemical shifts. In fact, the situation is more complex than this explanation suggests. By careful examination of Newman projections along the C2−C3 bond, it can be determined that protons H_a and H_b can never become equivalent with any degree of rotation.

C.   Diasteretopic Fluorines: 1-Bromo-2-Chloro-1,1,2-Trifluoroethane

To show that the idea of diastereotopic groups is quite general, let us examine one final example in which there are diastereotopic fluorine atoms. The ^1H proton spectrum of 1-bromo-2-chloro-1,1,2-trifluoroethane shows two doublets of doublets, one pair of doublets at about 5.7 ppm and the other at about 6.5 ppm. To save space, the spectrum is not included, but Figure 5.65 shows the splitting pattern and the Newman projection of this molecule. Remember that this is a proton spectrum, but the single proton is split by the three fluorines (spin = $\frac{1}{2}$).

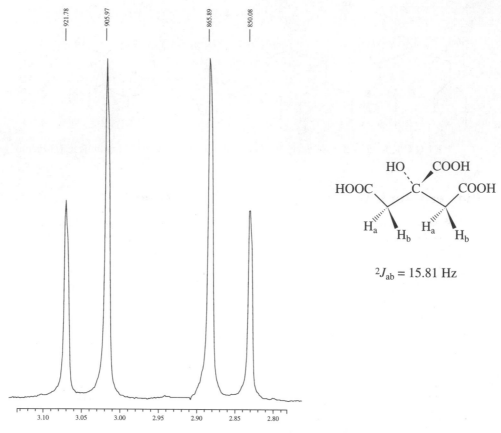

FIGURE 5.64   The 300-MHz ^{1}H spectrum of the diastereotopic methylene hydrogens in citric acid. While there is no adjacent stereocenter, restricted rotation renders these hydrogens nonequivalent. The resonances of the protons of the carboxylic acid and hydroxyl groups are not shown.

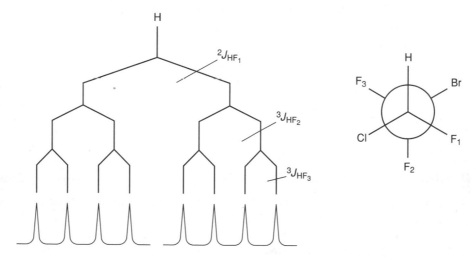

FIGURE 5.65   An analysis of the splitting caused by diastereotopic fluorine nuclei in the ^{1}H NMR spectrum of 1-bromo-2-chloro-1,1,2-trifluoroethane.

PROBLEMS

*1. Determine the coupling constants for the following compounds from their NMR spectra shown in this chapter. Draw tree diagrams for each of the protons.

(a) Vinyl acetate (Fig. 5.17).

(b) Crotonic acid (Fig. 5.21).

(c) 2-Nitrophenol (Fig. 5.49).

(d) 3-Nitrobenzoic acid (Fig. 5.50).

(e) Furfuryl alcohol (Fig. 5.51).

(f) 2-Picoline (2-methylpyridine) (Fig. 5.52).

*2. Estimate the expected splitting (J in Hertz) for the lettered protons in the following compounds; i.e., give J_{ab}, J_{ac}, J_{bc}, and so on. You may want to refer to the tables in Appendix 5.

(a)

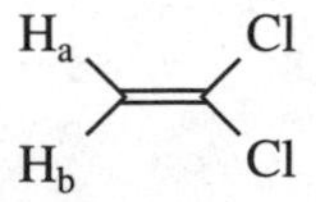

H_a ⟍ Cl
H_b ⟋ Cl

(b)

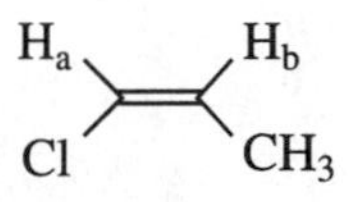

H_a ⟍ H_b
Cl ⟋ CH_3

(c)

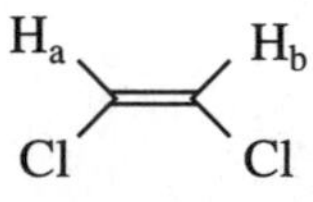

H_a ⟍ H_b
Cl ⟋ Cl

(d)

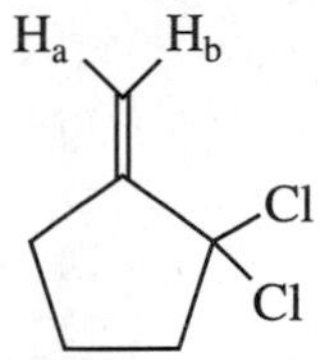

H_a H_b
      Cl
      Cl

(e)

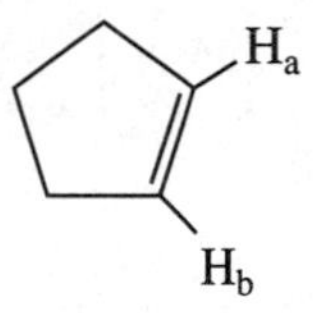

H_a
H_b

(f)

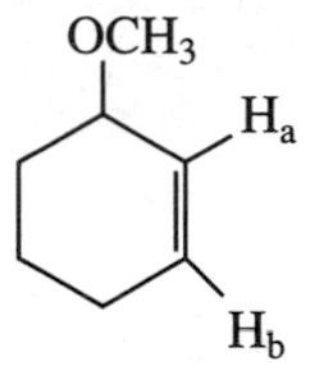

OCH_3
H_a
H_b

(g)

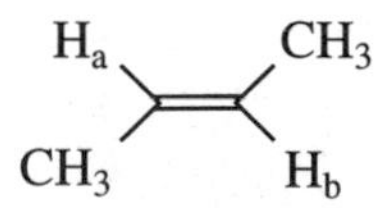

H_a ⟍ CH_3
CH_3 ⟋ H_b

(h)

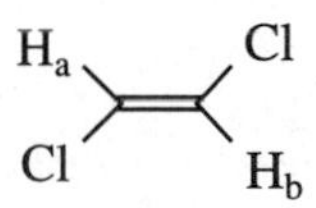

H_a ⟍ Cl
Cl ⟋ H_b

(i)

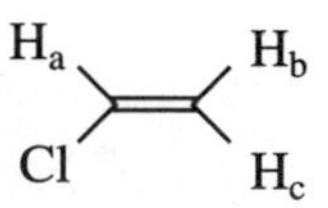

H_a ⟍ H_b
Cl ⟋ H_c

*3. Determine the coupling constants for methyl vinyl sulfone. Draw tree diagrams for each of the three protons shown in the expansions, using Figures 5.23–5.25 as examples. Assign the protons to the structure shown using the letters a, b, c, and d. Hertz values are shown above each of the peaks in the expansions.

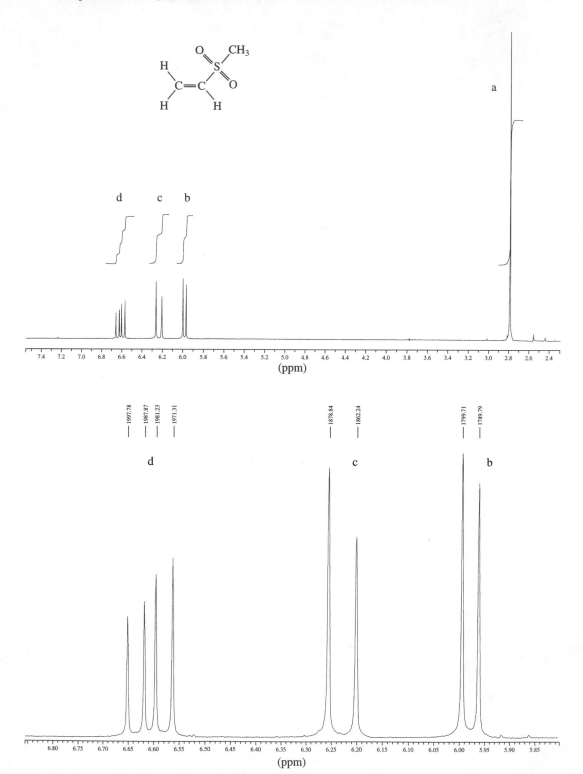

*4. The proton NMR spectrum shown in this problem is of *trans*-4-hexen-3-one. Expansions are shown for each of the five unique types of protons in this compound. Determine the coupling constants. Draw tree diagrams for each of the protons shown in the expansions and label them with the appropriate coupling constants. Also determine which of the coupling constants are ^3J and which are ^4J. Assign the protons to the structure using the letters a, b, c, d, and e. Hertz values are shown above each of the peaks in the expansions.

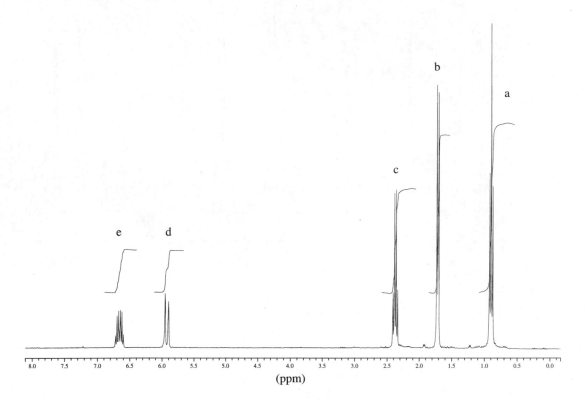

(ppm)

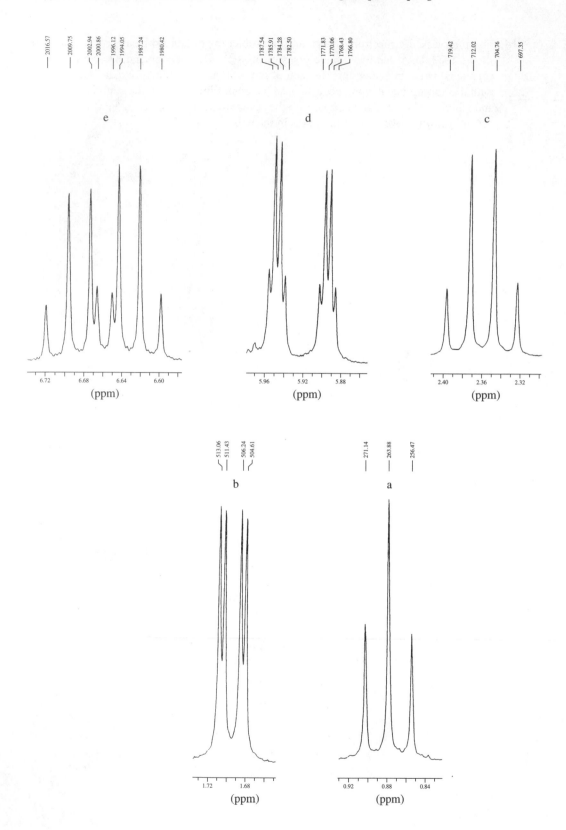

*5. The proton NMR spectrum shown in this problem is of *trans*-2-pentenal. Expansions are shown for each of the five unique types of protons in this compound. Determine the coupling constants. Draw tree diagrams for each of the protons shown in the expansions and label them with the appropriate coupling constants. Also determine which of the coupling constants are ^3J and which are ^4J. Assign the protons to the structure using the letters a, b, c, d, and e. Hertz values are shown above each of the peaks in the expansions.

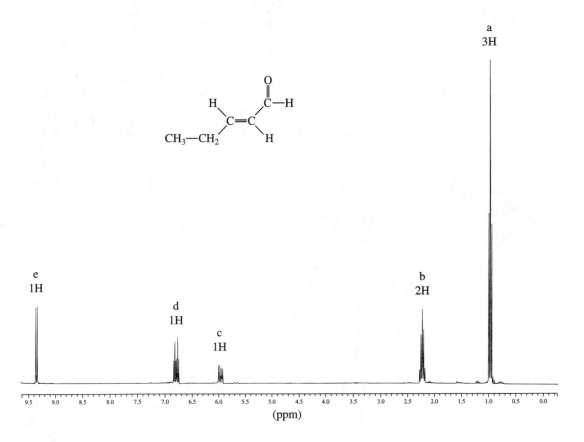

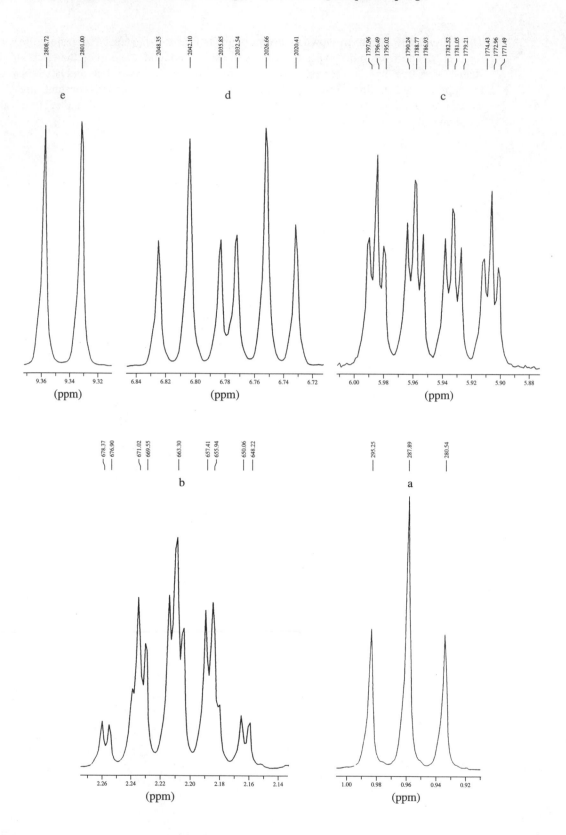

*6. In which of the following two compounds are you likely to see allylic (^4J) coupling?

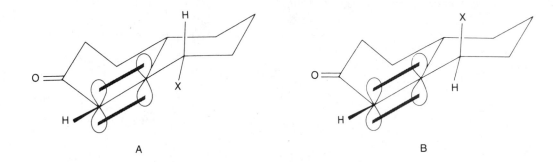

A   B

7. The reaction of dimethyl malonate with acetaldehyde (ethanal) under basic conditions yields a compound with formula $C_7H_{10}O_4$. The proton NMR is shown here. The normal carbon-13 and the DEPT experimental results are tabulated:

Normal Carbon	DEPT-135	DEPT-90
16 ppm	Positive	No peak
52.2	Positive	No peak
52.3	Positive	No peak
129	No peak	No peak
146	Positive	Positive
164	No peak	No peak
166	No peak	No peak

Determine the structure and assign the peaks in the proton NMR spectrum to the structure.

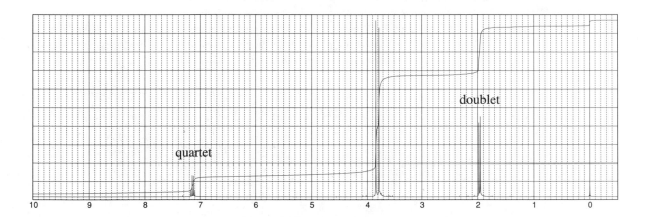

quartet

doublet

8. Diethyl malonate can be monoalkylated and dialkylated with bromoethane. The proton NMR spectra are provided for each of these alkylated products. Interpret each spectrum and assign an appropriate structure to each spectrum.

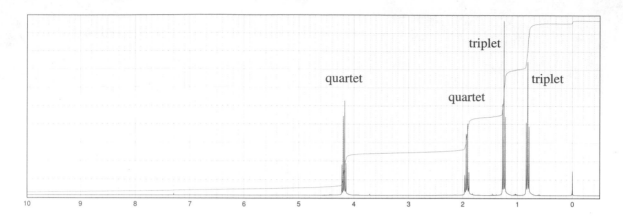

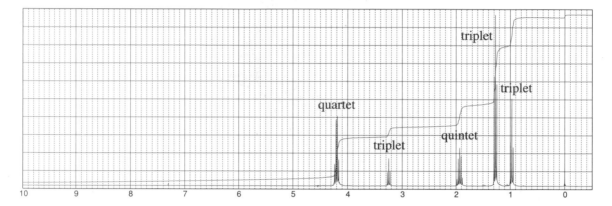

9. The proton NMR spectral information shown in this problem is for a compound with formula $C_{10}H_{10}O_3$. A disubstituted aromatic ring is present in this compound. Expansions are shown for each of the unique protons. Determine the *J* values and draw the structure of this compound. The doublets at 6.45 and 7.78 ppm provide an important piece of information. Likewise, the broad peak at about 12.3 ppm provides information on one of the functional groups present in this compound. Assign each of the peaks in the spectrum.

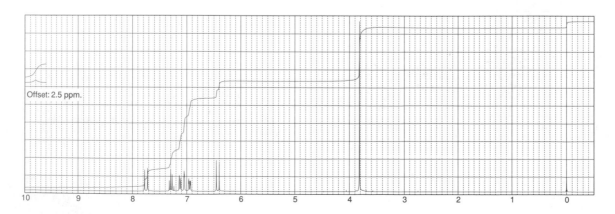

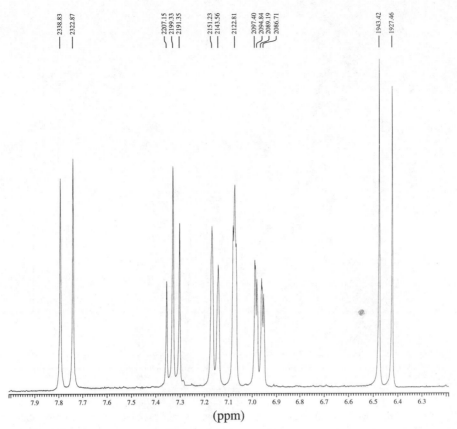

10. The proton NMR spectral information shown in this problem is for a compound with formula $C_8H_8O_3$. An expansion is shown for the region between 8.2 and 7.0 ppm. Analyze this region to determine the structure of this compound. A broad peak (1H) appearing near 12.0 ppm is not shown in the spectrum. Draw the structure of this compound and assign each of the peaks in the spectrum.

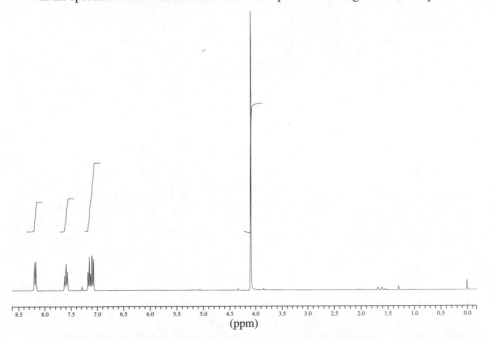

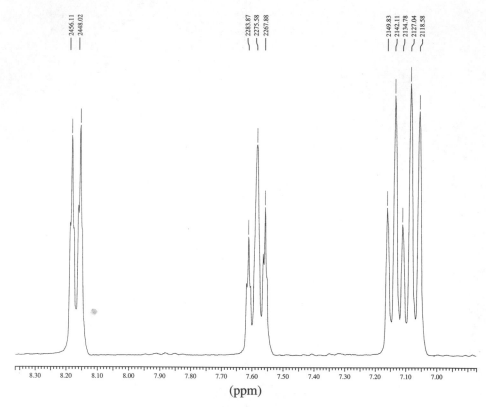

11. The proton NMR spectral information shown in this problem is for a compound with formula $C_{12}H_8N_2O_4$. An expansion is shown for the region between 8.3 and 7.2 ppm. No other peaks appear in the spectrum. Analyze this region to determine the structure of this compound. Strong bands appear at 1352 and 1522 cm^{-1} in the infrared spectrum. Draw the structure of this compound.

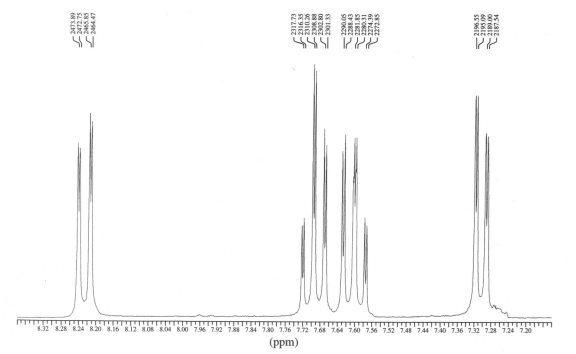

12. The proton NMR spectral information shown in this problem is for a compound with formula $C_9H_{11}NO$. Expansions of the protons appearing in the range 9.8 and 3.0 ppm are shown. No other peaks appear in the full spectrum. The usual aromatic and aliphatic C—H stretching bands appear in the infrared spectrum. In addition to the usual C—H bands, two weak bands also appear at 2720 and 2842 cm^{-1}. A strong band appears at 1661 cm^{-1} in the infrared spectrum. Draw the structure of this compound.

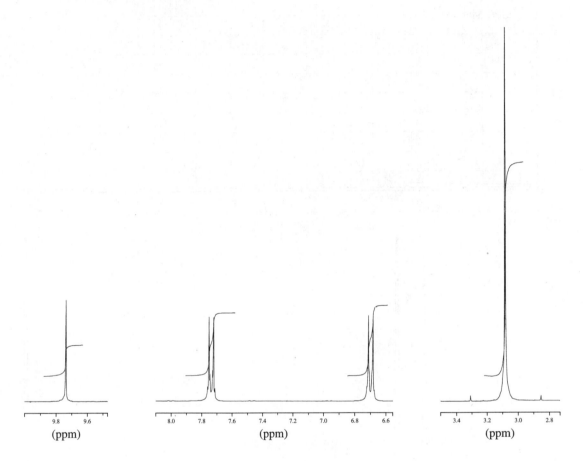

13. The fragrant natural product anethole ($C_{10}H_{12}O$) is obtained from anise by steam distillation. The proton NMR spectrum of the purified material follows. Expansions of each of the peaks are also shown, except for the singlet at 3.75 ppm. Deduce the structure of anethole, including stereochemistry, and interpret the spectrum.

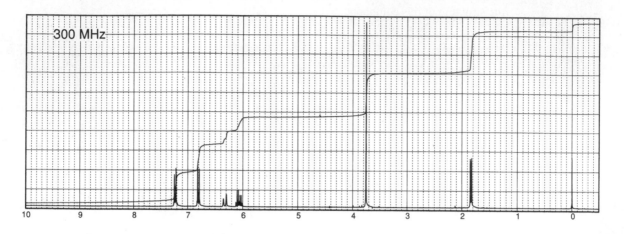

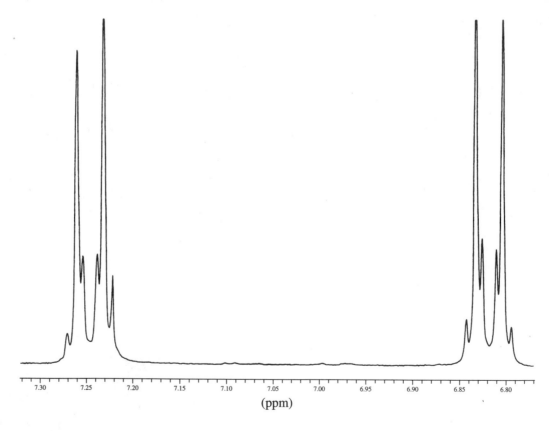

(ppm)

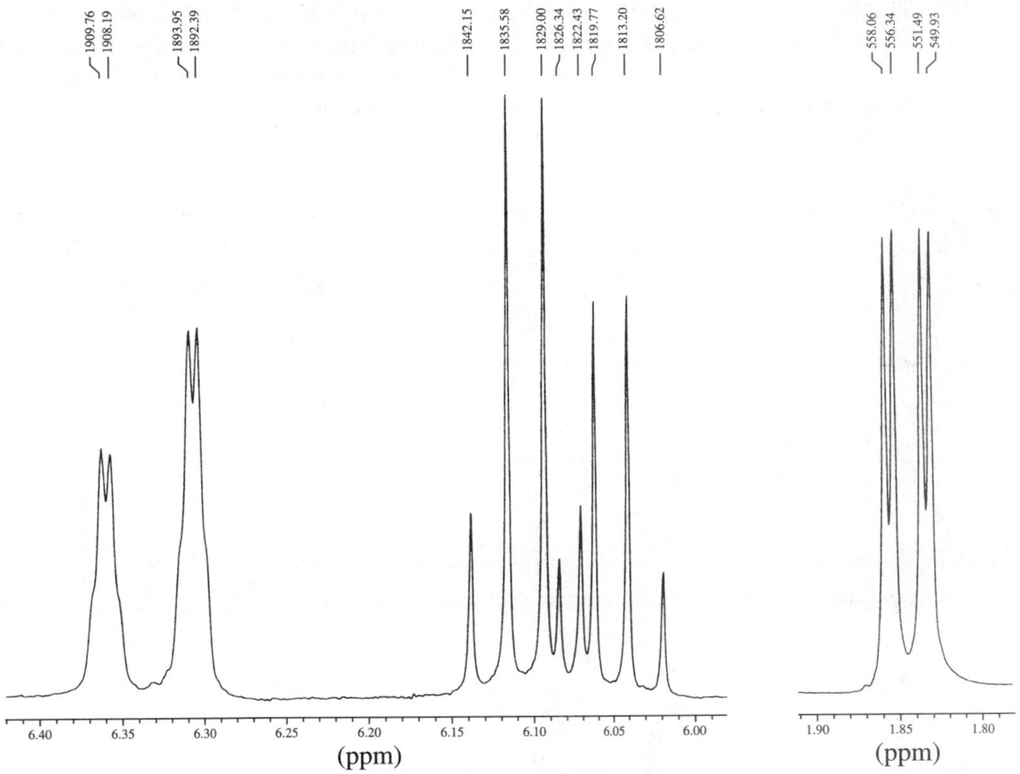
1909.76
1908.19
1893.95
1892.39
1842.15
1835.58
1829.00
1826.34
1822.43
1819.77
1813.20
1806.62
558.06
556.34
551.49
549.93
6.40
6.35
6.30
6.25
6.20
6.15
6.10
6.05
6.00
(ppm)
1.90
1.85
1.80
(ppm)

14. Determine the structure of the following aromatic compound with formula C_8H_7BrO.

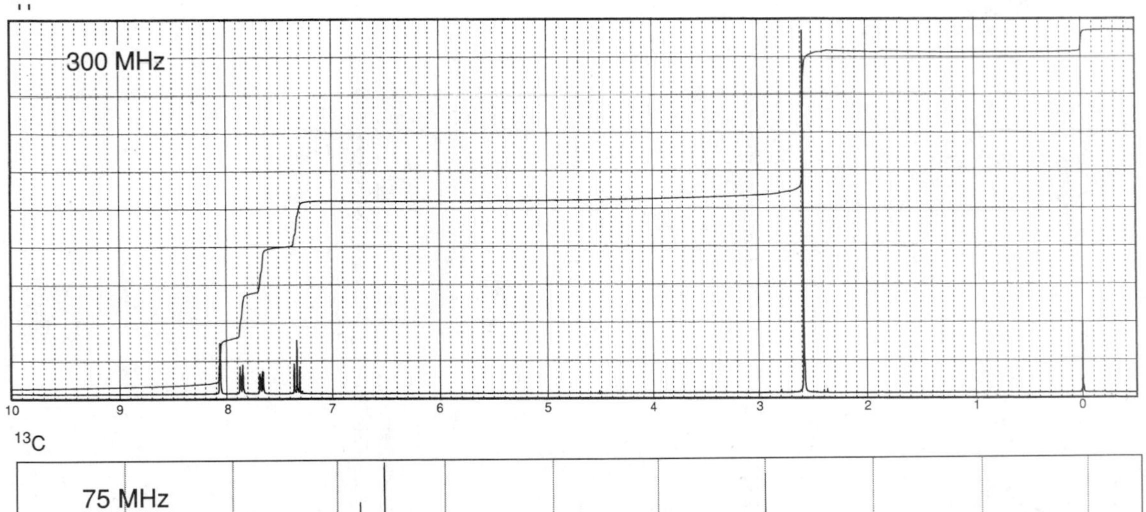
300 MHz
10
9
8
7
6
5
4
3
2
1
0

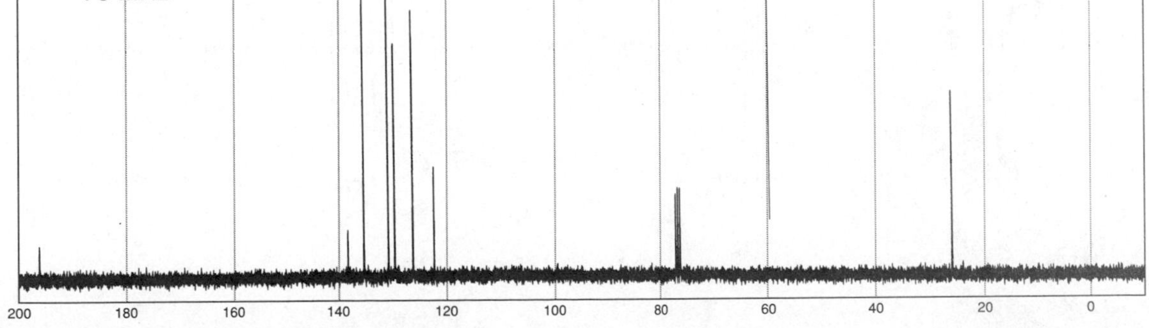
13C
75 MHz
200
180
160
140
120
100
80
60
40
20
0

*15. The following spectrum of a compound with formula $C_5H_{10}O$ shows interesting patterns at about 2.4 and 9.8 ppm. Expansions of these two sets of peaks are shown. Expansions of the other patterns (not shown) in the spectrum show the following patterns: 0.92 ppm (triplet), 1.45 ppm (sextet), and 1.61 ppm (quintet). Draw a structure of the compound. Draw tree diagrams of the peaks at 2.4 and 9.8 ppm, including coupling constants.

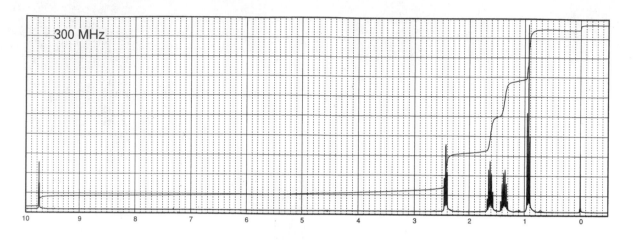

300 MHz

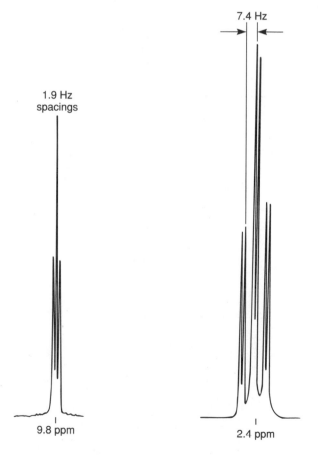

7.4 Hz

1.9 Hz
spacings

9.8 ppm

2.4 ppm

*16. The proton NMR spectral information shown in this problem is for a compound with formula $C_{10}H_{12}O_3$. A broad peak appearing at 12.5 ppm is not shown in the proton NMR reproduced here. The normal carbon-13 spectral results, including DEPT-135 and DEPT-90 results, are tabulated:

Normal Carbon	DEPT-135	DEPT-90
15 ppm	Positive	No peak
40	Negative	No peak
63	Negative	No peak
115	Positive	Positive
125	No peak	No peak
130	Positive	Positive
158	No peak	No peak
179	No peak	No peak

Draw the structure of this compound.

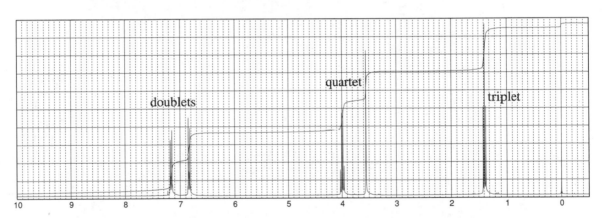

17. The proton NMR spectral information shown in this problem is for a compound with formula $C_{10}H_9N$. Expansions are shown for the region from 8.7 to 7.0 ppm. The normal carbon-13 spectral results, including DEPT-135 and DEPT-90 results, are tabulated:

Normal Carbon	DEPT-135	DEPT-90
19 ppm	Positive	No peak
122	Positive	Positive
124	Positive	Positive
126	Positive	Positive
128	No peak	No peak
129	Positive	Positive
130	Positive	Positive
144	No peak	No peak
148	No peak	No peak
150	Positive	Positive

Draw the structure of this compound and assign each of the protons in your structure. The coupling constants should help you to do this (see Appendix 5).

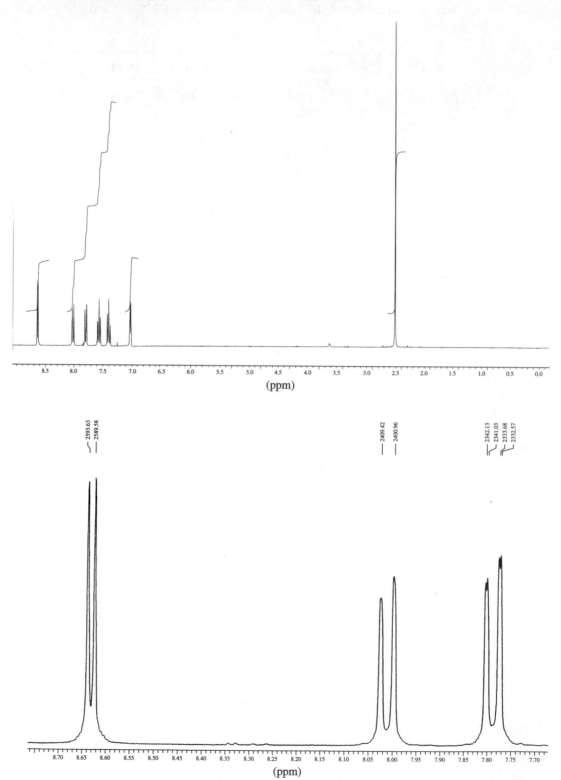

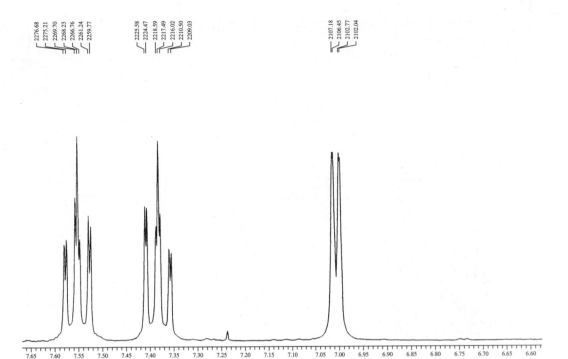

2276.68
2275.21
2269.70
2268.23
2266.76
2261.24
2259.77

2225.58
2224.47
2218.59
2217.49
2216.02
2210.50
2209.03

2107.18
2106.45
2102.77
2102.04

(ppm)

18. The proton NMR spectral information shown in this problem is for a compound with formula $C_9H_{14}O$. Expansions are shown for all the protons. The normal carbon-13 spectral results, including DEPT-135 and DEPT-90 results, are tabulated:

Normal Carbon	DEPT-135	DEPT-90
14 ppm	Positive	No peak
22	Negative	No peak
27.8	Negative	No peak
28.0	Negative	No peak
32	Negative	No peak
104	Positive	Positive
110	Positive	Positive
141	Positive	Positive
157	No peak	No peak

Draw the structure of this compound and assign each of the protons in your structure. The coupling constants should help you to do this (see Appendix 5).

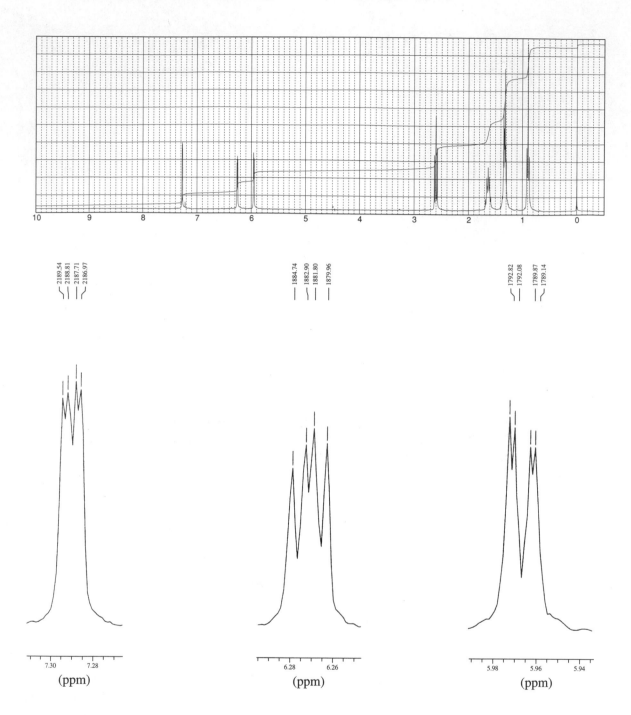

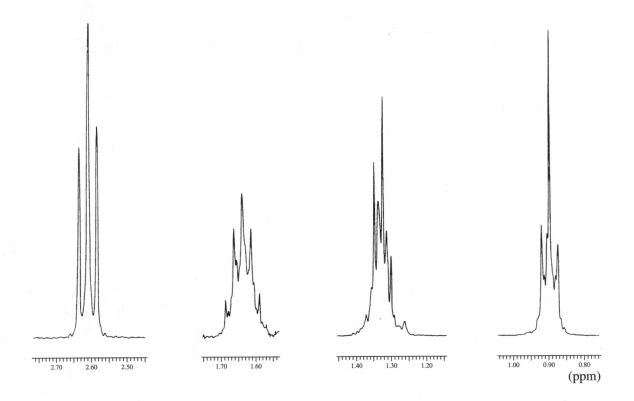

(ppm)

19. The proton NMR spectral information shown in this problem is for a compound with formula $C_{10}H_{12}O_2$. One proton, not shown, is a broad peak that appears at about 12.8 ppm. Expansions are shown for the protons absorbing in the region from 3.5 to 1.0 ppm. The monosubstituted benzene ring is shown at about 7.2 ppm, but is not expanded because it is uninteresting. The normal carbon-13 spectral results, including DEPT-135 and DEPT-90 results, are tabulated:

Normal Carbon	DEPT-135	DEPT-90
22 ppm	Positive	No peak
36	Positive	Positive
43	Negative	No peak
126.4	Positive	Positive
126.6	Positive	Positive
128	Positive	Positive
145	No peak	No peak
179	No peak	No peak

Draw the structure of this compound and assign each of the protons in your structure. Explain why the interesting pattern is obtained between 2.50 and 2.75 ppm. Draw tree diagrams as part of your answer.

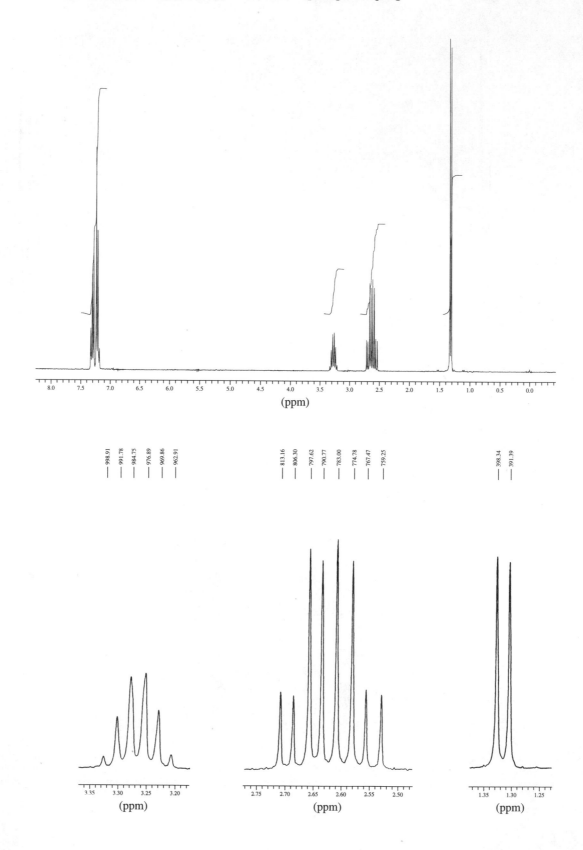

20. The spectrum shown in this problem is of 1-methoxy-1-buten-3-yne. Expansions are shown for each proton. Determine the coupling constants for each of the protons and draw tree diagrams for each. The interesting part of this problem is the presence of significant long-range coupling constants. There are ^3J, ^4J, and ^5J couplings in this compound. Be sure to include all of them in your tree diagram (graphical analysis).

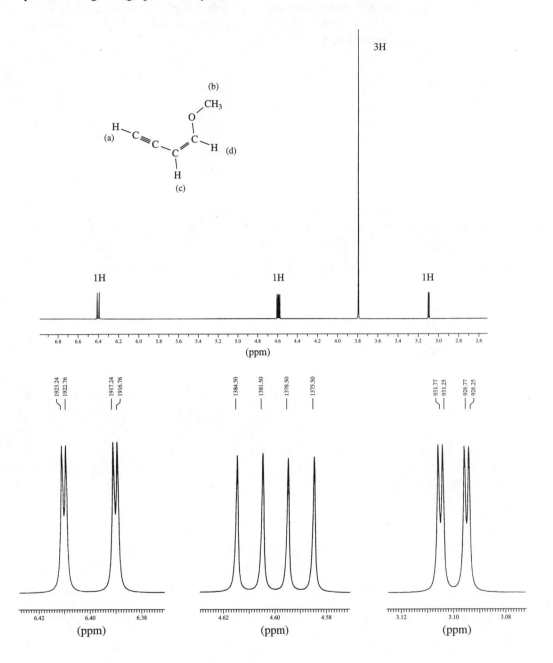

21. The partial proton NMR spectra (**A** and **B**) are given for the *cis* and *trans* isomers of the compound shown below (the bands for the three phenyl groups are not shown in either NMR). Draw the structures for each of the isomers and use the magnitude of the coupling constants to assign a structure to each spectrum. It may be helpful to use a molecular modeling program to determine the dihedral angles for each compound. The finely spaced doublet at 3.68 ppm in spectrum **A** is the band for the O—H peak. Assign each of the peaks in spectrum **A** to the structure. The O—H peak is not shown in spectrum **B**, but assign the pair of doublets to the structure using chemical shift information.

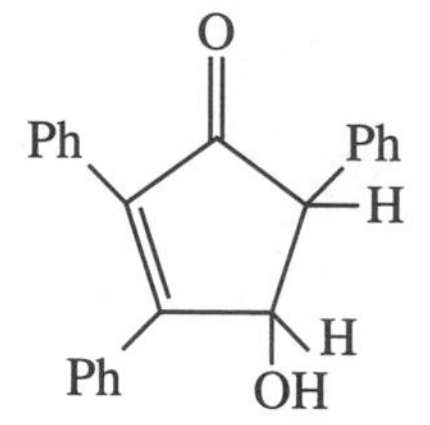
O
Ph
Ph
H
H
Ph
OH

Spectrum A

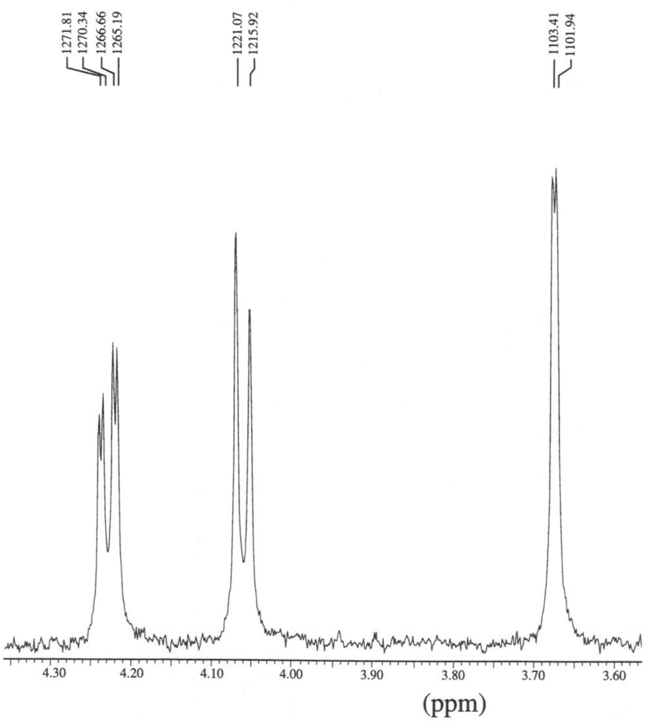
1271.81
1270.34
1266.66
1265.19
1221.07
1215.92
1103.41
1101.94
4.30
4.20
4.10
4.00
3.90
3.80
3.70
3.60
(ppm)

Spectrum B

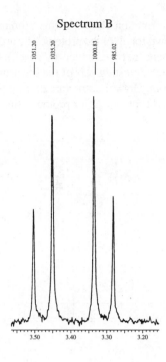

*22. Coupling constants between hydrogen and fluorine nuclei are often quite large: $^3J_{HF} \cong 3\text{–}25$ Hz and $^2J_{HF} \cong 44\text{–}81$ Hz. Since fluorine-19 has the same nuclear spin quantum number as a proton, we can use the $n + 1$ Rule with fluorine-containing organic compounds. One often sees larger H–F coupling constants, as well as smaller H–H couplings, in proton NMR spectra.

(a) Predict the appearance of the proton NMR spectrum of F–CH$_2$–O–CH$_3$.

(b) Scientists using modern instruments directly observe many different NMR-active nuclei by changing the frequency of the spectrometer. How would the fluorine NMR spectrum for F–CH$_2$–O–CH$_3$ appear?

*23. The proton NMR spectral information shown in this problem is for a compound with formula $C_9H_8F_4O$. Expansions are shown for all of the protons. The aromatic ring is disubstituted. In the region from 7.10 to 6.95 ppm, there are two doublets (1H each). One of the doublets is partially overlapped with a singlet (1H). The interesting part of the spectrum is the one proton pattern found in the region from 6.05 to 5.68 ppm. Draw the structure of the compound and draw a tree diagram for this pattern (see Appendix 5 and Problem 22 for proton to fluorine coupling constants).

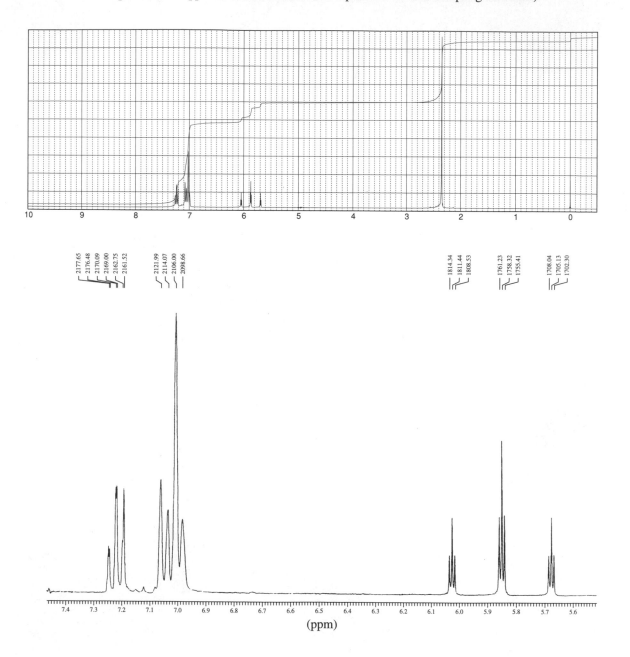

24. A compound with the formula C_2H_4BrF has the following NMR spectrum. Draw the structure for this compound. Using the Hertz values on the expansions, calculate the coupling constants. Completely explain the spectrum.

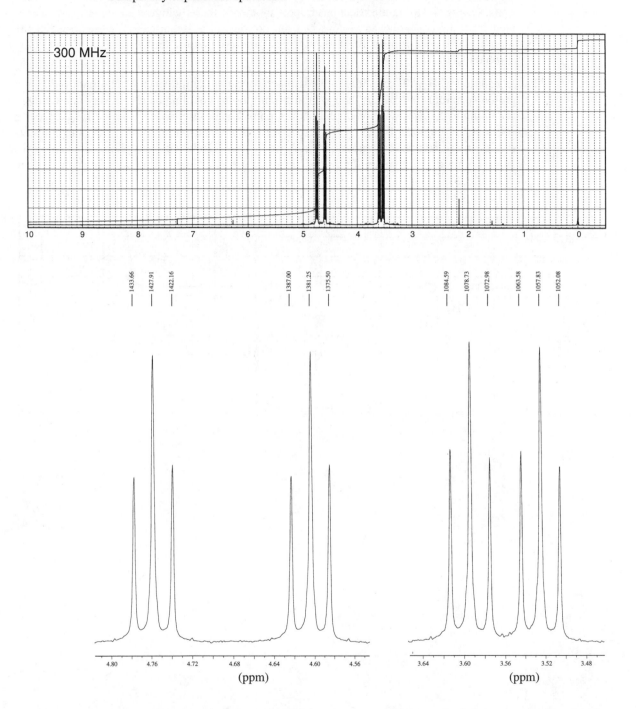

(ppm)   (ppm)

*25. Predict the proton and deuterium NMR spectra of $D-CH_2-O-CH_3$, remembering that the spin quantum number for deuterium = 1. Compare the proton spectrum to that of $F-CH_2-O-CH_3$ (Problem 22a).

*26. Although the nuclei of chlorine ($I = \frac{3}{2}$), bromine ($I = \frac{3}{2}$), and iodine ($I = \frac{5}{2}$) exhibit nuclear spin, the geminal and vicinal coupling constants, J_{HX} (vic) and J_{HX} (gem), are normally zero. These atoms are simply too large and diffuse to transmit spin information via their plethora of electrons. Owing to strong electrical quadrupole moments, these halogens are completely decoupled from directly attached protons or from protons on adjacent carbon atoms. Predict the proton NMR spectrum of $Br-CH_2-O-CH_3$ and compare it to that of $F-CH_2-O-CH_3$ (Problem 22a).

*27. In addition to $H-{}^{19}F$ coupling, it is possible to observe the influence of phosphorus-31 on a proton spectrum ($H-{}^{31}P$). Although proton–phosphorus coupling constants vary considerably according to the hybridization of phosphorus, phosphonate esters have ^2J and ^3J $H-P$ coupling constants of about 13 Hz and 8 Hz, respectively. Since phosphorus-31 has the same nuclear-spin quantum number as a proton, we can use the $n + 1$ Rule with phosphorus-containing organic compounds. Explain the following spectrum for dimethyl methylphosphonate (see Appendix 5).

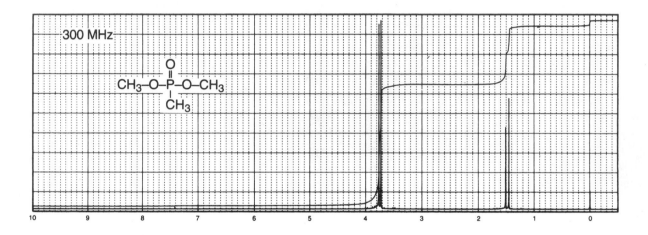

28. The proton NMR spectra for methyltriphenylphosphonium halide and its carbon-13 analogue are shown in this problem. Concentrating your attention on the doublet at 3.25 ppm and the pair of doublets between 2.9 and 3.5 ppm, interpret the two spectra. You may need to refer to Appendix 5 and Appendix 9. Estimate the coupling constants in the two spectra. Ignore the phenyl groups in your interpretation.

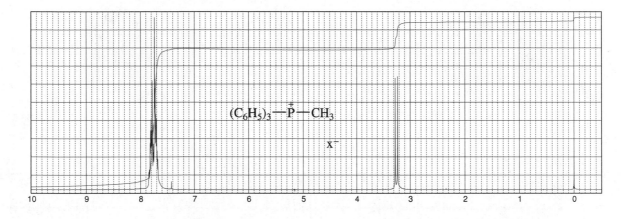

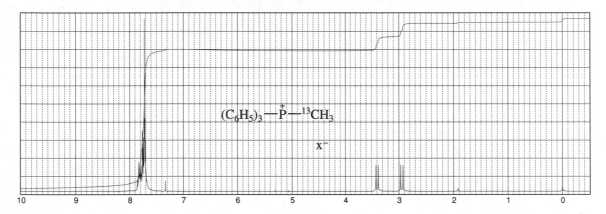

29. All three of the compounds, a, b, and c, have the same mass (300.4 amu). Identify each compound and assign as many peaks as you can, paying special attention to methyl and vinyl hydrogens. There is a small CHCl₃ peak near 7.3 ppm in each spectrum that should be ignored when analyzing the spectra.

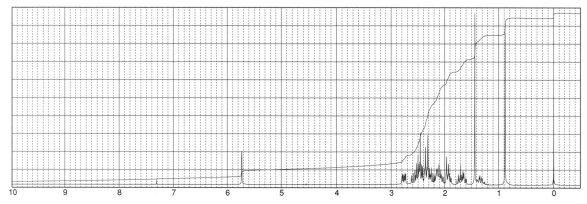

(a)   (b)   (c)

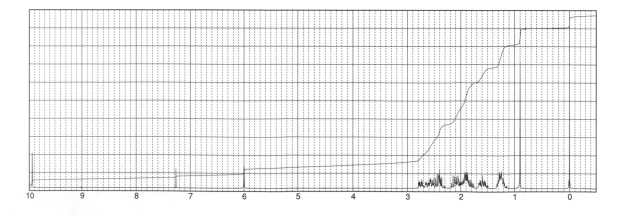

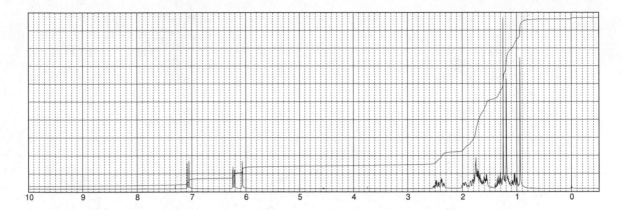

*30. Calculate the chemical shifts for the indicated protons, using Table 1 in Appendix 6.

(a) CH$_3$—$\overset{\overset{\textstyle O}{\|}}{C}$—CH$_3$
　　　↑

(b) CH$_3$—CH$_2$—$\overset{\overset{\textstyle O}{\|}}{C}$—CH$_2$—$\overset{\overset{\textstyle O}{\|}}{C}$—O—CH$_3$
　　　　　　　　↑　　　　　↑

(c) Cl—CH$_2$—$\overset{\overset{\textstyle O}{\|}}{C}$—O—CH$_3$
　　　　　　↑

(d) CH$_3$—CH$_2$—CH$_2$—C≡C—H
　　　　　　　　↑

(e) CH$_3$—CH$_2$—$\overset{\overset{\textstyle Cl}{|}}{\underset{\underset{\textstyle Cl}{|}}{C}}$—H
　　　　　　　　　↑

(f) CH$_2$=$\overset{\overset{\textstyle CH_3}{|}}{C}$—CH$_2$—O—H
　　　　　　　　↑

*31. Calculate the chemical shifts for the vinyl protons, using Table 2 in Appendix 6.

(a)
$$\underset{H}{\overset{H}{}}C=C\underset{CH_3}{\overset{C-O-CH_3}{}}$$
with C=O on the upper right carbon

(b)
$$\underset{H}{\overset{CH_3}{}}C=C\underset{\overset{\|}{O}}{\overset{H}{}}C-O-CH_3$$

(c)
$$\underset{H}{\overset{H}{}}C=C\underset{H}{\overset{C_6H_5}{}}$$

(d)
$$\underset{H}{\overset{C_6H_5}{}}C=C\underset{\overset{\|}{O}}{\overset{H}{}}C-CH_3$$

(e)
$$\underset{CH_3}{\overset{H}{}}C=C\underset{H}{\overset{CH_2-OH}{}}$$

(f)
$$\underset{CH_3}{\overset{CH_3}{}}C=C\underset{\overset{\|}{O}}{\overset{H}{}}C-CH_3$$

*32. Calculate the chemical shifts for the aromatic protons, using Table 3 in Appendix 6.

(a)

(b)

(c)

(d)

(e)

(f)

(g)

(h)

(i)

REFERENCES

Textbooks

Bovey, F. A., *NMR Spectroscopy*, Academic Press, New York, 1969.

Derome, A. E., *Modern NMR Techniques for Chemistry Research*, Pergamon Press, Oxford, 1987.

Friebolin, H., *Basic One- and Two-Dimensional NMR Spectroscopy*, 2nd. ed., VCH Publishers, New York, 1993.

Gunther, H., *NMR Spectroscopy*, 2nd ed., John Wiley and Sons, New York, 1995.

Jackman, L. M., and S. Sternhell, *Applications of Nuclear Magnetic Resonance Spectroscopy in Organic Chemistry*, 2d ed., Pergamon Press, London, 1969.

Macomber, R. S., *NMR Spectroscopy—Essential Theory and Practice*, College Outline Series, Harcourt, Brace Jovanovich, New York, 1988.

Macomber, R. S., *A Complete Introduction to Modern NMR Spectroscopy*, John Wiley and Sons, New York, 1997.

Pople, J. A., W. C. Schneider, and H. J. Bernstein, *High Resolution Nuclear Magnetic Resonance*, McGraw–Hill, New York, 1959.

Roberts, J. D., *Nuclear Magnetic Resonance: Applications to Organic Chemistry*, McGraw–Hill, New York, 1959.

Roberts, J. D., *An Introduction to the Analysis of Spin–Spin Splitting in High Resolution Nuclear Magnetic Resonance Spectra*, W. A. Benjamin, New York, 1962.

Sanders, J. K. M., and B. K. Hunter, *Modern NMR Spectroscopy—A Guide for Chemists*, 2d ed., Oxford University Press, Oxford, England, 1993.

Silverstein, R. M., and F. X. Webster, *Spectrometric Identification of Organic Compounds*, 6th ed., John Wiley and Sons, New York, 1998.

Wiberg, K. B., and B. J. Nist, *The Interpretation of NMR Spectra*, W. A. Benjamin, New York, 1962.

Yoder, C. H., and C. D. Schaeffer, *Introduction to Multinuclear NMR*, Benjamin-Cummings, Menlo Park, CA, 1987.

Compilations of Spectra

Ault, A., and M. R. Ault, *A Handy and Systematic Catalog of NMR Spectra*, 60 MHz with some 270 MHz, University Science Books, Mill Valley, CA, 1980.

Pouchert, C. J., and J. Behnke, *The Aldrich Library of ^{13}C and ^1H FT-NMR Spectra,* 300 MHz, Aldrich Chemical Company, Milwaukee, WI, 1993.

Computer Programs

Bell, H., "FIDWIN," IBM PC/Windows, Virginia Tech, Blacksburg, VA. (Dr. Bell has other programs as well, all available from *http://www.chem.vt.edu* or e-mail: hbell@chemserver.chem.vt.edu.)

"Felix," Silicon Graphics Computers, Biosym Technologies, Inc., 4 Century Drive, Parsippany, NJ 07054.

"HyperNMR," IBM PC/Windows, Hypercube, Inc., 419 Phillip Street, Waterloo, Ontario, Canada N2L 3X2.

NSF–Project Seraphim, University of Wisconsin, has several programs, with information available on *gopher.jchemed.chem.wisc.edu*; e-mail: jcesoft@macc. wisc.edu:

T. Farrar, "PC NMR4Windows"

J. Swartz, "PCNMR" and "Library of Spectra for PCNMR"

"NUTS," Acorn NMR, demo version from *ftp.netcom. com/pub/wp/woodyc*; e-mail: support@acornnmr.com.

Shatz, P., "RACCOON," NSF–Project Seraphim, University of Wisconsin.

"TurboNMR," Silicon Graphics Computers, Biosym Technologies, Inc., 4 Century Drive, Parsippany, NJ 07054.

"WIN-NMR," USA Bruker Instruments, Inc., Manning Park, Billerica, MA 01821. (Optional modules include NMR-SIM, WIN-DAISY, and WIN-FIT.)

Papers

Mann, B. "The Analysis of First-Order Coupling Patterns in NMR Spectra," *Journal of Chemical Education, 72* (1995): 614.

Hoye, T. R., P. R. Hanson, and J. R. Vyvyan, "A Practical Guide to First-Order Multiplet Analysis in ^{1}H NMR Spectroscopy," *Journal of Organic Chemistry 59* (1994): 4096.

Web sites

WWW NMR Webservers

Bruker–Germany *http://www.bruker.de/Bruker.html*

NMRFAM, Madison *http://gopher.nmrfam.wisc.edu/*

University of Florida *http://micro.ifas.ufl.edu/*

Wilson Lab *http://www.wilson.ucsd.edu/ education/spectroscopy/ spectroscopy.html*

Peter Lundberg, University of Umea, Sweden, has compiled a very complete list of educational NMR software. It is available from a number of sites, including: *http://atlas.chemistry.uakron.edu:8080/cdept.docs/ MAGNET/sware.html* and the Bruker Web sites.

http://www.aist.go.jp/RIODB/SDBS/menu-e.html

Integrated Spectral DataBase System for Organic Compounds, National Institute of Materials and Chemical Research, Tsukuba, Ibaraki 305-8565, Japan. This database includes infrared, mass spectra, and NMR data (proton and carbon-13) for a number of compounds.

http://www.chem.ucla.edu/~webnmr/

UCLA Department of Chemistry and Biochemistry, in connection with Cambridge University Isotope Laboratories, maintains a website, WebSpecta, that provides NMR and IR spectroscopy problems for students to interpret. They provide links to other sites with problems for students to solve.

http://www.nd.edu/~smithgrp/structure/workbook.html

Combined structure problems provided by the Smith group at Notre Dame University.

NUCLEAR MAGNETIC RESONANCE SPECTROSCOPY

Part Four: Other Topics in One-Dimensional NMR

In this chapter, we shall consider some additional topics in one-dimensional nuclear magnetic resonance spectroscopy. Among the topics that will be covered will be the variability in chemical shifts of protons attached to electronegative elements such as oxygen and nitrogen, the special characteristics of protons attached to nitrogen, the effects of solvent on chemical shift, lanthanide shift reagents, and spin decoupling experiments.

6.1 PROTONS ON OXYGEN: ALCOHOLS

Under the usual conditions of determining the NMR spectrum of an alcohol, and for most alcohols, no coupling is observed between the hydroxyl hydrogen and hydrogens on the carbon atom to which the hydroxyl group is attached ($-CH-OH$). Coupling actually occurs between these hydrogens, but, owing to other factors, spin–spin splitting is often not observed. Whether or not spin–spin splitting involving the hydroxyl hydrogen is observed in a given alcohol depends on several factors, including temperature, purity of the sample, and the solvent used. These factors are all related to the rate at which hydroxyl protons exchange with one another (or the solvent) in the solution. Under normal conditions, the rate of exchange of protons between alcohol molecules is faster than the rate at which the NMR spectrometer can respond.

$$R-O-H_a + R'-O-H_b \rightleftharpoons R-O-H_b + R'-O-H_a$$

About 10^{-2} to 10^{-3} sec is required for an NMR transitional event to occur and be recorded. At room temperature, a typical pure liquid alcohol sample undergoes intermolecular proton exchange at a rate of about 10^5 protons per second. This means that the average residence time of a single proton on the oxygen atom of a given alcohol before it exchanges is about 10^{-5} sec. This is a much shorter time than is required for a spin transition, and it leads to complications. For instance, a proton just starting to have resonance may suddenly begin to exchange; or one that is between two molecules may begin to have resonance and then suddenly affix itself to a molecule. Because the NMR spectrometer cannot respond rapidly to these situations, the net result, as far as it is concerned, is that the proton is unattached more frequently than it is attached to oxygen, and the spin interaction between the hydroxyl proton and any other proton in the molecule is effectively decoupled. *Rapid chemical exchange decouples spin interactions*, and the NMR spectrometer records only the *average environment* it senses for the exchanging proton. The hydroxyl proton, for instance, often exchanges between alcohol molecules so rapidly that it "sees" all the possible spin orientations of hydrogens attached to the α carbon as a single time-averaged spin configuration. Similarly, the α hydrogens see so many different protons on the hydroxyl oxygen (some with spin $+\frac{1}{2}$ some with spin $-\frac{1}{2}$) that the spin configuration they sense is an average or intermediate value between $+\frac{1}{2}$ and $-\frac{1}{2}$, that is, zero. In effect, the NMR spectrometer is like a camera with a slow shutter speed that is used to photograph a fast event. Events that are faster than the click of the shutter mechanism are either blurred or averaged.

If the rate of exchange in an alcohol can be slowed down to the point at which it approaches the "time scale of the NMR" (i.e., $<10^2$ to 10^3 exchanges per second), then coupling can be observed. For instance, the NMR spectrum of methanol at 25°C (ca. 300 K) consists of only two peaks, both unsplit singlets, integrating for one proton and three protons, respectively. However, at temperatures below −33°C (less than 240 K) the spectrum changes dramatically. The one-proton O−H resonance becomes a quartet ($^3J = 5$ Hz), and the three-proton methyl resonance becomes a doublet ($^3J = 5$ Hz). Clearly, at or below −33°C (less than 240 K) chemical exchange has slowed to the point at which it is on the time scale of the NMR spectrometer, and coupling to the hydroxyl proton is observed. At temperatures intermediate between 25°C and −33°C (300 K to 240 K), transitional spectra are seen. Figure 6.1 is a stacked plot of NMR spectra of methanol determined at a range of temperatures from 290 K to 200 K.

The room-temperature spectrum of an ordinary sample of ethanol (Fig. 6.2) shows no coupling of the hydroxyl proton to the methylene protons. Thus, the hydroxyl proton is seen as a singlet, and the methylene protons (split by the methyl group) are seen as a quartet. Apparently the rate of exchange in such a sample is faster than the NMR time scale, and decoupling of the hydroxyl and methylene protons is observed. However, if a sample of ethanol is purified in such a way as to eliminate all traces of impurity (especially of acid and water, thereby slowing the proton exchange rate), the hydroxyl–methylene coupling can be observed in the form of increased complexity of the

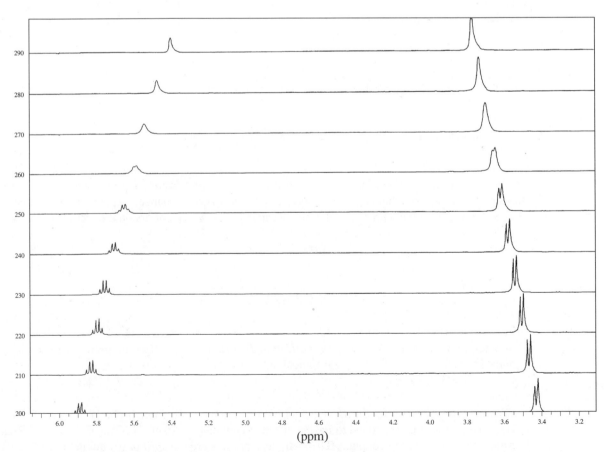

F I G U R E 6 . 1 Stacked plot of NMR spectra of methanol determined at a range of temperatures from 290 K to 200 K.

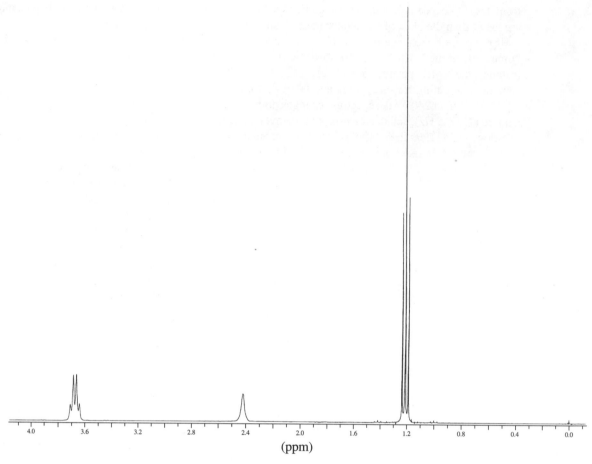

F I G U R E 6 . 2 The NMR spectrum of an ordinary ethanol sample

spin–spin splitting patterns. The hydroxyl absorption becomes a triplet, and the methylene absorptions are seen as an overlapping pair of quartets. The hydroxyl resonance is apparently split (just as the methyl group is, but with different *J*) into a triplet by its two neighbors on the methylene carbon.

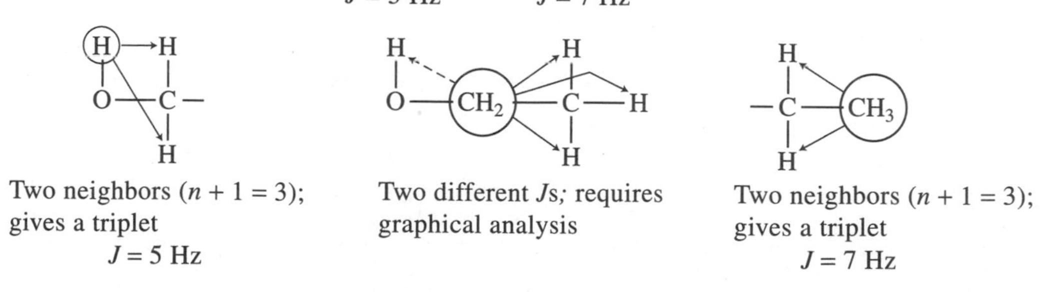

$J = 5$ Hz $J = 7$ Hz

Two neighbors ($n + 1 = 3$);
gives a triplet
$J = 5$ Hz

Two different *J*s; requires
graphical analysis

Two neighbors ($n + 1 = 3$);
gives a triplet
$J = 7$ Hz

The coupling constant for the methylene–hydroxyl interaction is found to be ^3J (CH_2, OH) = ca. 5 Hz. The methyl triplet is found to have a different coupling constant, ^3J (CH_3, CH_2) = ca. 7 Hz., for the methylene–methyl coupling. The methylene protons *are not split* into a quintet by their four neighbors, as the coupling constants for hydroxyl–methylene and methyl–methylene are different. As noted in Chapter 5, the $n + 1$ Rule does not apply in such an instance; each interaction occurs independently of the other, and a graphical analysis is required to approximate the correct pattern.

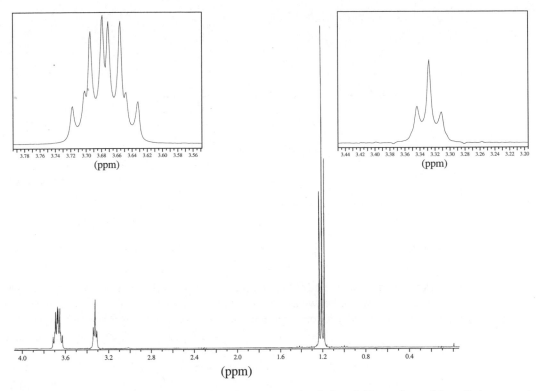

FIGURE 6.3 The NMR spectrum of an ultrapure sample of ethanol. Expansions of the splitting patterns are included.

Figure 6.3 shows the spectrum of ultrapure ethanol. Notice in the expanded splitting patterns that the methylene protons are split into two overlapping quartets (a doublet of quartets).[1,2] If even a drop of acid is added to the ultrapure ethanol sample, proton exchange becomes so fast that the methylene and hydroxyl protons are decoupled, and the simpler spectrum (Fig. 6.2) is obtained.

6.2 EXCHANGE IN WATER AND D$_2$O

A. Acid/Water and Alcohol/Water Mixtures

When two compounds, each of which contains an O—H group, are mixed, one often observes only a single NMR absorption due to O—H. For instance, consider first the spectra of (1) pure acetic acid, (2) pure water, and (3) a 1:1 mixture of acetic acid and water. Figure 6.4 indicates their general appearances.

Mixtures of acetic acid and water might be expected to show three peaks, since there are two distinct types of hydroxyl groups in the solutions—one on acetic acid and one on water. In addition, the methyl group on acetic acid should give an absorption peak. In actuality, however, mixtures of these two reagents produce only two peaks. The methyl peak occurs at its normal position in the mixture, but there is only a single hydroxyl peak *between* the hydroxyl positions of the pure substances. Apparently, exchange of the type

$$CH_3COOH_a + H-O-H_b \rightleftharpoons CH_3COOH_b + H-O-H_a$$

---

[1] This pattern could also be analyzed as a "quartet of doublets" by reversing the order in which the couplings are considered.

[2] Try drawing the tree diagram for this splitting. See Problem 1 at the end of this chapter.

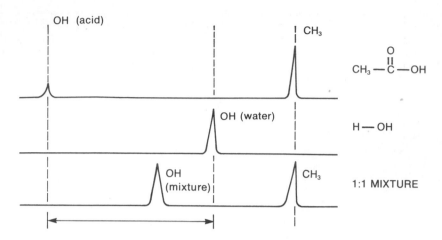

FIGURE 6.4   A comparison of the spectra of acetic acid, water, and a 1:1 mixture of the two.

occurs so rapidly that the NMR again "sees" the hydroxyl protons only in an averaged environment intermediate between the two extremes of the pure substances. The exact position of the O—H resonance depends on the relative amounts of acid and water. In general, if there is more acid than water, the resonance appears closer to the pure acid hydroxyl resonance. With the addition of more water, the resonance moves closer to that of pure water.

Samples of ethanol and water show a similar type of behavior, except that at low concentration of water in ethanol (~1%) both peaks are still often observed. As the amount of water is increased, however, the rate of exchange increases, and the peaks coalesce into the single averaged peak.

B.   Deuterium Exchange

When compounds with acidic hydrogen atoms are placed in D_2O, the acidic hydrogens exchange with deuterium. Sometimes a drop of acid or base catalyst is required, but frequently the exchange occurs spontaneously. The catalyst, however, allows a faster approach to equilibrium, a process that can require anywhere from several minutes to an hour or more.

Acids, phenols, alcohols, and amines are the functional groups that exchange most readily. Basic catalysis works best for acids and phenols, while acidic catalysis is most effective for alcohols and amines.

Basic catalyst

$$RCOOH + D_2O \rightleftharpoons RCOOD + DOH$$

$$ArOH + D_2O \rightleftharpoons ArOD + DOH$$

Acidic catalyst

$$ROH + D_2O \rightleftharpoons ROD + DOH$$

$$RNH_2 + D_2O \rightleftharpoons RND_2 + DOH$$

The result of each deuterium exchange is that the peaks due to the exchanged hydrogens "disappear" from the ^1H NMR spectrum. Notice that each of these exchange reactions produces the hybrid DOH molecule. In a molecule that has several exchangeable groups, several DOH groups are produced. Since all of the hydrogens end up in DOH molecules, the "lost" hydrogens generate a new peak, that of the hydrogen in DOH.

D_2O can be used as a solvent for NMR when it is difficult to dissolve polar compounds in the standard solvents. For instance, many acids are difficult to dissolve in $CDCl_3$. A basic solution of

D$_2$O with NaOD is easily produced by adding a small amount of sodium metal to D$_2$O. This solution readily dissolves carboxylic acids since it converts them to water-soluble (D$_2$O-soluble) salts. The peak due to the hydrogen of the carboxyl group is lost, but a new DOH peak appears.

$$2\,D_2O + 2\,Na \longrightarrow 2\,NaOD + D_2$$

$$RCOOH + NaOD \rightleftharpoons RCOONa + DOH$$

This D$_2$O/NaOD solvent mixture can also be used to exchange the α hydrogens in some ketones, aldehydes, and esters.

$$R{-}CH_2{-}\overset{\displaystyle O}{\overset{\|}{C}}{-}R + 2\,NaOD \rightleftharpoons R{-}CD_2{-}\overset{\displaystyle O}{\overset{\|}{C}}{-}R + 2\,NaOH$$

$$NaOH + D_2O \rightleftharpoons NaOD + DOH$$

Amines dissolve in solutions of D$_2$O to which the acid DCl has been added. The amino protons end up in the DOH peak.

$$R{-}NH_2 + 3\,DCl \rightleftharpoons R{-}ND_3{}^+\,Cl^- + 2\,HCl$$

$$HCl + D_2O \rightleftharpoons DCl + DOH$$

Note, however, that deuterium can actually complicate a proton spectrum. Since it has spin $= 1$, multiplets can end up with more peaks than they originally had. Consider the methine hydrogen in the following case. This hydrogen would be a triplet in the all-hydrogen compound, but it would be a five-line pattern in the deuterated compound. The proton-coupled ^{13}C spectrum would also show an increased complexity due to deuterium (see Section 4.13 p. 189).

$$-CH_2{-}\boxed{CH}{-}\qquad \text{versus} \qquad -CD_2{-}\boxed{CH}{-}$$

Triplet   Five lines

If the NMR spectrum of a particular substance is complicated by the presence of an $-$OH or $-$NH proton, it is possible to simplify the spectrum by removing the peak arising from the exchangeable protons: simply add a few drops of deuterium oxide to the NMR tube containing the solution of the compound being studied. After recapping and shaking the tube vigorously for a few seconds, return the sample tube to the probe and redetermine the spectrum. The peak arising from the exchangeable proton will either disappear or greatly diminish in intensity. The added deuterium oxide is immiscible with the NMR solvent and forms a layer on top of the solution. The presence of this layer, however, does not interfere in the determination of the spectrum. A new peak, owing to the presence of H$-$O$-$D, may be observed, generally around 4.5 to 5.0 ppm.

C.   Peak Broadening Due to Exchange

Rapid intermolecular proton exchange often (but not always!) leads to *peak broadening*. Rather than having a sharp and narrow shape, the peak sometimes increases in width at the base and loses height as a result of rapid exchange. An O$-$H peak can often be distinguished from all other singlets on the basis of this shape difference. Note the hydroxyl peak in Figure 6.2. Peak broadening is caused by factors that are rather complicated, and we will leave their explanation to more advanced texts. We note only that

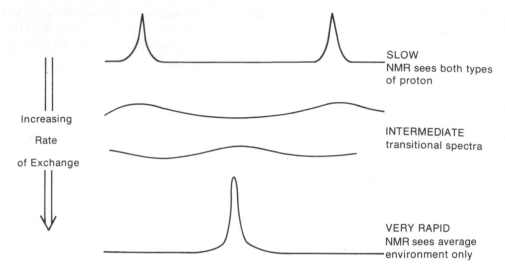

FIGURE 6.5 The effect of the rate of exchange on the NMR spectrum of a hydroxylic compound dissolved in water.

the phenomenon is *time dependent* and that the intermediate transitional stages of peak coalescence are sometimes seen in NMR spectra when the rate of exchange is neither slower nor faster than the NMR time scale but instead is on roughly the same order of magnitude. Figure 6.5 illustrates these situations.

Also, do not forget that when the spectrum of a pure acid or alcohol is determined in an inert solvent (e.g., $CDCl_3$ or CCl_4), the NMR absorption position is concentration dependent. You will recall that this is due to hydrogen-bonding differences. If you have not, now is a good time to reread Section 3.11C and 3.19F.

6.3 OTHER TYPES OF EXCHANGE: TAUTOMERISM

The exchange phenomena that have been presented thus far in this chapter have been essentially *intermolecular* in nature. They are examples of **dynamic NMR**, in which the NMR spectrometer is used to study processes that involve the rapid inverconversion of molecular species. The rates of these interconversions as a function of temperature can be studied, and they can be compared with the NMR time scale.

Molecules whose structures differ markedly in the arrangement of atoms, but that exist in equilibrium with one another, are called **tautomers**. The most common type of tautomerism is **keto–enol tautomerism**, in which the species differ mainly by the position of a hydrogen atom.

In general, the keto form is much more stable than the enol form, and the equilibrium lies strongly in favor of the keto form. Keto–enol tautomerism is generally considered an *intermolecular* process.

1,3-Dicarbonyl compounds are capable of exhibiting keto–enol tautomerism; this is illustrated for the case of **acetylacetone.**

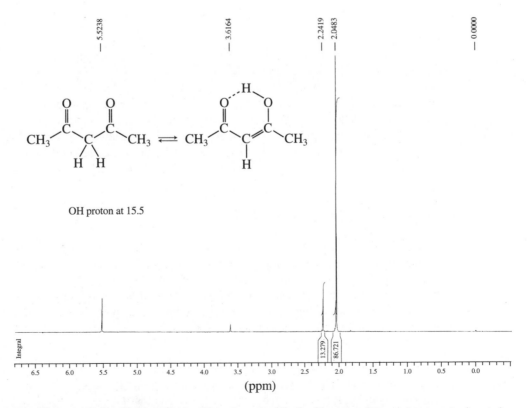

keto   enol

For most 1,3-dicarbonyl compounds, the equilibrium lies substantially to the right, favoring the *enol*. The enol form is stabilized through the formation of a strong intramolecular hydrogen bond.

The proton NMR spectrum of acetylacetone is shown in Figure 6.6. The O—H proton of the enol form can be seen very far downfield, at $\delta = 15.5$ ppm, as well as the vinyl C—H proton at $\delta = 5.5$ ppm. Note also the strong CH_3 peak from the enol from (2.0 ppm), and compare it with the much weaker CH_3 peak from the keto form (2.2 ppm). Also note that the CH_2 peak at 3.6 ppm is weak. Clearly, the enol form predominates in this equilibrium. The fact the we can see the spectra of both tautomeric forms, superimposed on each other, suggests that the rate of conversion of keto form to enol form, and vice versa, must be slow on the NMR time scale.

Another type of tautomerism, *intra*molecular in nature, is called **valence tautomerism** (or **valence isomerization**). Valence tautomers rapidly interconvert with one another, but the tautomeric forms differ principally by the positions of *covalent bonds*, rather than by the positions of protons.

There are many examples of valence tautomerism in the literature. An interesting example is the isomerization of **bullvalene**.

OH proton at 15.5

FIGURE 6.6 The NMR spectrum of acetylacetone. The O—H proton of the enol tautometer is not shown.

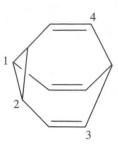

At low temperatures (below −85°C), the proton NMR spectrum of bullvalene consists of four complex multiplets (none of the protons is equivalent to any other proton in the molecule). As the temperature is raised, however, the multiplets broaden and move closer together. Eventually, at +120°C, the entire spectrum consists of *one* sharp singlet.

To explain the temperature-dependent behavior of the spectrum of bullvalene, chemists have determined that bullvalene rearranges through a series of isomerizations known as **Cope rearrangements**.

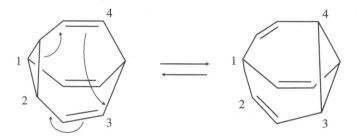

Notice the positions labeled **1** and **2** are part of a cyclopropane ring in one structure, but position **2** is part of a double bond in the second structure. If this rearrangement is repeated so that every bond becomes involved, all 10 hydrogens in bullvalene will become equivalent. An examination of the temperature at which the multiplets coalesce into one broad singlet (+15°C) allows the energy of activation, and thus the rate constant, for the isomerization to be determined. This process would be virtually impossible to study by any other technique except NMR spectroscopy.

There are many other examples of rearrangements, isomerizations, and equilibrations that lend themselves to study by NMR spectroscopy. This technique is used to study changes in conformations of cyclic molecules, proton-exchange reactions, and biochemical processes.

▢ 6.4 PROTONS ON NITROGEN: AMINES

In simple amines, as in alcohols, intermolecular proton exchange is usually fast enough to decouple spin–spin interactions between protons on nitrogen and those on the α carbon atom. Under such conditions, the amino hydrogens usually appear as a sharp singlet (unsplit), and in turn the hydrogens on the α carbon are also unsplit by amino hydrogens. The rate of exchange can be made slower by making the solution strongly acidic (pH <1) and forcing the protonation equilibrium to favor the quaternary ammonium cation rather than the free amine.

$$R-CH_2-NH_2 + H^+ \rightleftharpoons R-CH_2-\overset{\overset{\displaystyle H}{|}}{\underset{\underset{\displaystyle H}{|}}{N^+}}-H$$

excess
(pH <1)

Under these conditions, the predominant species in solution is the protonated amine, and intermolecular proton exchange is slowed, often allowing us to observe spin–spin coupling interactions that are decoupled and masked by exchange in the free amine.

In amides, which are less basic than amines, proton exchange is slow, and coupling is often observed between the protons on nitrogen and those on the α carbon of an alkyl substituent that is substituted on the same nitrogen.

$$^3J_{HH} \sim 7 \text{ Hz}$$

The spectra of *n*-butylamine (Fig. 6.7) and 1-phenylethylamine (Fig. 6.8) are examples of uncomplicated spectra (no NH−CH splitting). Unfortunately, the spectra of amines are not always this simple. Another factor can complicate the splitting patterns of both amines and amides: nitrogen itself has a nuclear spin, which is unity ($I = 1$). Nitrogen can therefore adopt spin states +1, 0, and −1. On the

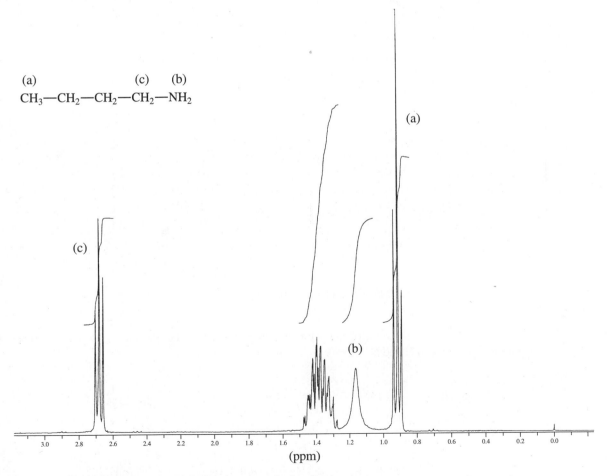

FIGURE 6.7 The NMR spectrum of *n*-butylamine.

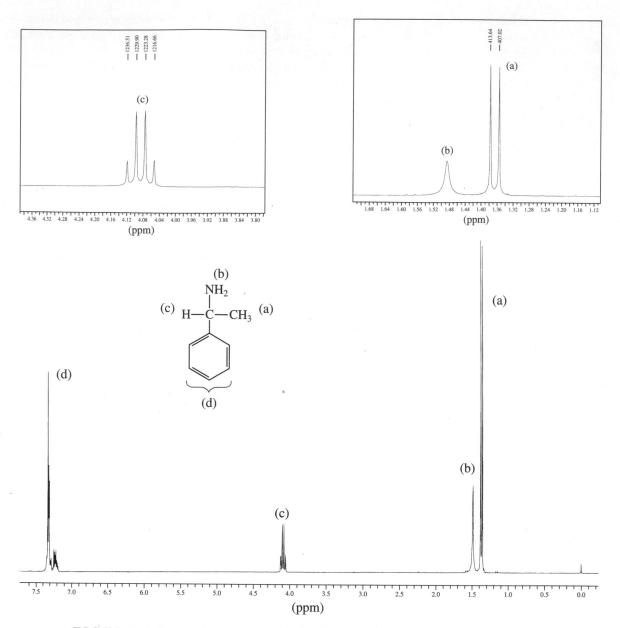

FIGURE 6.8 The NMR spectrum of 1-phenylethylamine

basis of what we know so far of spin–spin coupling, we can predict the following possible types of interaction between H and N.

Direct coupling
$^{1}J \sim 50$ Hz

Geminal coupling
^{2}J and $^{3}J \sim$ negligible
(i.e., almost always zero)

Vicinal coupling

Of these types of coupling, the geminal and vicinal types are very rarely seen, and we can dismiss them. Observation of direct coupling is infrequent but not unknown. Direct coupling is not observed if the hydrogen on the nitrogen is undergoing rapid exchange. The same conditions that decouple NH—CH or HO—CH proton–proton interactions also decouple N—H nitrogen–proton interactions. In cases where direct coupling is observed, the coupling constant is found to be quite large: $^1J \sim 50$ Hz.

One of the cases in which both N—H and CH—NH proton–proton coupling can be observed is the NMR spectrum of methylamine in aqueous hydrochloric acid solution (pH <1). The species actually being observed in this medium is methylammonium chloride, i.e., the hydrochloride salt of methylamine. Figure 6.9 simulates this spectrum. The peak at about 2.2 ppm is due to water (of which there is plenty in aqueous hydrochloric acid solution!). Figures 6.10 and 6.11 analyze the remainder of the spectrum.

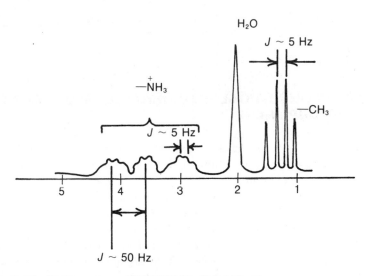

FIGURE 6.9 The NMR spectrum of $CH_3NH_3^+$ in H_2O (pH <1).

ANALYSIS OF THE PROTONS ON NITROGEN

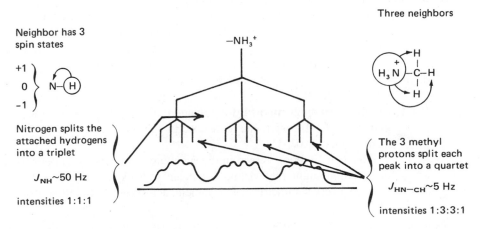

FIGURE 6.10 An analysis of the NMR spectrum of methylammonium chloride: The protons on nitrogen.

ANALYSIS OF THE METHYL PROTONS

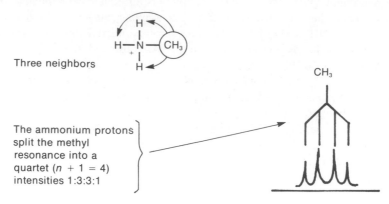

Three neighbors

The ammonium protons split the methyl resonance into a quartet ($n + 1 = 4$) intensities 1:3:3:1

FIGURE 6.11 An analysis of the NMR spectrum of methylammonium chloride: The methyl protons.

6.5 PROTONS ON NITROGEN: QUADRUPOLE BROADENING AND DECOUPLING

Elements that have $I \leq \frac{1}{2}$ have approximately spherical distributions of charge within their nuclei. Those that have $I > \frac{1}{2}$ have ellipsoidal distributions of charge within their nuclei, and as a result have a **quadrupole moment.**

Thus, a major factor determining the magnitude of a quadrupole moment is the symmetry about the nucleus. Unsymmetrical nuclei with a large quadrupole moment are very sensitive both to interaction with the magnetic field of the NMR spectrometer and to magnetic and electric perturbations of their valence electrons or their environment. Nuclei with large quadrupole moments undergo transitions at faster rates than nuclei with small moments and easily reach saturation—the condition in which both upward and downward transitions occur at a rapid rate. Rapid nuclear transitions lead to decoupling of the nucleus with a quadrupole moment from the adjacent NMR-active nuclei. These adjacent nuclei see a single averaged spin ($I = 0$) for the nucleus with the quadrupole moment, and no splitting occurs. Chlorine, bromine, and iodine have large quadrupole moments and are effectively decoupled from interaction with adjacent protons. Note, however, that fluorine ($I = \frac{1}{2}$) has no quadrupole moment, and it does couple with protons.

Nitrogen has a moderate-size quadrupole moment, and its transitions do not occur as rapidly as those in, say, bromine. Furthermore, the transitional rates and lifetimes of its excited spin states (i.e., its quadrupole moments) seem to vary slightly from one molecule to another. Solvent environment and temperature also seem to affect the quadrupole moment. As a result, three distinct situations are possible with a nitrogen atom:

1. **Small quadrupole moment for nitrogen.** In this case coupling is seen. An attached hydrogen (as in N—H) is split into three absorption peaks because of the three possible spin states of nitrogen (+1, 0, −1).

    This first situation is seen in the spectrum of methylammonium chloride (Figures 6.9 to 6.11). Ammonium, methylammonium, and tetraalkylammonium salts place the nitrogen nucleus in a very symmetrical environment, and $^1H - ^{14}N$ coupling is observed.[3]

---

[3] A similar case is seen in borohydride ion where $^1H - ^{11}B$ and $^1H - ^{10}B$ couplings are easily observed due to the symmetry

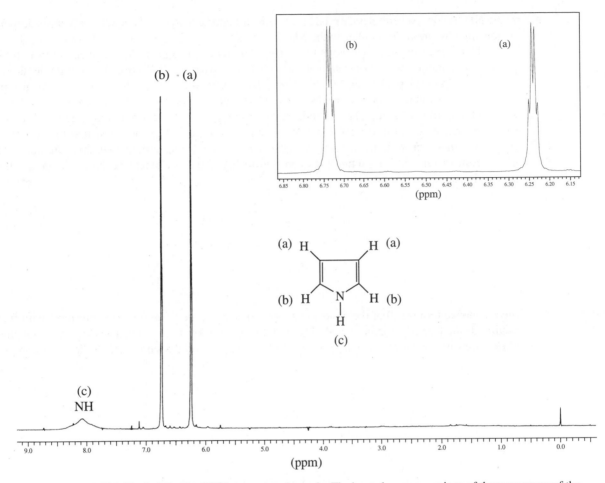

FIGURE 6.12 The NMR spectrum of pyrrole. The inset shows expansions of the resonances of the C—H protons of the ring.

2. **Large quadrupole moment for nitrogen.** In this case no coupling is seen. Due to rapid transitions among the three spin states of nitrogen, an attached proton (as in N—H) "sees" an averaged (zero) spin state for nitrogen. A singlet is observed for the hydrogen.

   This second situation is seen frequently in primary aromatic amines, such as substituted anilines.

3. **Moderate quadrupole moment for nitrogen.** This intermediate case leads to peak broadening, called **quadrupole broadening,** rather than splitting. The attached proton (as in N—H) is "not sure of what it sees."

Figure 6.12, the NMR spectrum of pyrrole, shows an extreme example of quadrupole broadening, where the NH absorption extends from 7.5 to 8.5 ppm.

6.6 AMIDES

Quadrupole broadening usually affects only the proton (or protons) attached directly to nitrogen. In the proton NMR spectrum of an amide, we expect to see the NH proton appear as a very broad singlet. In some instances, however, one might observe the protons on a carbon atom next to the nitrogen split by the NH proton. Nevertheless, the NH peak will still appear as a broad singlet; nuclear quadrupole broadening obscures any coupling to the NH.

While considering amides, note that groups attached to an amide nitrogen often exhibit different chemical shifts. For instance, the NMR spectrum of *N,N*-dimethylformamide shows two distinct methyl peaks (Fig. 6.13). Normally one would expect the two identical groups attached to nitrogen to be chemically equivalent because of free rotation around the C—N bond to the carbonyl group. This is indeed the case on the chemical reaction time scale. However, the rate of rotation around this bond is apparently slowed to the point that on the NMR time scale these groups are no longer equivalent. You may speculate that this circumstance arises from resonance interaction between the unshared pair of electrons on nitrogen and the carbonyl group.

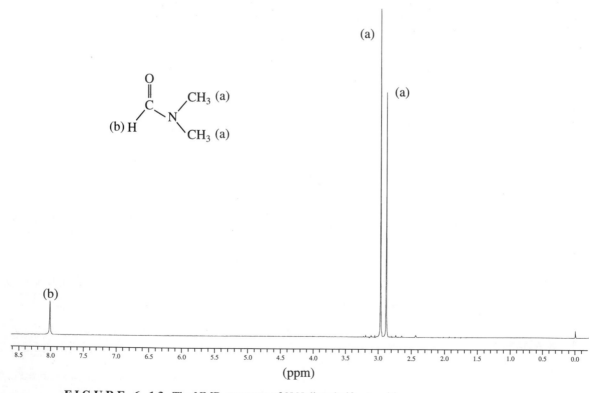

The resonance requires that the molecule adopt a planar geometry, and it thus interferes with free rotation. If the free rotation is slowed to the point where it takes longer than an NMR transition, the NMR spectrometer "sees" two different methyl groups, one on the same side of the C=N bond as

FIGURE 6.13 The NMR spectrum of *N,N*-dimethylformamide.

the carbonyl group and the other on the opposite side. Thus, the groups are in magnetically different environments and have slightly different chemical shifts.

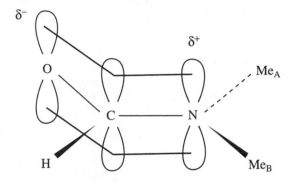

If one successively raises the temperature of the dimethylformamide sample and redetermines the spectrum, the two peaks first broaden (80 to 100°C), then merge to a single broad peak (~120°C), and finally give a sharp singlet (150°C). The increase of temperature apparently speeds up the rate of rotation to the point where the NMR spectrometer records an "average" methyl group. That is, the methyl groups exchange environments so rapidly that during the period of time required for the NMR excitation of one of the methyl protons, that proton is simultaneously experiencing all of its possible conformational positions. Figure 6.14 illustrates changes in the appearance of the methyl resonances of *N,N*-dimethylformamide with temperature.

In Figure 6.15, the spectrum of the amide of chloroacetic acid appears to show quadrupole broadening of the $-NH_2$ resonance. Also, notice that there are *two* N—H peaks. Quadrupole broadening may be present in this case, but other factors must also be considered. In amides, restricted rotation often occurs about the C—N bond, leading to nonequivalence of the two hydrogens on the nitrogen. Even in a substituted amide (RCONHR′), the single hydrogen could have two different absorptions.

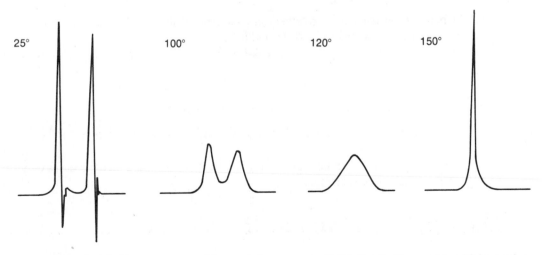

FIGURE 6.14 The appearance of the methyl resonances of *N,N*-dimethylformamide with increasing temperature.

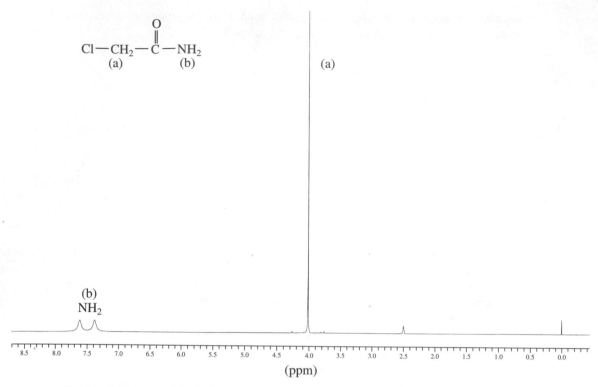

FIGURE 6.15 The NMR spectrum of chloroacetamide.

Depending on the rate of rotation, an averaging of the two NH absorptions could lead to peak broadening (see Sections 6.1, 6.2C, and 6.4).

Thus, in amides, three different peak-broadening factors must always be considered:

1. Quadrupole broadening

2. An intermediate rate of hydrogen exchange on nitrogen

3. Nonequivalence of the NH hydrogen(s) due to restricted rotation

The latter two effects should disappear at higher temperature, which increases either the rate of rotation or the rate of exchange.

▢ 6.7 THE EFFECT OF SOLVENT ON CHEMICAL SHIFT

Chemists generally obtain the NMR spectrum of a substance by following a typical routine. The substance must be dissolved in a solvent, and the solvent that is selected should have certain desirable properties. It should be cheap, it should dissolve a wide range of substances, and it should contain deuterium (for FT-NMR instruments). Deuteriochloroform (chloroform-d, $CDCl_3$) fulfills these requirements very well. This solvent works very well in most applications, and chemists frequently do not consider the role of the solvent in determining the spectrum beyond this point.

The observed chemical shifts, however, depend not only on the structure of the molecule being studied, but also on the interactions between the molecule and the surrounding solvent molecules. If the solvent consists of nonpolar molecules, such as hydrocarbons, there is only weak interaction between solute and solvent (van der Waals interactions or London forces), and the solvent has only a minimal effect on the observed chemical shift.

If the solvent that is selected is polar (e.g., acetone, chloroform, or dimethyl sulfoxide), there are stronger dipole interactions between solvent and solute, especially if the solute molecule also contains polar bonds. The interactions between the polar solvent and a polar solute are likely to be stronger than the interactions between the solvent and tetramethylsilane (TMS, which is nonpolar), and the result is that the observed chemical shift of the molecule of interest will be shifted with respect to the observed chemical shift in a nonpolar solvent. The magnitude of this **solvent-induced shift** can be on the order of several tenths of a part per million in a proton spectrum.

If the solvent has a strong diamagnetic anisotropy (e.g., benzene, pyridine, or nitromethane), the interaction between the solute and the anisotropic field of the solvent will give rise to significant chemical shifts. Again, the solvent will interact more strongly with the solute than it does with TMS. The result is a significant chemical shift for the solute molecules, with respect to the chemical shift of TMS. Solvents such as benzene and pyridine will cause the observed resonance of a given proton to be shifted to a higher field (smaller δ), while other solvents, such as acetonitrile, cause a shift in the opposite direction. This difference appears to be dependent on the shape of the solvent molecules. Aromatic solvents, such as benzene and pyridine, have a flat shape, while acetonitrile has a rod-like shape. The shape of the solvent molecule affects the nature of the solute–solvent complexes that are formed in solution.

The chemist can use these solvent-induced shifts to clarify complex spectra that feature overlapping multiplets. By adding a small amount (up to 20%) of a benzene-d_6 or pyridine-d_5 to the $CDCl_3$ solution of an unknown, a dramatic effect on the appearance of the spectrum can be observed. The chemical shifts of peaks in the proton spectrum can be shifted by as much as 1 ppm, with the result that overlapping multiplets may be separated from one another sufficiently to allow them to be analyzed. The use of this "benzene trick" is an easy way to simplify a crowded spectrum.

6.8 CHEMICAL SHIFT REAGENTS; HIGH-FIELD SPECTRA

Occasionally the 60-MHz spectrum of an organic compound, or a portion of it, is almost undecipherable because the chemical shifts of several groups of protons are all very similar. In such a case, all of the proton resonances occur in the same area of the spectrum, and often peaks overlap so extensively that individual peaks and splittings cannot be extracted. One of the ways in which such a situation can be simplified is by the use of a spectrometer that operates at a frequency higher than 60 MHz. Two common frequencies in use for this purpose in modern instruments are 300 and 400 MHz.

Although coupling constants do not depend on the frequency or the field strength of operation of the NMR spectrometer, chemical shifts in Hertz *are* dependent on these parameters (as Section 3.17 discussed). This circumstance can often be used to simplify an otherwise undecipherable spectrum.

Suppose, for instance, that a compound contains three multiplets: a quartet and two triplets derived from groups of protons with very similar chemical shifts. At 60 MHz, these peaks may overlap and simply give an unresolved envelope of absorption. In redetermining the spectrum at higher field strengths, the coupling constants do not change, but the chemical shifts in Hertz (not parts per million) of the proton groups (H_A, H_B, H_C) responsible for the multiplets do increase. At 300 MHz the individual multiplets are cleanly separated and resolved (see, for example, Fig. 3.35). Also remember that second-order effects disappear at higher field and that many spectra are simplified to first-order spectra at 300 MHz (Sections 5.12A and 5.12F).

Researchers have known for some time that interactions between molecules and solvents, such as those due to hydrogen bonding, can cause large changes in the resonance positions of certain types of protons (e.g., hydroxyl and amino). They have also known that changing from the usual NMR solvents such as CCl_4 and $CDCl_3$ to solvents such as benzene, which impose local anisotropic effects on surrounding molecules, can greatly affect the resonance positions of some groups of protons. In many cases it is possible to resolve partially overlapping multiplets by such a solvent change. However, the use of **chemical shift reagents**, an innovation dating from about 1969, allows a rapid and relatively inexpensive means of resolving overlapping multiplets in some spectra. Most of these chemical shift reagents are organic complexes of paramagnetic rare-earth metals from the lanthanide series. When such metal complexes are added to the compound whose spectrum is being determined, workers note profound shifts in the resonance positions of the various groups of protons. The direction of the shift (upfield or downfield) depends primarily on which metal is being used. Complexes of europium, erbium, thulium, and ytterbium shift resonances to lower field, while complexes of cerium, praseodymium, neodymium, samarium, terbium, and holmium generally shift resonances to higher field. The advantage of using such reagents is that shifts similar to those observed at higher field can be induced without the purchase of an expensive 300-MHz NMR instrument.

Of the lanthanides, europium is probably the most commonly used metal. Two of its widely used complexes are *tris*-(dipivalomethanato) europium and *tris*-(6,6,7,7,8,8,8-heptafluoro-2,2-dimethyl-3,5-octanedionato) europium, frequently abbreviated $Eu(dpm)_3$ and $Eu(fod)_3$, respectively. Their structures follow.

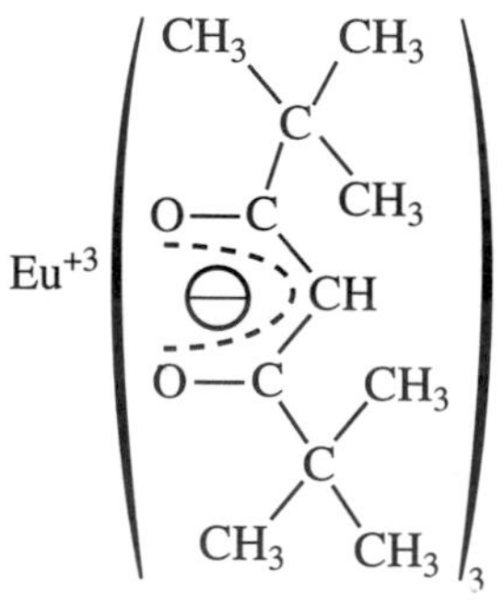

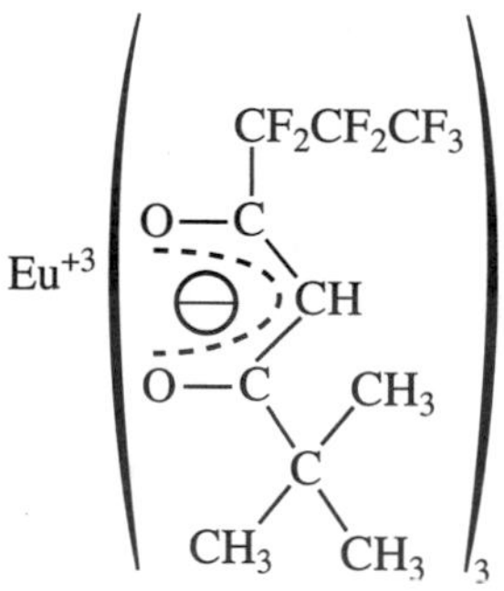

These lanthanide complexes produce spectral simplifications in the NMR spectrum of any compound with a relatively basic pair of electrons (an unshared pair) which can coordinate with Eu^{3+}. Typically, aldehydes, ketones, alcohols, thiols, ethers, and amines all interact:

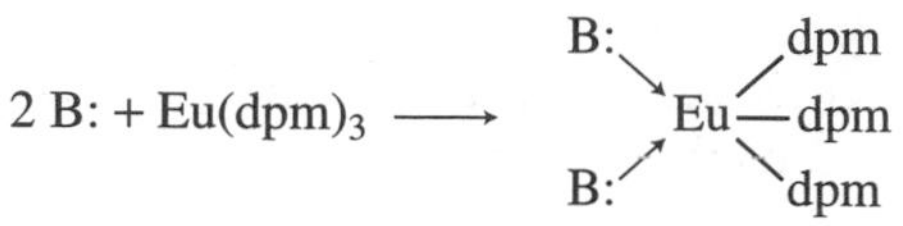

The amount of shift a given group of protons experiences depends on (1) the distance separating the metal (Eu^{3+}) and that group of protons and (2) the concentration of the shift reagent in the solu-

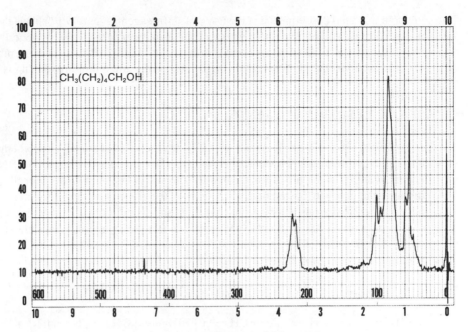

FIGURE 6.16 The normal 60-MHz NMR spectrum of 1-hexanol.

tion. Because of the latter dependence, it is necessary to include the number of mole equivalents of shift reagent used or its molar concentration when reporting a lanthanide-shifted spectrum.

The spectra of 1-hexanol (Figs. 6.16 and 6.17) beautifully illustrate the distance factor. In the absence of shift reagent, the spectrum shown in Figure 6.16 is obtained. Only the triplet of the terminal methyl group and the triplet of the methylene group next to the hydroxyl are resolved in the spectrum. The other protons (aside from O—H) are found together in a broad, unresolved group. With the shift reagent added (Fig. 6.17), each of the methylene groups is clearly separated and is resolved into the proper multiplet structure. The spectrum is in every sense *first-order* and thus simplified; all of the splittings are explained by the $n + 1$ Rule.

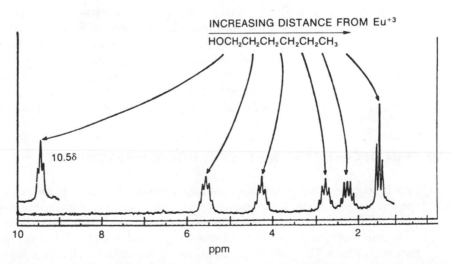

FIGURE 6.17 The 100-MHz NMR spectrum of 1-hexanol with 0.29-mole equivalent of Eu(dpm)$_3$ added. (From Sanders, J. K. M., and D. H. Williams, *Chem. Commun.*, 442 (1970). Reprinted by permission.)

Note one final consequence of the use of a shift reagent. Figure 6.17 shows that the multiplets are not as nicely resolved into sharp peaks as one usually expects. This is due to the fact that the shift reagents cause a small amount of line broadening. At high shift-reagent concentrations this problem becomes serious, but at most useful concentrations the amount of broadening is tolerable.

6.9 CHIRAL RESOLVING AGENTS

A group attached to a chiral carbon normally has the same chemical shift whether the chiral carbon has R or S configuration. However, the group can be made diastereotopic in the NMR (have different chemical shifts) when the racemic parent compound is treated with an optically pure **chiral resolving agent** to produce diastereomers. In this case, the group is no longer found in two enantiomers but in two different diastereomers, and its chemical shift is different in each environment.

For instance, if the partly resolved amine α-phenylethylamine, a mixture containing both the R and S enantiomers, is mixed with optically pure (S)-$(+)$-O-acetylmandelic acid in an NMR tube containing $CDCl_3$, two diastereomers form:

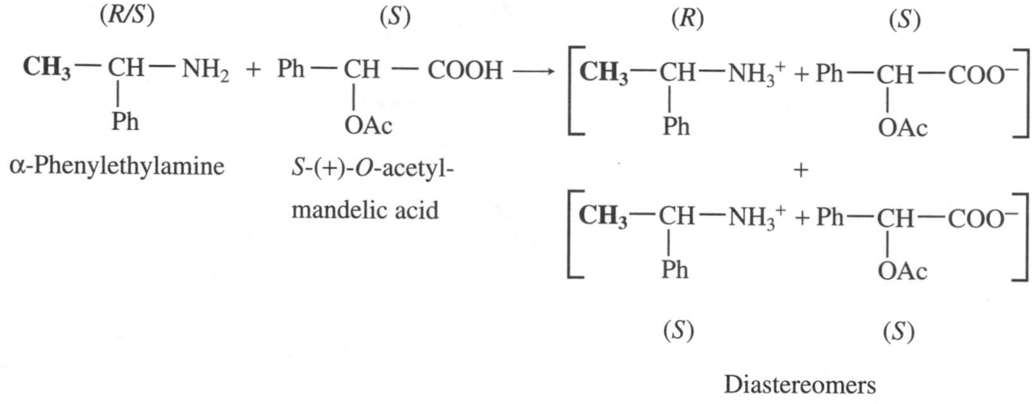

The methyl groups in the amine portions of the two diastereomeric salts are attached to a stereocenter, S in one case and R in the other. As a result, the methyl groups themselves become diastereomeric (diastereotopic), and they have different chemical shifts. In this case, the R isomer is downfield and the S isomer is upfield. These methyl groups appear at approximately (varying with concentrations) 1.1 and 1.2 ppm, respectively, in the proton NMR spectrum of the mixture. Since the methyl groups are adjacent to a methine (CH) group, they appear as doublets, as shown in Figure 6.18.

The doublets which appear may be integrated in order to determine the percentages of the R and S amines in the mixture. In the example, the NMR spectrum was determined with a mixture made by dissolving equal quantities (a 50/50 mixture) of the original unresolved $(\pm)$-α-phenyl-ethylamine and a student's resolved product, which contained predominantly (S)-$(-)$-α-phenyl-ethylamine.

6.10 SPIN DECOUPLING METHODS; DOUBLE RESONANCE

Spin decoupling, sometimes called **double resonance,** is another technique used to simplify NMR spectra. It is very useful for determining which spin–spin multiplets are actually linked with one another and for locating absorptions that may be hidden, or "buried," under other group absorptions.

In spin decoupling, a second radiofrequency signal is applied while the spectrum is scanned in the usual fashion. This radiofrequency can be set to irradiate a selected group of protons in the molecule being measured. Irradiation causes the selected group of protons to undergo rapid transitions

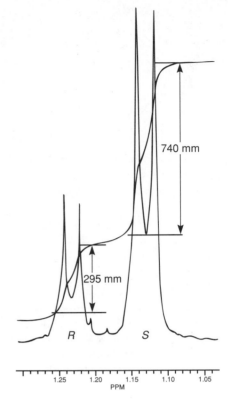

FIGURE 6.18 The 300-MHz ^{1}H spectrum of a 50–50 mixture of resolved and unresolved (racemic) α-phenylethylamine in CDCl$_3$. The chiral resolving agent (S)-(+)-O-acetylmandelic acid was added.

among their nuclear spin states. Just as with rapid exchange (Section 6.1) or with the nuclear quadrupole phenomenon (Section 6.5), these rapid nuclear transitions decouple the protons of this group from any spin–spin interactions, and the NMR spectrum is simplified.

As an example, consider propyl bromide. Figure 6.19A is its normal 60-MHz NMR spectrum. Figure 6.19B shows the effect of redetermining the NMR spectrum with simultaneous irradiation of the α CH$_2$ protons (the large arrow indicates the point of irradiation). Under these conditions, the resonance of the α CH$_2$ protons collapses and cannot be observed. The α CH$_2$ protons are caused to cycle through all of their spin states very rapidly and are thereby decoupled from all spin–spin interactions with the adjacent β CH$_2$ protons. Therefore, the β CH$_2$ protons couple only with the three methyl protons on the other side, and their resonance is simplified to a quartet rather than the original sextet. The methyl resonance, which does not couple with the α CH$_2$ protons even under nonirradiative conditions, is unchanged. The two β CH$_2$ protons split it into a triplet. Figures 6.19C and 6.19D show the effects of double irradiation (decoupling) of the β CH$_2$ and γ CH$_3$ protons, respectively.

The principal application of spin decoupling is as an aid in the interpretation of complicated spectra. For instance, the preceding experiments definitely establish the nature of the spin–spin interactions in propyl bromide. Spectrum B shows that the α CH$_2$ protons couple with the β CH$_2$ protons but not with the γ CH$_3$ protons. The resonance of the latter protons is left unchanged by irradiation of the α CH$_2$ group, while the resonance of the β CH$_2$ protons, which do couple, is simplified. Spectrum C shows that the α CH$_2$ and γ CH$_3$ protons do not interact but that the β CH$_2$ protons are responsible for the splitting of both groups in spectrum a. Similarly, spectrum D establishes that the γ CH$_3$ protons couple only with the β CH$_2$ protons. In other words, the decoupled spectra establish beyond a doubt which groups of protons interact to yield the multiplet structures in the propyl bromide spectrum. Although these relationships may have seemed obvious for propyl bromide, a relatively simple compound, one can easily see that such experiments could be extremely useful in the spectrum of a complicated compound.

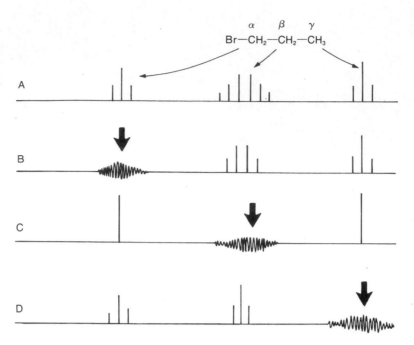

FIGURE 6.19 NMR spectra of propyl bromide: (A) the normal spectrum at 60 MHz; (B) the effect of decoupling (irradiating) the α CH$_2$ protons; (C) the effect of decoupling the β CH$_2$ protons; (D) the effect of decoupling the γCH$_3$ protons.

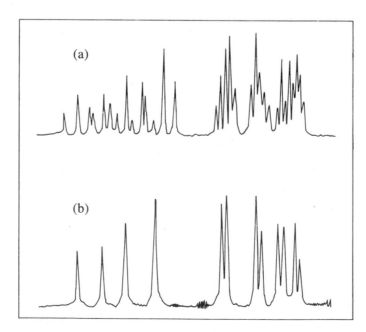

FIGURE 6.20 A spectrum of the vinyl protons of allyl bromide (CH$_2$=CH−CH$_2$−Br): (a) the normal spectrum; (b) the spectrum with the −CH$_2$−Br protons decoupled. (From McFarlane, W., *Chem. Britain*, 5 (1969):143. Reprinted by permission.)

Double resonance is often useful in the interpretation of difficult spectra that have extensive multiplet structures. By careful selection of the double resonance frequency, one can often simplify such a spectrum to the point where it is easy to interpret. Allyl bromide (Fig. 6.20) is an illustrative example. The resonances of the double-bond protons are very complex, owing principally to allylic couplings with the $-CH_2-$ group protons. If the allylic $-CH_2-Br$ protons are decoupled, however, the spectrum simplifies to the typical AMX pattern found for many alkenes (see vinyl acetate, Figs. 5.17 and 5.18). The lower trace of Figure 6.20 illustrates this AMX simplification. You can see that, while it would have been difficult to determine J values for the various vinyl interactions in the original spectrum, it is relatively easy to do so in the decoupled spectrum, which presents a much simpler splitting pattern. Today, spin-decoupling methods have in large part been replaced by 2D-NMR techniques, which are discussed in Chapter 10.

6.11 NOE DIFFERENCE SPECTRA

In many cases of interpretation of NMR spectra, it would be helpful to be able to distinguish protons by their *spatial* location within a molecule. For example, for alkenes it would be useful to determine whether two groups are *cis* to each other or whether they represent a *trans* isomer. In bicyclic molecules, the chemist may wish to know whether a substituent is in an *exo* or in an *endo* position. Many of these types of problems cannot be solved by an analysis of chemical shift or by examination of spin–spin splitting effects.

A handy method for solving these types of problems is **NOE difference spectroscopy**. This technique is based on the same phenomenon that gives rise to the nuclear Overhauser effect (Section 4.5), except that it uses *homonuclear*, rather than a heteronuclear, decoupling. In the discussion of the nuclear Overhauser effect, attention was focused on the case in which a hydrogen atom was directly bonded to a ^{13}C atom and the hydrogen nucleus was saturated by a broad-band signal. In fact, however, for two nuclei to interact via the nuclear Overhauser effect, the two nuclei do not need to be directly bonded; it is sufficient that they be *near* each other (generally within about 3 Å. Nuclei that are in close spatial proximity are capable of relaxing one another by a **dipolar** mechanism. If the magnetic moment of one nucleus, as it precesses in the presence of an applied magnetic field, happens to generate an oscillating field that has the same frequency as the resonance frequency of a nearby nucleus, the two affected nuclei will undergo a mutual exchange of energy, and they will relax one another. The two groups of nuclei that interact by this dipolar process must be very near each other; the magnitude of the effect decreases as r^{-6}, where r is the distance between the nuclei.

We can take advantage of this dipolar interaction with an appropriately timed application of a low-power decoupling pulse. If we irradiate one group of protons, any nearby protons that interact with it by a dipolar mechanism will experience an enhancement in signal *intensity*.

The typical NOE difference experiment consists of *two* separate spectra. In the first experiment, the decoupler frequency is tuned to match exactly the group of protons that we wish to irradiate. The second experiment is conducted under conditions identical to the first experiment, except that the frequency of the decoupler is adjusted to a value far away in the spectrum from any peaks (often the residual proton peak from the trace amount of $CHCl_3$ present in the $CDCl_3$ solvent is chosen for this second frequency). The two spectra are subtracted from each other (this is done by treating digitized data within the computer), and the *difference* spectrum is plotted.

The NOE difference spectrum thus obtained would be expected to show a *negative* signal for the group of protons that had been irradiated. *Positive* signals should be observed *only* for those nuclei that interact with the irradiated protons by means of a dipolar mechanism. In other words, only those nuclei that are located within about 2.5 to 3 Å of the irradiated protons will give rise to a positive signal. All other nuclei that are not affected by the irradiation will appear as very weak or absent signals.

The spectra presented in Figure 6.21 illustrate an NOE difference analysis of **ethyl methacrylate**.

$$\begin{array}{c} H_Z \qquad CH_3 \\ \diagdown \quad \diagup \\ C=C \\ \diagup \qquad \diagdown \\ H_E \qquad\quad C-O-CH_2-CH_3 \\ \| \\ O \end{array}$$

The upper spectrum shows the normal proton NMR spectrum of this compound. We see peaks arising from the two vinyl hydrogens at 5.5 to 6.1 ppm. It might be assumed that H_E should be shifted further downfield than H_Z owing to the "through space" deshielding effect of the carbonyl group. It is necessary, however, to confirm this prediction through experiment to determine unambiguously which of these peaks corresponds to H_Z and which corresponds to H_E.

The second spectrum was determined with the simultaneous irradiation of the methyl resonance at 1.9 ppm. We immediately see that the 1.9 ppm peak appears as a strongly negative peak. The only peak in the spectrum that appears as a positive peak is the vinyl proton peak at 5.5 ppm. The other vinyl peak at 6.1 ppm has nearly disappeared, as have most of the other peaks in the spectrum. The presence of a positive peak at 5.5 ppm confirms that this peak must come from proton H_Z; proton H_E is too far away from the methyl group to experience any dipolar relaxation effects.

The above result could have been obtained by conducting the experiment in the opposite direction. Irradiation of the vinyl proton at 5.5 ppm would have caused the methyl peak at 1.9 ppm to be positive. The results, however, would not be very dramatic; it is always more effective to irradiate the group with the larger number of equivalent hydrogens and observe the enhancement of the group with the smaller number of hydrogens, rather than vice versa.

Finally, the third spectrum was determined with the simultaneous irradiation of the H_E peak at 6.1 ppm. The only peak that appears as a positive peak is the H_Z peak at 5.5 ppm, as expected. The methyl peak at 1.9 ppm does not show any enhancement, confirming that the methyl group is distant from the proton responsible for the peak at 6.1 ppm.

This example is intended to illustrate how NOE difference spectroscopy can be used to solve complex structural problems. This technique is particularly well suited to the solution of problems involving the location of substituents around an aromatic ring and stereochemical differences in alkenes or in bicyclic compounds.

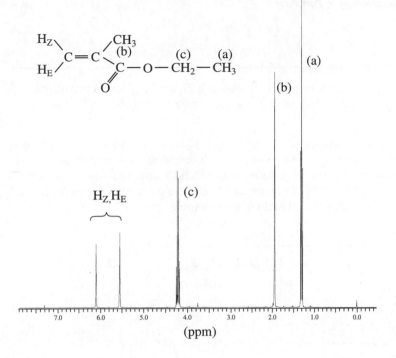

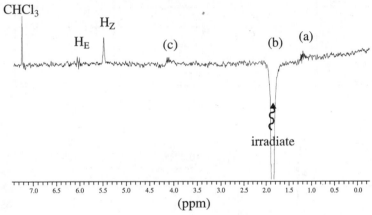

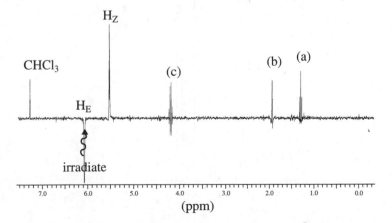

FIGURE 6.21 NOE Difference Spectrum of Ethyl Methacrylate. (Top Spectrum: Proton NMR Spectrum of Ethyl Methacrylate Without Decoupling; Middle Spectrum: NOE Difference Spectrum, With Irradiation at 1.9 ppm; Bottom Spectrum: NOE Difference Spectrum With Irradiation at 6.1 ppm)

PROBLEMS

*1. The spectrum of an ultrapure sample of ethanol is shown in Figure 6.3. Draw a tree diagram for the methylene groups in ethanol that takes into account the coupling to both the hydroxyl and methyl groups.

*2. The following spectrum is for a compound with the formula $C_5H_{10}O$. The peak at about 1.9 ppm is solvent and concentration dependent. Expansions are included, along with an indication of the spacing of the peaks in Hertz. The pairs of peaks at about 5.0 and 5.2 ppm have fine structure. How do you explain this small coupling? Draw the structure of the compound, assign the peaks, and include tree diagrams for the expanded peaks in the spectrum.

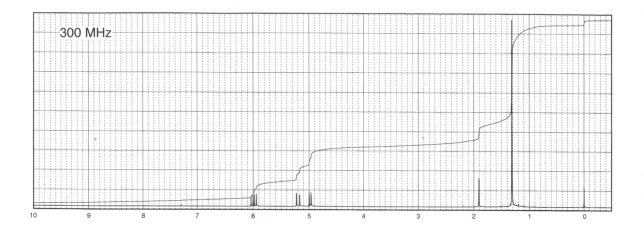

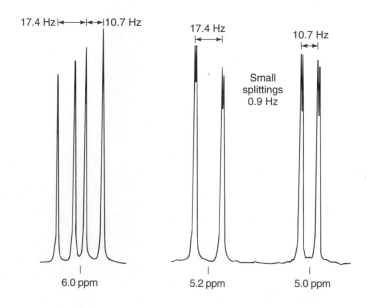

*3. Determine the structure of the aromatic compound with formula C_6H_5BrO. The peak at about 5.6 ppm is solvent dependent and shifts readily when the sample is diluted. The expansions that are provided show ^4J couplings of about 1.6 Hz.

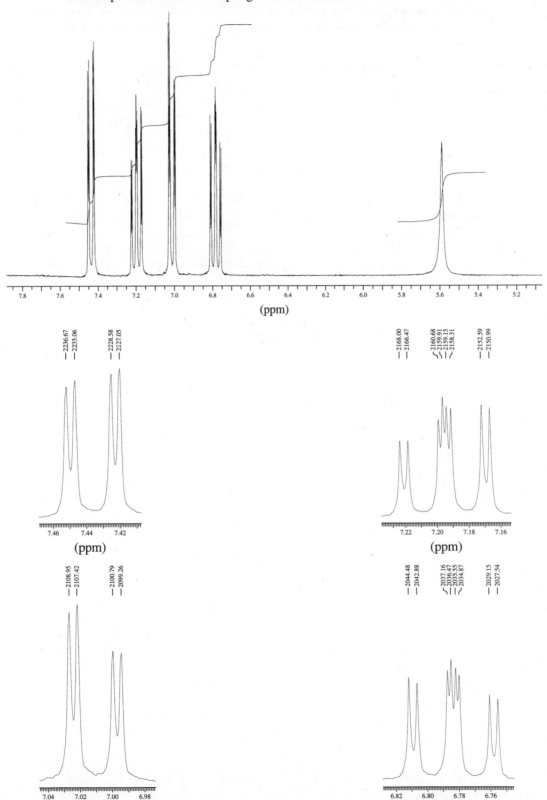

*4. The compound whose spectrum is shown is derived from 2-methylphenol. The formula of the product obtained is $C_9H_{10}O_2$. The infrared spectrum shows prominent peaks at 3136 and 1648 cm^{-1}. The broad peak at 8.16 ppm is solvent dependent. Determine the structure of this compound using the spectrum provided and calculations of the chemical shifts (see Appendix 6). The calculated values will be only approximate, but should allow you to determine the correct structure.

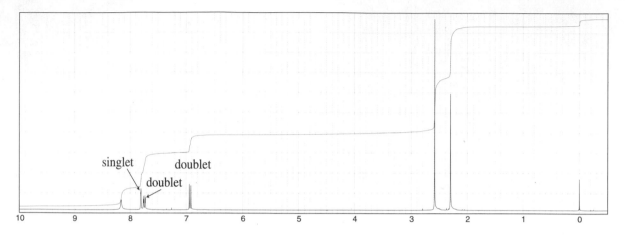

*5. The spectrum and expansions provided in this problem are for one of the compounds shown below. The broad peak at 5.25 ppm is solvent dependent. Using calculations of the *approximate* chemical shifts and the appearance and position of the peaks (singlet and doublets), determine the correct structure. The chemical shifts may be calculated from the information provided in Appendix 6. The calculated values will be only approximate, but should allow you to determine the correct structure.

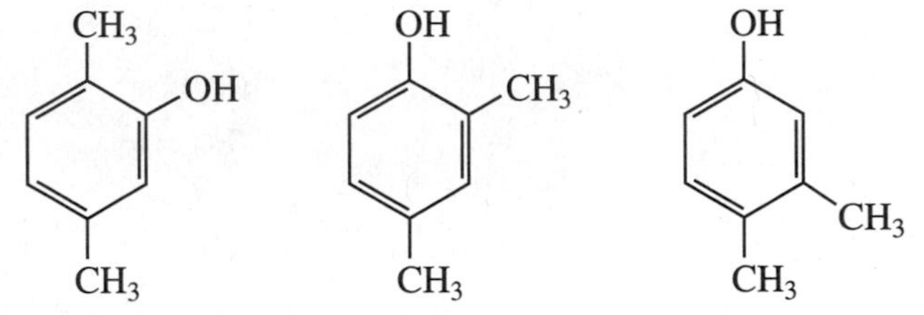

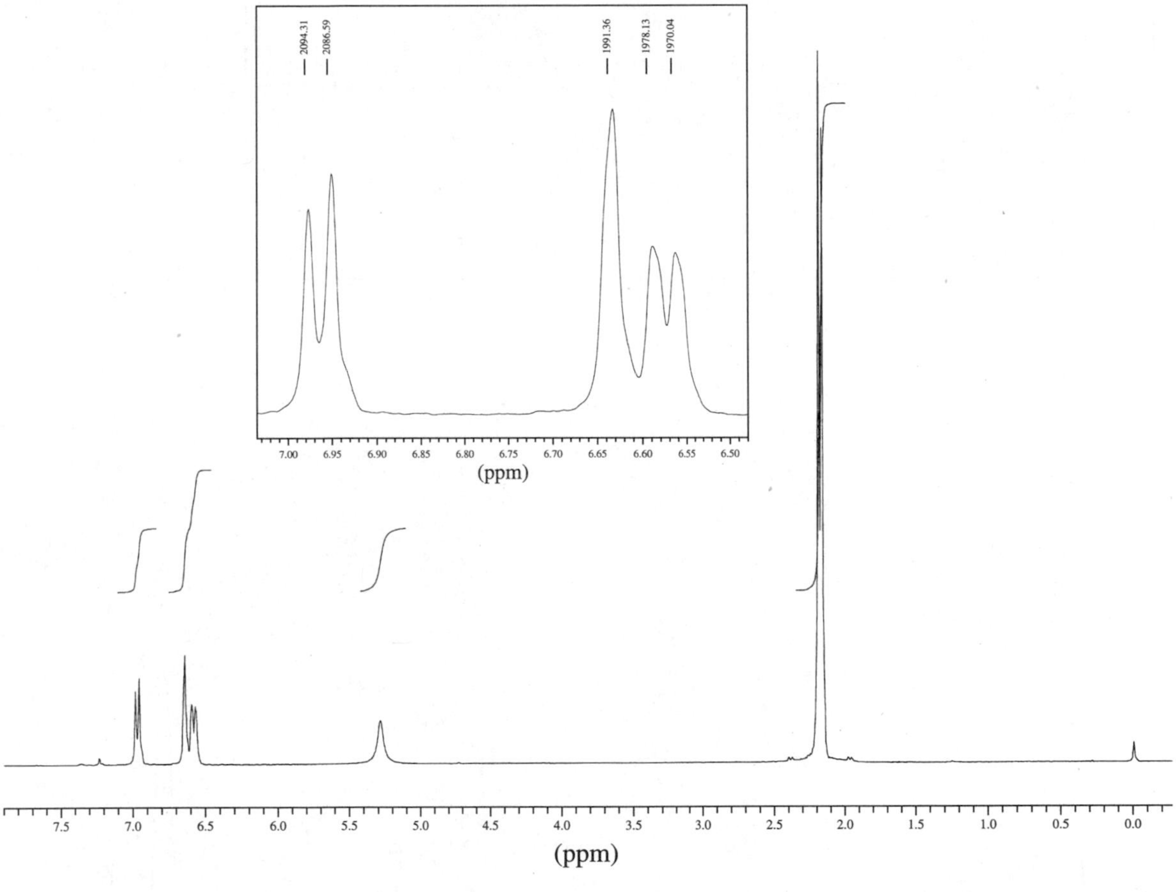

*6. The proton NMR spectrum for a compound with formula $C_5H_{10}O_2$ is shown. Determine the structure of this compound. The peak at 2.1 ppm is solvent dependent. Expansions are provided for some of the protons. Comment on the fine structure on the peak at 4.78 ppm. The normal carbon-13, DEPT-135, and DEPT-90 spectra data are tabulated.

Normal Carbon	DEPT-135	DEPT-90
22 ppm	Positive	No peak
41	Negative	No peak
60	Negative	No peak
112	Negative	No peak
142	No peak	No peak

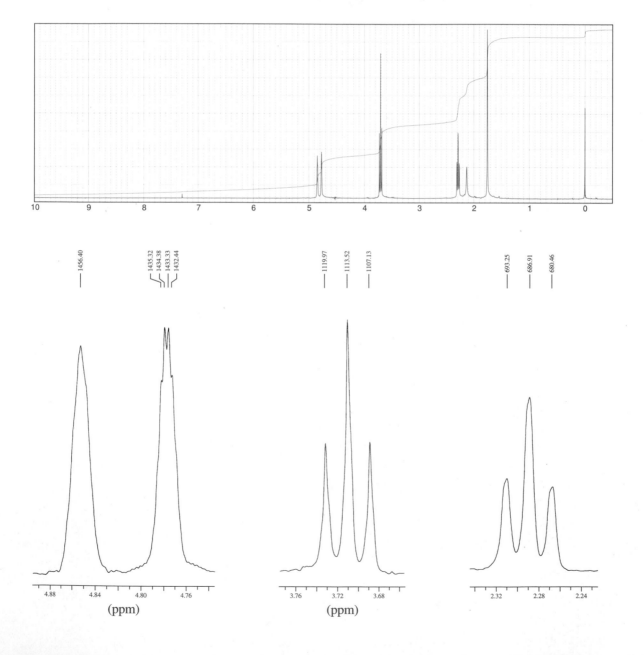

7. The proton NMR spectrum of a compound with formula $C_5H_{10}O_2$ is shown. The peak at 2.1 ppm is solvent dependent. The infrared spectrum shows a broad and strong peak at 3332 cm^{-1}. The normal carbon-13, DEPT-135, and DEPT-90 spectra data are tabulated.

Normal Carbon	DEPT-135	DEPT-90
11 ppm	Negative	No peak
18	No peak	No peak
21	Positive	No peak
71	Negative	No peak

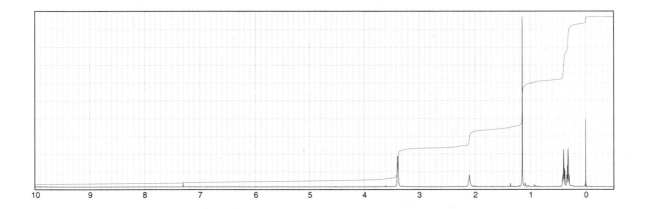

8. Determine the structure of the aromatic compound with formula $C_9H_9ClO_3$. The infrared spectrum shows a very broad band from 3300 to 2400 cm^{-1} and a strong band at 1714 cm^{-1}. The full proton NMR spectrum and expansions are provided. The compound is prepared by a nucleophilic substitution reaction of the sodium salt of 3-chlorophenol on a halogen-bearing substrate.

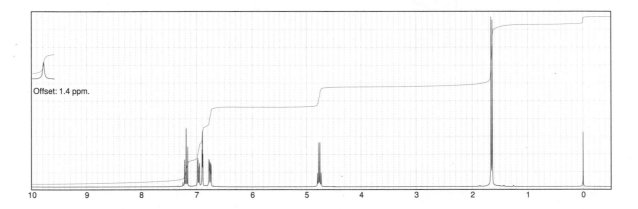

Offset: 1.4 ppm.

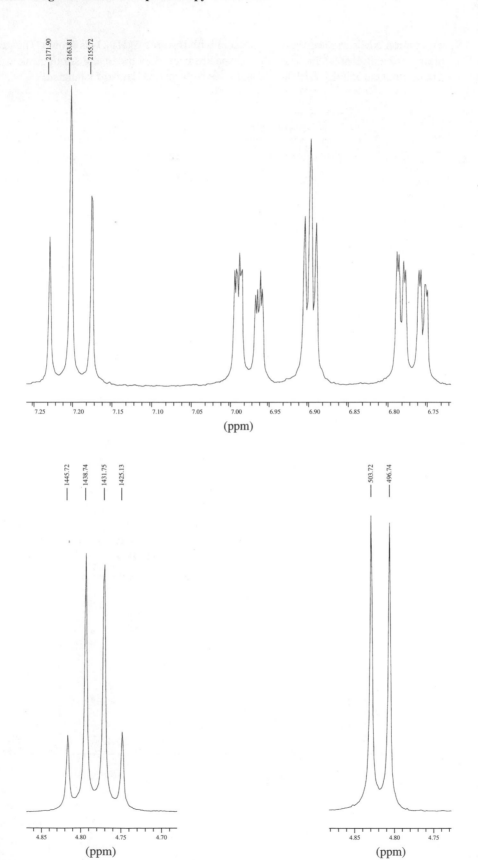

2171.90
2163.81
2155.72

(ppm)

1445.72
1438.74
1431.75
1425.13

503.72
496.74

(ppm)   (ppm)

*9. Determine the structure of a compound with formula $C_{10}H_{15}N$. The proton NMR spectrum is shown. The infrared spectrum has medium bands at 3420 and 3349 cm^{-1} and a strong band at 1624 cm^{-1}. The broad peak at 3.5 ppm in the NMR shifts when DCl is added while the other peaks stay in the same positions.

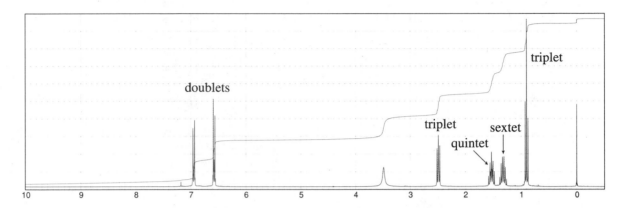

*10. Determine the structure of a compound with formula $C_6H_5Br_2N$. The proton NMR spectrum is shown. The infrared spectrum has medium bands at 3420 and 3315 cm^{-1} and a strong band at 1612 cm^{-1}. The normal carbon, DEPT-135, and DEPT-90 spectra data are tabulated.

Normal Carbon	DEPT-135	DEPT-90
109 ppm	No peak	No peak
119	Positive	Positive
132	Positive	Positive
142	No peak	No peak

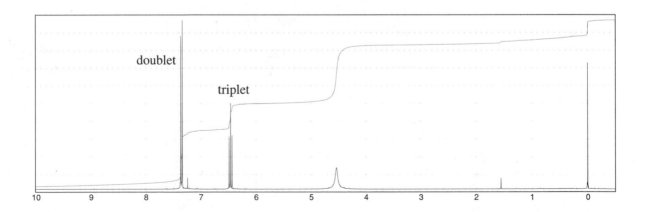

11. There are three spectra shown in this problem along with three structures of aromatic primary amines. Assign each spectrum to the appropriate structure. You should calculate the *approximate* chemical shifts (Appendix 6) and use these values along with the appearance and position of the peaks (singlet and doublets) to assign the correct structure.

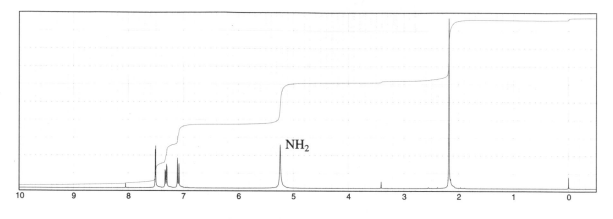

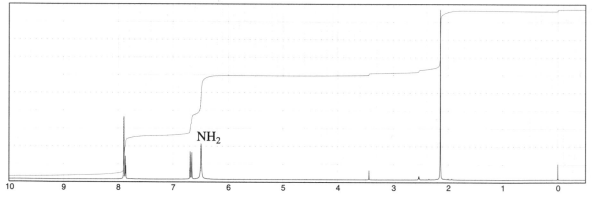

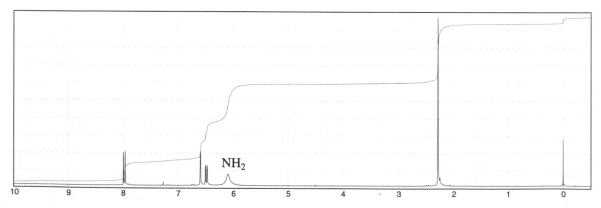

*12. When aniline is chlorinated, a product with the formula $C_6H_5NCl_2$ is obtained. The spectrum of this compound is shown. The expansions are labeled to indicate couplings, in Hertz. Determine the structure and substitution pattern of the compound, and assign each set of peaks. Explain the splitting patterns.

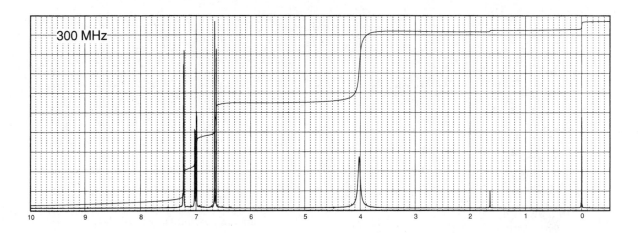

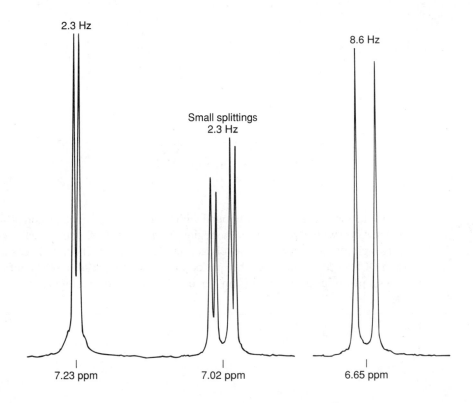

*13. A naturally occurring amino acid with the formula $C_3H_7NO_2$ gives the following proton NMR spectrum when determined in deuterium oxide solvent. The amino and carboxyl protons merge into a single peak at 4.9 ppm in the D_2O solvent (not shown); the peaks of each multiplet are separated by 7 Hz. Determine the structure of this amino acid.

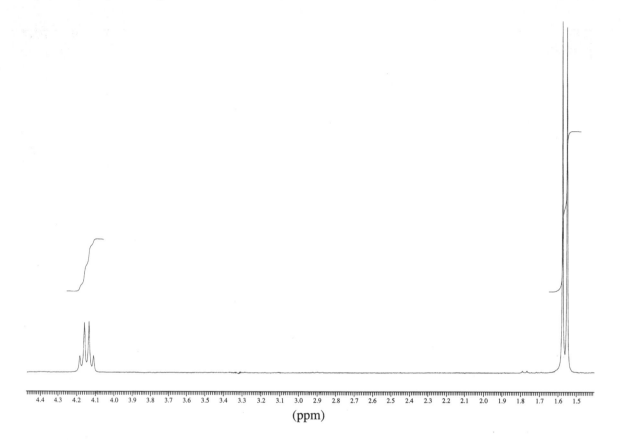

(ppm)

14. Determine the structure of a compound with formula C_7H_9N. The proton NMR spectrum is shown, along with expansions of the region from 7.10 to 6.60 ppm. The three-peak pattern for the two protons at about 7 ppm involves overlapping peaks. The broad peak at 3.5 ppm shifts when DCl is added while the other peaks stay in the same positions. The infrared spectrum shows a pair of peaks near 3400 cm^{-1} and an out-of-plane bending band at 751 cm^{-1}.

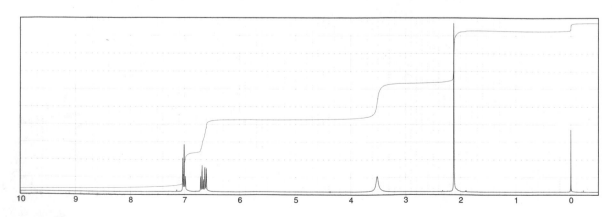

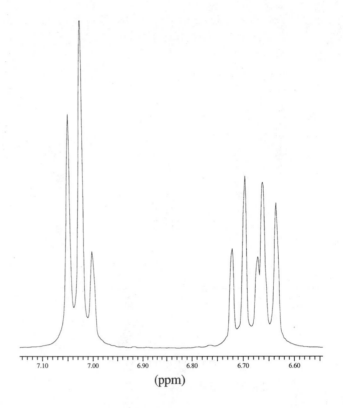

(ppm)

15. A naturally occurring amino acid with the formula $C_9H_{11}NO_3$ gives the following proton NMR spectrum when determined in deuterium oxide solvent with DCl added. The amino, carboxyl, and hydroxyl protons merge into a single peak at 5.1 ppm (4 H) in D_2O. Determine the structure of this amino acid and explain the pattern that appears in the range 3.17 to 3.40 ppm, including coupling constants.

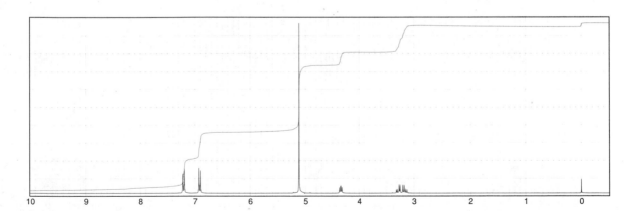

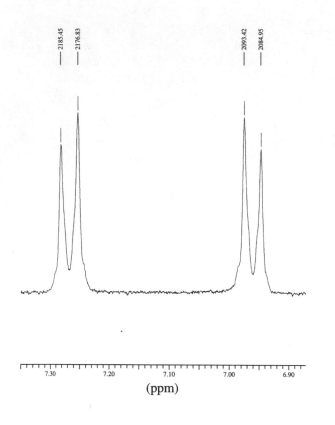

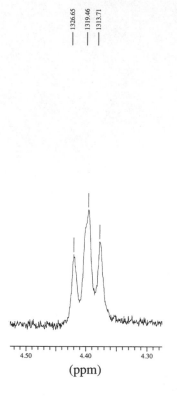

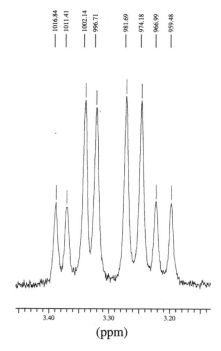

16. Determine the structure of a compound with formula $C_6H_{10}O_2$. The proton NMR spectrum with expansions is provided. Comment as to why the proton appearing at 6.91 ppm is a triplet of quartets, with spacing of 1.47 Hz. Also comment on the "singlet" at 1.83 that shows fine structure. The normal carbon, DEPT-135, and DEPT-90 spectral results are tabulated.

Normal Carbon	DEPT-135	DEPT-90
12 ppm	Positive	No peak
13	Positive	No peak
22	Negative	No peak
127	No peak	No peak
147	Positive	Positive
174	No peak	No peak

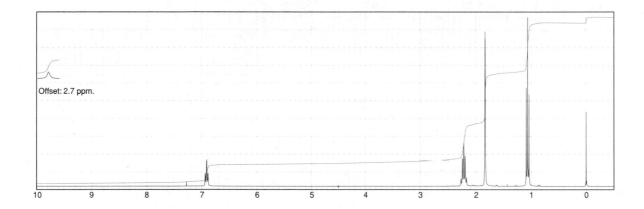

Offset: 2.7 ppm.

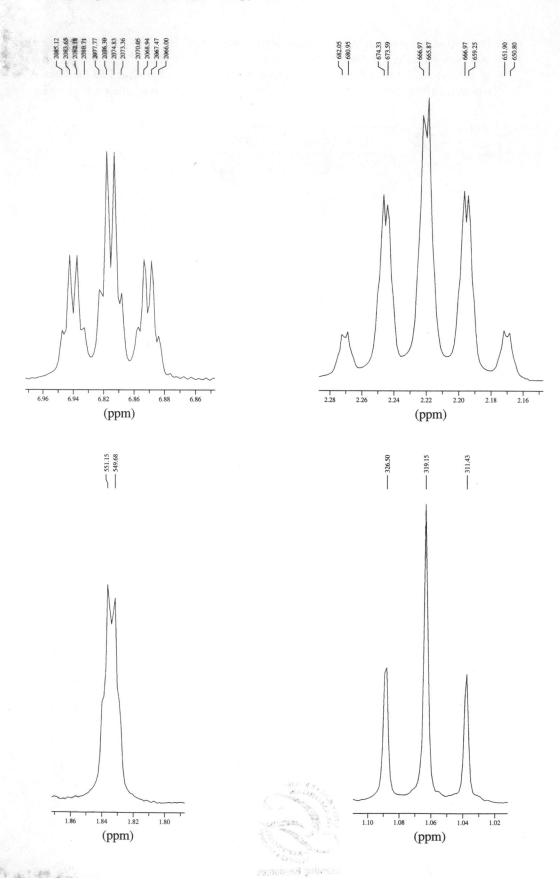

17. The following proton NMR spectrum is of a discontinued analgesic drug, phenacetin ($C_{10}H_{13}NO_2$). Phenacetin is structurally related to the very popular and current analgesic drug, acetaminophen. Phenacetin contains an amide functional group. Two tiny impurity peaks appear near 3.4 and 8.1 ppm. Give the structure of this compound, and interpret the spectrum.

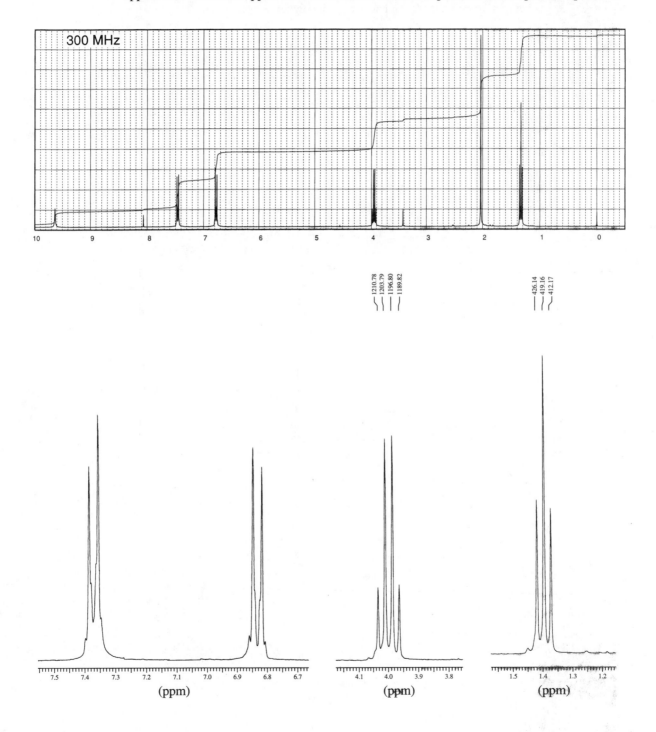

18. The proton NMR spectrum shown in this problem is for a common insect repellent, *N*,*N*-diethyl-*m*-toluamide, determined at 360 K. This problem also shows a stacked plot of this compound determined in the temperature range of 290 to 360 K (27 to 87°C). Explain why the spectrum changes from two pairs of broadened peaks near 1.2 and 3.4 ppm at low temperature to a triplet and quartet at the higher temperatures.

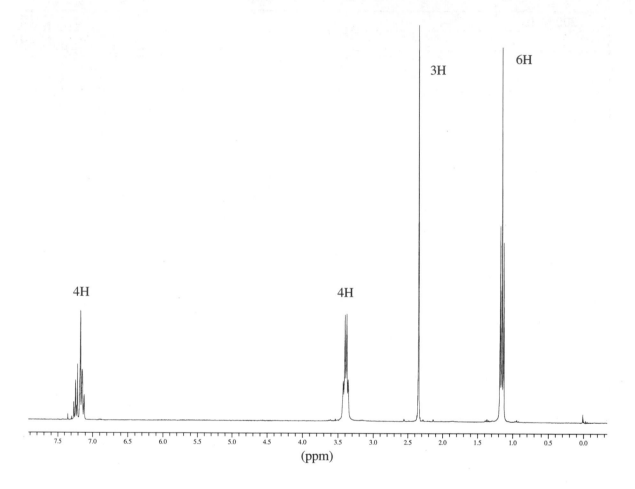

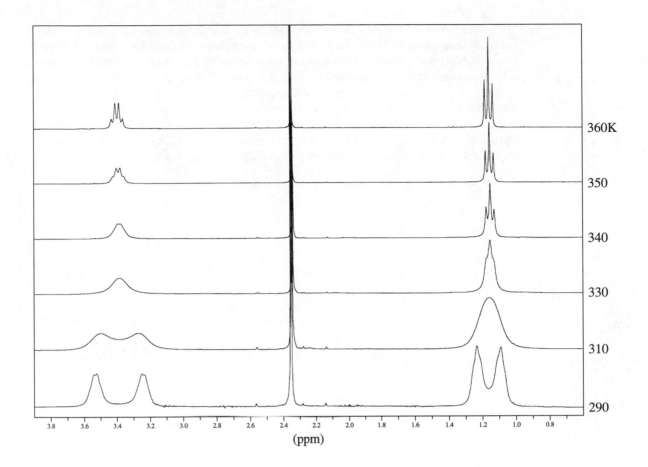

19. The proton NMR spectral information shown in this problem is for a compound with formula C_4H_7Cl. Expansions are shown for each of the unique protons. The original "quintet" pattern centering on 4.52 ppm is simplified to a doublet by irradiating the protons at 1.59 ppm (see Section 6.10). In another experiment, irradiation of the proton at 4.52 ppm simplifies the original pattern centering on 5.95 ppm to the four peak pattern shown. The doublet at 1.59 ppm becomes a singlet when the proton at 4.52 ppm is irradiated. Determine the coupling constants and draw the structure of this compound. Notice that there are ^2J, ^3J, and ^4J couplings present in this compound. Draw a tree diagram for the proton at 5.95 ppm (nondecoupled) and explain why irradiation of the proton at 4.52 ppm simplified the pattern. Assign each of the peaks in the spectrum.

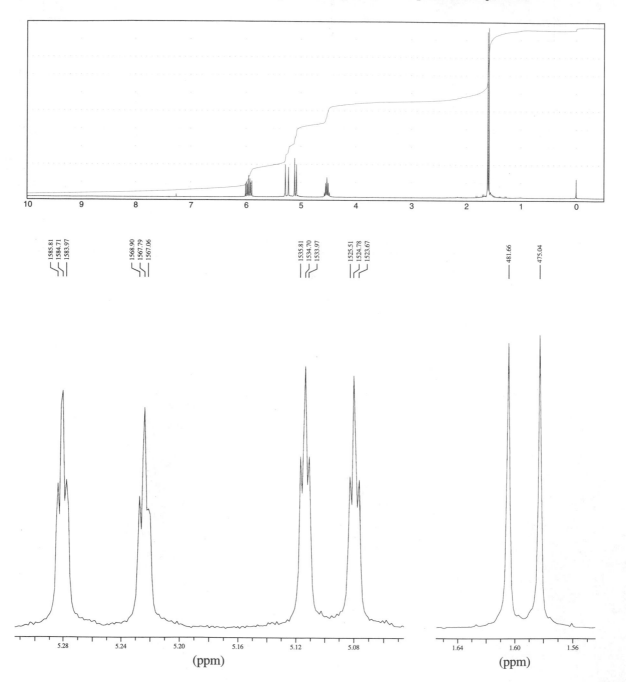

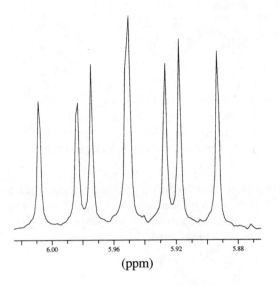

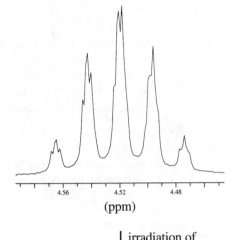

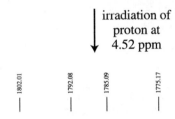

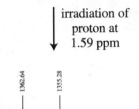

irradiation of
proton at
4.52 ppm

irradiation of
proton at
1.59 ppm

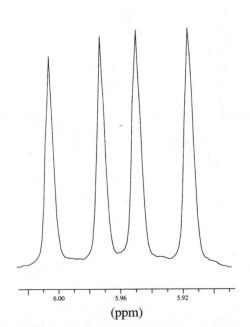

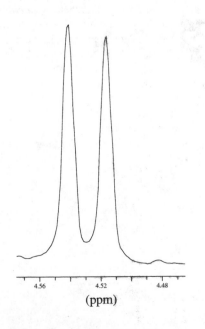

20. In Problem 11, calculations proved to be a good way of assigning structures to the spectra of some aromatic amines. Describe an experimental way of differentiating between the following amines.

*21. At room temperature, the NMR spectrum of cyclohexane shows only a single resonance peak. As the temperature of the sample is lowered, the sharp single peak broadens until at −66.7°C it begins to split into two peaks, both broad. As the temperature is lowered further to −100°C, each of the two broad bands begins to give a splitting pattern of its own. Explain the origin of these two families of bands.

*22. In *cis*-1-bromo-4-*tert*-butylcyclohexane, the proton on carbon-4 is found to give resonance at 4.33 ppm. In the *trans* isomer, the resonance of the C4 hydrogen is at 3.63 ppm. Explain why these compounds should have different chemical shift values for the C4 hydrogen. Can you explain the fact that this difference is not seen in the 4-bromomethylcyclohexanes except at very low temperature?

REFERENCES

Crews, P., J. Rodriguez, and M. Jaspars, *Organic Structure Analysis*, Oxford University Press, New York, 1998.

Friebolin, H., *Basic One- and Two-Dimensional NMR Spectroscopy*, 2nd ed., VCH Publishers, New York, 1993.

Gunther, H., *NMR Spectroscopy*, 2nd ed., John Wiley and Sons, New York, 1995.

Jackman, L. M., and S. Sternhell, *Applications of Nuclear Magnetic Resonance Spectroscopy in Organic Chemistry*, 2nd ed., Pergamon Press, London, 1969.

Lambert, J. B., H. F. Shurvell, D. A. Lightner, and R. G. Cooks, *Organic Structural Spectroscopy*, Prentice Hall, Upper Saddle River, NJ, 1998.

Macomber, R. S., *NMR Spectroscopy—Essential Theory and Practice*, College Outline Series, Harcourt, Brace Jovanovich, New York, 1988.

Macomber, R. S., *A Complete Introduction to Modern NMR Spectroscopy*, John Wiley and Sons, New York, 1997.

Pople, J. A., W. C. Schneider, and H. J. Bernstein, *High Resolution Nuclear Magnetic Resonance*, McGraw–Hill, New York, 1969.

Sanders, J. K. M., and B. K. Hunter, *Modern NMR Spectroscopy—A Guide for Chemists*, 2nd ed., Oxford University Press, Oxford, England, 1993.

Silverstein, R. M., F. X. Webster, *Spectrometric Identification of Organic Compounds*, 6th ed., John Wiley and Sons, New York, 1998.

Yoder, C. H., and C. D. Schaeffer, *Introduction to Multinuclear NMR*, Benjamin-Cummings, Menlo Park, CA, 1987.

In addition to these references, also consult textbook references, compilations of spectra, computer programs, and NMR-related Internet addresses cited at the end of Chapter 5

CHAPTER 7

ULTRAVIOLET SPECTROSCOPY

Most organic molecules and functional groups are transparent in the portions of the electromagnetic spectrum which we call the **ultraviolet (UV)** and **visible (VIS)** regions—that is, the regions where wavelengths range from 190 nm to 800 nm. Consequently, absorption spectroscopy is of limited utility in this range of wavelengths. However, in some cases we can derive useful information from these regions of the spectrum. That information, when combined with the detail provided by infrared and nuclear magnetic resonance spectra, can lead to valuable structural proposals.

7.1 THE NATURE OF ELECTRONIC EXCITATIONS

When continuous radiation passes through a transparent material, a portion of the radiation may be absorbed. If that occurs, the residual radiation, when it is passed through a prism, yields a spectrum with gaps in it, called an **absorption spectrum.** As a result of energy absorption, atoms or molecules pass from a state of low energy (the initial, or **ground, state**) to a state of higher energy (the **excited state**). Figure 7.1 depicts this excitation process, which is quantized. The electromagnetic radiation that is absorbed has energy exactly equal to the energy *difference* between the excited and ground states.

In the case of ultraviolet and visible spectroscopy, the transitions that result in the absorption of electromagnetic radiation in this region of the spectrum are transitions between **electronic** energy levels. As a molecule absorbs energy, an electron is promoted from an occupied orbital to an unoccupied orbital of greater potential energy. Generally, the most probable transition is from the **highest occupied molecular orbital (HOMO)** to the **lowest unoccupied molecular orbital (LUMO)**. The energy differences between electronic levels in most molecules vary from 125 to 650 kJ/mole (kilojoules per mole).

For most molecules, the lowest-energy occupied molecular orbitals are the σ orbitals, which correspond to σ bonds. The π orbitals lie at somewhat higher energy levels, and orbitals that hold unshared pairs, the **nonbonding (n) orbitals**, lie at even higher energies. The unoccupied, or **antibonding orbitals** (π^* and σ^*), are the orbitals of highest energy. Figure 7.2a shows a typical progression of electronic energy levels.

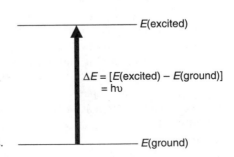

$$\Delta E = [E(\text{excited}) - E(\text{ground})]$$
$$= h\upsilon$$

E(excited)

E(ground)

FIGURE 7.1 The excitation process.

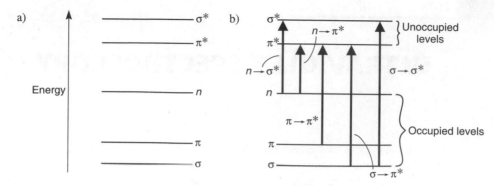

FIGURE 7.2 Electronic energy levels and transitions.

In all compounds other than alkanes, the electrons may undergo several possible transitions of different energies. Some of the most important transitions are:

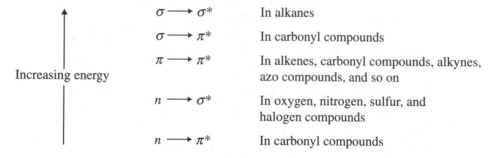

Figure 7.2b illustrates these transitions. Electronic energy levels in aromatic molecules are more complicated than the ones depicted here. Section 7.14 will describe the electronic transitions of aromatic compounds.

Clearly, the energy required to bring about transitions from the highest occupied energy level (HOMO) in the ground state to the lowest unoccupied energy level (LUMO) is less than the energy required to bring about a transition from a lower occupied energy level. Thus, in Figure 7.2b an $n \rightarrow \pi^*$ transition would have a lower energy than a $\pi \rightarrow \pi^*$ transition. For many purposes, the transition of lowest energy is the most important.

Not all of the transitions that at first sight appear possible are observed. Certain restrictions, called **selection rules,** must be considered. One important selection rule states that transitions that involve a change in the spin quantum number of an electron during the transition are not allowed to take place; they are called **"forbidden" transitions.** Other selection rules deal with the numbers of electrons that may be excited at one time, with symmetry properties of the molecule and of the electronic states, and with other factors that need not be discussed here. Transitions that are formally forbidden by the selection rules are often not observed. However, theoretical treatments are rather approximate, and in certain cases forbidden transitions *are* observed, although the intensity of the absorption tends to be much lower than for transitions that are **allowed** by the selection rules. The $n \rightarrow \pi^*$ transition is the most common type of forbidden transition.

7.2 THE ORIGIN OF UV BAND STRUCTURE

For an atom that absorbs in the ultraviolet, the absorption spectrum sometimes consists of very sharp lines, as would be expected for a quantized process occurring between two discrete energy levels. For molecules, however, the UV absorption usually occurs over a wide range of wavelengths, because molecules (as opposed to atoms) normally have many excited modes of vibration and rotation at room temperature. In fact, the vibration of molecules cannot be completely "frozen out" even at absolute zero. Consequently, a collection of molecules generally has its members in many states of vibrational and rotational excitation. The energy levels for these states are quite closely spaced, corresponding to energy differences considerably smaller than those of electronic levels. The rotational and vibrational levels are thus "superimposed" on the electronic levels. A molecule may therefore undergo electronic and vibrational–rotational excitation simultaneously, as shown in Figure 7.3.

Because there are so many possible transitions, each differing from the others by only a slight amount, each electronic transition consists of a vast number of lines spaced so closely that the spectrophotometer cannot resolve them. Rather, the instrument traces an "envelope" over the entire pattern. What is observed from these types of combined transitions is that the UV spectrum of a molecule usually consists of a broad **band** of absorption centered near the wavelength of the major transition.

7.3 PRINCIPLES OF ABSORPTION SPECTROSCOPY

The greater the number of molecules capable of absorbing light of a given wavelength, the greater the extent of light absorption. Furthermore, the more effectively a molecule absorbs light of a given wavelength, the greater the extent of light absorption. From these guiding ideas, the following empirical expression, known as the **Beer–Lambert Law,** may be formulated.

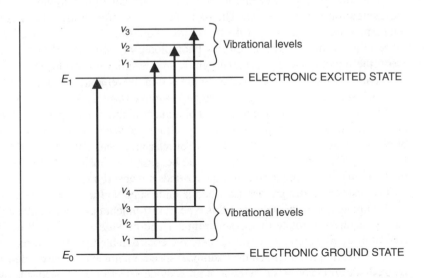

FIGURE 7.3 Electronic transitions with vibrational transitions superimposed. (Rotational levels, which are very closely spaced within the vibrational levels, are omitted for clarity.)

$$A = \log(I_0/I) = \varepsilon c l \text{ for a given wavelength} \qquad \textbf{Equation 7.1}$$

A = absorbance

I_0 = intensity of light incident upon sample cell

I = intensity of light leaving sample cell

c = molar concentration of solute

l = length of sample cell (cm)

ε = molar absorptivity

The term $\log(I_0/I)$ is also known as the **absorbance** (or the **optical density** in older literature) and may be represented by A. The **molar absorptivity** (formerly known as the **molar extinction coefficient**) is a property of the molecule undergoing an electronic transition and is not a function of the variable parameters involved in preparing a solution. The size of the absorbing system and the probability that the electronic transition will take place control the absorptivity, which ranges from 0 to 10^6. Values above 10^4 are termed **high-intensity absorptions,** while values below 10^3 are **low-intensity absorptions.** Forbidden transitions (see Section 7.1) have absorptivities in the range from 0 to 1000.

The Beer–Lambert Law is rigorously obeyed when a *single species* gives rise to the observed absorption. The law may not be obeyed, however, when different forms of the absorbing molecule are in equilibrium, when solute and solvent form complexes through some sort of association, when *thermal* equilibrium exists between the ground electronic state and a low-lying excited state, or when fluorescent compounds or compounds changed by irradiation are present.

7.4 INSTRUMENTATION

The typical ultraviolet–visible spectrophotometer consists of a **light source,** a **monochromator,** and a **detector.** The light source is usually a deuterium lamp, which emits electromagnetic radiation in the ultraviolet region of the spectrum. A second light source, a tungsten lamp, is used for wavelengths in the visible region of the spectrum. The monochromator is a diffraction grating; its role is to spread the beam of light into its component wavelengths. A system of slits focuses the desired wavelength on the sample cell. The light that passes through the sample cell reaches the detector, which records the intensity of the transmitted light (I). The detector is generally a photomultiplier tube, although in modern instruments photodiodes are also used. In a typical double-beam instrument, the light emanating from the light source is split into two beams, the **sample beam** and the **reference beam.** When there is no sample cell in the reference beam, the detected light is taken to be equal to the intensity of light entering the sample (I_0).

The sample cell must be constructed of a material that is transparent to the electromagnetic radiation being used in the experiment. For spectra in the visible range of the spectrum, cells composed of glass or plastic are generally suitable. For measurements in the ultraviolet region of the spectrum, however, glass and plastic cannot be used because they absorb ultraviolet radiation. Instead, cells made of quartz must be used, since quartz does not absorb radiation in this region.

The instrument design just described is quite suitable for measurement at only one wavelength. If a complete spectrum is desired, this type of instrument has some deficiencies. A mechanical system is required to rotate the monochromator and provide a scan of all desired wavelengths. This type of system operates slowly, and therefore considerable time is required to record a spectrum.

A modern improvement on the traditional spectrophotometer is the **diode-array spectrophotometer.** A diode array consists of a series of photodiode detectors positioned side by side on a silicon crystal. Each diode is designed to record a narrow band of the spectrum. The diodes are connected so that the entire spectrum is recorded at once. This type of detector has no moving parts and can record spectra very quickly. Furthermore, its output can be passed to a computer, which can process the information and provide a variety of useful output formats. Since the number of photodiodes is limited, the speed and convenience described here are obtained at some small cost in resolution. For many applications, however, the advantages of this type of instrument outweigh the loss of resolution.

7.5 PRESENTATION OF SPECTRA

The ultraviolet–visible spectrum is generally recorded as a plot of absorbance versus wavelength. It is customary to then replot the data with either ε or log ε plotted on the ordinate and wavelength plotted on the abscissa. Figure 7.4, the spectrum of benzoic acid, is typical of the manner in which spectra are displayed. However, very few electronic spectra are reproduced in the scientific literature; most are described by indications of the wavelength maxima and absorptivities of the principal absorption peaks. For benzoic acid, a typical description might be

$$\lambda_{max} = 230 \text{ nm} \qquad \log \varepsilon = 4.2$$
$$272 \qquad \qquad 3.1$$
$$282 \qquad \qquad 2.9$$

Figure 7.4 is the actual spectrum that corresponds to these data.

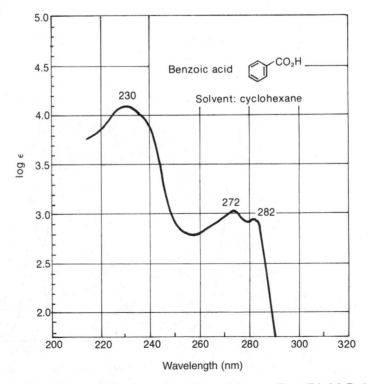

FIGURE 7.4 Ultraviolet spectrum of benzoic acid in cyclohexane. (From Friedel, R. A., and M. Orchin, *Ultraviolet Spectra of Aromatic Compounds*, John Wiley and Sons, New York, 1951. Reprinted by permission.)

7.6 SOLVENTS

The choice of the solvent to be used in ultraviolet spectroscopy is quite important. The first criterion for a good solvent is that it should not absorb ultraviolet radiation in the same region as the substance whose spectrum is being determined. Usually solvents that do not contain conjugated systems are most suitable for this purpose, although they vary as to the shortest wavelength at which they remain transparent to ultraviolet radiation. Table 7.1 lists some common ultraviolet spectroscopy solvents and their cutoff points, or minimum regions of transparency.

Of the solvents listed in Table 7.1, water, 95% ethanol, and hexane are most commonly used. Each is transparent in the regions of the ultraviolet spectrum where interesting absorption peaks from sample molecules are likely to occur.

A second criterion for a good solvent is its effect on the fine structure of an absorption band. Figure 7.5 illustrates the effects of polar and nonpolar solvents on an absorption band. A nonpolar solvent does not hydrogen bond with the solute, and the spectrum of the solute closely approximates the spectrum that would be produced in the gaseous state, where fine structure is often observed. In a polar solvent, the hydrogen bonding forms a solute–solvent complex, and the fine structure may disappear.

TABLE 7.1
SOLVENT CUTOFFS

Acetonitrile	190 nm	*n*-Hexane	201 nm
Chloroform	240	Methanol	205
Cyclohexane	195	Isooctane	195
1,4-Dioxane	215	Water	190
95% Ethanol	205	Trimethyl phosphate	210

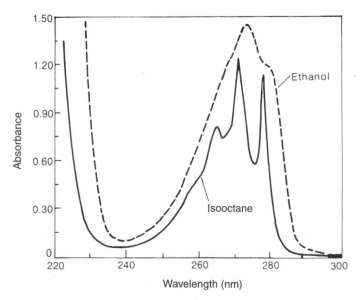

FIGURE 7.5 Ultraviolet spectra of phenol in ethanol and in isooctane. (From Coggeshall, N. D., and E. M. Lang, *J. Am. Chem. Soc., 70* [1948]: 3288. Reprinted by permission.)

TABLE 7.2
SOLVENT SHIFTS ON THE $n \rightarrow \pi^*$ TRANSITION OF ACETONE

Solvent	H_2O	CH_3OH	C_2H_5OH	$CHCl_3$	C_6H_{14}
λ_{max} (nm)	264.5	270	272	277	279

A third criterion for a good solvent is its ability to influence the wavelength of ultraviolet light that will be absorbed via stabilization of either the ground or the excited state. Polar solvents do not form hydrogen bonds as readily with the excited states of polar molecules as with their ground states, and these polar solvents increase the energies of electronic transitions in the molecules. Polar solvents shift transitions of the $n \rightarrow \pi^*$ type to shorter wavelengths. On the other hand, in some cases the excited states may form stronger hydrogen bonds than the corresponding ground states. In such a case, a polar solvent shifts an absorption to longer wavelength, since the energy of the electronic transition is decreased. Polar solvents shift transitions of the $\pi \rightarrow \pi^*$ type to longer wavelengths. Table 7.2 illustrates typical effects of a series of solvents on an electronic transition.

7.7 WHAT IS A CHROMOPHORE?

Although the absorption of ultraviolet radiation results from the excitation of electrons from ground to excited states, the nuclei that the electrons hold together in bonds play an important role in determining which wavelengths of radiation are absorbed. The nuclei determine the strength with which the electrons are bound and thus influence the energy spacing between ground and excited states. Hence, the characteristic energy of a transition and the wavelength of radiation absorbed are properties of a group of atoms rather than of electrons themselves. The group of atoms producing such an absorption is called a **chromophore.** As structural changes occur in a chromophore, the exact energy and intensity of the absorption are expected to change accordingly. Very often, it is extremely difficult to predict from theory how the absorption will change as the structure of the chromophore is modified, and it is necessary to apply empirical working guides to predict such relationships.

Alkanes. For molecules, such as alkanes, which contain nothing but single bonds and lack atoms with unshared electron pairs, the only electronic transitions possible are of the $\sigma \rightarrow \sigma^*$ type. These transitions are of such a high energy that they absorb ultraviolet energy at very short wavelengths—shorter than the wavelengths that are experimentally accessible using typical spectrophotometers. Figure 7.6 illustrates this type of transition. The excitation of the σ-bonding electron to the σ^*-antibonding orbital is depicted at the right.

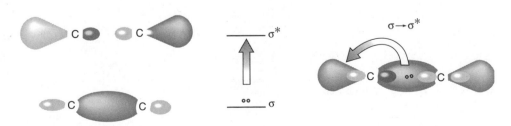

FIGURE 7.6 $\sigma \rightarrow \sigma^*$ transition.

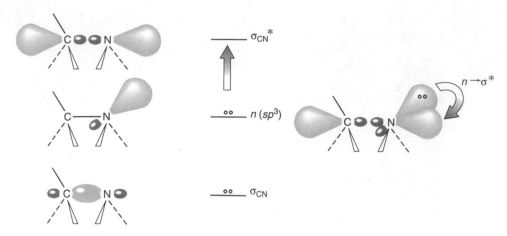

FIGURE 7.7 $n \rightarrow \sigma^*$ transition.

Alcohols, Ethers, Amines, and Sulfur Compounds. In saturated molecules that contain atoms bearing nonbonding pairs of electrons, transitions of the $n \rightarrow \sigma^*$ type become important. They are also rather high-energy transitions, but they do absorb radiation that lies within an experimentally accessible range. Alcohols and amines absorb in the range from 175 to 200 nm, while organic thiols and sulfides absorb between 200 and 220 nm. Most of the absorptions are below the cutoff points for the common solvents, and so they are not observed in solution spectra. Figure 7.7 illustrates an $n \rightarrow \sigma^*$ transition for an amine. The excitation of the nonbonding electron to the antibonding orbital is shown at the right.

Alkenes and Alkynes. With unsaturated molecules, $\pi \rightarrow \pi^*$ transitions become possible. These transitions are of rather high energy as well, but their positions are sensitive to the presence of substitution, as will be clear later. Alkenes absorb around 175 nm, and alkynes absorb around 170 nm. Figure 7.8 shows this type of transition.

Carbonyl Compounds. Unsaturated molecules that contain atoms such as oxygen or nitrogen may also undergo $n \rightarrow \pi^*$ transitions. These are perhaps the most interesting and most studied transitions, particularly among carbonyl compounds. These transitions are also rather sensitive to substitution on the chromophoric structure. The typical carbonyl compound undergoes an $n \rightarrow \pi^*$ transition around 280 to 290 nm ($\varepsilon = 15$). Most $n \rightarrow \pi^*$ transitions are forbidden and hence are of low intensity. Carbonyl compounds also have a $\pi \rightarrow \pi^*$ transition at about 188 nm ($\varepsilon = 900$). Figure 7.9 shows the $n \rightarrow \pi^*$ and $\pi \rightarrow \pi^*$ transitions of the carbonyl group.[1]

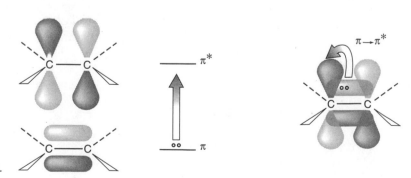

FIGURE 7.8 $\pi \rightarrow \pi^*$ transition.

---

[1] Contrary to what you might expect from simple theory, the oxygen atom of the carbonyl group is *not sp²* hybridized. Spectroscopists have shown that although the carbon atom is *sp²* hybridized, the hybridization of the oxygen atom more closely approximates *sp*.

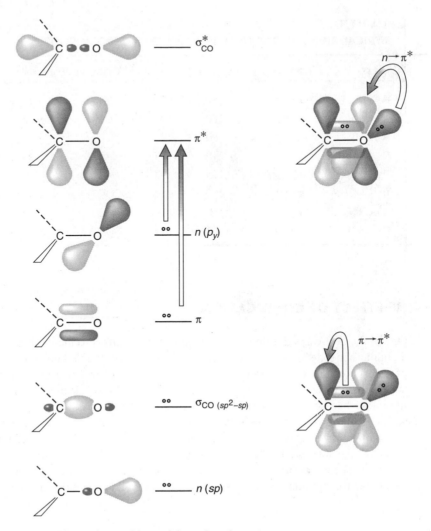

FIGURE 7.9 Electronic transitions of the carbonyl group.

Table 7.3 lists typical absorptions of simple isolated chromophores. You may notice that these *simple* chromophores nearly all absorb at approximately the same wavelength (160 to 210 nm).

The attachment of substituent groups in place of hydrogen on a basic chromophore structure changes the position and intensity of an absorption band of the chromophore. The substituent groups may not give rise to the absorption of the ultraviolet radiation themselves, but their presence modifies the absorption of the principal chromophore. Substituents that increase the intensity of the absorption, and possibly the wavelength, are called **auxochromes.** Typical auxochromes include methyl, hydroxyl, alkoxy, halogen, and amino groups.

Other substituents may have any of four kinds of effects on the absorption:

1. **Bathochromic shift** (red shift)—a shift to lower energy or longer wavelength.

2. **Hypsochromic shift** (blue shift)—a shift to higher energy or shorter wavelength.

3. **Hyperchromic effect**—an increase in intensity.

4. **Hypochromic effect**—a decrease in intensity.

TABLE 7.3
TYPICAL ABSORPTIONS OF SIMPLE ISOLATED CHROMOPHORES

Class	Transition	λ_{max} (nm)	log ε	Class	Transition	λ_{max} (nm)	log ε
R—OH	$n \rightarrow \sigma^*$	180	2.5	R—NO$_2$	$n \rightarrow \pi^*$	271	<1.0
R—O—R	$n \rightarrow \sigma^*$	180	3.5	R—CHO	$\pi \rightarrow \pi^*$	190	2.0
R—NH$_2$	$n \rightarrow \sigma^*$	190	3.5		$n \rightarrow \pi^*$	290	1.0
R—SH	$n \rightarrow \sigma^*$	210	3.0	R$_2$CO	$\pi \rightarrow \pi^*$	180	3.0
R$_2$C=CR$_2$	$\pi \rightarrow \pi^*$	175	3.0		$n \rightarrow \pi^*$	280	1.5
R—C≡C—R	$\pi \rightarrow \pi^*$	170	3.0	RCOOH	$n \rightarrow \pi^*$	205	1.5
R—C≡N	$n \rightarrow \pi^*$	160	<1.0	RCOOR$'$	$n \rightarrow \pi^*$	205	1.5
R—N=N—R	$n \rightarrow \pi^*$	340	<1.0	RCONH$_2$	$n \rightarrow \pi^*$	210	1.5

7.8 THE EFFECT OF CONJUGATION

One of the best ways to bring about a bathochromic shift is to increase the extent of conjugation in a double-bonded system. In the presence of conjugated double bonds, the electronic energy levels of a chromophore move closer together. As a result, the energy required to produce a transition from an occupied electronic energy level to an unoccupied level decreases, and the wavelength of the light absorbed becomes longer. Figure 7.10 illustrates the bathochromic shift that is observed in a series of conjugated polyenes as the length of the conjugated chain is increased.

Conjugation of two chromophores not only results in a bathochromic shift but increases the intensity of the absorption. These two effects are of prime importance in the use and interpretation of electronic spectra of organic molecules, because conjugation shifts the selective light absorption of isolated chromophores from a region of the spectrum that is not readily accessible to a region that is

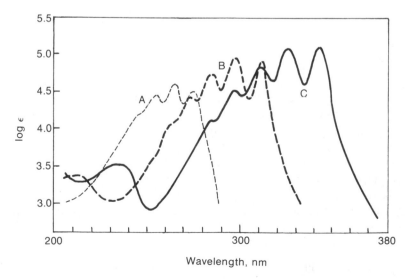

FIGURE 7.10 CH$_3$—(CH=CH)$_n$—CH$_3$ Ultraviolet spectra of dimethylpolyenes. (a) $n = 3$; (b) $n = 4$; (c) $n = 5$. (From Nayler, P., and M. C. Whiting, *J. Chem. Soc.* 1955: 3042.)

TABLE 7.4
EFFECT OF CONJUGATION ON ELECTRONIC TRANSITIONS

	λ_{max} (nm)	ε
Alkenes		
Ethylene	175	15,000
1,3-Butadiene	217	21,000
1,3,5-Hexatriene	258	35,000
β-Carotene (11 double bonds)	465	125,000
Ketones		
Acetone		
$\pi \rightarrow \pi^*$	189	900
$n \rightarrow \pi^*$	280	12
3-Buten-2-one		
$\pi \rightarrow \pi^*$	213	7,100
$n \rightarrow \pi^*$	320	27

easily studied with commercially available spectrophotometers. The exact position and intensity of the absorption band of the conjugated system can be correlated with the extent of conjugation in the system. Table 7.4 illustrates the effect of conjugation on some typical electronic transitions.

7.9 THE EFFECT OF CONJUGATION ON ALKENES

The bathochromic shift that results from an increase in the length of a conjugated system implies that an increase in conjugation decreases the energy required for electronic excitation. This is true and can be explained most easily by the use of molecular orbital theory. According to molecular orbital (MO) theory, the atomic p orbitals on each of the carbon atoms combine to make π molecular orbitals. For instance, in the case of ethylene (ethene), we have two atomic p orbitals, ϕ_1 and ϕ_2. From these two p orbitals we form two π molecular orbitals, ψ_1 and ψ_2^*, by taking linear combinations. The bonding orbital, ψ_1, results from the addition of the wave functions of the two p orbitals, and the antibonding orbital, ψ_2^*, results from the subtraction of these two wave functions. The new bonding orbital, a **molecular orbital,** has an energy lower than that of either of the original p orbitals; likewise, the antibonding orbital has an elevated energy. Figure 7.11 illustrates this diagrammatically.

Notice that *two* atomic orbitals were combined to build the molecular orbitals, and as a result, *two* molecular orbitals were formed. There were also two electrons, one in each of the atomic p orbitals. As a result of combination, the new π system contains *two* electrons. Because we fill the lower-energy orbitals first, these electrons end up in ψ_1, the bonding orbital, and they constitute a new π bond. Electronic transition in this system is a $\pi \rightarrow \pi^*$ transition from ψ_1 to ψ_2^*.

Now, moving from this simple two-orbital case, consider 1,3-butadiene, which has *four* atomic p orbitals that form its π system of two conjugated double bonds. Since we had four atomic orbitals with which to build, *four* molecular orbitals result. Figure 7.12 represents the orbitals of ethylene on the same energy scale as the new orbitals, for the sake of comparison.

Notice that the transition of lowest energy in 1,3-butadiene, $\psi_2 \rightarrow \psi_3^*$, is a $\pi \rightarrow \pi^*$ transition and that it has a *lower energy* than the corresponding transition in ethylene, $\psi_1 \rightarrow \psi_2^*$. This result is general. As we increase the number of p orbitals making up the conjugated system, the transition from the **highest occupied molecular orbital (HOMO)** to the **lowest unoccupied molecular orbital (LUMO)**

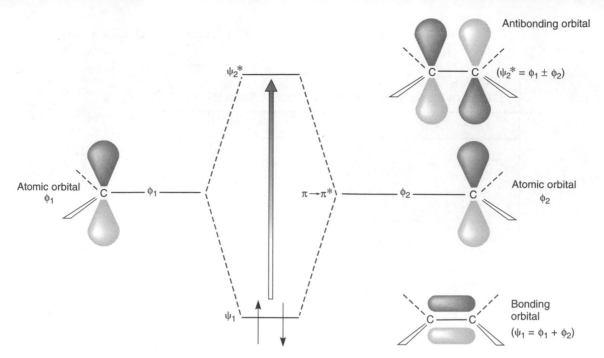

FIGURE 7.11 Formation of the molecular orbitals for ethylene.

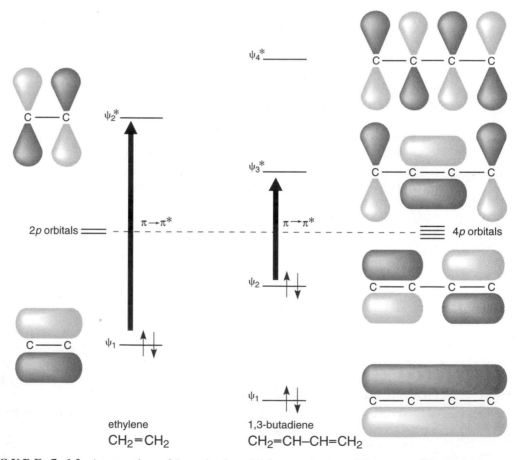

FIGURE 7.12 A comparison of the molecular orbital energy levels and the energy of the $\pi \rightarrow \pi^*$ transitions in ethylene and 1,3-butadiene.

has progressively lower energy. The energy gap dividing the bonding and antibonding orbitals becomes progressively smaller with increasing conjugation. Figure 7.13 plots the molecular orbital energy levels of several conjugated polyenes of increasing chain length on a common energy scale. Arrows indicate the HOMO–LUMO transitions. The increased conjugation shifts the observed wavelength of the absorption to higher values.

In a qualitatively similar fashion, many auxochromes exert their bathochromic shifts by means of an extension of the length of the conjugated system. The strongest auxochromes invariably possess a pair of unshared electrons on the atom attached to the double-bond system. Resonance interaction of this lone pair with the double bond(s) increases the length of the conjugated system.

$$\left[\ce{>C=C-\ddot{B}} \longleftrightarrow \ce{>\ddot{C}-C=B+} \right]$$

As a result of this interaction, as just shown, the nonbonded electrons become part of the π system of molecular orbitals, increasing its length by one extra orbital. Figure 7.14 depicts this interaction for ethylene and an unspecified atom, B, with an unshared electron pair. However, any of the typical auxochromic groups, $-OH$, $-OR$, $-X$, or $-NH_2$, could have been illustrated specifically.

In the new system, the transition from the highest occupied orbital, ψ_2, to the antibonding orbital, ψ_3^*, always has lower energy than the $\pi \rightarrow \pi^*$ transition would have in the system without the interaction. Although MO theory can explain this general result, it is beyond the scope of this book.

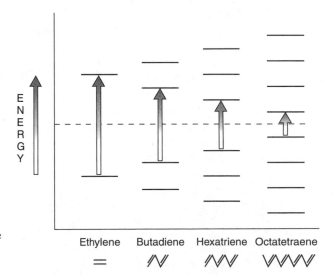

FIGURE 7.13 A comparison of the $\pi \rightarrow \pi^*$ energy gap in a series of polyenes of increasing chain length.

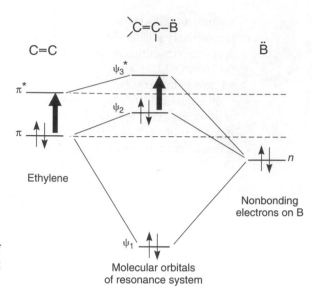

FIGURE 7.14 Energy relationships of the new molecular orbitals and the interacting π system and its auxochrome.

In similar fashion, methyl groups also produce a bathochromic shift. However, as methyl groups do not have unshared electrons, the interaction is thought to result from overlap of the C—H bonding orbitals with the π system, as follows.

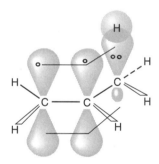

This type of interaction is often called **hyperconjugation.** Its net effect is an extension of the π system.

7.10 THE WOODWARD–FIESER RULES FOR DIENES

In butadiene, two possible $\pi \longrightarrow \pi^*$ transitions can occur: $\psi_2 \longrightarrow \psi_3^*$ and $\psi_2 \longrightarrow \psi_4^*$. We have already discussed the easily observable $\psi_2 \longrightarrow \psi_3^*$ transition (see Fig. 7.12). The $\psi_2 \longrightarrow \psi_4^*$ transition is not often observed, for two reasons. First, it lies near 175 nm for butadiene; second, it is a forbidden transition for the *s-trans* conformation of double bonds in butadiene.

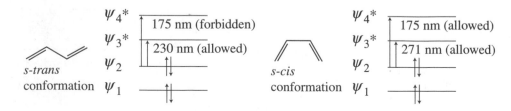

A transition at 175 nm lies below the cutoff points of the common solvents used to determine UV spectra (Table 7.1) and therefore is not easily detectable. Furthermore, the *s-trans* conformation is more favorable for butadiene than is the *s-cis* conformation. Therefore, the 175-nm band is not usually detected.

In general, conjugated dienes exhibit an intense band (ε = 20,000 to 26,000) in the region from 217 to 245 nm, owing to a $\pi \rightarrow \pi^*$ transition. The position of this band appears to be quite insensitive to the nature of the solvent.

Butadiene and many simple conjugated dienes exist in a planar *s-trans* conformation, as noted. Generally, alkyl substitution produces bathochromic shifts and hyperchromic effects. However, with certain patterns of alkyl substitution, the wavelength increases but the intensity decreases. The 1,3-dialkylbutadienes possess too much crowding between alkyl groups to permit them to exist in the *s-trans* conformation. They convert, by rotation around the single bond, to an *s-cis* conformation, which absorbs at longer wavelengths but with lower intensity than the corresponding *s-trans* conformation.

s-trans   *s-cis*

In cyclic dienes, where the central bond is a part of the ring system, the diene chromophore is usually held rigidly in either the *s-trans* (transoid) or the *s-cis* (cisoid) orientation. Typical absorption spectra follow the expected pattern:

Homoannular diene (cisoid or *s-cis*)
Less intense, ε = 5,000–15,000
λ longer (273 nm)

Heteroannular diene (transoid or *s-trans*)
More intense, ε = 12,000–28,000
λ shorter (234 nm)

By studying a vast number of dienes of each type, Woodward and Fieser devised an empirical correlation of structural variations that enables us to predict the wavelength at which a conjugated diene will absorb. Table 7.5 summarizes the rules. Following are a few sample applications of these rules. Notice that the pertinent parts of the structures are shown in bold face.

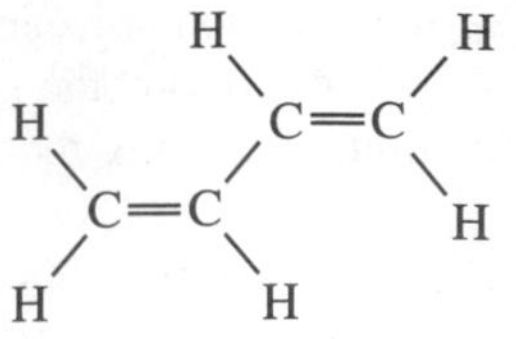

Transsoid: 214 nm
Observed: 217 nm

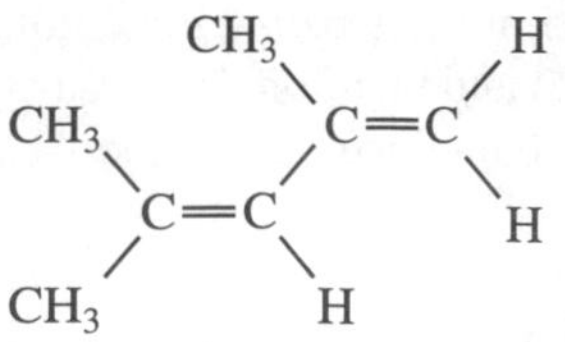

Transsoid: 214 nm
Alkyl groups: $3 \times 5 =$ 15
                                 229 nm

Observed: 228 nm

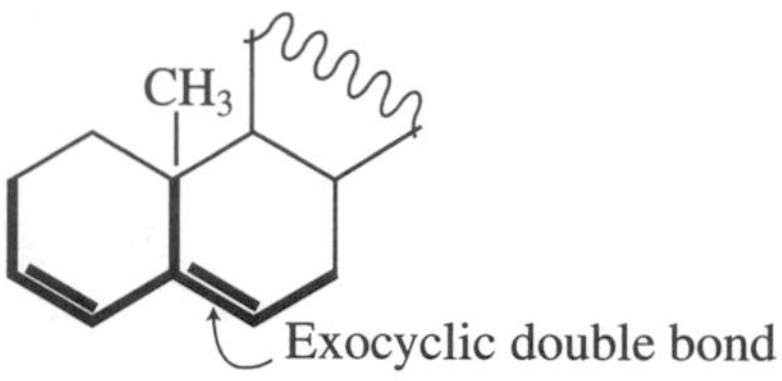

Exocyclic double bond

Transsoid: 214 nm
Ring residues: $3 \times 5 =$ 15
Exocyclic double bond: 5
                                234 nm

Observed: 235 nm

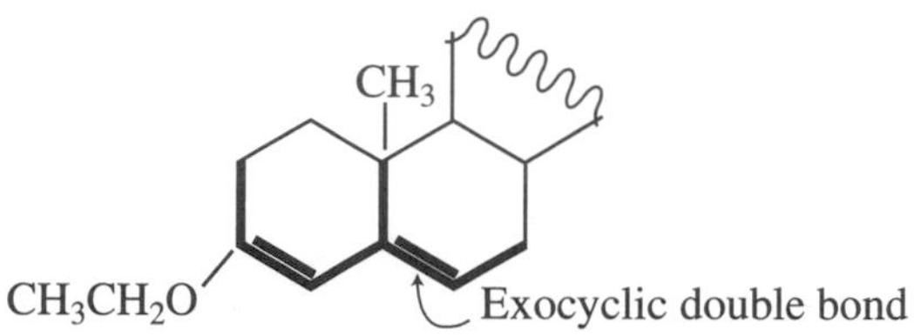

Exocyclic double bond

Transsoid: 214 nm
Ring residues: $3 \times 5 =$ 15
Exocyclic double bond: 5
—OR: 6
                                240 nm

Observed: 241 nm

TABLE 7.5
EMPIRICAL RULES FOR DIENES

	Homoannular (cisoid)	Heteroannular (transoid)
Parent	$\lambda = 253$ nm	$\lambda = 214$ nm
Increments for:		
Double-bond-extending conjugation	30	30
Alkyl substituent or ring residue	5	5
Exocyclic double bond	5	5
Polar groupings:		
$-OCOCH_3$	0	0
$-OR$	6	6
$-Cl, -Br$	5	5
$-NR_2$	60	60

In this context, an "exocyclic double bond" is a double bond that lies outside of a given ring. Notice that the exocyclic bond may lie within one ring even though it is outside another ring. Often, an exocyclic double bond will be found at a junction point on rings. Here is an example of a compound with the exocyclic double bonds labeled with asterisks:

Three exocyclic double bonds = 3 × 5 = 15 nm

Cisoid: 253 nm
Alkyl substituent: 5
Ring residues: 3 × 5 =   15
Exocyclic double bond: 5
                               278 nm

Observed: 275 nm

Cisoid: 253 nm
Ring residues: 5 × 5 =   25
Double-bond-extending conjugation: 2 × 30 =   60
Exocyclic double bond: 3 × 5 =   15
CH₃COO—:   0
                               353 nm

Observed: 355 nm

7.11 CARBONYL COMPOUNDS; ENONES

As discussed in Section 7.7, carbonyl compounds have two principal UV transitions, the allowed $\pi \rightarrow \pi^*$ transition and the forbidden $n \rightarrow \pi^*$ transition.

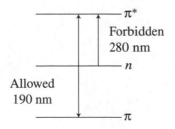

Of these, only the $n \rightarrow \pi^*$ transition, although it is weak (forbidden), is commonly observed above the usual cutoff points of solvents. Substitution on the carbonyl group by an auxochrome with a lone pair of electrons, such as $-NR_2$, $-OH$, $-OR$, $-NH_2$, or $-X$, as in amides, acids, esters, or acid chlorides, gives a pronounced hypsochromic effect on the $n \rightarrow \pi^*$ transition and a lesser, bathochromic effect on the $\pi \rightarrow \pi^*$ transition. Such bathochromic shifts are caused by resonance interaction similar to that discussed in Section 7.9. Seldom, however, are these effects large enough to bring the $\pi \rightarrow \pi^*$ band into the region above the solvent cutoff point. Table 7.6 lists the hypsochromic effects of an acetyl group on the $n \rightarrow \pi^*$ transition.

The hypsochromic shift of the $n \rightarrow \pi^*$ is due primarily to the inductive effect of the oxygen, nitrogen, or halogen atoms. They withdraw electrons from the carbonyl carbon, causing the lone pair of electrons on oxygen to be held more firmly than they would be in the absence of the inductive effect.

If the carbonyl group is part of a conjugated system of double bonds, both the $n \rightarrow \pi^*$ and the $\pi \rightarrow \pi^*$ bands are shifted to longer wavelengths. However, the energy of the $n \rightarrow \pi^*$ transition does not decrease as rapidly as that of the $\pi \rightarrow \pi^*$ band, which is more intense. If the conjugated chain becomes long enough, the $n \rightarrow \pi^*$ band is "buried" under the more intense $\pi \rightarrow \pi^*$ band. Figure 7.15 illustrates this effect for a series of polyene aldehydes.

Figure 7.16 shows the molecular orbitals of a simple enone system, along with those of the noninteracting double bond and the carbonyl group.

TABLE 7.6

HYPSOCHROMIC EFFECTS OF LONE-PAIR AUXOCHROMES ON THE $n \rightarrow \pi^*$ TRANSITION OF A CARBONYL GROUP

	λ_{max}	ε_{max}	Solvent
$CH_3-\overset{\displaystyle O}{\overset{\|}{C}}-H$	293 nm	12	Hexane
$CH_3-\overset{\displaystyle O}{\overset{\|}{C}}-CH_3$	279	15	Hexane
$CH_3-\overset{\displaystyle O}{\overset{\|}{C}}-Cl$	235	53	Hexane
$CH_3-\overset{\displaystyle O}{\overset{\|}{C}}-NH_2$	214	—	Water
$CH_3-\overset{\displaystyle O}{\overset{\|}{C}}-OCH_2CH_3$	204	60	Water
$CH_3-\overset{\displaystyle O}{\overset{\|}{C}}-OH$	204	41	Ethanol

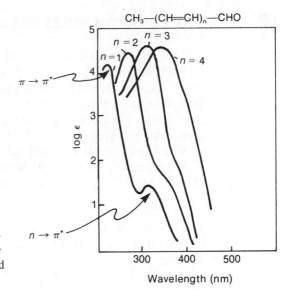

FIGURE 7.15 The spectra of a series of polyene aldehydes. (From Murrell, J. N., *The Theory of the Electronic Spectra of Organic Molecules*, Methuen and Co., Ltd., London, 1963. Reprinted by permission.)

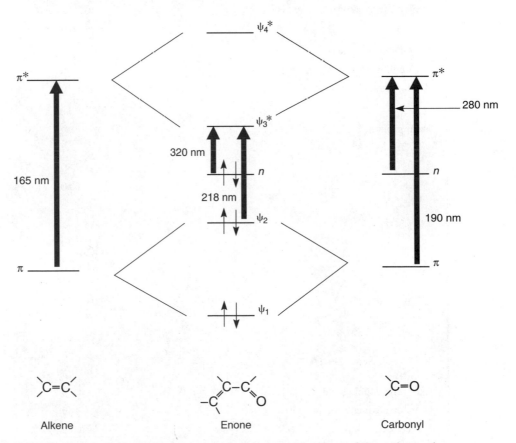

FIGURE 7.16 The orbitals of an enone system compared to those of the noninteracting chromophores.

7.12 WOODWARD'S RULES FOR ENONES

The conjugation of a double bond with a carbonyl group leads to intense absorption ($\varepsilon = 8{,}000$ to $20{,}000$) corresponding to a $\pi \rightarrow \pi^*$ transition of the carbonyl group. The absorption is found between 220 and 250 nm in simple enones. The $n \rightarrow \pi^*$ transition is much less intense ($\varepsilon = 50$ to 100) and appears at 310 to 330 nm. Although the $\pi \rightarrow \pi^*$ transition is affected in predictable fashion by structural modifications of the chromophore, the $n \rightarrow \pi^*$ transition does not exhibit such predictable behavior.

Woodward examined the ultraviolet spectra of numerous enones and devised a set of empirical rules that enable us to predict the wavelength at which the $\pi \rightarrow \pi^*$ transition occurs in an unknown enone. Table 7.7 summarizes these rules.

TABLE 7.7
EMPIRICAL RULES FOR ENONES

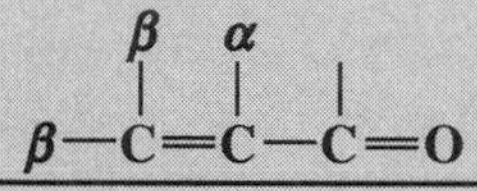

$$\overset{\beta}{|} \quad \overset{\alpha}{|}$$
$$\beta - C = C - C = O \qquad\qquad \overset{\delta}{|} \quad \overset{\gamma}{|} \quad \overset{\beta}{|} \quad \overset{\alpha}{|}$$
$$\delta - C = C - C = C - C = O$$

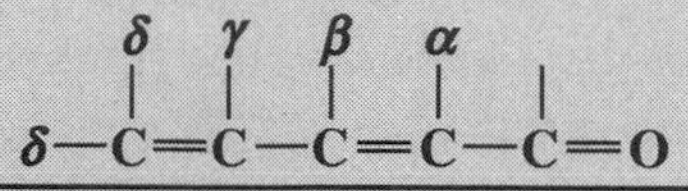

Base values:		
Six-membered ring or acyclic parent enone		= 215 nm
Five-membered ring parent enone		= 202 nm
Acyclic dienone		= 245 nm
Increments for:		
Couble-bond-extending conjugation		30
Alkyl group or ring residue	α	10
	γ and higher	18
Polar groupings:		
$-$OH	α	35
	β	30
	δ	50
$-$OCOCH$_3$	α,β,δ	6
$-$OCH$_3$	α	35
	β	30
	γ	17
	δ	31
$-$Cl	α	15
	β	30
$-$Br	α	25
	β	30
$-$Nr$_2$	β	95
Exocyclic double bond		5
Homocyclic diene component		39
Solvent correction		Variable
	λ_{max}^{EtOH} (calc) = Total	

Following are a few sample applications of these rules. The pertinent parts of the structures are shown in bold face.

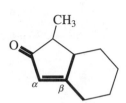

Acyclic enone:	215 nm
α-CH$_3$:	10
β-CH$_3$: 2 × 12 =	24
	249 nm
Observed:	249 nm

Six-membered enone:	215 nm
Double-bond-extending conjugation:	30
Homocyclic diene:	39
δ-Ring residue:	18
	302 nm
Observed:	300 nm

Five-membered enone:	202 nm
β-Ring residue: 2 × 12 =	24
Exocyclic double-bond:	5
	231 nm
Observed:	226 nm

Five-membered enone:	202 nm
α-Br:	25
β-Ring residue: 2 × 12 =	24
Exocyclic double bond:	5
	256 nm
Observed:	251 nm

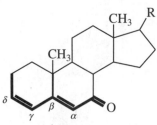

Six-membered enone:	215 nm
Double-bond-extending conjugation:	30
β-Ring residue:	12
δ-Ring residue:	18
Exocyclic double-bond:	5
	280 nm
Observed:	280 nm

7.13 α,β-UNSATURATED ALDEHYDES, ACIDS, AND ESTERS

α,β-Unsaturated aldehydes generally follow the same rules as enones (see the preceding section) except that their absorptions are displaced by about 5 to 8 nm toward shorter wavelength than those of the corresponding ketones. Table 7.8 lists the empirical rules for unsaturated aldehydes.

Nielsen developed a set of rules for *α,β*-unsaturated acids and esters that are similar to those for enones (Table 7.9).

Consider 2-cyclohexenoic and 2-cycloheptenoic acids as examples.

COOH	*α,β*-dialkyl	217 nm calc.
	Double bond is in a six-membered ring, adds nothing	217 nm obs.
COOH	*α,β*-dialkyl	217 nm
	Double bond is in a seven-membered ring	+ 5
		222 nm calc.
		222 nm obs.

TABLE 7.8
EMPIRICAL RULES FOR UNSATURATED ALDEHYDES

$$\beta \diagdown \underset{\beta \diagup}{C} = \underset{\alpha}{C} - \underset{\diagdown O}{C} \diagup ^{H}$$

Parent	208 nm
With *α* or *β* alkyl groups	220
With *α,β* or *β,β* alkyl groups	230
With *α,β,β* alkyl groups	242

7.14 AROMATIC COMPOUNDS

The absorptions that result from transitions within the benzene chromophore can be quite complex. The ultraviolet spectrum contains three absorption bands, which sometimes contain a great deal of fine structure. The electronic transitions are basically of the $\pi \rightarrow \pi*$ type, but their details are not as simple as in the cases of the classes of chromophores described in earlier sections of this chapter.

Figure 7.17a shows the molecular orbitals of benzene. If you were to attempt a simple explanation for the electronic transitions in benzene, you would conclude that there are four possible transitions but each transition has the same energy. You would predict that the ultraviolet spectrum of benzene consists of one absorption peak. However, owing to electron–electron repulsions and symmetry considerations, the actual energy states from which electronic transitions occur are somewhat modified. Figure 7.17b shows the energy-state levels of benzene. Three electronic transitions take

TABLE 7.9
EMPIRICAL RULES FOR UNSATURATED ACIDS AND ESTERS

Base values for:

With α or β alkyl group	208 nm
With α,β or β,β alkyl groups	217
With α,β,β alkyl groups	225
For an exocyclic α,β double bond	Add 5 nm
For an endocyclic α,β double bond in a five- or seven-membered ring	Add 5 nm

place to these excited states. Those transitions, which are indicated in Figure 7.17b, are the so-called **primary bands** at 184 and 202 nm and the **secondary** (fine-structure) **band** at 255 nm. Figure 7.18 is the spectrum of benzene. Of the primary bands, the 184-nm band (the **second primary band**) has a molar absorptivity of 47,000. It is an allowed transition. Nevertheless, this transition is not observed under usual experimental conditions, because absorptions at this wavelength are in the vacuum ultraviolet region of the spectrum, beyond the range of most commercial instruments. In polycyclic aromatic compounds, the second primary band is often shifted to longer wavelengths, in

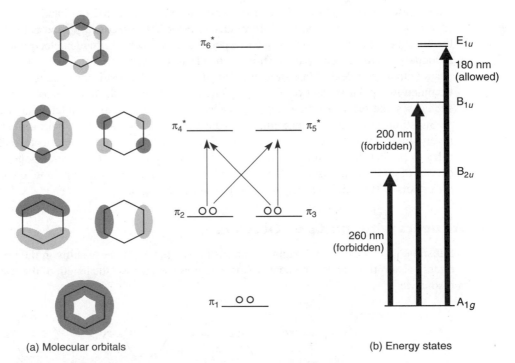

(a) Molecular orbitals   (b) Energy states

FIGURE 7.17 Molecular orbitals and energy states for benzene.

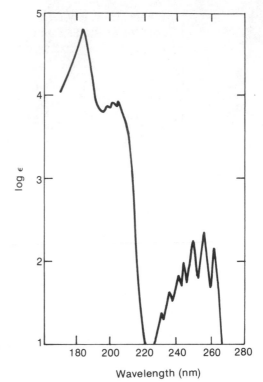

FIGURE 7.18 Ultraviolet spectrum of benzene. (From Petruska, J., *J. Chem. Phys., 34* [1961]: 1121. Reprinted by permission.)

which case it can be observed under ordinary conditions. The 202-nm band is much less intense ($\varepsilon = 7400$), and it corresponds to a forbidden transition. The secondary band is the least intense of the benzene bands ($\varepsilon = 230$). It also corresponds to a symmetry-forbidden electronic transition. The secondary band, caused by interaction of the electronic energy levels with vibrational modes, appears with a great deal of fine structure. This fine structure is lost if the spectrum of benzene is determined in a polar solvent or if a single functional group is substituted onto the benzene ring. In such cases, the secondary band appears as a broad peak, lacking in any interesting detail.

Substitution on the benzene ring can cause bathochromic and hyperchromic shifts. Unfortunately, these shifts are difficult to predict. Consequently, it is impossible to formulate empirical rules to predict the spectra of aromatic substances as was done for dienes, enones, and the other classes of compounds discussed earlier in this chapter. You may gain a qualitative understanding of the effects of substitution by classifying substituents into groups.

A.   Substituents with Unshared Electrons

Substituents that carry nonbonding electrons (*n* electrons) can cause shifts in the primary and secondary absorption bands. The nonbonding electrons can increase the length of the π system through resonance.

The more available these *n* electrons are for interaction with the π system of the aromatic ring, the greater the shifts will be. Examples of groups with *n* electrons are the amino, hydroxyl, and methoxy groups, as well as the halogens.

Interactions of this type between the *n* and π electrons usually cause shifts in the primary and secondary benzene absorption bands to longer wavelength (extended conjugation). In addition, the presence of *n* electrons in these compounds gives the possibility of $n \rightarrow \pi^*$ transitions. If an *n* electron is excited into the extended π^* chromophore, the atom from which it was removed becomes electron deficient, while the π system of the aromatic ring (which also includes atom Y) acquires an extra electron. This causes a separation of charge in the molecule and is generally represented as regular resonance, as was shown earlier. However, the extra electron in the ring is actually in a π^* orbital and would be better represented by structures of the following type, with the asterisk representing the excited electron.

Such an excited state is often called a **charge-transfer** or an **electron-transfer** excited state.

In compounds that are acids or bases, pH changes can have very significant effects on the positions of the primary and secondary bands. Table 7.10 illustrates the effects of changing the pH of the solution on the absorption bands of various substituted benzenes. In going from benzene to phenol, notice the shift from 203.5 to 210.5 nm—a 7-nm shift—in the primary band. The secondary band shifts from 254 to 270 nm—a 16-nm shift. However, in phenoxide ion, the conjugate base of phenol, the primary band shifts from 203.5 to 235 nm (a 31.5-nm shift), and the secondary band shifts from 254 to 287 nm (a 33-nm shift). The intensity of the secondary band also increases. In phenoxide ion there are more *n* electrons, and they are more available for interaction with the aromatic π system, than in phenol.

TABLE 7.10
pH EFFECTS ON ABSORPTION BANDS

Substituent	Primary		Secondary	
	λ(nm)	ε	λ(nm)	ε
$\bigcirc$—H	203.5	7,400	254	204
—OH	210.5	6,200	270	1,450
—O$^-$	235	9,400	287	2,600
—NH$_2$	230	8,600	280	1,430
—NH$_3^+$	203	7,500	254	169
—COOH	230	11,600	273	970
—COO$^-$	224	8,700	268	560

The comparison of aniline and anilinium ion illustrates a reverse case. Aniline exhibits shifts similar to those of phenol. From benzene to aniline, the primary band shifts from 203.5 to 230 nm (a 26.5-nm shift), and the secondary band shifts from 254 to 280 nm (a 26-nm shift). However, these large shifts are not observed in the case of anilinium ion, the conjugate acid of aniline. For anilinium ion, the primary and secondary bands do not shift at all. The quaternary nitrogen of anilinium ion has no unshared pairs of electrons to interact with the benzene π system. Consequently, the spectrum of anilinium ion is almost identical to that of benzene.

B.   Substituents Capable of π-Conjugation

Substituents that are themselves chromophores usually contain π electrons. Just as in the case of n electrons, interaction of the benzene-ring electrons and the π electrons of the substituent can produce a new electron transfer band. At times, this new band may be so intense as to obscure the secondary band of the benzene system. Notice that this interaction induces the opposite polarity; the ring becomes electron deficient.

Table 7.10 demonstrates the effect of acidity or basicity of the solution on such a chromophoric substituent group. In the case of benzoic acid, the primary and secondary bands are shifted substantially from those noted for benzene. However, the magnitudes of the shifts are somewhat smaller in the case of benzoate ion, the conjugate base of benzoic acid. The intensities of the peaks are lower than for benzoic acid, as well. We expect electron transfer of the sort just shown to be less likely when the functional group already bears a negative charge.

C.   Electron-Releasing and Electron-Withdrawing Effects

Substituents may have differing effects on the positions of absorption maxima, depending upon whether they are electron releasing or electron withdrawing. Any substituent, regardless of its influence on the electron distribution elsewhere in the aromatic molecule, shifts the primary absorption band to longer wavelength. Electron-withdrawing groups have essentially no effect on the position of the secondary absorption band unless, of course, the electron-withdrawing group is also capable of acting as a chromophore. However, electron-releasing groups increase both the wavelength and the intensity of the secondary absorption band. Table 7.11 summarizes these effects, with electron-releasing and electron-withdrawing substituents grouped together.

D.   Disubstituted Benzene Derivatives

With disubstituted benzene derivatives, it is necessary to consider the effect of each of the two substituents. For *para*-disubstituted benzenes, two possibilities exist. If both groups are electron releasing or if they are both electron withdrawing, they exert effects similar to those observed with monosubstituted benzenes. The group with the stronger effect determines the extent of shifting of

TABLE 7.11
ULTRAVIOLET MAXIMA FOR VARIOUS AROMATIC COMPOUNDS

Substituent		Primary		Secondary	
		λ(nm)	ε	λ(nm)	ε
	—H	203.5	7,400	254	204
Electron-releasing substituents	—CH$_3$	206.5	7,000	261	225
	—Cl	209.5	7,400	263.5	190
	—Br	210	7,900	261	192
	—OH	210.5	6,200	270	1,450
	—OCH$_3$	217	6,400	269	1,480
	—NH$_2$	230	8,600	280	1,430
Electron-withdrawing substituents	—CN	224	13,000	271	1,000
	—COOH	230	11,600	273	970
	—COCH$_3$	245.5	9,800		
	—CHO	249.5	11,400		
	—NO$_2$	268.5	7,800		

the primary absorption band. If one of the groups is electron releasing while the other is electron withdrawing, the magnitude of the shift of the primary band is greater than the sum of the shifts due to the individual groups. The enhanced shifting is due to resonance interactions of the following type.

If the two groups of a disubstituted benzene derivative are either *ortho* or *meta* to each other, the magnitude of the observed shift is approximately equal to the sum of the shifts caused by the individual groups. With substitution of these types, there is no opportunity for the kind of direct resonance interaction between substituent groups that is observed with *para* substituents. In the case of *ortho* substituents, the steric inability of both groups to achieve coplanarity inhibits resonance.

For the special case of substituted benzoyl derivatives, an empirical correlation of structure with the observed position of the primary absorption band has been developed (Table 7.12). It provides a means of estimating the position of the primary band for benzoyl derivatives within about 5 nm.

TABLE 7.12
EMPIRICAL RULES FOR BENZOYL DERIVATIVES

Parent chromophore:

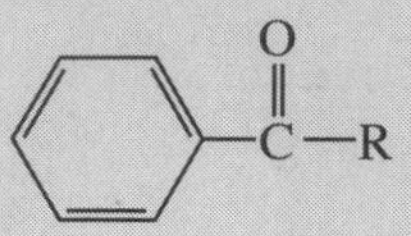

R = alkyl or ring residue		246
R = H		250
R = OH or OAlkyl		230

Increment for each substituent:

—Alkyl or ring residue	*o, m*	3
	p	10
—OH, —OCH₃, or —OAlkyl	*o, m*	7
	p	25
—O⁻	*o*	11
	m	20
	p	78
—Cl	*o, m*	0
	p	10
—Br	*o, m*	2
	p	15
—NH₂	*o, m*	13
	p	58
—NHCOCH₃	*o, m*	20
	p	45
—NHCH₃	*p*	73
—N(CH₃)₂	*o, m*	20
	p	85

Following are two sample applications of these rules.

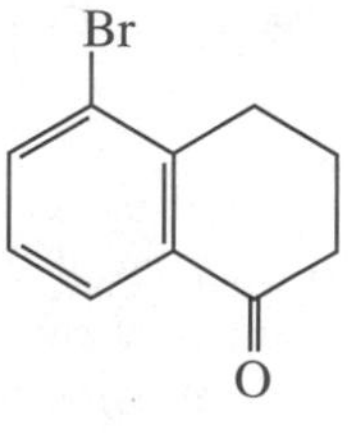

Parent chromophore:	246 nm
o-Ring residue:	3
m-Br:	2
	251 nm
Observed:	253 nm

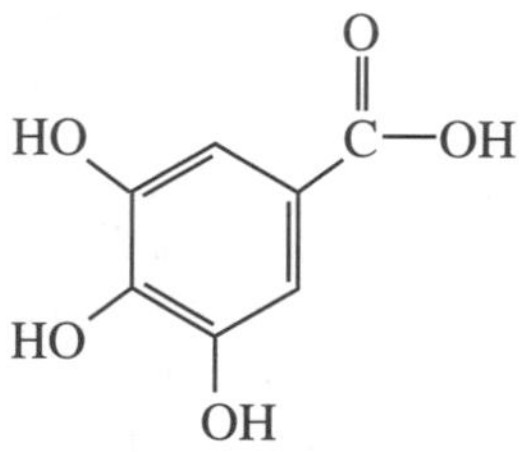

Parent chromophore:	230 nm
m-OH: 2 × 7 =	14
p-OH:	25
	269 nm
Observed:	270 nm

E.   Polynuclear Aromatic Hydrocarbons and Heterocyclic Compounds

Researchers have observed that the primary and secondary bands in the spectra of polynuclear aromatic hydrocarbons shift to longer wavelength. In fact, even the second primary band, which appears at 184 nm for benzene, is shifted to a wavelength within the range of most UV spectrophotometers. This band lies at 220 nm in the spectrum of naphthalene. As the extent of conjugation increases, the magnitude of the bathochromic shift also increases.

The ultraviolet spectra of the polynuclear aromatic hydrocarbons possess characteristic shapes and fine structure. In the study of spectra of substituted polynuclear aromatic derivatives, it is common practice to compare them with the spectrum of the unsubstituted hydrocarbon. The nature of the chromophore can be identified on the basis of similarity of peak shapes and fine structure. This technique involves the use of model compounds. Section 7.15 will discuss it further.

Figure 7.19 shows the ultraviolet spectra of naphthalene and anthracene. Notice the characteristic shape and fine structure of each spectrum, as well as the effect of increased conjugation on the positions of the absorption maxima.

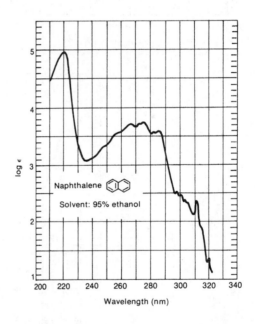

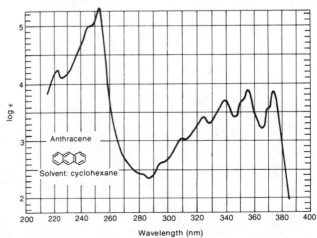

FIGURE 7.19 Ultraviolet spectra of naphthalene and anthracene. (From Friedel, R. A., and M. Orchin, *Ultraviolet Spectra of Aromatic Compounds*, John Wiley and Sons, New York, 1951. Reprinted by permission.)

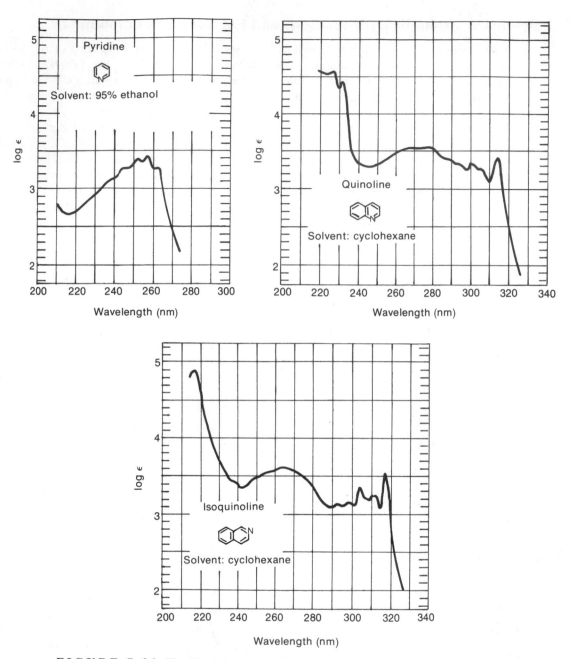

FIGURE 7.20 The ultraviolet spectra of pyridine, quinoline, and isoquinoline. (From Friedel, R. A., and M. Orchin, *Ultraviolet Spectra of Aromatic Compounds*, John Wiley and Sons, New York, 1951. Reprinted by permission.)

Heterocyclic molecules have electronic transitions that include combinations of $\pi \rightarrow \pi^*$ and $n \rightarrow \pi^*$ transitions. The spectra can be rather complex, and analysis of the transitions involved will be left to more advanced treatments. The common method of studying derivatives of heterocyclic molecules is to compare them to the spectra of the parent heterocyclic systems. Section 7.15 will further describe the use of model compounds in this fashion.

Figure 7.20 includes the ultraviolet spectra of pyridine, quinoline, and isoquinoline. You may wish to compare the spectrum of pyridine with that of benzene (Fig. 7.18) and the spectra of quinoline and isoquinoline with the spectrum of naphthalene (Fig. 7.19).

7.15 MODEL COMPOUND STUDIES

Very often, the ultraviolet spectra of several members of a particular class of compounds are very similar. Unless you are thoroughly familiar with the spectroscopic properties of each member of the class of compounds, it is very difficult to distinguish the substitution patterns of individual molecules by their ultraviolet spectra. You can, however, determine the gross nature of the chromophore of an unknown substance by this method. Then, based on knowledge of the chromophore, you can employ the other spectroscopic techniques described in this book to elucidate the precise structure and substitution of the molecule.

This approach—the use of model compounds—is one of the best ways to put the technique of ultraviolet spectroscopy to work. By comparing the UV spectrum of an unknown substance with that of a similar but less highly substituted compound, you can determine whether or not they contain the same chromophore. Many of the books listed in the references at the end of this chapter contain large collections of spectra of suitable model compounds, and with their help you can establish the general structure of the part of the molecule that contains the π electrons. You can then utilize infrared or NMR spectroscopy to determine the detailed structure.

As an example, consider an unknown substance that has the molecular formula $C_{15}H_{12}$. A comparison of its spectrum (Fig. 7.21) with that of anthracene (Fig. 7.19) shows that the two spectra are nearly identical. Disregarding minor bathochromic shifts, the same general peak shape and fine structure appear in the spectra of both the unknown and anthracene, the model compound. You may then conclude that the unknown is a substituted anthracene derivative. Further structure determination reveals that the unknown is 9-methylanthracene. The spectra of model compounds can be obtained from published catalogues of ultraviolet spectra. In cases in which a suitable model compound is not available, a model compound can be synthesized and its spectrum determined.

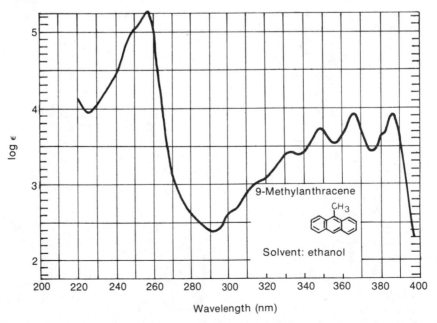

FIGURE 7.21 The ultraviolet spectrum of 9-methylanthracene. (From Friedel, R. A., and M. Orchin, *Ultraviolet Spectra of Aromatic Compounds*, John Wiley and Sons, New York, 1951. Reprinted by permission.)

7.16   VISIBLE SPECTRA: COLOR IN COMPOUNDS

The portion of the electromagnetic spectrum lying between about 400 and 750 nm is the **visible** region. Light waves with wavelengths between these limits appear colored to the human eye. As anyone who has seen light diffracted by a prism or the diffraction effect of a rainbow knows, one end of the visible spectrum is violet and the other is red. Light with wavelengths near 400 nm is violet, while that with wavelengths near 750 nm is red.

The phenomenon of color in compounds, however, is not as straightforward as the preceding discussion would suggest. If a substance absorbs visible light, it appears to have a color; if not, it appears white. However, compounds that absorb light in the visible region of the spectrum do not possess the color corresponding to the wavelength of the absorbed light. Rather, there is an inverse relationship between the observed color and the color absorbed.

When we observe light **emitted** from a source, as from a lamp or an emission spectrum, we observe the color corresponding to the wavelength of the light being emitted. A light source emitting violet light emits light at the high-energy end of the visible spectrum. A light source emitting red light emits light at the low-energy end of the spectrum.

However, when we observe the color of a particular object or substance, we do not observe that object or substance emitting light. (Certainly the substance does not glow in the dark.) Rather, we observe the light that is being **reflected.** The color that our eye perceives is not the color corresponding to the wavelength of the light absorbed but its **complement.** When white light falls on an object, light of a particular wavelength is absorbed. The remainder of the light is reflected. The eye and brain register all of the reflected light as the color complementary to the color that was absorbed.

In the case of transparent objects or solutions, the eye receives the light that is **transmitted.** Again, light of a particular wavelength is absorbed, and the remaining light passes through to reach the eye. As before, the eye registers this transmitted light as the color complementary to the color that was absorbed. Table 7.13 illustrates the relationship between the wavelength of light absorbed by a substance and the color perceived by an observer.

Some familiar compounds may serve to underscore these relationships between the absorption spectrum and the observed color. The structural formulas of these examples are shown. Notice that each of these substances has a highly extended conjugated system of electrons. Such extensive conjugation shifts their electronic spectra to such long wavelengths that they absorb visible light and appear colored.

TABLE 7.13

RELATIONSHIP BETWEEN THE COLOR OF LIGHT ABSORBED BY A COMPOUND AND THE OBSERVED COLOR OF THE COMPOUND

Color of Light Absorbed	Wavelength of Light Absorbed (nm)	Observed Color
Violet	400	Yellow
Blue	450	Orange
Blue-green	500	Red
Yellow-green	530	Red-violet
Yellow	550	Violet
Orange-red	600	Blue-green
Red	700	Green

β-Carotene (pigment from carrots): $\lambda_{max} = 452$ nm, **orange**
Cyanidin (blue pigment of cornflower): $\lambda_{max} = 545$ nm, **blue**
Malachite green (a triphenylmethane dye): $\lambda_{max} = 617$ nm, **green**

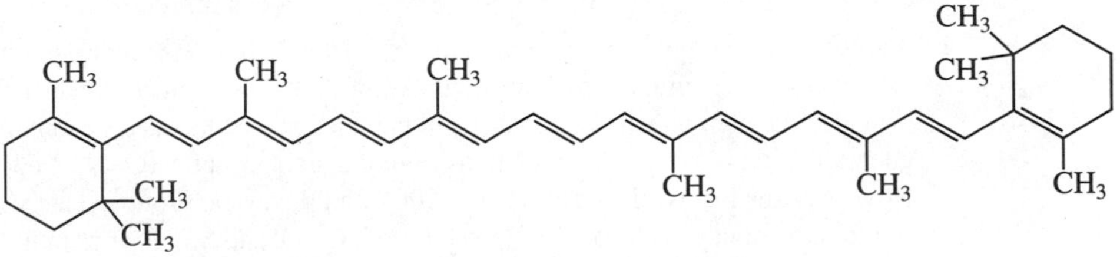

β-Carotene (a carotenoid, which is a class of plant pigments)
$\lambda_{max} = 452$ nm

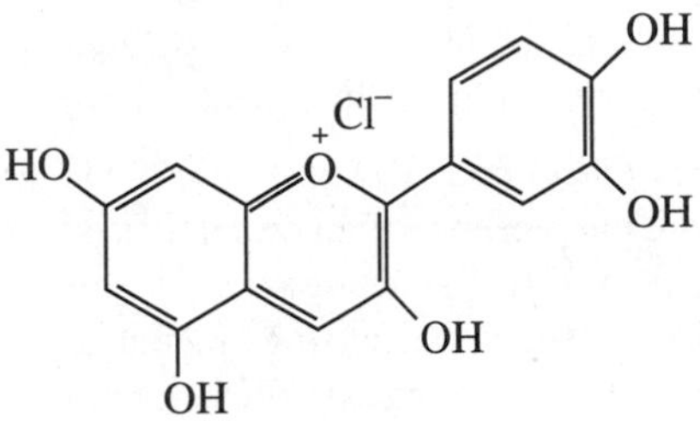

Cyanidin chloride (an anthocyanin, another class of plant pigments)
$\lambda_{max} = 545$ nm

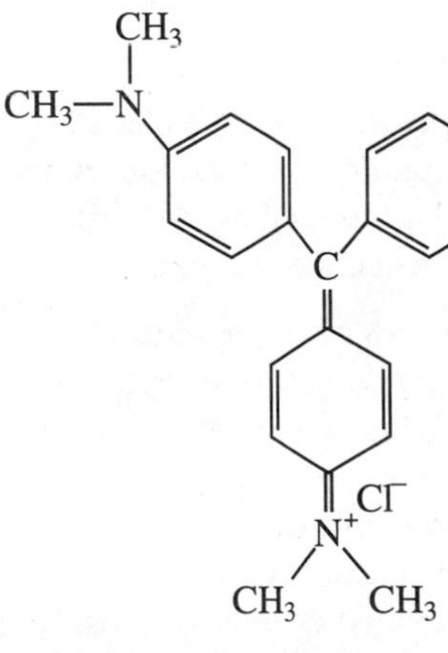

Malachite green (a triphenylmethane dye)
$\lambda_{max} = 617$ nm

7.17 WHAT TO LOOK FOR IN AN ULTRAVIOLET SPECTRUM: A PRACTICAL GUIDE

It is often difficult to extract a great deal of information from a UV spectrum used by itself. It should be clear by now that a UV spectrum is most useful when at least a general idea of the structure is already known; in this way, the various empirical rules can be applied. Nevertheless, several generalizations can serve to guide our use of UV data. These generalizations are a good deal more meaningful when combined with infrared and NMR data—which can, for instance, definitely identify carbonyl groups, double bonds, aromatic systems, nitro groups, nitriles, enones, and other important chromophores. In the absence of infrared or NMR data, the following observations should be taken only as guidelines.

1. *A single band of low to medium intensity ($\varepsilon = 100$ to $10,000$) at wavelengths less than 220 nm* usually indicates an $n \rightarrow \sigma^*$ transition. Amines, alcohols, ethers, and thiols are possibilities, provided the nonbonded electrons are not included in a conjugated system. An exception to this generalization is that the $n \rightarrow \pi^*$ transition of cyano groups ($-C\equiv N :$) appears in this region. However, this is a weak transition ($\varepsilon < 100$), and the cyano group is easily identified in the infrared. Do not neglect to look for N—H, O—H, C—O, and S—H bands in the infrared spectrum.

2. *A single band of low intensity ($\varepsilon = 10$ to 100) in the region 250 to 360 nm, with no major absorption at shorter wavelengths (200 to 250 nm),* usually indicates an $n \rightarrow \pi^*$ transition. Since the absorption does not occur at long wavelength, a simple, or unconjugated, chromophore is indicated, generally one that contains an O, N, or S atom. Examples of this may include C=O, C=N, N=N, $-NO_2$, $-COOR$, $-COOH$, or $-CONH_2$. Once again, infrared and NMR spectra should help a great deal.

3. *Two bands of medium intensity ($\varepsilon = 1,000$ to $10,000$), both with λ_{max} above 200 nm,* generally indicate the presence of an aromatic system. If an aromatic system is present, there may be a good deal of fine structure in the longer-wavelength band (in nonpolar solvents only). Substitution on the aromatic rings increases the molar absorptivity above 10,000, particularly if the substituent increases the length of the conjugated system.

   In polynuclear aromatic substances, a third band appears near 200 nm, a band that in simpler aromatics occurs below 200 nm, where it cannot be observed. Most polynuclear aromatics (and heterocyclic compounds) have very characteristic intensity and band-shape (fine-structure) patterns, and they may often be identified via comparison to spectra that are available in the literature. The textbooks by Jaffé and Orchin and by Scott, which are listed in the references at the end of this chapter, are good sources of spectra.

4. *Bands of high intensity ($\varepsilon = 10,000$ to $20,000$) that appear above 210 nm* generally represent either an α,β-unsaturated ketone (check the infrared spectrum), a diene, or a polyene. The greater the length of the conjugated system, the longer the observed wavelength. For dienes, the λ_{max} may be calculated using the Woodward–Fieser Rules (Section 7.10).

5. *Simple ketones, acids, esters, amides, and other compounds containing both π systems and unshared electron pairs show two absorptions:* an $n \rightarrow \pi^*$ transition at longer wavelengths (>300 nm, low intensity) and a $\pi \rightarrow \pi^*$ transition at shorter wavelengths (<250 nm, high intensity). With conjugation (enones), the λ_{max} of the $\pi \rightarrow \pi^*$ band moves to longer wavelengths and can be predicted by Woodward's Rules (Section 7.12). The ε value usually rises above 10,000 with conjugation, and, as it is very intense, it may obscure or bury the weaker $n \rightarrow \pi^*$ transition.

   For α,β-unsaturated esters and acids, Nielsen's Rules (Section 7.13) may be used to predict the position of λ_{max} with increasing conjugation and substitution.

6. *Compounds that are highly colored* (have absorption in the visible region) are likely to contain a long-chain conjugated system or a polycyclic aromatic chromophore. Benzenoid compounds may be colored if they have enough conjugating substituents. For nonaromatic systems, usually a minimum of four to five conjugated chromophores are required to produce absorption in the visible region. However, some simple nitro, azo, nitroso, α-diketo, polybromo, and polyiodo compounds may also exhibit color, as may many compounds with quinoid structures.

PROBLEMS

*1. The ultraviolet spectrum of benzonitrile shows a primary absorption band at 224 nm and a secondary band at 271 nm.

  (a) If a solution of benzonitrile in water, with a concentration of 1×10^{-4} molar, is examined at a wavelength of 224 nm, the absorbance is determined to be 1.30. The cell length is 1 cm. What is the molar absorptivity of this absorption band?

  (b) If the same solution is examined at 271 nm, what will be the absorbance reading ($\varepsilon = 1000$)? What will be the intensity ratio, I_0/I?

*2. Draw structural formulas that are consistent with the following observations.

  (a) An acid, $C_7H_4O_2Cl_2$, shows a UV maximum at 242 nm.

  (b) A ketone, $C_8H_{14}O$, shows a UV maximum at 248 nm.

  (c) An aldehyde, $C_8H_{12}O$, absorbs in the UV with $\lambda_{max} = 244$ nm.

*3. Predict the UV maximum for each of the following substances.

  (a)

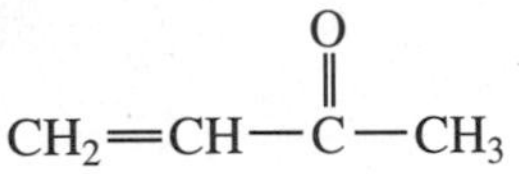

  (b)

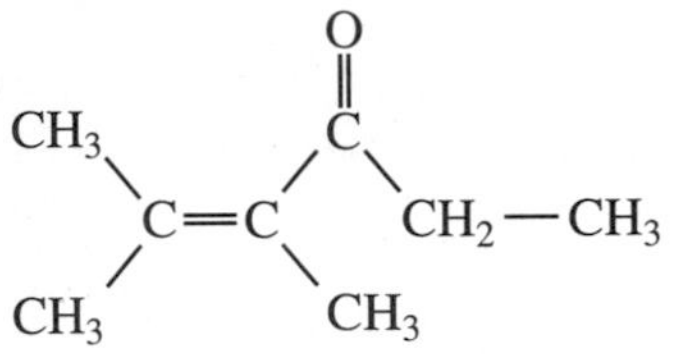

  (c)

  (d)

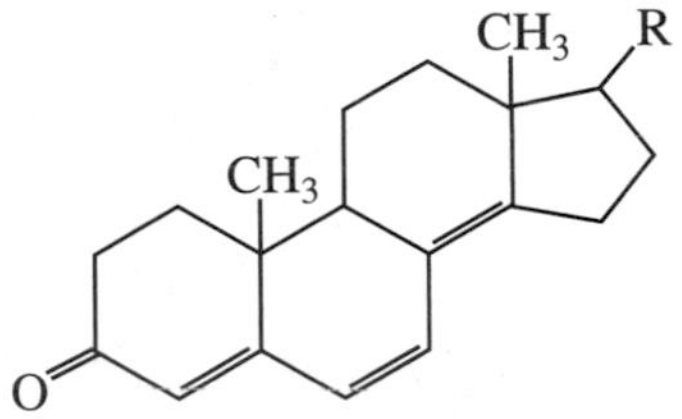

  (e)

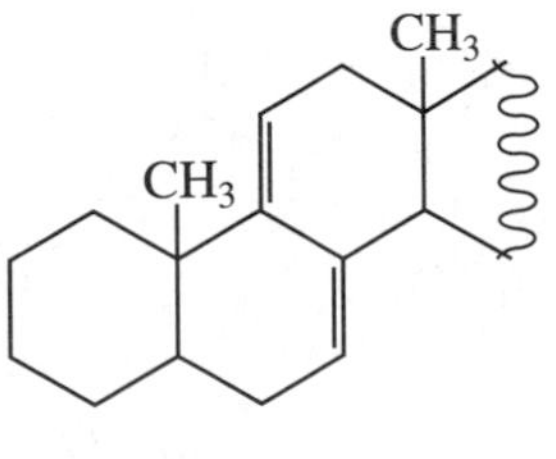

  (f)

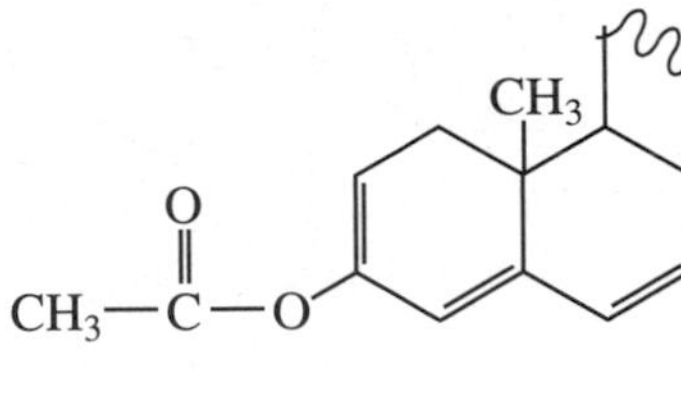

  (g)

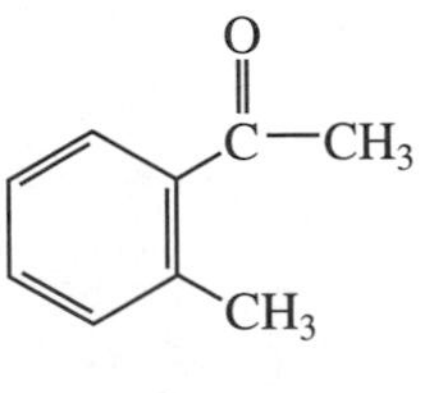

  (h)

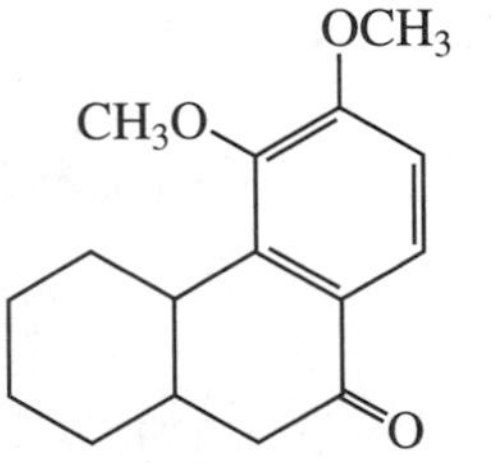

  (i)

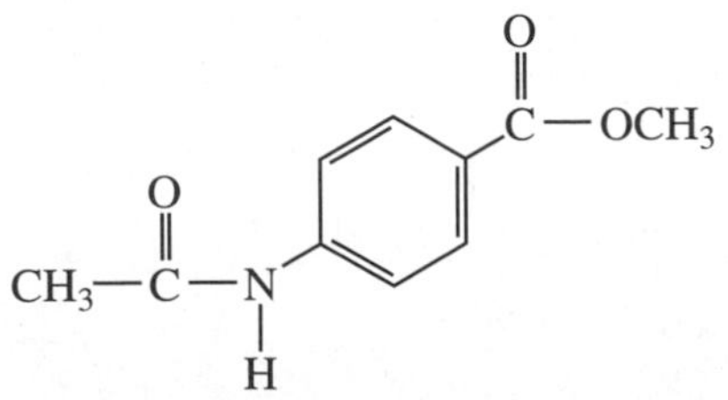

  (j)

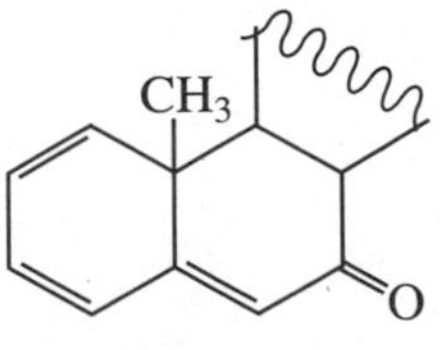

*4. The UV spectrum of acetone shows absorption maxima at 166, 189, and 279 nm. What type of transition is responsible for each of these bands?

*5. Chloromethane has an absorption maximum at 172 nm, bromomethane shows an absorption at 204 nm, and iodomethane shows a band at 258 nm. What type of transition is responsible for each band? How can the trend of absorptions be explained?

*6. What types of electronic transitions are possible for each of the following compounds?

    (a) Cyclopentene

    (b) Acetaldehyde

    (c) Dimethyl ether

    (d) Methyl vinyl ether

    (e) Triethylamine

    (f) Cyclohexane

7. Predict and explain whether UV/visible spectroscopy can be used to distinguish between the following pairs of compounds. Where possible, support your answers with calculations.

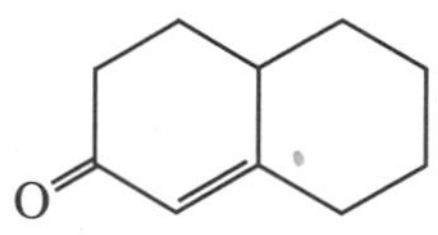

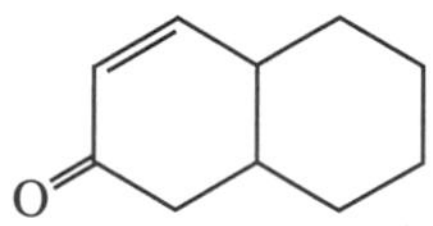

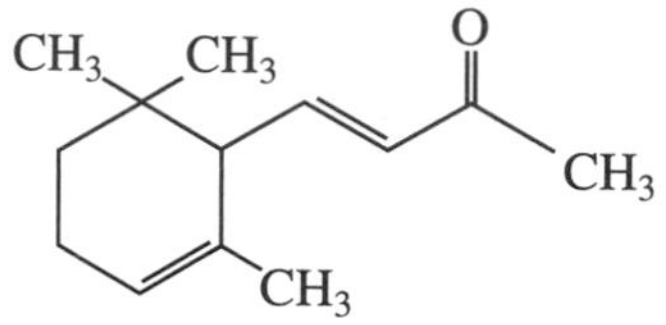

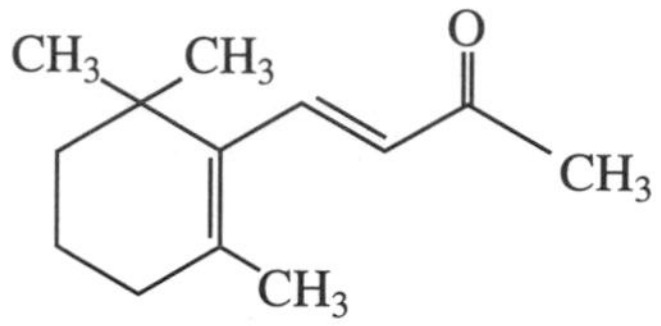

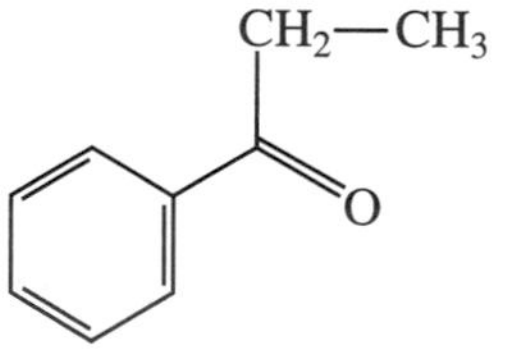

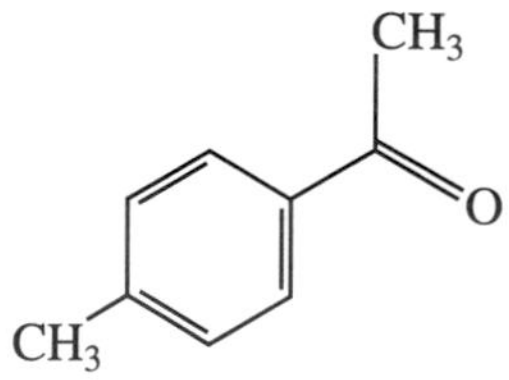

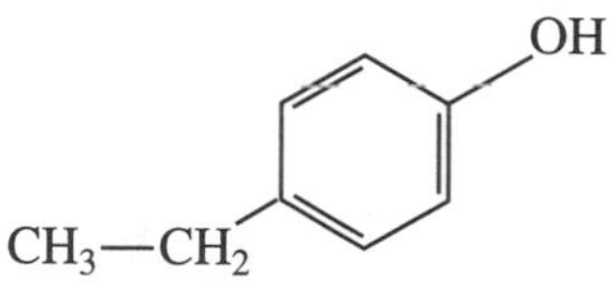

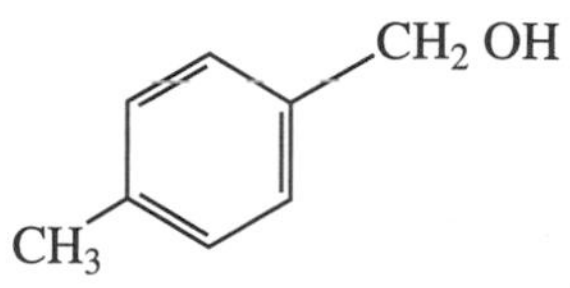

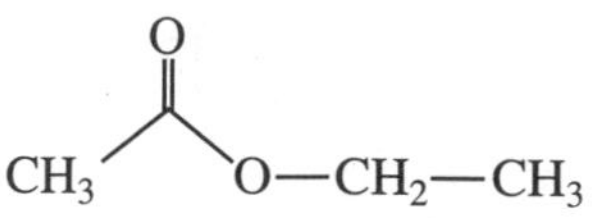

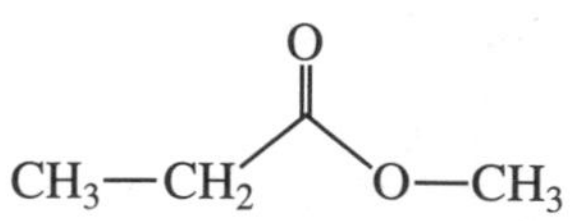

8. (a) Predict the UV maximum for the reactant and product of the following photochemical reaction.

(b) Is UV spectroscopy a good way to distinguish the reactant from the product?

(c) How would you use infrared spectroscopy to distinguish between the reactant and the product?

(d) How would you use proton NMR to distinguish between the reactant and the product (two ways)?

(e) How could you distinguish between the reactant and the product by using DEPT NMR (see Chapter 10)?

REFERENCES

American Petroleum Institute Research Project 44, *Selected Ultraviolet Spectral Data*, Vols. I–IV, Thermodynamics Research Center, Texas A&M University, College Station, Texas, 1945–1977.

Friedel, R. A., and M. Orchin, *Ultraviolet Spectra of Aromatic Compounds*, John Wiley and Sons, New York, 1951.

Graselli, J. G., and W. M. Ritchey, eds., *Atlas of Spectral Data and Physical Constants*, CRC Press, Cleveland, OH, 1975.

Hershenson, H. M., *Ultraviolet Absorption Spectra: Index for 1954–1957*, Academic Press, New York, 1959.

Jaffé, H. H., and M. Orchin, *Theory and Applications of Ultraviolet Spectroscopy*, John Wiley and Sons, New York, 1964.

Parikh, V. M., *Absorption Spectroscopy of Organic Molecules*, Addison–Wesley Publishing Co., Reading, MA, 1974, Chapter 2.

Scott, A. I., *Interpretation of the Ultraviolet Spectra of Natural Products*, Pergamon Press, New York, 1964.

Silverstein, R. M., and Webster, F. X. *Spectrometric Identification of Organic Compounds*, 6th ed., John Wiley and Sons, New York, 1998, Chapter 7.

Stern, E. S., and T. C. J. Timmons, *Electronic Absorption Spectroscopy in Organic Chemistry*, St. Martin's Press, New York, 1971.

Websites:

http://webbook.nist.gov/chemistry/

The National Institute of Standards and Technology (NIST) has developed the WebBook. This site includes UV/visible spectra, gas phase infrared spectra, and mass spectral data for compounds.

CHAPTER 8

MASS SPECTROMETRY

The principles that underlie mass spectrometry predate all of the other instrumental techniques described in this book. The fundamental principles date to 1898. In 1911, J. J. Thomson used a mass spectrum to demonstrate the existence of neon-22 in a sample of neon-20, thereby establishing that elements could have isotopes. The earliest mass spectrometer, as we know it today, was built in 1918. However, the method of mass spectrometry did not come into common use until quite recently, when reasonably inexpensive and reliable instruments became available. With the advent of commercial instruments that can be maintained fairly easily, are priced within reason for many industrial and academic laboratories, and provide high resolution, the technique has become quite important in structure elucidation studies.

8.1 THE MASS SPECTROMETER

In its simplest form, the mass spectrometer performs three essential functions. First, it subjects molecules to bombardment by a stream of high-energy electrons, converting some of the molecules to ions, which are accelerated in an electric field. Second, the accelerated ions are separated according to their mass-to-charge ratios in a magnetic or electric field. Finally, the ions that have a particular mass-to-charge ratio are detected by a device which can count the number of ions striking it. The detector's output is amplified and fed to a recorder. The trace from the recorder is a **mass spectrum**—a graph of the number of particles detected as a function of mass-to-charge ratio.

When we examine each function in detail, we see that the mass spectrometer is actually somewhat more complex than just described. Before the ions can be formed, a stream of molecules must be introduced into the **ionization chamber** where the ionization takes place. A **sample inlet system** provides this stream of molecules.

A sample studied by mass spectrometry may be a gas, a liquid, or a solid. Enough of the sample must be converted to the vapor state to obtain the stream of molecules that must flow into the ionization chamber. With gases, of course, the substance is already vaporized, so a simple inlet system can be used. This inlet system is only partially evacuated so that the ionization chamber itself is at a lower pressure than the sample inlet system. The sample is introduced into a larger reservoir, from which the molecules of vapor can be drawn into the ionization chamber, which is at low pressure. To ensure that a steady stream of molecules is passing into the ionization chamber, the vapor travels through a small pinhole, called a **molecular leak,** before entering the chamber. The same system can be used for volatile liquids or solids. For less volatile materials, the system can be designed to fit within an oven, which can heat the sample to provide a greater vapor pressure. Care must be taken not to heat any sample to a temperature at which it might decompose.

With rather nonvolatile solids, a direct-probe method of introducing the sample may be used. The sample is placed on the tip of the probe, which is then inserted through a vacuum lock into the ionization chamber. The sample is placed very close to the ionizing beam of electrons. The probe can be heated, thus causing vapor from the sample to be evolved in proximity to the beam of electrons. A system such as this can be used to study samples of molecules with vapor pressures lower than 10^{-9} mm Hg at room temperature.

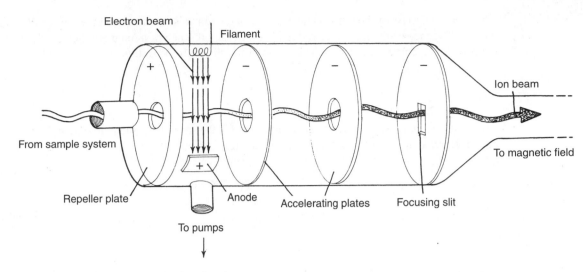

FIGURE 8.1 Ionization chamber.

Once the stream of sample molecules has entered the ionization chamber, a beam of high-energy electrons bombards it. This process converts the molecules to ions. The ions are then accelerated in an electric field. Figure 8.1 is a diagram of a typical ionization chamber.

In the ionization chamber, the beam of high-energy electrons is emitted from a **filament** that is heated to several thousand degrees Celsius. In normal operation, the emitted electrons have an energy of about 70 electron volts (eV). These high-energy electrons strike the stream of molecules that has been admitted from the sample system and ionize the molecules in the stream by removing electrons from them; the molecules are thus converted to positive ions. A **repeller plate,** which carries a positive electrical potential, directs the newly created ions toward a series of **accelerating plates.** A large potential difference, ranging from 1 to 10 kilovolts (kV), applied across these accelerating plates produces a beam of rapidly traveling positive ions. One or more **focusing slits** direct the ions into a uniform beam.

Most of the sample molecules are not ionized at all but are continuously drawn off by vacuum pumps that are connected to the ionization chamber. Some of the molecules are converted to negative ions through the absorption of electrons. The repeller plate absorbs these negative ions. A small proportion of the positive ions that are formed may have a charge greater than one (a loss of more than one electron). These are accelerated in the same way as the singly charged positive ions.

The energy required to remove an electron from an atom or molecule is its **ionization potential.** Most organic compounds have ionization potentials ranging between 8 and 15 eV. However, a beam of electrons does not create ions with high efficiency until it strikes the stream of molecules with a potential of 50 to 70 eV. To produce reproducible spectra, electrons of this energy range are generally used to ionize the sample.

From the ionization chamber, the beam of ions passes through a short field-free region. From there it enters the **mass analyzer,** the region where the ions are separated according to their mass-to-charge ratios.

The kinetic energy of an accelerated ion is equal to

$$\frac{1}{2}mv^2 = eV$$

where m is the mass of the ion, v is the velocity of the ion, e is the charge on the ion, and V is the potential difference of the ion-accelerating plates. In the presence of a magnetic field, a charged particle describes a curved flight path. The equation which yields the radius of curvature of this path is

$$r = \frac{mv}{eH}$$

where r is the radius of curvature of the path and H is the strength of the magnetic field. If these two equations are combined to eliminate the velocity term, the result is

$$\frac{m}{e} = \frac{H^2 r^2}{2V}$$

Equation 8.1

This is the important equation that governs the behavior of an ion in the mass-analyzer portion of the mass spectrometer.

As can be seen from Equation 8.1, the greater the value of m/e, the larger the radius of the curved path. The analyzer tube of the instrument is constructed to have a fixed radius of curvature. A particle with the correct m/e ratio can negotiate the curved analyzer tube and reach the detector. Particles with m/e ratios that are either too large or too small strike the sides of the analyzer tube and do not reach the detector. The method would not be very interesting if ions of only one mass could be detected. Therefore, either the accelerating voltage or the magnetic field strength is continuously varied in order that all of the ions produced in the ionization chamber can be detected. The record produced from the detector system is in the form of a plot of the numbers of ions versus their values of m/e.

An important consideration in mass spectrometry is **resolution,** defined according to the relationship

$$R = \frac{M}{\Delta M}$$

Equation 8.2

where R is the resolution, M is the mass of the particle, and ΔM is the difference in mass between a particle of mass M and the particle of next higher mass that can be resolved by the instrument. Low-resolution instruments have R values ranging as high as 2000. For some applications, resolutions 5 to 10 times that amount are required.

To obtain higher resolutions, modifications of this basic instrument design are used. Since the particles leaving the ionization chamber do not all have precisely the same velocity, a **double-focusing mass spectrometer** may be used. In such an instrument, the beam of ions passes through an electric field region before entering the magnetic field. The particles describe a curved path in each of these regions. In the presence of an electric field, the particles all travel at the same velocity, so the resolution of the magnetic field region improves.

The **detector** of a typical instrument consists of a counter that produces a current that is proportional to the number of ions that strike it. Through the use of electron multiplier circuits, this current can be measured so accurately that the current caused by just one ion striking the detector can be measured. The signal from the detector is fed to a **recorder,** which produces the mass spectrum. In modern instruments, the output of the detector is fed through an interface to a computer. The computer can store the data, provide the output in both tabular and graphic forms, and compare the data to standard spectra, which are contained in spectra libraries that are also stored in the computer. In some older instruments, the electron current from the detector is fed to a series of five galvanometers of varying sensitivity. The galvanometers frequently have sensitivities in the ratios 1, 1/3, 1/10, 1/30, and 1/100. Each is capable of recording a spectrum with five simultaneous traces at different sensitivities. By using the five traces, it is possible to record the weakest peaks while still keeping the strongest peaks on scale.

8.2 GAS CHROMATOGRAPHY–MASS SPECTROMETRY

A very useful innovation in sample introduction systems is the use of a gas chromatograph coupled to a mass spectrometer. In effect, the mass spectrometer acts in the role of detector. In this technique, known as **gas chromatography–mass spectrometry (GC-MS),** the gas stream emerging from the gas chromatograph is admitted through a valve into a tube, where it passes over a molecular leak. Some of the gas stream is thus admitted into the ionization chamber of the mass spectrometer. In this way it is possible to obtain the mass spectrum of every component in a mixture being injected into the gas chromatograph.

A drawback of this method involves the need for rapid scanning by the mass spectrometer. The instrument must determine the mass spectrum of each component in the mixture before the next component exits from the gas chromatography column, in order that one substance is not contaminated by the next fraction before its spectrum has been obtained.

Since high-efficiency capillary columns are used in the gas chromatograph, in most cases compounds are completely separated before the gas stream is analyzed. The instrument must have the capability of obtaining at least one scan per second in the range of 10 to 300 amu. Even more scans are necessary if a narrower range of masses is to be analyzed.

The effluent from the gas-chromatograph part of the instrument can also be directed into an FT-IR instrument so that infrared spectra rather than mass spectra can be obtained. In that case, the infrared spectrophotometer acts as the detector for the gas chromatograph.

The mass spectrometer which is coupled to the gas chromatograph should be relatively compact and capable of high resolution. In many instruments, a highly useful modification has been to replace the magnetic field region with a quadrupole system. In a **quadrupole mass spectrometer** (Fig. 8.2), a set of four solid rods is arranged parallel to the direction of the ion beam. The rods should be hyperbolic in cross section, although cylindrical rods may be used. A direct-current voltage and a radiofrequency are applied to the rods, generating an oscillating electrostatic field in the region between the rods. Depending upon the ratio of the radiofrequency amplitude to the direct-current voltage, ions acquire an oscillation in this electrostatic field. Ions of an incorrect mass-to-charge ratio (too small or too large) undergo an unstable oscillation. The amplitude of the oscillation continues to increase until the particle strikes one of the rods. Ions of the correct mass-to-charge ratio undergo a stable oscillation of constant amplitude. These ions do not strike the rods but pass through the analyzer to reach the detector. The resolution of the system can be adjusted by varying the ratio of radiofrequency amplitude to DC voltage. Resolutions as high as 10,000 may be obtained with this type of mass analyzer.

With a GC-MS system, one can also analyze a mixture and conduct a library search on each component of the mixture. If the components are known compounds, they can be identified tentatively by

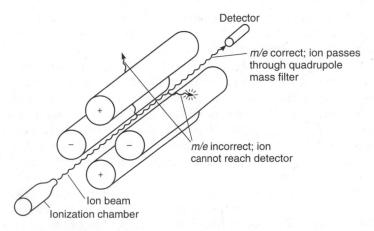

FIGURE 8.2 Quadrupole mass analyzer.

comparisons with compounds found in the computer library. In this way, a "hit list" can be generated which reports .on the probability that the compound in the library matches the known substance (see Section 8.8).

A new technique that resembles the GC-MS technique described here is **high-performance liquid chromatography–mass spectrometry (HPLC-MS).** An HPLC instrument is coupled through a special interface to a mass spectrometer. The substances that elute from the HPLC column are detected by the mass spectrometer, and their mass spectra can be displayed, analyzed, and compared with standard spectra found in the computer library built into the instrument.

8.3 THE MASS SPECTRUM

The mass spectrum is a plot of ion abundance versus m/e ratio. Figure 8.3 is a portion of a typical mass spectrum—that of dopamine, a substance that acts as a neurotransmitter in the central ner-vous system. The upper part of Figure 8.3 shows the five traces that are obtained with a multiple-galvanometer detection system. Because a spectrum of this sort is difficult to use, usually this information is converted to the form of a bar graph. Such a graph appears beneath the spectrum in Figure 8.3. With a modern instrument, in which the signal from the detector is digitized and stored in the instrument's computer, it is possible to obtain a direct plot of the data in a form essentially identical to the bar graph in Figure 8.3. Commonly, mass spectral results may also be presented in tabular form, as in Table 8.1.

The most abundant ion formed in the ionization chamber gives rise to the tallest peak in the mass spectrum, called the **base peak.** In the mass spectrum of dopamine, the base peak is indicated at an m/e value of 124. The relative abundances of all the other peaks in the spectrum are reported as percentages of the abundance of the base peak.

As we mentioned earlier, the beam of electrons in the ionization chamber converts some of the sample molecules to positive ions. Some of these types of ions are of sufficient importance to warrant further examination. The simple removal of an electron from a molecule yields an ion whose weight is the actual molecular weight of the original molecule. This ion is the **molecular ion,** which is frequently symbolized by M^+. The value of m/e at which the molecular ion appears on the mass spectrum, assuming that the ion has only one electron missing, gives the molecular weight of the original molecule. If you can identify the molecular ion peak in the mass spectrum, you will be able to use the spectrum to determine the molecular weight of an unknown substance. Ignoring heavy isotopes for the moment, the molecular ion peak is the heaviest peak in the mass spectrum; it is indicated in the graphic presentation in Figure 8.3 ($m/e = 153$). Strictly speaking, the molecular ion is a **radical-cation** since it contains an unpaired electron as well as a positive charge.

Molecules in nature do not occur as isotopically pure species. Virtually all atoms have heavier isotopes that occur in characteristic natural abundances. Hydrogen occurs largely as ^1H, but about 0.02% of hydrogen atoms are the isotope ^2H. Carbon normally occurs as ^{12}C, but about 1.1% of carbon atoms are the heavier isotope ^{13}C. With the possible exception of fluorine and a few additional elements, most other elements have a certain percentage of naturally occurring heavier isotopes. Peaks caused by ions bearing those heavier isotopes also appear in mass spectra. The relative abundances of such isotopic peaks are proportional to the abundances of the isotopes in nature. Most often, the isotopes occur one or two mass units above the mass of the "normal" atom. Therefore, besides looking for the molecular ion (M^+) peak, one would also attempt to locate the $M + 1$ and $M + 2$ peaks. As Section 8.5 will demonstrate, the relative abundances of the $M + 1$ and $M + 2$ peaks can be used to determine the molecular formula of the substance being studied. In Figure 8.3, the isotopic peaks are the low-intensity peaks at m/e values higher (154 and 155) than that of the molecular ion peak.

We have seen that the beam of electrons in the ionization chamber can produce the molecular ion. This beam is also sufficiently powerful to break some of the bonds in the molecule, producing a series of molecular fragments. The positively charged fragments are also accelerated in the ionization

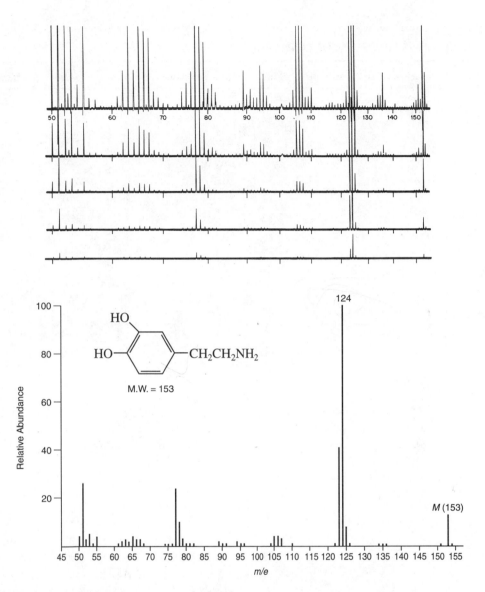

FIGURE 8.3 Partial mass spectrum of dopamine.

chamber, sent through the analyzer, detected, and recorded on the mass spectrum. These **fragment ions** appear at *m/e* values corresponding to their individual masses. Very often a fragment ion, rather than the parent ion, is the most abundant ion produced in the mass spectrum. A second means of producing fragment ions exists if the molecular ion, once it is formed, is so unstable that it disintegrates before it can pass into the accelerating region of the ionization chamber. Lifetimes less than 10^{-6} sec are typical in this type of fragmentation. The fragments that are charged then appear as fragment ions in the mass spectrum. A great deal of structural information about a substance can be determined from an examination of the fragmentation pattern in the mass spectrum. Section 8.6 will examine some fragmentation patterns for common classes of compounds.

Ions with lifetimes on the order of 10^{-6} sec are accelerated in the ionization chamber before they have an opportunity to disintegrate. These ions may disintegrate into fragments while they are passing into the analyzer region of the mass spectrometer. The fragment ions formed at that point have considerably lower energy than normal ions, since the uncharged portion of the original ion carries away

TABLE 8.1
PARTIAL MASS SPECTRUM OF DOPAMINE

m/e	Relative Abundance	m/e	Relative Abundance	m/e	Relative Abundance
50	4.00	76	1.48	114	0.05
50.5	0.05	77	24.29	115	0.19
51	25.71	78	10.48	116	0.24
51.5	0.19	79	2.71	117	0.24
52	3.00	80	0.81	118	0.14
52.5	0.62	81	1.05	119	0.19
53	5.43	82	0.67	120	0.14
53.5	0.19	83	0.14	121	0.24
54	1.00	84	0.10	122	0.71
55	4.00	85	0.10	123	41.43
56	0.43	86	0.14	124	100.00 (base peak)
56.5	0.05 (metastable peak)	87	0.14	125	7.62
57	0.33	88	0.19	126	0.71
58	0.10	89	1.57	127	0.10
58.5	0.05	89.7	0.10 (metastable peak)	128	0.10
59	0.05	90	0.57	129	0.10
59.5	0.05	90.7	0.10 (metastable peak)	131	0.05
60	0.10	91	0.76	132	0.19
60.5	0.05	92	0.43	133	0.14
61	0.52	93	0.43	134	0.52
61.5	0.10	94	1.76	135	0.52
62	1.57	95	1.43	136	1.48
63	3.29	96	0.52	137	0.33
64	1.57	97	0.14	138	0.10
65	3.57	98	0.05	139	0.10
65.5	0.05	99	0.05	141	0.19
66	3.14	100.6	0.19 (metastable peak)	142	0.05
66.5	0.14	101	0.10	143	0.05
67	2.86	102	0.14	144	0.05
67.5	0.10	103	0.24	145	0.05
68	0.67	104	0.76	146	0.05
69	0.43	105	4.29	147	0.05
70	0.24	106	4.29	148	0.10
71	0.19	107	3.29	149	0.24
72	0.05	108	0.43	150	0.33
73	0.14	109	0.48	151	1.00
74	0.67	110	0.86	152	0.38
74.5	0.05	111	0.10	153	13.33 (molecular ion)
75	1.00	112	0.05	154	1.48
75.5	0.14	113	0.05	155	0.19

some of the energy that the ion received as it was accelerated. As a result, the fragment ion produced in the analyzer follows an abnormal flight path on its way to the detector. This ion appears at an m/e ratio that depends on its mass as well as the mass of the original ion from which it formed. Such an ion gives rise to what is termed a **metastable ion peak** in the mass spectrum. Metastable ion peaks are usually broad peaks, and they frequently appear at nonintegral values of m/e. The equation that relates the position of the metastable ion peak in the mass spectrum to the mass of the original ion is

$$m_1^+ \longrightarrow m_2^+ + \text{fragment}$$

$$m^* = \frac{(m_2)^2}{m_1}$$

where m^* is the apparent mass of the metastable ion in the mass spectrum, m_1 is the mass of the original ion from which the fragment formed, and m_2 is the mass of the new fragment ion. A metastable ion peak is useful in some applications, since its presence definitely links two peaks together. Metastable ion peaks can be used to prove a proposed fragmentation pattern or to aid in the solution of structure proof problems.

8.4 DETERMINATION OF MOLECULAR WEIGHT

Section 8.1 showed that when a beam of high-energy electrons impinges upon a stream of sample molecules, ionization of electrons from the molecules takes place. The resulting ions, called **molecular ions,** are then accelerated, sent through a magnetic field, and detected. If these molecular ions have lifetimes of at least 10^{-5} sec, they reach the detector without breaking into fragments. The user then observes the m/e ratio that corresponds to the molecular ion in order to determine the molecular weight of the sample molecule.

In practice, molecular weight determination is not quite as easy as the preceding paragraph suggests. First you must understand that the value of the mass of any ion accelerated in a mass spectrometer is its true mass, and not its molecular weight obtained through the use of chemical atomic weights. The chemical scale of atomic weights is based on weighted averages of the weights of all of the isotopes of a given element. The mass spectrometer can distinguish between masses of particles bearing the most common isotopes of the elements and particles bearing heavier isotopes. Consequently, the masses that are observed for molecular ions are the masses of the molecules in which every atom is present as its most common isotope.

In the second place, molecules subjected to bombardment by electrons may break apart into fragment ions. As a result of this fragmentation, mass spectra can be quite complex, with peaks appearing at a variety of m/e ratios. You must be quite careful to be certain that the suspected peak is indeed that of the molecular ion and not that of a fragment ion. This distinction becomes particularly crucial when the abundance of the molecular ion is low, as when the molecular ion is rather unstable and fragments easily.

The masses of the ions detected in the mass spectrum must be measured accurately. An error of only one mass unit in the assignment of mass spectral peaks can render determination of structure impossible.

One method of confirming that a particular peak corresponds to a molecular ion is to vary the energy of the ionizing electron beam. If the energy of the beam is lowered, the tendency of the molecular ion to fragment lessens. As a result, the intensity of the molecular ion peak should increase with decreasing electron potential, while the intensities of the fragment ion peaks should decrease.

Certain facts must apply to a molecular ion peak:

1. The peak must correspond to the ion of highest mass in the spectrum, excluding isotopic peaks that occur at even higher masses. The isotopic peaks are usually of much lower intensity than the molecular ion peak. At the sample pressures used in most spectral studies, the probability that ions and molecules will collide to form heavier particles is quite low.

2. The ion must have an odd number of electrons. When a molecule is ionized by an electron beam, it loses one electron to become a radical-cation. The charge on such an ion is one, thus making it an ion with an odd number of electrons.

3. The ion must be capable of forming the important fragment ions in the spectrum, particularly the fragments of relatively high mass, by loss of logical neutral fragments. Section 8.6 will explain these fragmentation processes in detail.

The observed abundance of the suspected molecular ion must correspond to expectations based on the assumed molecule structure. Highly branched substances undergo fragmentation very easily. Observation of an intense molecular ion peak for a highly branched molecule thus would be unlikely. The lifetimes of molecular ions vary according to the following generalized sequence.

Aromatic compounds > conjugated alkenes > alicyclic compounds > organic sulfides > unbranched hydrocarbons > mercaptans > ketones > amines > esters > ethers > carboxylic acids > branched hydrocarbons > alcohols

Another rule that is sometimes used to verify that a given peak corresponds to the molecular ion is the so-called **Nitrogen Rule.** This rule states that if a compound has an even number of nitrogen atoms (or no nitrogen atoms), its molecular ion will appear at an even mass value. On the other hand, a molecule with an odd number of nitrogen atoms will form a molecular ion with an odd mass. The Nitrogen Rule stems from the fact that nitrogen, although it has an even mass, has an odd-numbered valence. Consequently, an extra hydrogen atom is included as a part of the molecule, giving it an odd mass. To picture this effect, consider ethylamine, $C_2H_5NH_2$. This substance has one nitrogen atom, and its mass is an odd number (45), whereas ethylenediamine, $H_2N-CH_2-CH_2-NH_2$, has two nitrogen atoms, and its mass is an even number (60).

You must be careful when studying molecules containing chlorine or bromine atoms, since these elements have two commonly occurring isotopes. Chlorine has isotopes of 35 (relative abundance = 75.77%) and 37 (relative abundance = 24.23%); bromine has isotopes of 79 (relative abundance = 50.5%) and 81 (relative abundance = 49.5%). When these elements are present, take special care not to confuse the molecular ion peak with a peak corresponding to the molecular ion with a heavier halogen isotope present.

In many of the cases that you are likely to encounter in mass spectrometry, the molecular ion can be observed in the mass spectrum. Once you have identified that peak in the spectrum, the problem of molecular weight determination is solved. However, with molecules that form unstable molecular ions, you may not observe the molecular ion peak. Molecular ions with lifetimes less than 10^{-5} sec break up into fragments before they can be accelerated. The only peaks that are observed in such cases are those due to fragment ions. You will be obliged to deduce the molecular weight of the substance from the fragmentation pattern on the basis of known patterns of fragmentation for certain classes of compounds.

Alcohols undergo dehydration very easily. Consequently, the molecular ion loses water (mass = 18) as a neutral fragment before it can be accelerated. To determine the mass of an alcohol molecular ion, you must locate the heaviest fragment and keep in mind that it may be necessary to add 18 to its mass. Acetates undergo loss of acetic acid (mass = 60) easily. If acetic acid is lost, the weight of the molecular ion is 60 mass units higher than the mass of the heaviest fragment.

The conjugate acids of oxygen and nitrogen compounds are reasonably stable. Spectra run with sample pressures higher than 0.5 mm Hg may show peaks due to ion–molecule collisions. In such collisions, a hydrogen atom is transferred from a molecule to an ion. The resulting ion is then accelerated. Since oxygen compounds form fairly stable oxonium ions and nitrogen compounds form ammonium ions, ion–molecule collisions form peaks in the mass spectrum that appear one mass unit *higher* than the mass of the molecular ion. At times, the formation of ion–molecule products may be helpful in the determination of the molecular weight of an oxygen or nitrogen compound.

Additional techniques, such as **field ionization** and **chemical ionization,** are useful in the study of unstable molecular ions. Section 8.7 will discuss these methods in detail.

8.5 DETERMINATION OF MOLECULAR FORMULAS

A. Precise Mass Determination

Perhaps the most important application of high-resolution mass spectrometers is the determination of very precise molecular weights of substances. We are accustomed to thinking of atoms as having integral atomic masses—for example, H = 1, C = 12, and O = 16. However, if we determine atomic masses with sufficient precision, we find that this is not true. In 1923 Aston discovered that every isotopic mass is characterized by a small "mass defect." The mass of each atom actually differs from a whole mass number by an amount known as the nuclear packing fraction. Table 8.2 gives the actual masses of some atoms.

Depending upon the atoms contained in a molecule, it is possible for particles of the same nominal mass to have slightly different measured masses when precise determinations are possible. To illustrate, a molecule with a molecular weight of 60 could be C_3H_8O, $C_2H_8N_2$, $C_2H_4O_2$, or CH_4N_2O. These species have the following precise masses:

$$
\begin{array}{ll}
C_3H_8O & 60.05754 \\
C_2H_8N_2 & 60.06884 \\
C_2H_4O_2 & 60.02112 \\
CH_4N_2O & 60.03242
\end{array}
$$

TABLE 8.2
PRECISE MASSES OF SOME COMMON ELEMENTS

Element	Atomic Weight	Nuclide	Mass
Hydrogen	1.00797	^1H	1.00783
		^2H	2.01410
Carbon	12.01115	^{12}C	12.0000
		^{13}C	13.00336
Nitrogen	14.0067	^{14}N	14.0031
		^{15}N	15.0001
Oxygen	15.9994	^{16}O	15.9949
		^{17}O	16.9991
		^{18}O	17.9992
Fluorine	18.9984	^{19}F	18.9984
Silicon	28.086	^{28}Si	27.9769
		^{29}Si	28.9765
		^{30}Si	29.9738
Phosphorus	30.974	^{31}P	30.9738
Sulfur	32.064	^{32}S	31.9721
		^{33}S	32.9715
		^{34}S	33.9679
Chlorine	35.453	^{35}Cl	34.9689
		^{37}Cl	36.9659
Bromine	79.909	^{79}Br	78.9183
		^{81}Br	80.9163
Iodine	126.904	^{127}I	126.9045

Observation of a molecular ion with a mass of 60.058 would establish that the unknown molecule is C_3H_8O. An instrument with a resolution of about 5320 would be required to distinguish among these peaks. That is well within the capability of modern mass spectrometers, which can attain resolutions greater than one part in 20,000. A precisely determined molecular weight can also provide information about the molecular formula of a substance under study. (See Appendix 7.)

In a modern, state-of-the-art mass spectrometry laboratory, the determination of the molecular formula of an unknown substance is done using exact masses. The spectroscopist must be careful, however, as the method cannot be used if the molecular ion peak is very weak or absent.

It is interesting to compare the precision of molecular weight determinations by mass spectrometry with the chemical methods described in Chapter 1, Section 1.2. Chemical methods give results which are accurate to only two or three significant figures (± 0.1 to 1%). Molecular weights determined by mass spectrometry have an accuracy of about $\pm 0.005\%$. Clearly, mass spectrometry is much more precise than chemical methods of determining molecular weight.

B. Isotope Ratio Data

The preceding section described a method of determining molecular formulas using data from high-resolution mass spectrometers. Another method of determining molecular formulas is to examine the relative intensities of the peaks due to the molecular ion and related ions that bear one or more heavy isotopes. The advantage of this latter method is that it is much less costly. A high-resolution instrument is not required. This method, however, is not typical of the way that a modern research scientist would determine the molecular formula of an unknown substance. Unfortunately, the isotopic peaks may be difficult to locate in the mass spectrum. Furthermore, this method is useless when the molecular ion peak is very weak or does not appear. The results obtained by this method may at times be somewhat ambiguous.

The example of ethane can illustrate the determination of a molecular formula from a comparison of the intensities of mass spectral peaks of the molecular ion and the ions bearing heavier isotopes. Ethane, C_2H_6, has a molecular weight of 30 when it contains the most common isotopes of carbon and hydrogen. Its molecular ion peak should appear at a position in the spectrum corresponding to a mass of 30. Occasionally, however, a sample of ethane yields a molecule in which one of the carbon atoms is a heavy isotope of carbon, ^{13}C. This molecule would appear in the mass spectrum at a mass of 31. The relative abundance of ^{13}C in nature is 1.08% of the ^{12}C atoms. In the tremendous number of molecules in a sample of ethane gas, either of the carbon atoms of ethane will turn out to be a ^{13}C atom 1.08% of the time. Since there are two carbon atoms in ethane, a molecule of mass 31 will turn up (2×1.08) or 2.16% of the time. Thus, we would expect to observe a peak of mass 31 with an intensity of 2.16% of the molecular ion peak intensity. This mass 31 peak is called the $M + 1$ peak, since its mass is one unit higher than that of the molecular ion.

You may notice that a particle of mass 31 could form in another manner. If a deuterium atom, ^2H, replaced one of the hydrogen atoms of ethane, the molecule would also have a mass of 31. The natural abundance of deuterium is only 0.016% of the abundance of ^1H atoms. The intensity of the $M + 1$ peak would be (6×0.016) or 0.096% of the intensity of the molecular ion peak, if we consider only contributions due to deuterium. When we add these contributions to those of ^{13}C, we obtain the observed intensity of the $M + 1$ peak, which is 2.26% of the intensity of the molecular ion peak.

A peak of mass 32 can form if both of the carbon atoms are replaced by ^{13}C atoms simultaneously. The probability that a molecule of formula $^{13}C_2H_6$ will appear in a natural sample of ethane is (1.08×1.08)/100, or 0.01%. A peak that appears two mass units higher than the mass of the molecular ion peak is called the $M + 2$ peak. The intensity of the $M + 2$ peak of ethane is only 0.01% of the intensity of the molecular ion peak. The contribution due to two deuterium atoms replacing hydrogen atoms would be (0.016×0.016)/100 = 0.00000256%, a negligible amount.[1] To assist in the determination of the ratios of molecular ion, $M + 1$, and $M + 2$ peaks, Table 8.3 lists the natural abundances of some

TABLE 8.3
NATURAL ABUNDANCES OF COMMON ELEMENTS AND THEIR ISOTOPES

Element		Relative Abundance					
Hydrogen	^1H	100	^2H	0.016			
Carbon	^{12}C	100	^{13}C	1.08			
Nitrogen	^{14}N	100	^{15}N	0.38			
Oxygen	^{16}O	100	^{17}O	0.04	^{18}O	0.20	
Fluorine	^{19}F	100					
Silicon	^{28}Si	100	^{29}Si	5.10	^{30}Si	3.35	
Phosphorus	^{31}P	100					
Sulfur	^{32}S	100	^{33}S	0.78	^{34}S	4.40	
Chlorine	^{35}Cl	100			^{37}Cl	32.5	
Bromine	^{79}Br	100			^{81}Br	98.0	
Iodine	^{127}I	100					

common elements and their isotopes. In this table, the relative abundances of the isotopes of each element are calculated by setting the abundances of the most common isotopes equal to 100.

To demonstrate how the intensities of the $M + 1$ and $M + 2$ peaks provide a unique value for a given molecular formula, consider two molecules of mass 42, propene (C_3H_6) and diazomethane (CH_2N_2). For propene, the intensity of the $M + 1$ peak should be $(3 \times 1.08) + (6 \times 0.016) = 3.34\%$, and the intensity of the $M + 2$ peak should be 0.05%. The natural abundance of ^{15}N isotopes of nitrogen is 0.38% of the abundance of ^{14}N atoms. In diazomethane, we expect the relative intensity of the $M + 1$ peak to be $1.08 + (2 \times 0.016) + (2 \times 0.38) = 1.87\%$ of the intensity of the molecular ion peak, and the intensity of the $M + 2$ peak would be 0.01% of the intensity of the molecular ion peak.

---

[1] The formula for calculating the intensity of the $M + 1$ peak is as follows.

$$\%(M + 1) = 100 \frac{(M+1)}{M} = 1.1 \times \text{number of C atoms}$$

$$+ 0.016 \times \text{number of H atoms}$$

$$+ 0.38 \times \text{number of N atoms} + \ldots$$

The formula for calculating the approximate intensity of the $M + 2$ peak is as follows.

$$\%(M + 2) = 100 \frac{(M+2)}{M} \cong \frac{(1.1 \times \text{number of C atoms})^2}{200}$$

$$+ \frac{(0.016 \times \text{number of H atoms})^2}{200}$$

$$+ 0.20 \times \text{number of O atoms}$$

These formulas are of limited usefulness unless one is already certain of the actual molecular formula. In practice, contributions due to hydrogen isotopes are not included in the calculation. The exact formula for calculating the intensity of the $M + 2$ peak is more complex than shown here (see Beynon). The formula given above tends to have better agreement with the observed intensities for compounds of high mass.

TABLE 8.4
ISOTOPE RATIOS FOR PROPENE AND DIAZOMETHANE

Compound	Molecular Mass	Relative Intensities		
		M	M + 1	M + 2
C_3H_6	42	100	3.34	0.05
CH_2N_2	42	100	1.87	0.01

Table 8.4 summarizes these intensity ratios. It shows that the two molecules have nearly the same molecular weight, but the relative intensities of the $M + 1$ and $M + 2$ peaks that they yield are quite different. As an additional illustration, Table 8.5 compares the ratios of the molecular ion, $M + 1$, and $M + 2$ peaks for three substances of mass 28: carbon monoxide, nitrogen, and ethene. Again, notice that the relative intensities of the $M + 1$ and $M + 2$ peaks provide a means of distinguishing among these molecules.

As molecules become larger and more complex, the number of possible combinations that yield $M + 1$ and $M + 2$ peaks grows. For a particular combination of atoms, the intensities of these peaks relative to the intensity of the molecular ion peak are unique. Thus, the isotope ratio method can be used to establish the molecular formula of a compound. Tables of possible combinations of carbon, hydrogen, oxygen, and nitrogen and intensity ratios for the $M + 1$ and $M + 2$ peaks for each combination have been developed. Appendices 10 an 11 and the book by Beynon (in the reference list at the end of this chapter) contain extensive tables of this sort. For a given molecular weight, you can examine the tables to find the molecular formula that corresponds to the isotope ratios observed.

For atoms other than carbon, hydrogen, oxygen, and nitrogen, it is necessary to calculate the expected intensities of the $M + 1$ and $M + 2$ peaks for a particular molecular formula. You must establish the presence of these other elements by means other than mass spectrometry.

When chlorine or bromine is present, the $M + 2$ peak becomes very significant. The heavy isotope of each of these elements is two mass units heavier than the lighter isotope. The natural abundance of ^{37}Cl is 32.5% that of ^{35}Cl; the natural abundance of ^{81}Br is 98.0% that of ^{79}Br. When either of these elements is present, the $M + 2$ peak becomes quite intense. If a compound contains two chlorine or bromine atoms, a quite distinct $M + 4$ peak, as well as an intense $M + 2$ peak, should be observed. In such a case, it is important to exercise caution in identifying the molecular ion peak in the mass spectrum. Section 8.6N will discuss the mass spectral properties of the organic halogen compounds in greater detail. Table 8.6 gives the relative intensities of isotope peaks for various combinations of bromine and chlorine atoms, and Figure 8.4 illustrates them.

TABLE 8.5
ISOTOPE RATIOS FOR CO, N_2, AND C_2H_4

Compound	Molecular Mass	Relative Intensities		
		M	M + 1	M + 2
CO	28	100	1.12	0.2
N_2	28	100	0.76	
C_2H_4	28	100	2.23	0.01

TABLE 8.6
RELATIVE INTENSITIES OF ISOTOPE PEAKS FOR VARIOUS COMBINATIONS OF BROMINE AND CHLORINE

Halogen	Relative Intensities			
	M	$M + 2$	$M + 4$	$M + 6$
Br	100	97.7		
Br_2	100	195.0	95.4	
Br_3	100	293.0	286.0	93.4
Cl	100	32.6		
Cl_2	100	65.3	10.6	
Cl_3	100	97.8	31.9	3.47
BrCl	100	130.0	31.9	
Br_2Cl	100	228.0	159.0	31.2
Cl_2Br	100	163.0	74.4	10.4

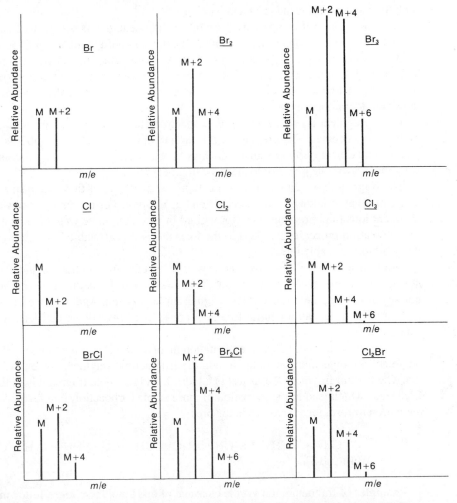

FIGURE 8.4 Mass spectra expected for various combinations of bromine and chlorine.

Examination of the intensity of the $M + 2$ peak is also useful for obtaining information about elements which may be present in the molecular formula. An unusually intense $M + 2$ peak can indicate that sulfur or silicon is present in the unknown substance. The relative abundances of ^{34}S and ^{30}Si are 4.40 and 3.35, respectively. A trained chemist knows that a larger-than-normal $M + 2$ peak can be the first hint that sulfur is present.

8.6 SOME FRAGMENTATION PATTERNS

When a molecule has been bombarded by high-energy electrons in the ionization chamber of a mass spectrometer, besides losing one electron to form an ion, the molecule also absorbs some of the energy transferred in its collision with the incident electrons. This extra energy places the molecular ion in an excited vibrational state. The vibrationally excited molecular ion may be unstable, and it may lose some of its extra energy by breaking apart into fragments. If the lifetime of the molecular ion is greater than 10^{-5} sec, a peak corresponding to the molecular ion appears in the mass spectrum. However, molecular ions with lifetimes less than 10^{-5} sec break apart into fragments before they are accelerated within the ionization chamber. In such cases, peaks corresponding to the mass-to-charge ratios for these fragments appear in the mass spectrum. For a given compound, not all of the molecular ions formed by ionization have precisely the same lifetime; some may have shorter lifetimes than others. As a result, in a spectrum one usually observes peaks corresponding to both the molecular ion and the fragments.

For most classes of compounds, the mode of fragmentation is somewhat characteristic and hence predictable. This section discusses some of the more important modes of fragmentation.

It is helpful to begin by describing some general principles that govern fragmentation processes. The ionization of the sample molecule forms a molecular ion that not only carries a positive charge but also has an unpaired electron. The molecular ion, then, is actually a radical-cation, and it contains an odd number of electrons.

When fragment ions form in the mass spectrometer, they almost always do so by means of unimolecular processes. The pressure of the sample in the ionization chamber is too low to permit a significant number of bimolecular collisions to occur. The unimolecular processes that are energetically most favorable give rise to the most fragment ions.

The fragment ions thus formed are cations. A great deal of their chemistry can be explained in terms of what is known about carbocations in solution. For example, alkyl substitution stabilizes fragment ions (and promotes their formation) in much the same way that it stabilizes carbocations. Fragmentation processes that lead to the formation of more stable ions are favored over processes that lead to less stable ions.

Often, fragmentation involves the loss of an electrically neutral fragment. This fragment does not appear in the mass spectrum, but its existence can be deduced by noting the difference in masses of the fragment ion and the original molecular ion. Again, processes that lead to the formation of a more stable neutral fragment are favored over those that lead to less stable neutral fragments.

The most common mode of fragmentation involves the cleavage of one bond. In this process, the odd-electron molecular ion yields an odd-electron neutral fragment and an even-electron fragment ion. The neutral fragment that is lost is a radical, and the ionic fragment is of the carbocation type. Cleavages which lead to the formation of more stable carbocations are favored. Thus, ease of fragmentation to form ions increases in the order

$$CH_3^+ < RCH_2^+ < R_2CH^+ < R_3C^+ < CH_2=CH-CH_2^+ < C_6H_5-CH_2^+$$
DIFFICULT   EASY

Examples of fragmentation via the cleavage of one bond include the following.

$$\left[R\!-\!\!\vdots\!-\!CH_3\right]^{\stackrel{+}{\cdot}} \longrightarrow R^+ + \cdot CH_3$$

$$\left[\begin{array}{c}O\\\parallel\\R\!-\!\!\vdots\!-\!C\!-\!R'\end{array}\right]^{\stackrel{+}{\cdot}} \longrightarrow \begin{array}{c}O\\\parallel\\{}^+C\!-\!R'\end{array} + \cdot R$$

$$\left[R\!-\!\!\vdots\!-\!X\right]^{\stackrel{+}{\cdot}} \longrightarrow R^+ + X\cdot$$

X = halogen, OR, SR, or NR$_2$
where R = H, alkyl, or aryl

The next most important type of fragmentation involves the cleavage of two bonds. In this process, the odd-electron molecular ion yields an odd-electron fragment ion and an even-electron neutral fragment, usually a small molecule of some type. Examples of this type of cleavage include the following.

$$\left[\begin{array}{cc}H & OH\\ \vdots & \vdots \\ R\!-\!CH\!-\!CH\!-\!R'\end{array}\right]^{\stackrel{+}{\cdot}} \longrightarrow [R\!-\!CH\!=\!CH\!-\!R']^{\stackrel{+}{\cdot}} + H_2O$$

$$\left[\begin{array}{cc}CH_2\!-\!CH_2\\ \vdots \quad\quad \vdots \\ R\!-\!CH\!-\!CH_2\end{array}\right]^{\stackrel{+}{\cdot}} \longrightarrow [R\!-\!CH\!=\!CH_2]^{\stackrel{+}{\cdot}} + CH_2\!=\!CH_2$$

$$\left[\begin{array}{c}O\\ \parallel \\ R\!-\!CH\!-\!CH_2\!-\!O\!-\!C\!-\!CH_3\\ \vdots\\ H\end{array}\right]^{\stackrel{+}{\cdot}} \longrightarrow [R\!-\!CH\!=\!CH_2]^{\stackrel{+}{\cdot}} + \begin{array}{c}O\\\parallel\\HO\!-\!C\!-\!CH_3\end{array}$$

In addition to these processes, fragmentation processes involving rearrangements, migrations of groups, and secondary fragmentations of fragment ions are also possible. These latter modes of fragmentation occur less often than the two cases already described, and additional discussion of them will be reserved for the compounds in which they are important.

To assist you in identifying possible fragment ions, Appendix 12 provides a table that lists the molecular formulas for common fragments with masses less than 105. More complete tables may be found in the books by Beynon and by Silverstein and Webster listed in the references at the end of this chapter.

A. Alkanes

For saturated hydrocarbons and organic structures containing large saturated hydrocarbon skeletons, the methods of fragmentation are quite predictable. What is known about the stabilities of carbocations in solution can be used to help us understand the fragmentation patterns of alkanes. The mass spectra of alkanes are characterized by strong molecular ion peaks and a regular series of fragment ion peaks separated by 14 amu.

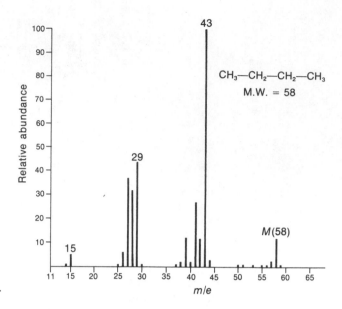

FIGURE 8.5 Mass spectrum of butane.

SPECTRAL ANALYSIS BOX — Alkanes

MOLECULAR ION

Strong M⁺

FRAGMENT IONS

Loss of CH_2 units in a series: $M - 14$, $M - 28$, $M - 42$, etc.

For a straight-chain, or "normal," alkane, a peak corresponding to the molecular ion can be observed, as in the mass spectra of butane (Fig. 8.5) and octane (Fig. 8.6). As the carbon skeleton becomes more highly branched, the intensity of the molecular ion peak decreases. You will see this effect easily if you compare the mass spectrum of butane with that of isobutane (Fig. 8.7). The mol-

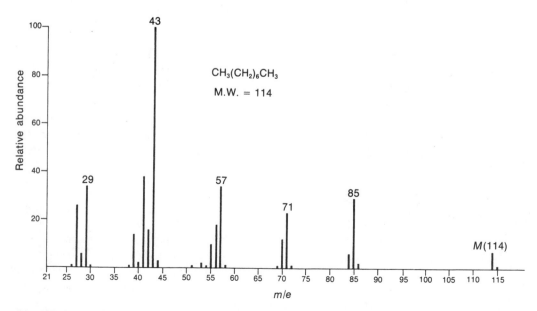

FIGURE 8.6 Mass spectrum of octane.

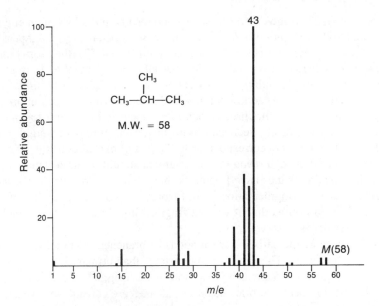

FIGURE 8.7 Mass spectrum of isobutane.

ecular ion peak in isobutane is much less intense than that in butane. Comparison of the mass spectra of octane and 2,2,4-trimethylpentane (Fig. 8.8) provides a more dramatic illustration of the effect of chain branching on the intensity of the molecular ion peak. The molecular ion peak in 2,2,4-trimethylpentane is too weak to be observed, while the molecular ion peak in its straight-chain isomer is quite readily observed.

The effect of chain branching on the intensity of the molecular ion peak can be understood by examining the method by which hydrocarbons undergo fragmentation. Straight-chain hydrocarbons appear to undergo fragmentation by breaking carbon–carbon bonds, resulting in a homologous series of fragmentation products. For example, in the case of butane, cleavage of the C1-to-C2 bond results in the loss of a methyl radical and the formation of the propyl carbocation

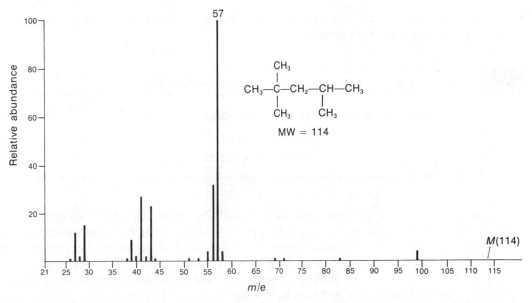

FIGURE 8.8 Mass spectrum of 2,2,4-trimethylpentane ("isooctane").

(*m/e* = 43). Cleavage of the C2-to-C3 bond results in the loss of an ethyl radical and the formation of the ethyl carbocation (*m/e* = 29). In the case of octane, fragment peaks due to the hexyl ion (*m/e* = 85), the pentyl ion (*m/e* = 71), the butyl ion (*m/e* = 57), the propyl ion (*m/e* = 43), and the ethyl ion (*m/e* = 29) are observed. Notice that alkanes fragment to form clusters of peaks that are 14 mass units (corresponding to one CH_2 group) apart from each other. Other fragments within each cluster correspond to additional losses of one or two hydrogen atoms. As is evident in the mass spectrum of octane, the three-carbon ions appear to be the most abundant, with the intensities of each cluster uniformly decreasing with increasing fragment weight. Interestingly, for long-chain alkanes, the fragment corresponding to the loss of one carbon atom is generally absent. In the mass spectrum of octane, a seven-carbon fragment should occur at a mass of 99, but it is not observed. Straight-chain alkanes have fragments that are always primary carbocations. Since these ions are rather unstable, fragmentation is not favored. A significant number of the original molecules survive electron bombardment without fragmenting. Consequently, a molecular ion peak of significant intensity is observed.

Cleavage of the carbon–carbon bonds of branched-chain alkanes can lead to secondary or tertiary carbocations. These ions are more stable than primary ions, so fragmentation becomes a more favorable process. A greater proportion of the original molecules undergo fragmentation, so the molecular ion peaks of branched-chain alkanes are considerably weaker or even absent. In isobutane, cleavage of a carbon–carbon bond yields an isopropyl carbocation, which is more stable than a normal propyl ion. Isobutane undergoes fragmentation more easily than butane because of the increased stability of its fragmentation products. With 2,2,4-trimethylpentane, the cleavage shown leads to the formation of a *tert*-butyl carbocation. Since tertiary carbocations are the most stable of the saturated alkyl carbocations, this cleavage is particularly favorable and accounts for the intense fragment peak at *m/e* = 57.

B.   Cycloalkanes

Cycloalkanes generally form strong molecular ion peaks. Fragmentation via the loss of a molecule of ethene (*M*–28) is common.

SPECTRAL ANALYSIS BOX — Cycloalkanes

MOLECULAR ION	FRAGMENT IONS
Strong M⁺	*M* – 28
	A series of peaks: *M* – 15, *M* – 29, *M* – 43, *M* – 57, etc.

The typical mass spectrum for a cycloalkane shows a relatively intense molecular ion peak. Fragmentation of ring compounds requires the cleavage of two carbon–carbon bonds, which is a more difficult process than cleavage of one such bond. Therefore, a larger proportion of cycloalkane molecules than of acyclic alkane molecules survive electron bombardment without undergoing fragmentation. In the mass spectra of cyclopentane (Fig. 8.9) and methylcyclopentane (Fig. 8.10), strong molecular ion peaks can be observed.

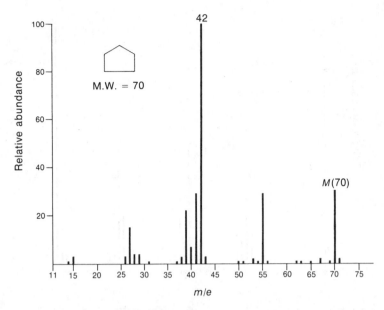

FIGURE 8.9 Mass spectrum of cyclopentane.

The fragmentation patterns of cycloalkanes may show mass clusters arranged in a homologous series, as in the alkanes. However, the most significant mode of cleavage of the cycloalkanes involves the loss of a molecule of ethene, either from the parent molecule or from intermediate radical-ions. The peak at $m/e = 42$ in cyclopentane and the peak at $m/e = 56$ in methylcyclopentane result from the loss of ethene from the parent molecule. Each of these fragment peaks is the most intense in the mass spectrum.

When the cycloalkane bears a side chain, loss of that side chain is a favorable mode of fragmentation. The fragment peak at $m/e = 69$ in the mass spectrum of methylcyclopentane is due to the loss of the CH_3 side chain. A secondary carbocation results from the loss of the methyl group.

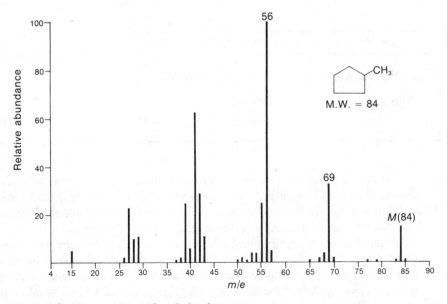

FIGURE 8.10 Mass spectrum of methylcyclopentane.

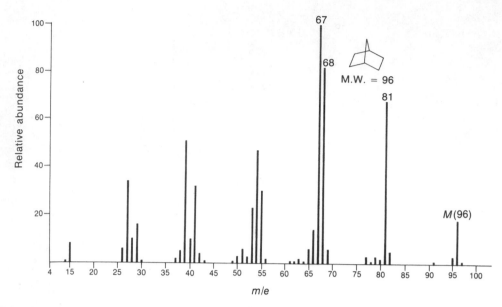

FIGURE 8.11 Mass spectrum of bicyclo[2.2.1]heptane (norbornane).

Applying these pieces of information to the mass spectrum of bicyclo[2.2.1]heptane (Fig. 8.11), we can identify fragment peaks due to the loss of the side chain (the one-carbon bridge, plus an additional hydrogen atom) at $m/e = 81$ and the loss of ethene at $m/e = 68$. The fragment ion peak at $m/e = 67$ is due to the loss of ethene plus an additional hydrogen atom.

C.   Alkenes

The mass spectra of most alkenes show distinct molecular ion peaks. Apparently, electron bombardment removes one of the electrons in the π bond, leaving the carbon skeleton relatively undisturbed. Fragmentation to form an allyl cation ($m/e = 41$) is favored.

SPECTRAL ANALYSIS BOX — Alkenes

MOLECULAR ION	FRAGMENT IONS
Strong M$^+$	$m/e = 41$
	A series of peaks: $M - 15, M - 29, M - 43, M - 57$, etc.

The double bond of an alkene is capable of absorbing substantial energy. As a result, the mass spectra of alkenes generally show a strong molecular ion peak. A characteristic of the mass spectra of alkenes is that the mass of the molecular ion should correspond to a molecular formula with an index of hydrogen deficiency equal to at least *one* (see Chapter 1).

When alkenes undergo fragmentation processes, the resulting fragment ions have formulas corresponding to $C_nH_{2n}{}^+$ and $C_nH_{2n-1}{}^+$. It is very difficult to locate double bonds in alkenes, since they migrate readily. The similarity of the mass spectra of alkene isomers is readily seen in the mass spectra of two isomers of the formula C_4H_8 (Figs. 8.12 and 8.13). The mass spectra are very nearly identical. Furthermore, *cis*- and *trans*-isomers produce essentially identical mass spectra.

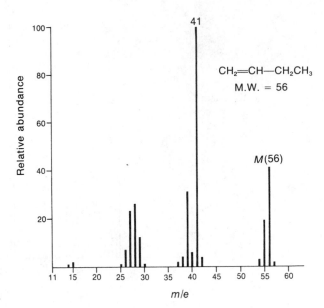

FIGURE 8.12 Mass spectrum of 1-butene.

An important fragment in the mass spectra of terminal alkenes, the allyl carbocation, occurs at an *m/e* value of 41. Its formation is due to cleavage of the type

$$\left[R \!\!\mid\!\! CH_2\!-\!CH\!\!=\!\!CH_2 \right]^{+} \longrightarrow R\cdot\ +\ \left[\overset{+}{C}H_2\!-\!CH\!\!=\!\!CH_2 \longleftrightarrow CH_2\!\!=\!\!CH\!-\!\overset{+}{C}H_2 \right]$$

The mass spectra of cycloalkenes show quite distinct molecular ion peaks. For many cycloalkenes, migration of bonds gives virtually identical mass spectra. Consequently, it may be impossible to locate the position of the double bond in a cycloalkene, particularly a cyclopentene or a cycloheptene.

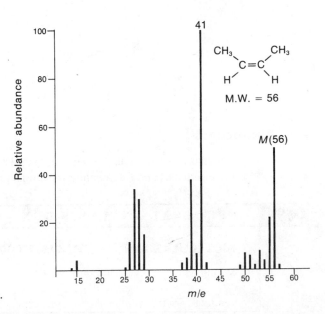

FIGURE 8.13 Mass spectrum of *cis*-2-butene.

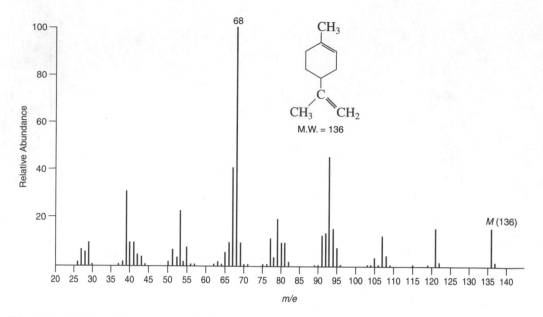

FIGURE 8.14 Mass spectrum of limonene.

Cyclohexenes do have a characteristic fragmentation pattern that corresponds to a reverse Diels–Alder reaction. This cleavage can be illustrated as

In the mass spectrum of limonene (Fig. 8.14), the intense peak at $m/e = 68$ corresponds to the diene fragment arising from the type of cleavage just described.

D. Alkynes

The mass spectra of alkynes are very similar to those of alkenes. The molecular ion peaks tend to be rather intense, and fragmentation patterns parallel those of the alkenes.

SPECTRAL ANALYSIS BOX — Alkynes	
MOLECULAR ION	**FRAGMENT IONS**
Strong M^+	$m/e = 39$
	Strong $M - 1$ peak

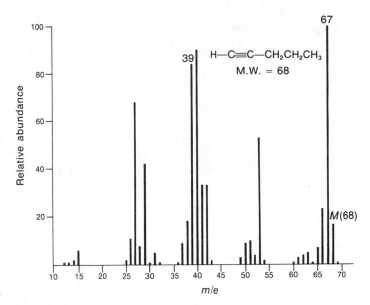

FIGURE 8.15 Mass spectrum of 1-pentyne.

As can be seen from the mass spectrum of 1-pentyne (Fig. 8.15), fragmentation of the type

$$\left[H\!-\!C\!\equiv\!C\!-\!CH_2\!-\!R\right]^{\dot{+}} \longrightarrow R\cdot + \left[H\!-\!C\!\equiv\!C\!-\!\overset{+}{C}H_2 \longleftrightarrow H\!-\!\overset{+}{C}\!=\!C\!=\!CH_2\right]$$

to yield the propargyl ion ($m/e = 39$) is important. The formation of the propargyl ion from alkynes is not as important as the formation of the allyl ions from alkenes, since the allyl ion is more stable than the propargyl ion.

Another important mode of fragmentation for terminal alkynes is the loss of the terminal hydrogen, yielding a strong $M - 1$ peak. This peak appears as the base peak ($m/e = 67$) in the spectrum of 1-pentyne.

$$\left[H\!-\!C\!\equiv\!C\!-\!R\right]^{\dot{+}} \longrightarrow H\cdot + {}^{+}C\!\equiv\!C\!-\!R$$

E. Aromatic Hydrocarbons

The mass spectra of most aromatic hydrocarbons show very intense molecular ion peaks. As is evident from the mass spectrum of benzene (Fig. 8.16), fragmentation of the benzene ring requires a great deal of energy. Such fragmentation is not observed to any significant extent. When a benzene ring contains side chains, however, a favored mode of fragmentation is cleavage of the side chain to form a **benzyl cation**, which spontaneously rearranges to a **tropylium ion**. When the side chain attached to a benzene ring contains three or more carbons, ions formed by a **McLafferty rearrangement** can be observed.

SPECTRAL ANALYSIS BOX—Aromatic Hydrocarbons

MOLECULAR ION	FRAGMENT IONS
Strong M$^+$	$m/e = 91$
	$m/e = 92$

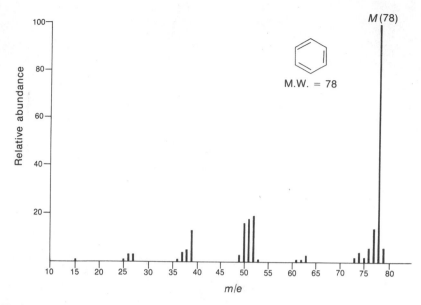

FIGURE 8.16 Mass spectrum of benzene.

When an alkyl group is attached to a benzene ring, preferential fragmentation occurs at a benzylic position to form a fragment ion of the formula $C_7H_7^+$ ($m/e = 91$). In the mass spectrum of toluene (Fig. 8.17), loss of hydrogen from the molecular ion gives a strong peak at $m/e = 91$. Although it might be expected that this fragment ion peak is due to the benzyl carbocation, evidence has accumulated that suggests that the benzyl carbocation actually rearranges to form the **tropylium ion.** Isotope-labeling experiments tend to confirm the formation of the tropylium ion.

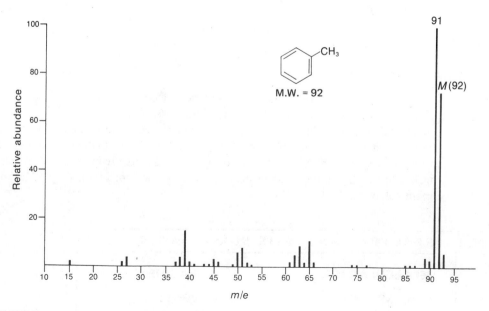

FIGURE 8.17 Mass spectrum of toluene.

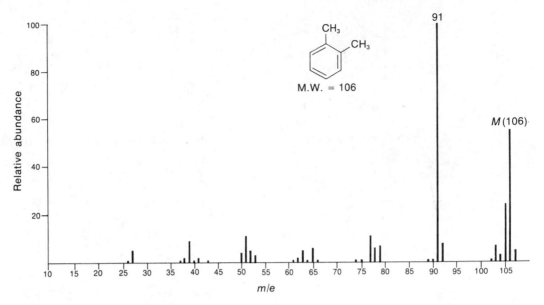

FIGURE 8.18 Mass spectrum of *ortho*-xylene.

Benzyl carbocation   Tropylium ion

The mass spectra of the xylene isomers (see Fig. 8.18 for an example) show a medium peak at $m/e = 105$, which is due to the methyltropylium ion. More importantly, xylene loses one methyl group to form the unsubstituted $C_7H_7^+$ ion ($m/e = 91$). The mass spectra of *ortho-*, *meta-*, and *para-* disubstituted aromatic rings are essentially identical. As a result, the positions of substitution of polyalkyl-substituted benzenes cannot be determined by mass spectrometry.

The formation of a substituted tropylium ion is typical for alkyl-substituted benzenes. In the mass spectrum of isopropylbenzene (Fig. 8.19), a strong peak appears at $m/e = 105$. This peak corresponds to loss of a methyl group to form a methyl-substituted tropylium ion.

$+ \cdot CH_3$ $m/e = 105$

Again in the mass spectrum of propylbenzene (Fig. 8.20), a strong peak due to the tropylium ion appears at $m/e = 91$.

When the alkyl group attached to the benzene ring is a propyl group or larger, an important type of rearrangement, called the **McLafferty rearrangement,** occurs. Using butylbenzene as an example, this arrangement may be depicted as

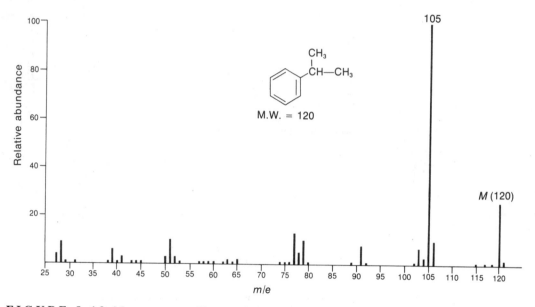

and leads to a peak at *m/e* = 92. This peak also appears in the mass spectrum of propylbenzene.

FIGURE 8.19 Mass spectrum of isopropylbenzene (cumene).

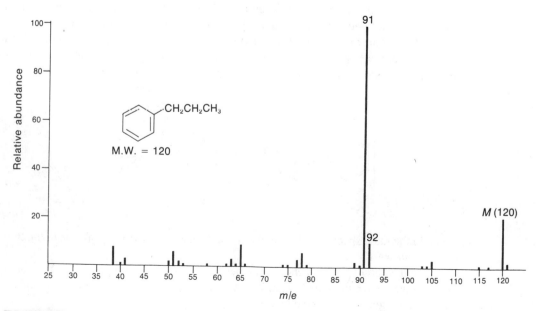

FIGURE 8.20 Mass spectrum of propylbenzene.

F.   Alcohols and Phenols

The intensity of the molecular ion peak in the mass spectrum of a primary or secondary alcohol is usually rather low. The molecular ion peak may be entirely absent in the mass spectrum of a tertiary alcohol. Fragmentation involves the loss of an alkyl group or the loss of a molecule of water.

SPECTRAL ANALYSIS BOX — Alcohols

MOLECULAR ION	FRAGMENT IONS
M^+ weak or absent	Loss of alkyl group
	$M - 18$

The mass spectrum of 1-butanol (Fig. 8.21) shows a very weak molecular ion peak at $m/e = 74$, while the mass spectrum of 2-butanol (Fig. 8.22) has a molecular ion peak ($m/e = 74$) that is too weak to be detected. The molecular ion peak for tertiary alcohol, 2-methyl-2-propanol (Fig. 8.23), is entirely absent.

The most important fragmentation reaction for alcohols is the loss of an alkyl group:

$$\left[R-\underset{R''}{\overset{R'}{C}}-OH \right]^{+\cdot} \longrightarrow R\cdot \; + \; \underset{R''}{\overset{R'}{C}}=\overset{+}{O}H$$

The largest alkyl group is most readily lost. In the spectrum of 1-butanol (Fig. 8.21), the intense peak at $m/e = 31$ is due to the loss of a propyl group to form an $H_2C=OH^+$ ion. 2-Butanol (Fig. 8.22) loses an ethyl group to form the $CH_3CH=OH^+$ fragment at $m/e = 45$. 2-Methyl-2-propanol (Fig. 8.23) loses a methyl group to form the $(CH_3)_2C=OH^+$ fragment at $m/e = 59$.

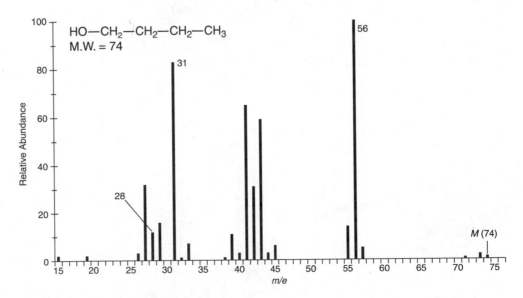

FIGURE 8.21 Mass spectrum of 1-butanol.

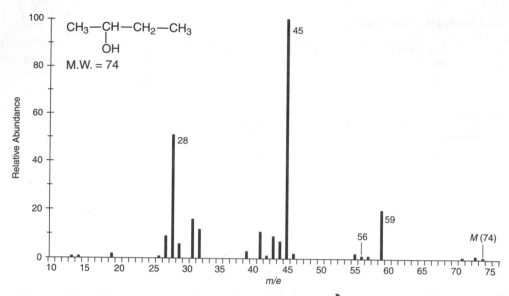

FIGURE 8.22 Mass spectrum of 2-butanol.

A second common mode of fragmentation involves dehydration. The importance of dehydration increases as the chain length of the alcohol increases. While the fragment ion peak resulting from dehydration ($m/e = 56$) is very intense in 1-butanol, it is very weak in the other butanol isomers. However, in the mass spectra of the five-carbon alcohols, this peak due to dehydration of the molecular ion is quite important.

Dehydration may occur by either of two mechanisms. The hot surfaces of the inlet system may stimulate dehydration of the alcohol molecule before the molecule comes in contact with the electrons. In this case, the dehydration is a **1,2-elimination** of water. However, the molecular ion, once it is formed, may also lose water. In that case the dehydration is a **1,4-elimination** of water via a cyclic mechanism:

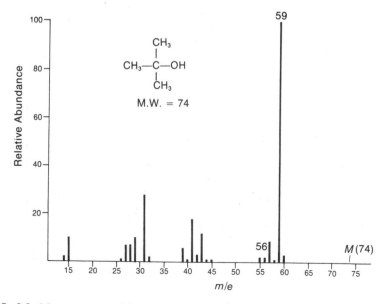

FIGURE 8.23 Mass spectrum of 2-methyl-2-propanol.

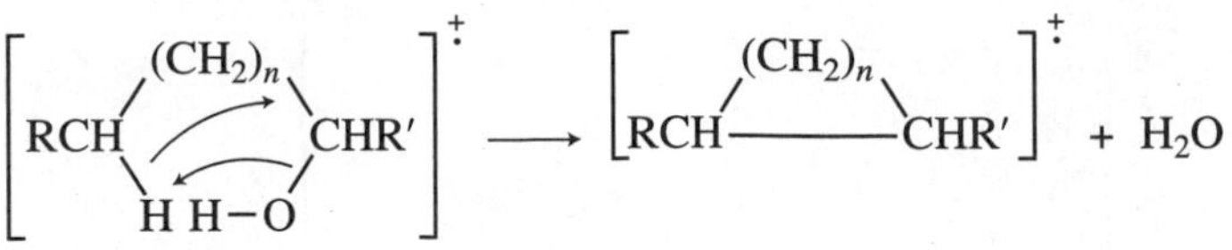

$$\left[RCH \overset{(CH_2)_n}{\underset{H\ H-O}{\diagup \diagdown}} CHR' \right]^{\ddagger} \longrightarrow \left[RCH \overset{(CH_2)_n}{\underset{}{\diagup \diagdown}} CHR' \right]^{\ddagger} + H_2O$$

$$n = 1 \text{ or } 2$$

Alcohols containing four or more carbons may undergo the simultaneous loss of both water and ethylene:

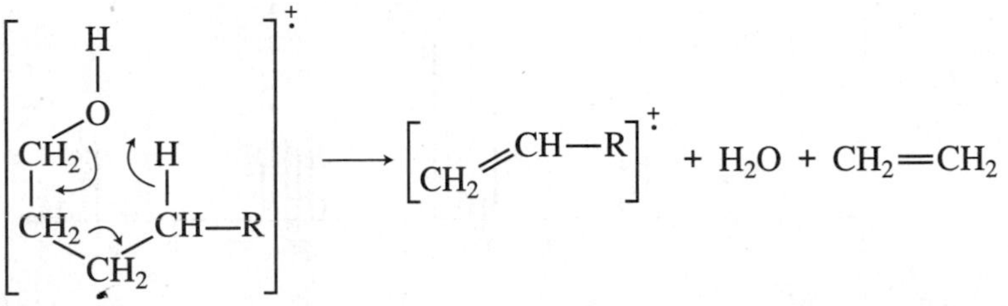

$$\left[\begin{array}{c} H \\ | \\ O \\ CH_2 \quad H \\ | \quad | \\ CH_2 \quad CH-R \\ CH_2 \end{array} \right]^{\ddagger} \longrightarrow \left[CH_2 \overset{}{=} CH-R \right]^{\ddagger} + H_2O + CH_2 \!\!=\!\! CH_2$$

In the case of 1-butanol, this fragment ion is responsible for a rather weak peak at $m/e = 28$. However, with 1-pentanol, the corresponding peak is the most intense in the spectrum.

Cyclic alcohols may undergo fragmentation by at least three different pathways:

(1)

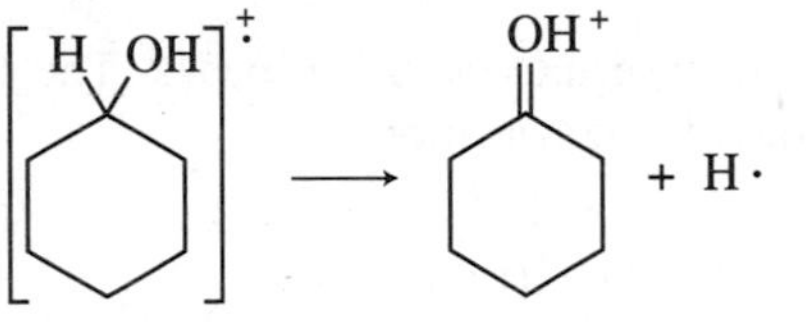

$$m/e = 99$$

(2)

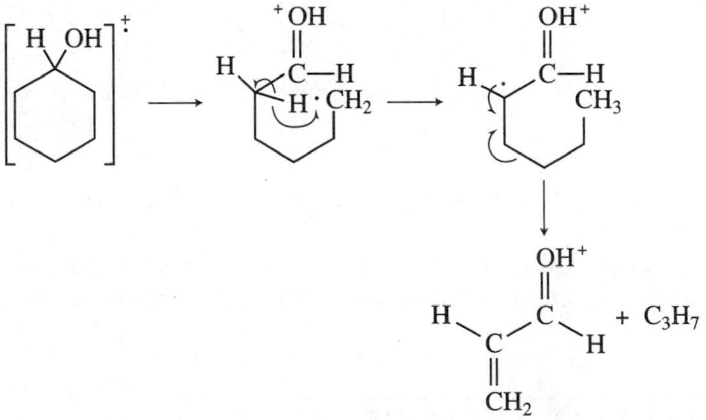

$$m/e = 57$$

(3)

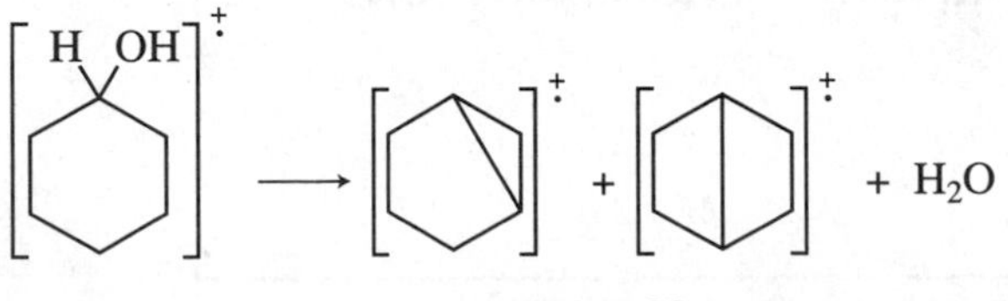

$$m/e = 82$$

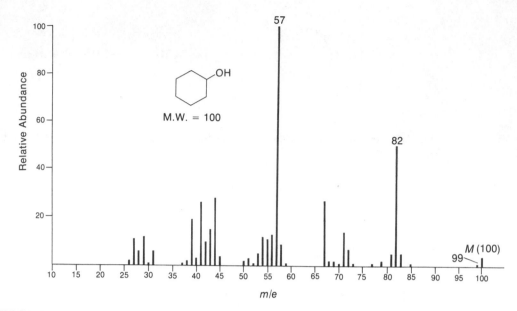

FIGURE 8.24 Mass spectrum of cyclohexanol.

A peak corresponding to each of these fragment ions can be observed in the mass spectrum of cyclohexanol (Fig. 8.24).

Benzylic alcohols exhibit intense molecular ion peaks. The following sequence of reactions illustrates their principal modes of fragmentation.

$$
\left[\begin{array}{c} CH_2OH \\ \bigcirc \end{array} \right]^{+\cdot} \longrightarrow \underset{\substack{m/e = 107 \\ + \text{ H·}}}{\bigcirc^{OH\cdot}} \longrightarrow \underset{\substack{m/e = 79 \\ + \text{ CO}}}{\bigcirc^{H\;H}} \longrightarrow \underset{m/e = 77}{C_6H_5{}^+ \; + \; H_2}
$$

Peaks arising from these fragment ions can be observed in the mass spectrum of benzyl alcohol (Fig. 8.25).

The mass spectra of phenols show strong molecular ion peaks. Favored modes of fragmentation involve loss of a hydrogen atom, loss of a molecular of carbon monoxide, and loss of a formyl radical.

SPECTRAL ANALYSIS BOX — Phenols

MOLECULAR ION	FRAGMENT IONS
M⁺ strong	$M - 1$
	$M - 28$
	$M - 29$

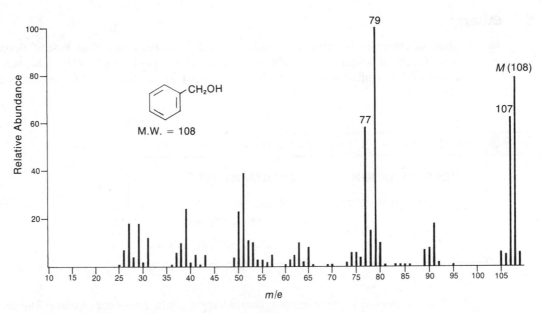

FIGURE 8.25 Mass spectrum of benzyl alcohol.

Phenols typically lose the elements of carbon monoxide to give strong peaks at *m/e* values that are 28 mass units below the value for the molecular ion. Such a peak is designated an $M - 28$ peak due to the relationship between its *m/e* value and the *m/e* value of the molecular ion. Phenols also lose the elements of the formyl radical (HCO·) to give strong $M - 29$ peaks (the fragment ion appears 29 mass units below the molecular ion). An exception appears to be the cresols, for which these fragment peaks are rather weak. Nevertheless, the $M - 28$ peak (*m/e* = 80) and the $M - 29$ peak (*m/e* = 79) appear in the mass spectrum of 2-methylphenol (Fig. 8.26).

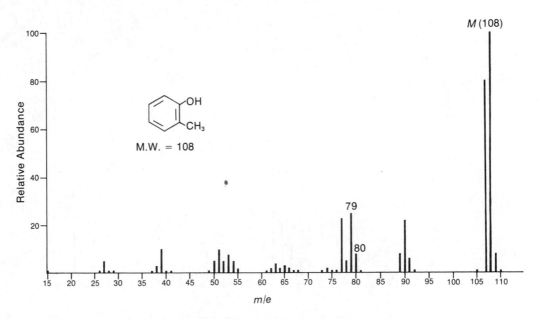

FIGURE 8.26 Mass spectrum of 2-methylphenol (*ortho*-cresol).

G.   Ethers

Aliphatic ethers tend to exhibit molecular ion peaks that are stronger than those of alcohols with the same molecular weights. Nevertheless, the molecular ion peaks of ethers are rather weak. Principal modes of fragmentation include α-cleavage, formation of carbocation fragments, and loss of an alkoxy group.

SPECTRAL ANALYSIS BOX — Ethers

MOLECULAR ION	**FRAGMENT IONS**
M^+ weak, but observable	α-Cleavage
	$m/e = 43, 59, 73$, etc.
	$M - 31, M - 45, M - 59$, etc.

The fragmentation of the ethers is somewhat similar to that of the alcohols. The carbon–carbon bond to the α carbon may be broken to yield a fragment ion that bears a positive charge on the oxygen.

$$\left[R \!\!\mid\!\! \overset{\alpha}{C}H_2 \!-\! OR \right]^{\cdot +} \longrightarrow CH_2 \!=\! \overset{+}{O}R + R \cdot$$

In the mass spectrum of diisopropyl ether (Fig. 8.27), this fragmentation gives rise to a peak at $m/e = 87$, due to the loss of a methyl group.

A second mode of fragmentation involves cleavage of the carbon–oxygen bond of an ether to yield a carbocation. Cleavage of this type in diisopropyl ether is responsible for the $C_3H_7^+$ fragment at $m/e = 43$.

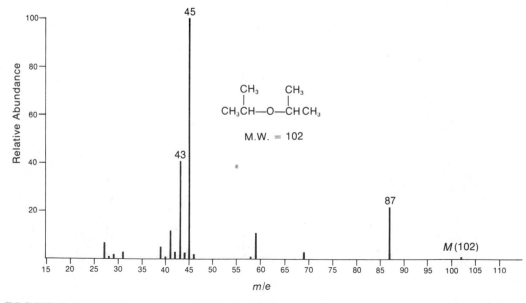

FIGURE 8.27 Mass spectrum of diisopropyl ether.

A third type of fragmentation occurs as a rearrangement reaction taking place on one of the fragment ions, rather than on the molecular ion itself. The rearrangement may be illustrated as follows.

$$R-CH=\overset{+}{O}+CH-\underset{\beta}{CH_2} \longrightarrow R-CH=\overset{+}{O}H + CH=CH_2$$

This type of rearrangement is particularly favored when the α carbon of the ether is branched. In the case of diisopropyl ether, this rearrangement gives rise to a $C_2H_5O^+$ fragment ($m/e = 45$).

As a result of these fragmentation processes, the fragment ion peaks of ethers often fall into the series 31, 45, 59, 73, and so on. These fragments are of either the RO^+ or the $ROCH_2^+$ type.

Acetals and ketals behave very similarly to ethers. However, fragmentation is even more favorable in acetals and ketals than in ethers, so the molecular ion peak of an acetal or ketal may be either extremely weak or totally absent.

Aromatic ethers may undergo cleavage reactions that involve loss of the alkyl group to form $C_6H_5O^+$ ions. These fragment ions then lose the elements of carbon monoxide to form $C_5H_5^+$ ions. In addition, an aromatic ether may lose the entire alkoxy group to yield ions of the types $C_6H_6^+$ and $C_6H_5^+$.

H. Aldehydes

The molecular ion peak of an aliphatic aldehyde is usually observable, although at times it may be fairly weak. Principal modes of fragmentation include α-cleavage and β-cleavage. If the carbon chain attached to the carbonyl group contains at least three carbons, McLafferty rearrangement is also observed.

SPECTRAL ANALYSIS BOX—Aldehydes

MOLECULAR ION	FRAGMENT IONS
M^+ weak, but observable (aliphatic)	Aliphatic:
M^+ strong (aromatic)	$m/e = 29$
	$M - 29$
	$M - 43$
	$m/e = 44$
	Aromatic:
	$M - 1$
	$M - 29$

Cleavage of one of the two bonds to the carbonyl group, sometimes called **α-cleavage,** occurs very commonly. It may be outlined as follows.

$$[R-CHO]^{\ddot{+}} \longrightarrow R-C\equiv O^+ + H\cdot$$

$$[R-CHO]^{\ddot{+}} \longrightarrow H-C\equiv O^+ + R\cdot$$

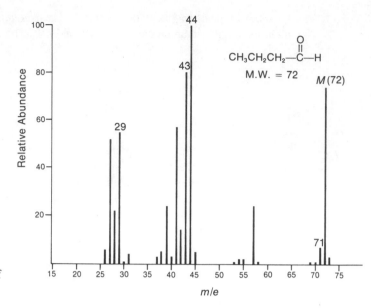

FIGURE 8.28 Mass spectrum of butyraldehyde.

The peak due to the loss of one hydrogen atom is very characteristic of aldehydes. This peak is observed at $m/e = 71$ in the mass spectrum of butyraldehyde (Fig. 8.28). The peak due to the formation of HCO^+ can be observed at $m/e = 29$; this is also a very characteristic peak in the mass spectra of aldehydes.

The second important mode of fragmentation for aldehydes is known as **β-cleavage.**

$$\left[R\!\!\stackrel{|}{\dotplus}\!\!CH_2\!-\!CHO \right]^{+\cdot} \longrightarrow R^+ + CH_2\!\!=\!\!CH\!-\!O\cdot$$

Coincidentally, the R^+ fragment ion also occurs at $m/e = 29$ in the mass spectrum of butyraldehyde. However, in higher aldehydes it occurs at higher m/e values. At any rate, the mass of this peak is 43 mass units less than that of the molecular ion.

The third major fragmentation pathway for aldehydes is the McLafferty rearrangement (see Section 8.6E).

$$\left[\begin{array}{c} R \diagdown_{CH} \diagup^H \quad \diagdown O \\ | \qquad \qquad \| \\ R \diagdown_{CH} \diagdown_{CH_2} \diagup_C \diagdown H \end{array} \right]^{+\cdot} \longrightarrow \begin{array}{c} R \diagdown_C \diagup^H \\ \| \\ R \diagdown_C \diagdown H \end{array} + [CH_2\!\!=\!\!CH\!-\!OH]^{+\cdot} \\ \qquad \qquad \qquad \qquad m/e = 44$$

The fragment ion formed in this rearrangement has $m/e = 44$. A peak at $m/e = 44$ is considered to be quite characteristic of the mass spectra of aldehydes. It is the most intense peak in the mass spectrum of butyraldehyde. Note that this rearrangement occurs only if the chain attached to the carbonyl group has three or more carbons.

Aromatic aldehydes exhibit intense molecular ion peaks. The loss of one hydrogen atom via α-cleavage is a very favorable process. The resulting $M - 1$ peak may be more intense than the molecular ion peak. In the mass spectrum of benzaldehyde (Fig. 8.29), the $M - 1$ peak appears at $m/e = 105$. You may also notice a peak at $m/e = 77$, which is due to the loss of the $-CHO$ group to give $C_6H_5^+$.

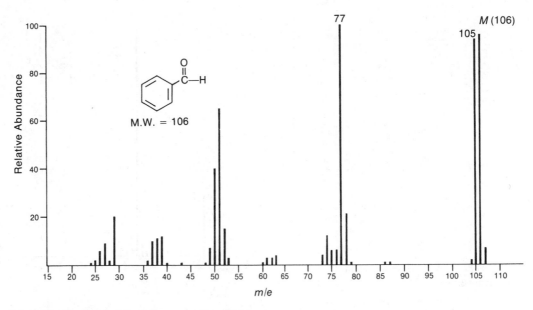

FIGURE 8.29 Mass spectrum of benzaldehyde.

I. Ketones

The mass spectra of ketones show an intense molecular ion peak. Loss of the alkyl groups attached to the carbonyl group is one of the most important fragmentation processes. The pattern of fragmentation is similar to that of aldehydes.

SPECTRAL ANALYSIS BOX — Ketones

MOLECULAR ION	FRAGMENT IONS
M^+ strong	Aliphatic:
	$M - 15, M - 29, M - 43$, etc.
	$m/e = 43$
	$m/e = 58, 72, 86$, etc.
	$m/e = 42, 83$
	Aromatic:
	$m/e = 105, 120$

Loss of alkyl groups by means of α-cleavage is an important mode of fragmentation. The larger of the two alkyl groups attached to the carbonyl group appears more likely to be lost. In the mass spectrum of 2-butanone (Fig. 8.30), the peak at $m/e = 43$, due to the loss of the ethyl group, is more intense than the peak at $m/e = 57$, which is due to the loss of the methyl group. Similarly, in the mass spectrum of 2-octanone (Fig. 8.31), loss of the hexyl group, giving a peak at $m/e = 43$, is more important than loss of the methyl group, which gives the weak peak at $m/e = 113$.

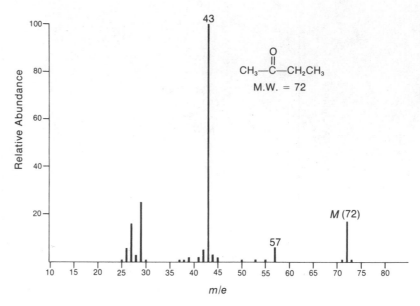

FIGURE 8.30 Mass spectrum of 2-butanone.

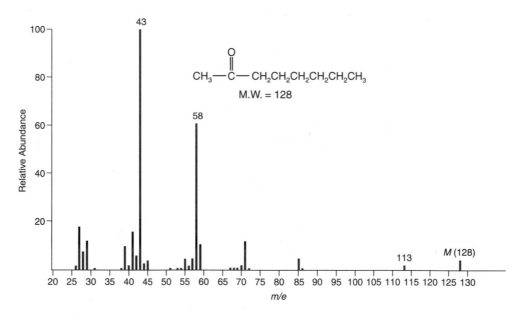

FIGURE 8.31 Mass spectrum of 2-octanone.

When the carbonyl group of a ketone has attached to it at least one alkyl group that is three or more carbon atoms in length, a McLafferty rearrangement is possible. It may be described as follows.

The peak at $m/e = 58$ in the mass spectrum of 2-octanone is due to the fragment ion that results from this rearrangement.

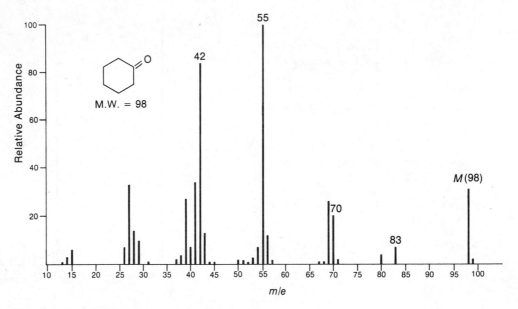

FIGURE 8.32 Mass spectrum of cyclohexanone.

Cyclic ketones may undergo a variety of fragmentation and rearrangement processes. Outlines of these processes for the case of cyclohexanone follow. A fragment ion peak corresponding to each process appears in the mass spectrum of cyclohexanone (Fig. 8.32).

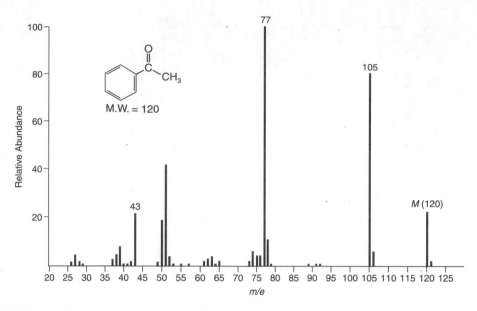

FIGURE 8.33 Mass spectrum of acetophenone.

Aromatic ketones undergo α-cleavage to lose the alkyl group and form the $C_6H_5CO^+$ ion ($m/e = 105$). This ion loses carbon monoxide to form the $C_6H_5^+$ ion ($m/e = 77$). These peaks appear prominently in the mass spectrum of acetophenone (Fig. 8.33). With larger alkyl groups attached to the carbonyl group of an aromatic ketone, a rearrangement of the McLafferty type is possible, as follows.

$$m/e = 120$$

The $m/e = 120$ fragment ion may undergo additional α-cleavage to yield the $C_6H_5CO^+$ ion at $m/e = 105$.

J. Esters

Even though fragmentation readily occurs, it is usually possible to observe weak molecular ion peaks in the mass spectra of methyl esters. The esters of alcohols higher than methanol form much weaker molecular ion peaks. Esters of alcohols larger than four carbons may form molecular ion peaks that are too weak to be observed.

The most important of the α-cleavage reactions involves the loss of the alkoxy group from an ester to form the corresponding acylium ion, RCO^+. The acylium ion peak appears at $m/e = 71$ in the mass spectrum of methyl butyrate (Fig. 8.34). The acylium ion peak is a useful diagnostic peak in the mass spectra of esters. A second useful peak results from the loss of the alkyl group from the acyl portion of the ester molecule, leaving a fragment $CH_3-O-C\!=\!O^+$ that appears at $m/e = 59$. This peak can also be observed in the mass spectrum of methyl butyrate. Again, this $m/e = 59$ peak, though less intense than the acylium ion peak, is a useful diagnostic peak for methyl esters. Other fragment ion peaks include the OCH_3^+ fragment and the R^+ fragment from the acyl portion of the ester molecule. These latter ions are much less important than the former two.

SPECTRAL ANALYSIS BOX—Esters

MOLECULAR ION	**FRAGMENT IONS**
M^+ weak, but generally observable	Methyl esters:
	$M - 31$
	$m/e = 59, 74$
	Higher esters:
	$M - 45, M - 59, M - 73$
	$m/e = 73, 87, 101$
	$m/e = 88, 102, 116$
	$m/e = 61, 75, 89$
	$m/e = 77, 105, 108$
	$M - 32, M - 46, M - 60$

The most important β-cleavage reaction of methyl esters is the McLafferty rearrangement:

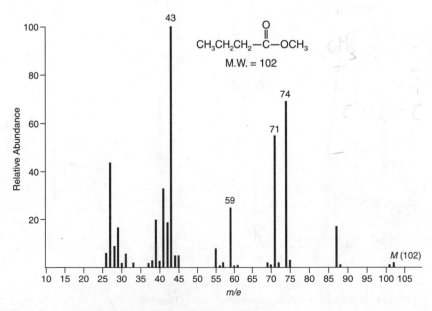

$$m/e = 74$$

This fragment ion peak is very important in the mass spectra of methyl esters, as is evident from its prominence in the mass spectrum of methyl butyrate.

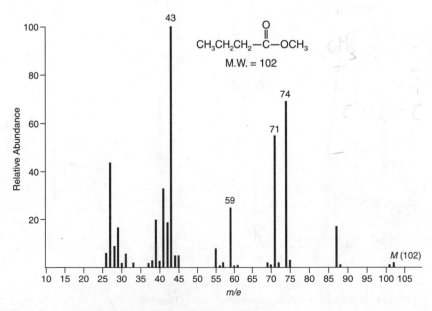

FIGURE 8.34 Mass spectrum of methyl butyrate.

Ethyl, propyl, butyl, and higher alkyl esters also undergo the α-cleavage and McLafferty rearrangements typical of the methyl esters. In addition, however, these esters may undergo a rearrangement of the alkyl portion of the molecule, in which a hydrogen atom from the alkyl portion is transferred to the carbonyl oxygen of the acyl portion of the ester. This rearrangement results in fragments of the type

$$R-C \begin{matrix} \overset{\cdot +}{O}H \\ \\ OH \end{matrix}$$

which appear in the series $m/e = 61, 75, 89$, and so on. These ions may also appear without the extra hydrogen as $RCOOH^+$ fragment ions.

Benzyl esters undergo rearrangement to eliminate the neutral ketene molecule.

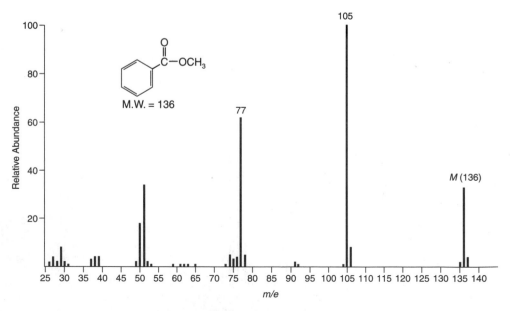

The resulting ion is often the most intense peak in the mass spectrum of such a compound.

Alkyl benzoate esters prefer to lose the alkoxy group to form the $C_6H_5CO^+$ ion ($m/e = 105$). This ion may lose carbon monoxide to form the $C_6H_5^+$ ion at $m/e = 77$. Each of these peaks appears in the mass spectrum of methyl benzoate (Fig. 8.35).

Alkyl substitution on benzoate esters appears to have little effect on the mass spectral results unless the alkyl group is in the *ortho* position with respect to the ester functional group. In this case, the alkyl group can interact with the ester function, with the elimination of a molecule of alcohol.

FIGURE 8.35 Mass spectrum of methyl benzoate.

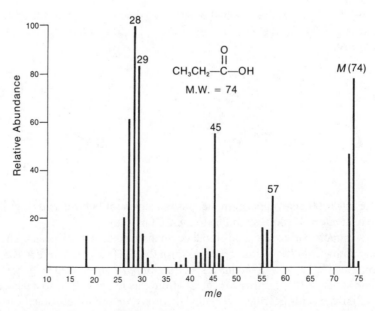

K. Carboxylic Acids

Aliphatic carboxylic acids generally show weak, but observable, molecular ion peaks. Aromatic carboxylic acids, on the other hand, show strong molecular ion peaks. The principal modes of fragmentation resemble those of the methyl esters.

S P E C T R A L   A N A L Y S I S   B O X — Carboxylic Acids

MOLECULAR ION	FRAGMENT IONS
Aliphatic carboxylic acids:	Aliphatic carboxylic acids:
M^+ weak, but observable	$M - 17, M - 45$
	$m/e = 45, 60$
Aromatic carboxylic acids:	Aromatic carboxylic acids:
M^+ strong	$M - 17, M - 45$
	$M - 18$

With short-chain acids, the loss of OH and COOH through α-cleavage on either side of the C=O group may be observed. In the mass spectrum of propionic acid (Fig. 8.36), loss of OH

FIGURE 8.36 Mass spectrum of propionic acid.

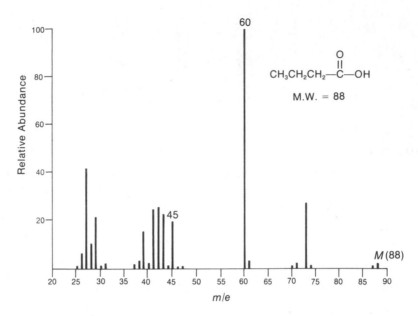

FIGURE 8.37 Mass spectrum of butyric acid.

gives rise to a peak at $m/e = 57$. Loss of COOH gives rise to a peak at $m/e = 29$. The intense peak at $m/e = 28$ is due to further fragmentation of the alkyl portion of the acid molecule. Loss of the alkyl group as a free radical, leaving the $COOH^+$ ion ($m/e = 45$), also occurs. This fragment ion peak appears in the mass spectrum, as well, and is characteristic of the mass spectra of carboxylic acids.

With acids containing γ hydrogens, the principal pathway for fragmentation is the McLafferty rearrangement. In the case of carboxylic acids, this rearrangement produces a prominent peak at $m/e = 60$.

The $m/e = 60$ peak appears in the mass spectrum of butyric acid (Fig. 8.37). You may also notice a peak at $m/e = 45$, corresponding to the $COOH^+$ ion.

Aromatic carboxylic acids produce intense molecular ion peaks. The most important fragmentation pathway involves loss of OH to form the $C_6H_5CO^+$ ion ($m/e = 105$), followed by loss of CO to form the $C_6H_5^+$ ion. In the mass spectrum of *para*-anisic acid (Fig. 8.38), loss of OH gives rise to a peak at $m/e = 135$. Further loss of CO from this ion gives rise to a peak at $m/e = 107$.

Benzoic acids bearing *ortho* alkyl, hydroxy, or amino substituents undergo loss of water through a rearrangement reaction similar to that observed for *ortho*-substituted benzoate esters, as illustrated at the end of Section 8.6J.

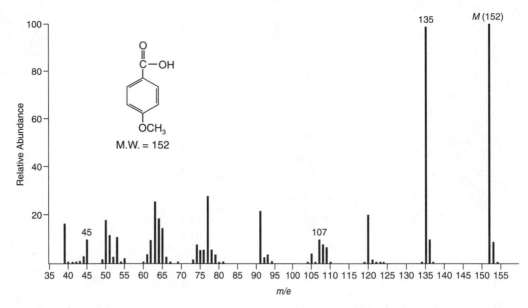

FIGURE 8.38 Mass spectrum of *para*-anisic acid.

L.   Amines

The value of the mass of the molecular ion can be of great help in identifying a substance as an amine. As stated in Section 8.4, a compound with an odd number of nitrogen atoms must have an odd-numbered molecular weight. On this basis, it is possible to quickly determine whether a substance could be an amine. Unfortunately, in the case of aliphatic amines, the molecular ion peak may be very weak or even absent.

SPECTRAL ANALYSIS BOX — Amines	
MOLECULAR ION	**FRAGMENT IONS**
M^+ weak or absent	α-Cleavage
Nitrogen Rule obeyed	$m/e = 30$

The most intense peak in the mass spectrum of an aliphatic amine arises from α-cleavage:

$$\left[R\!-\!\underset{|}{\overset{|}{C}}\!-\!\overset{/}{\underset{\backslash}{N}} \right]^{+\cdot} \longrightarrow R\cdot + \overset{\backslash}{\underset{/}{C}}\!=\!\overset{+}{\underset{\backslash}{N}}{\overset{/}{}}$$

When there is a choice of R groups to be lost through this process, the largest R group is lost preferentially. For primary amines that are not branched at the carbon next to the nitrogen, the most intense peak in the spectrum occurs at $m/e = 30$. It arises from α-cleavage:

$$\left[R\!-\!CH_2\!-\!NH_2 \right]^{+\cdot} \longrightarrow R\cdot + CH_2\!=\!\overset{+}{N}H_2$$
$$m/e = 30$$

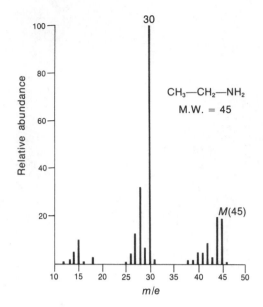

FIGURE 8.39 Mass spectrum of ethylamine.

The presence of this peak is good, although not conclusive, evidence that the test substance is a primary amine. The peak may arise from secondary fragmentation of ions formed from the fragmentation of secondary or tertiary amines, as well. In the mass spectrum of ethylamine (Fig. 8.39), the $m/e = 30$ peak can be seen clearly.

The same β-cleavage peak can also occur for long-chain primary amines. Further fragmentation of the R group of the amine leads to clusters of fragments 14 mass units apart, due to sequential loss of CH_2 units from the R group. Long-chain primary amines can also undergo fragmentation, via the process

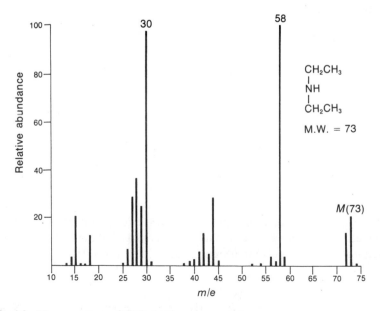

FIGURE 8.40 Mass spectrum of diethylamine.

$$\left[\text{R--CH}_2 \overset{\frown}{\underset{(CH_2)_n}{}} \text{NH}_2 \right]^{\overset{+}{\cdot}} \longrightarrow \text{R} \cdot + \underset{(CH_2)_n}{\text{CH}_2 \text{--} \overset{+}{\text{NH}}_2}$$

This is particularly favorable when $n = 4$, since a stable six-membered ring results. In this case, the fragment ion appears at $m/e = 86$.

Secondary and tertiary amines also undergo fragmentation processes as described earlier. The most important fragmentation is β-cleavage. In the mass spectrum of diethylamine (Fig. 8.40), the intense peak at $m/e = 58$ is due to loss of a methyl group. Again in the mass spectrum of triethylamine (Fig. 8.41), loss of methyl produces the most intense peak in the spectrum, at $m/e = 86$. In each case, further fragmentation of this initially formed fragment ion produces a peak at $m/e = 30$.

Cyclic aliphatic amines usually produce intense molecular ion peaks. Their principal modes of fragmentation are as follows.

$$\left[\overset{\displaystyle\bigcirc}{\underset{\text{CH}_3}{\overset{|}{\text{N}}}} \right]^{\overset{+}{\cdot}} \longrightarrow \underset{\text{CH}_3 + \text{H}\cdot}{\overset{\displaystyle\bigcirc}{\underset{|}{\overset{+}{\text{N}}}}} \longrightarrow \text{CH}_3\text{--}\overset{+}{\text{N}}\equiv\text{CH} + \cdot\text{CH}_2\text{CH}_2\text{CH}_2 \cdot$$

$m/e = 85 \qquad\qquad m/e = 84 \qquad\qquad\qquad m/e = 42$

$$\left[\overset{\displaystyle\bigcirc}{\underset{\text{CH}_3}{\overset{|}{\text{N}}}} \right]^{\overset{+}{\cdot}} \longrightarrow \underset{\text{CH}_3}{\overset{\cdot\text{CH}_2 \diagup \diagup \text{CH}_2}{\overset{|}{\overset{+}{\text{N}}}}} + \text{CH}_2{=}\text{CH}_2 \longrightarrow \text{CH}_2{=}\overset{+}{\text{N}}{=}\text{CH}_2 + \text{CH}_3 \cdot$$

$m/e = 57 \qquad\qquad\qquad\qquad m/e = 42$

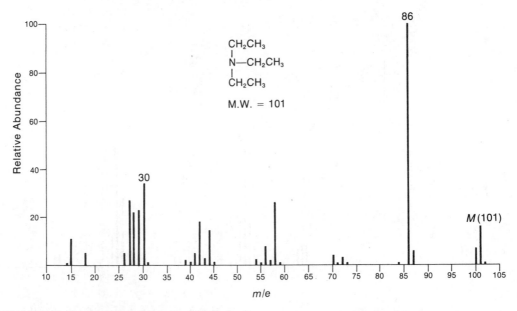

FIGURE 8.41 Mass spectrum of triethylamine.

Aromatic amines show intense molecular ion peaks. A moderately intense peak may appear at an m/e value one mass unit less than that of the molecular ion, due to loss of a hydrogen atom. The fragmentation of aromatic amines can be illustrated for the case of aniline:

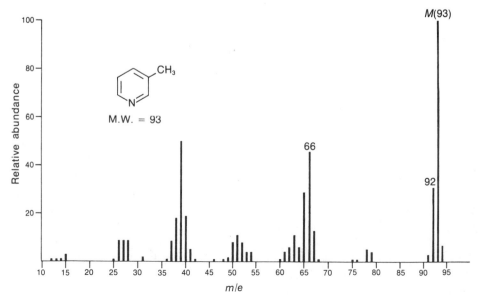

Very intense molecular ion peaks characterize substituted pyridines. Frequently, loss of a hydrogen atom to produce a peak at an m/e value one mass unit less than the molecular ion is also observed.

The most important fragmentation process for the pyridine ring is loss of the elements of hydrogen cyanide. This produces a fragment ion which is 27 mass units lighter than the molecular ion. In the mass spectrum of 3-methylpyridine (Fig. 8.42), you can see the peak due to loss of hydrogen ($m/e = 92$) and the peak due to loss of hydrogen cyanide ($m/e = 66$).

When the alkyl side chain attached to a pyridine ring contains three or more carbons arranged linearly, fragmentation via the McLafferty rearrangement can also occur.

FIGURE 8.42 Mass spectrum of 3-methylpyridine.

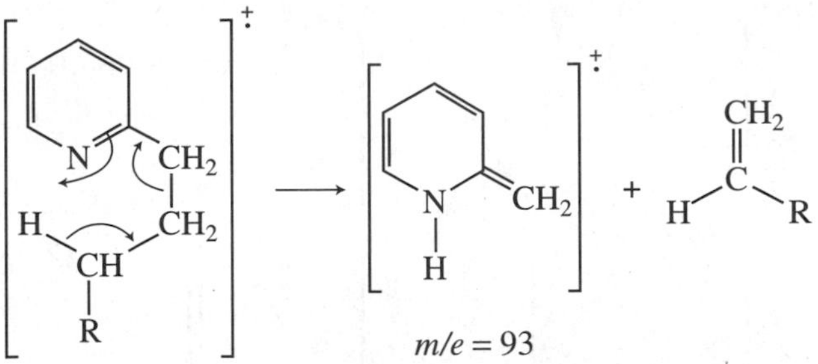

$m/e = 93$

This mode of cleavage is most important for substituents attached to the number 2 position of the ring.

M. Selected Nitrogen and Sulfur Compounds

As is true of amines, nitrogen-bearing compounds such as amides, nitriles, and nitro compounds must follow the Nitrogen Rule (explained more completely in Section 8.4): If they contain an odd number of nitrogen atoms, they must have an odd-numbered molecular weight.

Amides

The mass spectra of amides usually show observable molecular ion peaks. The fragmentation patterns of amides are quite similar to those of the corresponding esters and acids. The presence of a strong fragment ion peak at $m/e = 44$ is usually indicative of a primary amide. This peak arises from α-cleavage of the following sort.

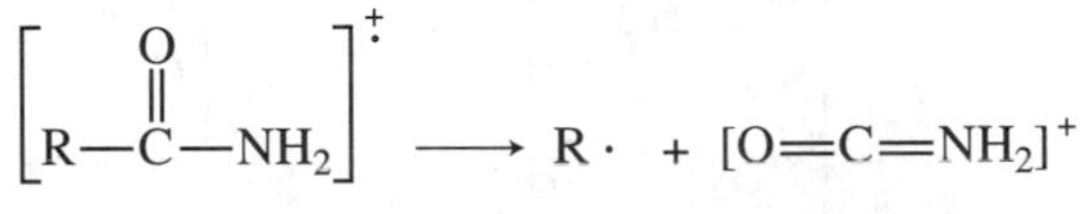

$$\left[R-\overset{\overset{\displaystyle O}{\|}}{C}-NH_2 \right]^{\ddagger} \longrightarrow R\cdot + [O{=}C{=}NH_2]^+$$

$$m/e = 44$$

Once the carbon chain in the acyl moiety of an amide becomes long enough to permit the transfer of a hydrogen attached to the γ position, McLafferty rearrangements become possible. For primary amides, the McLafferty rearrangement gives rise to a fragment ion peak at $m/e = 59$. For N-alkylamides, analogous peaks at m/e values of 63, 77, 91, and so on often appear.

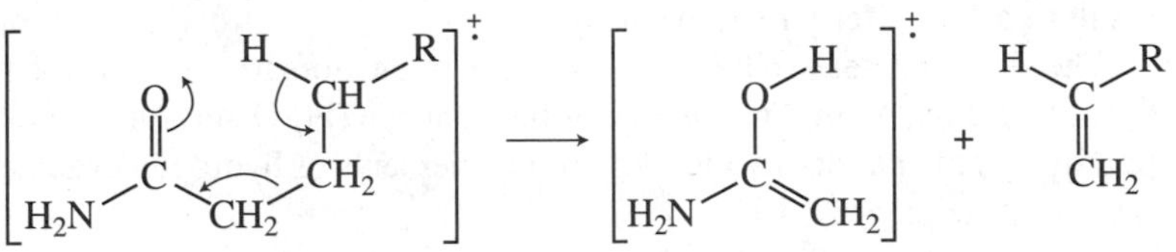

$m/e = 59$

Nitriles

Aliphatic nitriles usually undergo fragmentation so readily that the molecular ion peak is too weak to be observed. However, most nitriles form a peak due to the loss of one hydrogen atom, producing an ion of the type $R-CH{=}C{=}N^+$. Although this peak may be weak, it is a useful diagnostic peak in characterizing nitriles. In the mass spectrum of hexanenitrile (Fig. 8.43), this peak appears at $m/e = 96$.

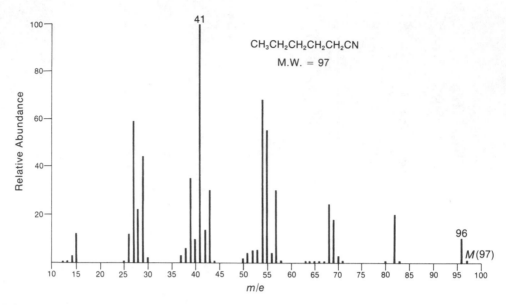

FIGURE 8.43 Mass spectrum of hexanenitrile.

When the alkyl group attached to the nitrile functional group is a propyl group or some longer hydrocarbon group, the most intense peak in the mass spectrum results from a McLafferty rearrangement:

$$m/e = 41$$

This peak, which appears in the mass spectrum of hexanenitrile, can be quite useful in characterizing an aliphatic nitrile. Unfortunately, as the alkyl group of a nitrile becomes longer, the probability of formation of the $C_3H_5^+$ ion, which also appears at $m/e = 41$, increases. With high-molecular-weight nitriles, most of the fragment ions of mass 41 are $C_3H_5^+$ ions, rather than ions formed as a result of a McLafferty rearrangement.

The strongest peak in the mass spectrum of an aromatic nitrile is the molecular ion peak. Loss of cyanide occurs, giving, in the case of benzonitrile (Fig. 8.44), the $C_6H_5^+$ ion at $m/e = 77$. More important fragmentation involves loss of the elements of hydrogen cyanide. In benzonitrile, this gives rise to a peak at $m/e = 76$.

Nitro Compounds

The molecular ion peak for an aliphatic nitro compound is seldom observed. The mass spectrum is the result of fragmentation of the hydrocarbon part of the molecule. However, the mass spectra of nitro compounds may show a moderate peak at $m/e = 30$, corresponding to the NO^+ ion, and a weaker peak at $m/e = 46$, corresponding to the NO_2^+ ion. These peaks appear in the mass spectrum of 1-nitropropane (Fig. 8.45). The intense peak at $m/e = 43$ is due to the $C_3H_7^+$ ion.

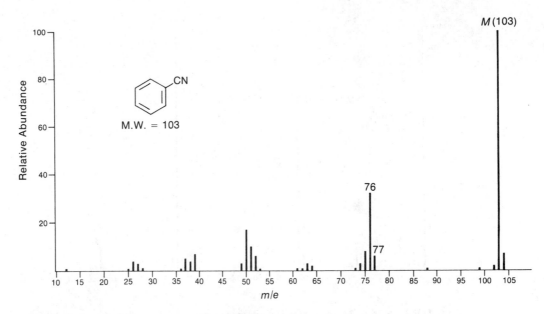

FIGURE 8.44 Mass spectrum of benzonitrile.

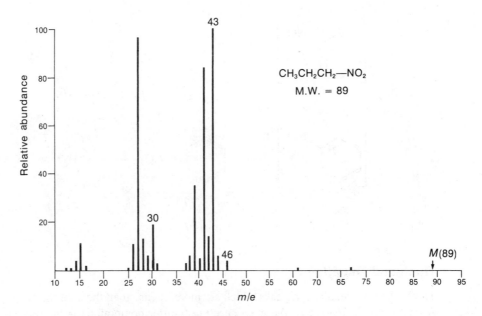

FIGURE 8.45 Mass spectrum of 1-nitropropane.

Aromatic nitro compounds show intense molecular ion peaks. The characteristic NO^+ ($m/e = 30$) and NO_2^+ ($m/e = 46$) peaks appear in the mass spectrum. The principal fragmentation pattern, however, involves loss of all or part of the nitro group. Using nitrobenzene (Fig. 8.46) as an example, this fragmentation pattern may be described as follows.

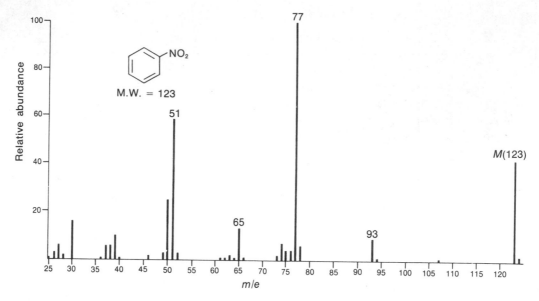

FIGURE 8.46 Mass spectrum of nitrobenzene.

Thiols and Thioethers

Thiols show molecular ion peaks which are more intense than those of the corresponding alcohols. A characteristic feature of the mass spectra of sulfur compounds is the presence of a significant $M + 2$ peak. This peak arises from the presence of the heavy isotope, ^{34}S, which has a natural abundance of 4.4%.

The fragmentation patterns of the thiols are very similar to those of the alcohols. As alcohols tend to undergo dehydration under some conditions, thiols tend to lose the elements of hydrogen sulfide, giving rise to an $M - 34$ peak.

Thioethers show mass spectral patterns which are very similar to those of the ethers. As in the case of the thiols, thioethers show molecular ion peaks that tend to be more intense than those of the corresponding ethers.

N. Alkyl Chlorides and Alkyl Bromides

The most dramatic feature of the mass spectra of alkyl chlorides and alkyl bromides is the presence of an important *M* + 2 peak. This peak arises because both chlorine and bromide are present in nature in two isotopic forms, each with a significant natural abundance.

For aliphatic halogen compounds, the molecular ion peak is strongest with alkyl iodides, less strong for alkyl bromides, weaker for alkyl chlorides, and weakest for alkyl fluorides. Furthermore, as the alkyl group increases in size or as the amount of branching at the [α] position increases, the intensity of the molecular ion peak decreases.

SPECTRAL ANALYSIS BOX — Alkyl Halides

MOLECULAR ION	FRAGMENT IONS
Strong *M* + 2 peak	Loss of Cl or Br
(for Cl, $M/M + 2 = 3:1$;	Loss of HCl
for Br, $M/M + 2 = 1:1$)	α-Cleavage

There are several important fragmentation mechanisms for the alkyl halides. Perhaps the most important is the simple loss of the halogen atom, leaving a carbocation. This fragmentation is most important where the halogen is a good leaving group. Therefore, this type of fragmentation is most prominent in the mass spectra of the alkyl iodides and the alkyl bromides. In the mass spectrum of 1-bromohexane (Fig. 8.47), the peak at $m/e = 85$ is due to the formation of the hexyl ion. This ion undergoes further fragmentation to form a $C_3H_7^+$ ion at $m/e = 43$. The corresponding heptyl ion peak at $m/e = 99$ in the mass spectrum of 2-chloroheptane (Fig. 8.48) is quite weak.

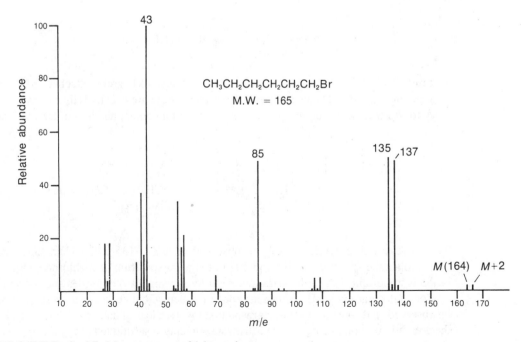

FIGURE 8.47 Mass spectrum of 1-bromohexane.

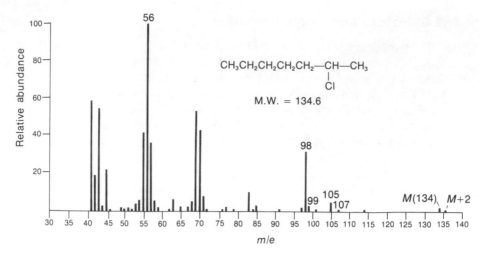

FIGURE 8.48 Mass spectrum of 2-chloroheptane.

Alkyl halides may also lose a molecule of hydrogen halide according to the process

$$[R—CH_2—CH_2—X]^{\ddot{+}} \longrightarrow [R—CH=CH_2]^{\ddot{+}} + HX$$

This mode of fragmentation is most important for alkyl fluorides and chlorides and is less important for alkyl bromides and iodides. In the mass spectrum of 1-bromohexane, the peak corresponding to the loss of hydrogen bromide at $m/e = 84$ is very weak. However, for 2-chloroheptane, the peak corresponding to the loss of hydrogen chloride at $m/e = 98$ is quite intense.

A less important mode of fragmentation is α-cleavage, for which a fragmentation mechanism might be

$$\left[R\dot{+}CH_2—X\right]^{\ddot{+}} \longrightarrow R\cdot + CH_2=X^+$$

In cases where the α position is branched, the heaviest alkyl group attached to the α carbon is lost with greatest facility. The peaks arising from α-cleavage are usually rather weak.

A fourth fragmentation mechanism involves rearrangement and loss of an alkyl radical:

$$\left[\begin{array}{cc} R—CH_2 & X \\ CH_2 & CH_2 \\ & CH_2 \end{array}\right]^{\ddot{+}} \longrightarrow \begin{array}{cc} CH_2——X^+ \\ CH_2 & CH_2 \\ CH_2 \end{array} + R\cdot$$

The corresponding cyclic ion can be observed at $m/e = 135$ and 137 in the mass spectrum of 1-bromohexane and at $m/e = 105$ and 107 in the mass spectrum of 2-chloroheptane. Such fragmentation is important only in the mass spectra of long-chain alkyl chlorides and bromides.

The molecular ion peaks in the mass spectra of benzyl halides are usually of sufficient intensity to be observed. The most important fragmentation involves loss of halogen to form the $C_7H_7^+$ ion. When the aromatic ring of a benzyl halide carries substituents, a substituted phenyl cation may also appear.

The molecular ion peak of an aromatic halide is usually quite intense. The most important mode of fragmentation involves loss of halogen to form the $C_6H_5^+$ ion.

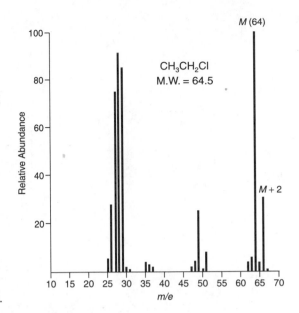

FIGURE 8.49 Mass spectrum of ethyl chloride.

Although the fragmentation patterns we have described are well characterized, the most interesting feature of the mass spectra of chlorine- and bromine-containing compounds is the presence of two molecular ion peaks. As Section 8.4 indicated, chlorine occurs naturally in two isotopic forms. The natural abundance of chlorine of mass 37 is 32.5% that of chlorine of mass 35. The natural abundance of bromine of mass 81 is 98.0% that of ^{79}Br. Therefore, the intensity of the $M + 2$ peak in a chlorine-containing compound should be 32.5% of the intensity of the molecular ion peak, and the intensity of the $M + 2$ peak in a bromine-containing compound should be almost equal to the intensity of the molecular ion peak. These pairs of molecular ion peaks (sometimes called doublets) appear in the mass spectra of ethyl chloride (Fig. 8.49) and ethyl bromide (Fig. 8.50).

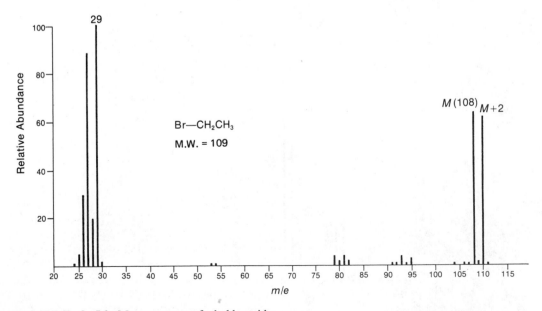

FIGURE 8.50 Mass spectrum of ethyl bromide.

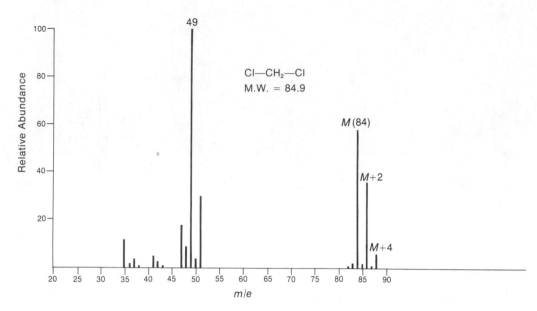

FIGURE 8.51 Mass spectrum of dichloromethane.

Table 8.6 in Section 8.5 can be used to determine what the ratio of the intensities of the molecular ion and isotopic peaks should be when more than one chlorine or bromine is present in the same molecule. The mass spectra of dichloromethane (Fig. 8.51), dibromomethane (Fig. 8.52), and 1-bromo-2-chloroethane (Fig. 8.53) are included here to illustrate some of the combinations of halogens listed in Table 8.6.

Unfortunately, it is not always possible to take advantage of these characteristic patterns to identify halogen compounds. Frequently the molecular ion peaks are too weak to permit accurate measurement of the ratio of the intensities of the molecular ion and isotopic peaks. However, it is often possible to make such a comparison on certain fragment ion peaks in the mass spectrum of a halogen compound. The mass spectrum of 1-bromohexane (Fig. 8.47) may be used to illustrate this method. The presence of bromine can be determined using the fragment ion peaks at m/e values of 135 and 137.

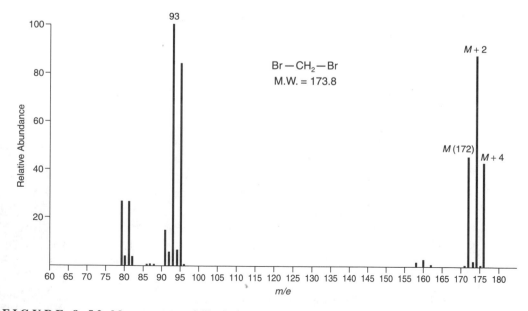

FIGURE 8.52 Mass spectrum of dibromomethane.

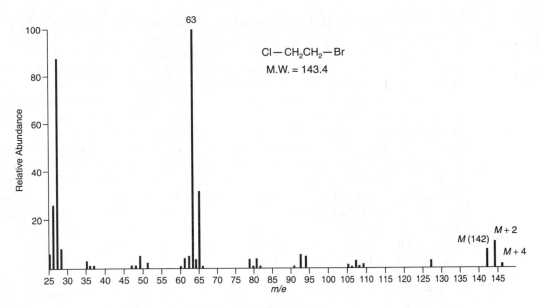

FIGURE 8.53 Mass spectrum of 1-bromo-2-chloroethane.

Since iodine and fluorine exist in nature in the form of only one isotope each, their mass spectra do not show isotopic peaks. The presence of halogen must be deduced either by noting the unusually weak $M + 1$ peak or by observing the mass difference between the fragment ions and the molecular ion.

8.7 ADDITIONAL TOPICS

As you may have observed from the discussions in Section 8.6, the molecular ions of many classes of compounds are so unstable that they decompose before they can reach the detector of the mass spectrometer. As a result, the molecular ion peaks for these classes of compounds are either very weak or totally absent. In such cases, two specialized techniques are useful for studying the molecular ion: field ionization and chemical ionization.

Field ionization involves passing the sample molecules very close to a thin wire that carries a high electrical potential. The strong electric field in the vicinity of this wire ionizes the molecules of sample. Molecular ions formed in this manner do not possess the high degree of vibrational energy found in molecular ions formed by electron impact. Consequently, the ions formed in field ionization are considerably more stable. The molecular ions formed in this manner are much more abundant than those formed by other means of ionization, and the molecular ion peak is usually fairly intense. In some cases, a molecular ion peak may be observed easily through the use of field ionization, whereas no peak would be observed after electron bombardment.

In **chemical ionization,** the sample is introduced into the ionization chamber along with 1 or 2 mm Hg of some reagent gas, usually methane. Essentially all of the electrons ionize methane molecules rather than sample molecules. Once the methane molecules are ionized, a series of ion–molecule collisions yields, among other species, the ions CH_5^+, $C_2H_5^+$, and $C_3H_5^+$. These ions act as strong Lewis acids and can react with sample molecules to produce their corresponding conjugate acids. A sample chemical equation illustrates this behavior:

$$CH_5^+ + R{-}H \longrightarrow RH_2^+ + CH_4$$

These protonated molecules, being cations, are accelerated in the usual way, giving rise to peaks with masses one unit higher than those of the expected molecular ions. Chemical ionization mass spectra show peaks one mass unit higher than those expected in electron impact spectra. Occasionally, it is possible for peaks to appear one mass unit *lower* than expected, due to the loss of hydrogen molecules from the protonated molecular ions. Nevertheless, chemical ionization methods are useful in the study of molecular ions, since the stability of the $M + 1$ peak is usually greater than that of the molecular ion. The $M + 1$ peak can be observed in cases in which the molecular ion peak is either very weak or totally absent.

A special type of mass spectrometer, called a **time-of-flight mass spectrometer,** may be used for certain applications. The time-of-flight instrument does not use a magnetic field to separate ions of varying masses. In this method, ions are produced from sample molecules by means of electron bombardment, as in most mass spectrometric methods. However, rather than using a continuous beam of high-energy electrons, the time-of-flight mass spectrometer passes the electrons through a control grid that has an oscillating positive potential applied across it. The positive potential oscillates at about 10,000 Hz, producing pulses of electrons that last about 0.25 μsec. The ions that are produced by these pulses of electrons are accelerated by an electric field that is pulsed at the same frequency as the control grid, but lagging behind the ionization pulse. In this acceleration, all ions obtain the same kinetic energy, and they reach velocities that depend upon their individual mass-to-charge ratios.

Rather than being focused by a magnetic field, the ions travel down a field-free, evacuated drift tube about 1 to 3 meters long. Because the ions travel at different velocities, they arrive at the detector at the end of the drift tube at different times. The output of the detector is amplified and recorded on an oscilloscope (although a chart recorder may also be used).

Because all of the ions that were created by a single electron pulse are collected and recorded on the oscilloscope, it is possible to produce a mass spectrum far more quickly than by conventional methods. This makes the time-of-flight technique particularly useful in the study of short-lived phenomena. The spectrum produced on the oscilloscope is a plot of the number of ions reaching the detector versus mass-to-charge ratio, as usual. However, in this case the mass-to-charge ratio is determined by an accurate electronic measurement of each ion's flight time from the ion source to the detector.

Time-of-flight instruments are incapable of the high resolution that conventional instruments can provide. Resolution of about 200 is possible, with a mass range of 1 to about 5000. Ions with m/e values greater than 5000 travel so slowly down the drift tube that ions of $m/e = 1$ from the next burst catch up to them before they can reach the detector.

Time-of-flight mass spectrometers are relatively simple, which makes it possible to use them in the field. Because of their value for studying short-lived species, they are particularly useful in kinetic studies, especially with applications to very fast reactions. Very rapid reactions such as combustion and explosions can be investigated with this technique.

Finally, consider an interesting nonlaboratory application of mass spectrometry. During the Gulf War of 1991, concern arose that Iraqi troops might be releasing chemical warfare agents against American troops. To guard against that possibility, the U.S. Army deployed a number of tracked vehicles, each equipped with a mass spectrometer. The mass spectrometer was used to sample the air and provide advance warning should any poisonous gases be released into the air.

8.8 COMPUTERIZED MATCHING OF SPECTRA WITH SPECTRAL LIBRARIES

A modern mass spectrometer is generally equipped with an on-line computer that can be used to collect, process, and display the mass spectral data. In addition, the computer often has one or more spectral libraries included in its memory. In some cases the instrument is equipped with libraries

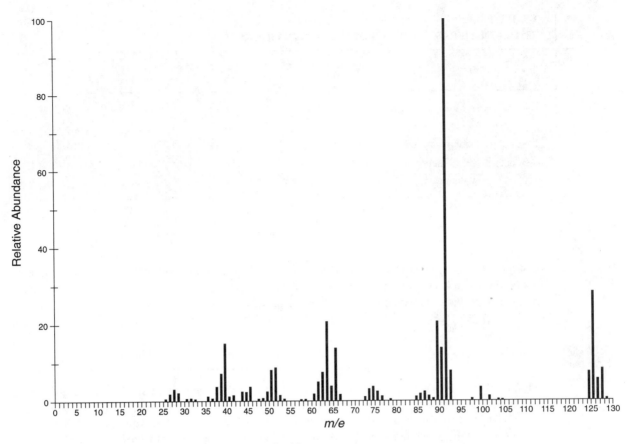

FIGURE 8.54 Mass spectrum of an unknown liquid.

when it is purchased; otherwise it is possible to purchase additional libraries. The computer can compare a mass spectrum it has determined with the spectra in these databases. The output of such a library search is a table that lists the names of possible compounds, their molecular formulas, and an indicator of the probability that the spectrum of the test compound matches the spectrum in the database. The probability is determined by the number of peaks (and their intensities) that can be matched. This type of table is often called a "hit list."

Figure 8.54 is the mass spectrum of an unknown liquid substance with an observed boiling point of 158 to 159°C. Table 8.7 reproduces the type of information that the computer would display as a hit list. Notice that the information includes the name of each compound that the computer has used for matching, its molecular weight and molecular formula, and its Chemical Abstracts Service (CAS) Registry number.

The information in Table 8.7 confirms that the unknown liquid is most likely **1-chloro-2-methylbenzene,** since the probability of a correct match is placed at 94%. It is interesting to note

TABLE 8.7
RESULT OF LIBRARY SEARCH FOR UNKNOWN LIQUID

Name	Molecular Weight	Formula	Probability	CAS No.
1. Benzene, 1-chloro-2-methyl-	126	C_7H_7Cl	94	000095-49-8
2. Benzene, 1-chloro-3-methyl-	126	C_7H_7Cl	70	000108-41-8
3. Benzene, 1-chloro-4-methyl-	126	C_7H_7Cl	60	000106-43-4
4. Benzene, (chloromethyl)-	126	C_7H_7Cl	47	000100-44-7
5. 1,3,5-Cycloheptatriene, 1-chloro-	126	C_7H_7Cl	23	032743-66-1

that the *meta* and *para* isomers show probabilities of 70% and 60%, respectively. Based on this illustration, we may conclude that computer matching of mass spectra is a very sensitive and accurate method of proving the structure of an unknown substance.

PROBLEMS

*1. A low-resolution mass spectrum of the alkaloid vobtusine showed the molecular weight to be 718. This molecular weight is correct for the molecular formulas $C_{43}H_{50}N_4O_6$ and $C_{42}H_{46}N_4O_7$. A high-resolution mass spectrum provided a molecular weight of 718.3743. Which of the possible molecular formulas is the correct one for vobtusine?

*2. A tetramethyltriacetyl derivative of oregonin, a diarylheptanoid xyloside found in red alder, was found by low-resolution mass spectrometry to have a molecular weight of 660. Possible molecular formulas include $C_{32}H_{36}O_{15}$, $C_{33}H_{40}O_{14}$, $C_{34}H_{44}O_{13}$, $C_{35}H_{48}O_{12}$, $C_{32}H_{52}O_{14}$, and $C_{33}H_{56}O_{13}$. High-resolution mass spectrometry indicated that the precise molecular weight was 660.278. What is the correct molecular formula for this derivative of oregonin?

*3. An unknown substance shows a molecular ion peak at $m/e = 170$ with a relative intensity of 100. The $M + 1$ peak has an intensity of 13.2, and the $M + 2$ peak has an intensity of 1.00. What is the molecular formula of the unknown?

*4. An unknown hydrocarbon has a molecular ion peak at $m/e = 84$, with a relative intensity of 31.3. The $M+1$ peak has a relative intensity of 2.06, and the $M + 2$ peak has a relative intensity of 0.08. What is the molecular formula for this substance?

*5. An unknown substance has a molecular ion peak at $m/e = 107$, with a relative intensity of 100. The relative intensity of the $M + 1$ peak is 8.00, and the relative intensity of the $M + 2$ peak is 0.30. What is the molecular formula for this unknown?

*6. The mass spectrum of an unknown liquid shows a molecular ion peak at $m/e = 78$, with a relative intensity of 23.6. The relative intensities of the isotopic peaks are as follows.

$m/e = 79$	Relative intensity $= 1.00$
80	7.55
81	0.25

What is the molecular formula of this unknown?

7. Assign a structure that would be expected to give rise to each of the following mass spectra. *Note*: Some of these problems may have more than one reasonable answer. In some cases, infrared spectral data have been included in order to make the solution to the problems more reasonable. We recommend that you review the index of hydrogen deficiency (Section 1.4) and the Rule of Thirteen (Section 1.5) and apply those methods to each of the following problems. To help you, we have provided an example problem with solution.

■ SOLVED EXAMPLE

An unknown compound has the mass spectrum shown. The infrared spectrum of the unknown shows significant peaks at

3102 cm^{-1}	3087	3062	3030	1688
1598	1583	1460	1449	1353
1221	952	746	691	

There is also a band from aliphatic C—H stretching from 2879 to 2979 cm^{-1}.

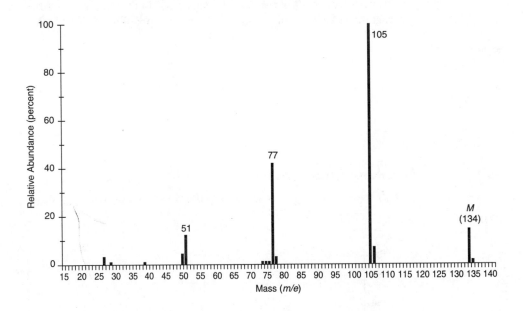

■ SOLUTION

1. The molecular ion appears at an *m/e* value of 134. Applying the Rule of Thirteen gives the following possible molecular formulas:

$$C_{10}H_{14} \qquad U = 4$$
$$C_9H_{10}O \qquad U = 5$$

2. The infrared spectrum shows a C=O peak at 1688 cm^{-1}. The position of this peak, along with the C—H stretching peaks in the 3030–3102 cm^{-1} range and C=C stretching peaks in the 1449–1598 cm^{-1} range, suggests a ketone where the carbonyl group is conjugated with a benzene ring. Such a structure would be consistent with the second molecular formula and with the index of hydrogen deficiency.

3. The base peak in the mass spectrum appears at $m/e = 105$. This peak is likely due to the formation of a benzoyl cation.

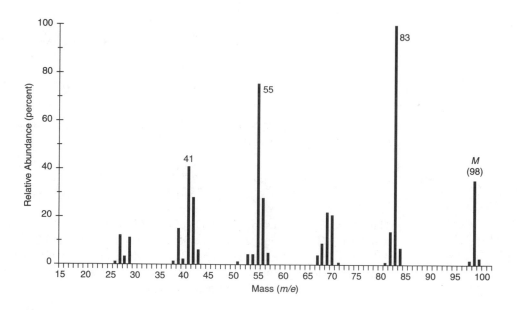

Subtracting the mass of the benzoyl ion from the mass of the molecular ion gives a difference of 29, suggesting that an ethyl group is attached to the carbonyl carbon. The peak appearing at $m/e = 77$ arises from the phenyl cation.

4. If we assemble all of the "pieces" suggested in the data, as described above, we conclude that the unknown compound must have been **propiophenone (1-phenyl-1-propanone).**

Problem 7 (continued)

*(a) The infrared spectrum has no interesting features except aliphatic C—H stretching and bending.

*(b) The infrared spectrum has a medium-intensity peak at about 1650 cm^{-1}. There is also a C—H out-of-plane bending peak near 880 cm^{-1}.

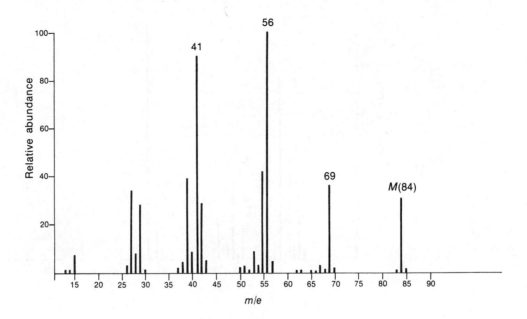

*(c) The infrared spectrum of this unknown has a prominent, broad peak at 3370 cm^{-1}. There is also a strong peak at 1159 cm^{-1}. The mass spectrum of this unknown does not show a molecular ion peak. You will have to deduce the molecular weight of this unknown from the heaviest fragment ion peak, which arises from the loss of a methyl group from the molecular ion.

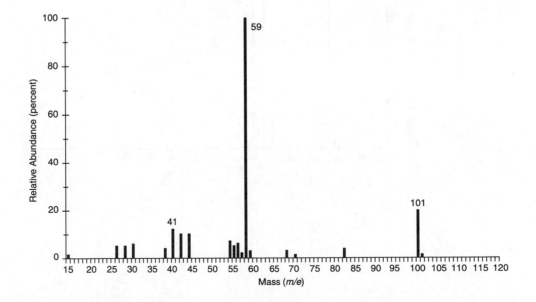

*(d) This unknown contains oxygen, but it does not show any significant infrared absorption peaks above 3000 cm^{-1}.

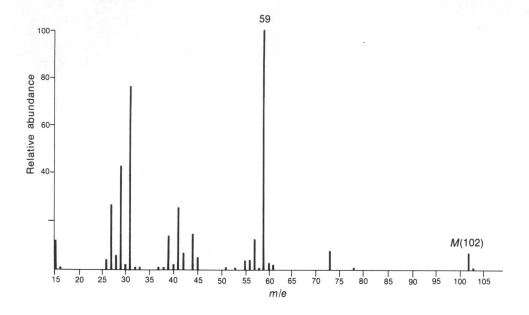

*(e) The infrared spectrum of this unknown shows a strong peak near 1725 cm^{-1}.

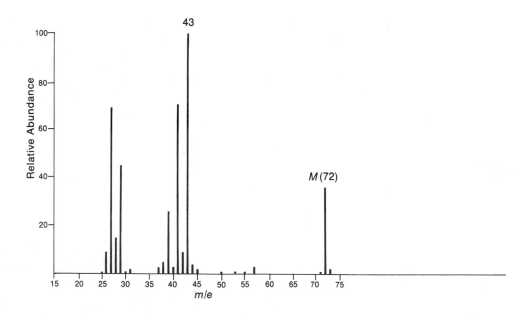

*(f) The infrared spectrum of this unknown shows a strong peak near 1715 cm⁻¹.

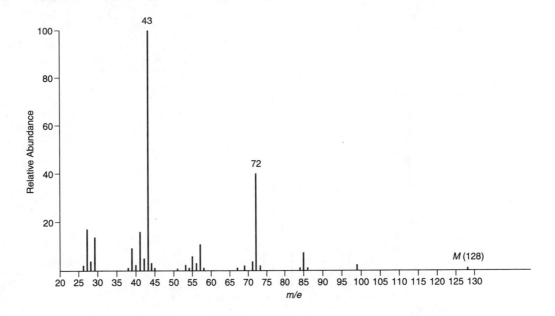

*(g) The infrared spectrum of this compound lacks any significant absorption above 3000 cm⁻¹.
There is a prominent peak near 1740 cm⁻¹ and another strong peak near 1200 cm⁻¹.

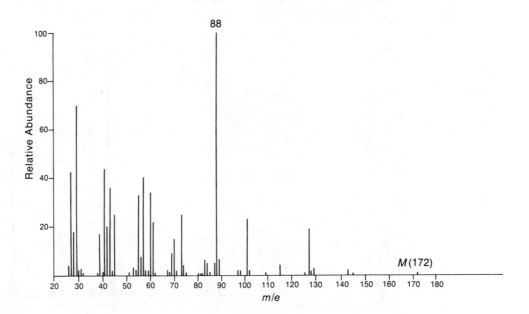

*(h) The infrared spectrum of this substance shows a very strong, broad peak in the range of 2500–3000 cm^{-1}, as well as a strong, somewhat broadened peak at about 1710 cm^{-1}.

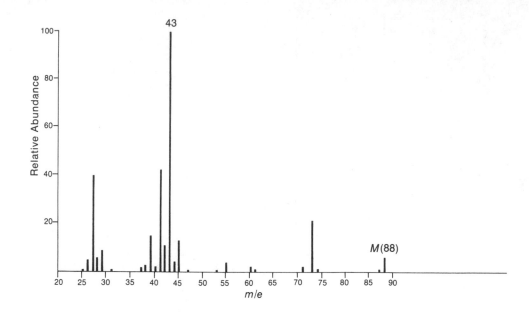

*(i) The ^{13}C NMR spectrum of this unknown shows only four peaks in the region 125–145 ppm. The infrared spectrum shows a very strong, broad peak extending from 2500 to 3500 cm^{-1}, as well a strong and somewhat broadened peak at 1680 cm^{-1}.

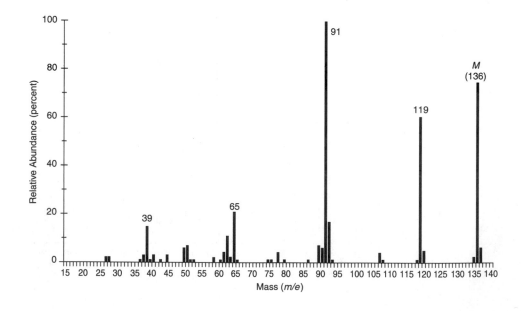

*(j) Note the odd value of mass for the molecular ion in this substance.

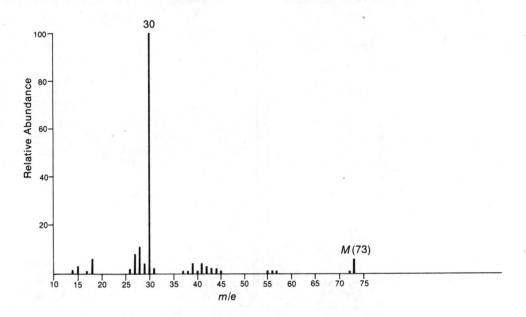

*(k) Notice the **M + 2** peak in the mass spectrum.

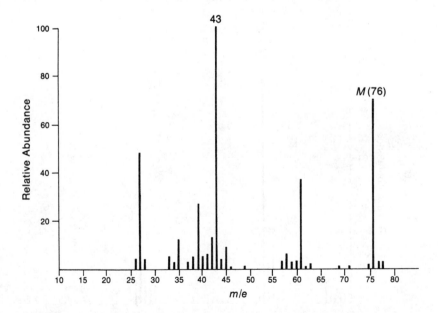

*(l)   The infrared spectrum of this unknown shows two strong peaks, one near 1350 cm^{-1} and the other near 1550 cm^{-1}. Notice that the mass of the molecular ion is *odd*.

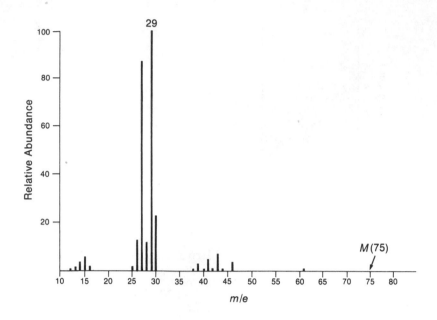

*(m)   There is a sharp peak of medium intensity near 2250 cm^{-1} in the infrared spectrum of this compound.

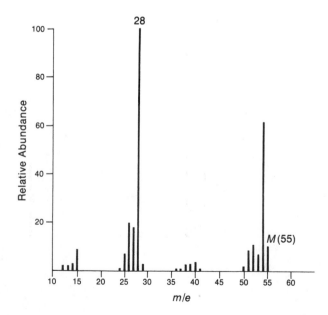

*(n) Consider the fragment ions at *m/e* = 127 and 128. From what ions might these peaks arise?

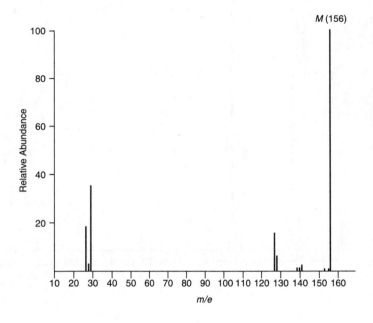

*(o)

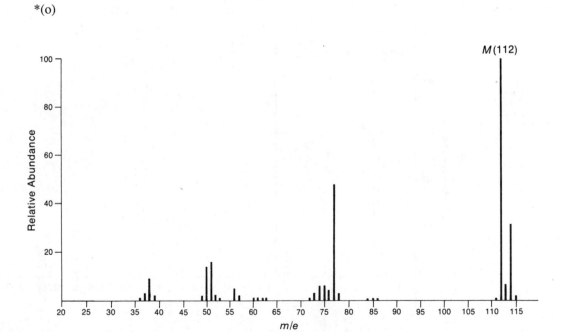

*(p)

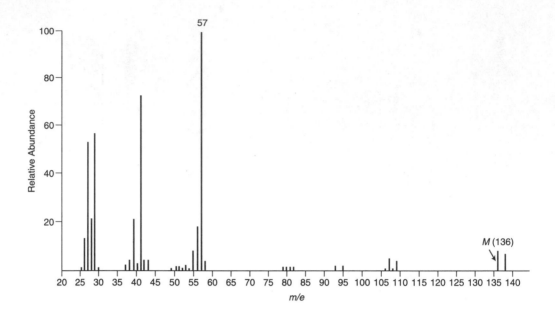

*(q)

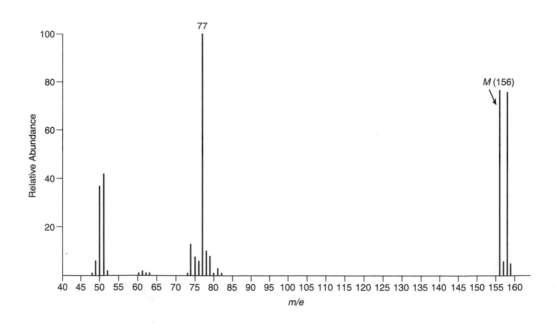

*(r)

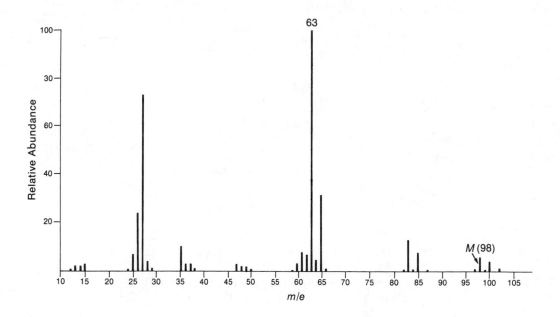

*(s) The infrared spectrum of this unknown shows a sharp peak at 3087 cm^{-1}, and a sharp peak at 1612 cm^{-1}, in addition to other absorptions. The unknown contains chlorine atoms, but some of the isotopic peaks ($M + n$) are too weak to be seen.

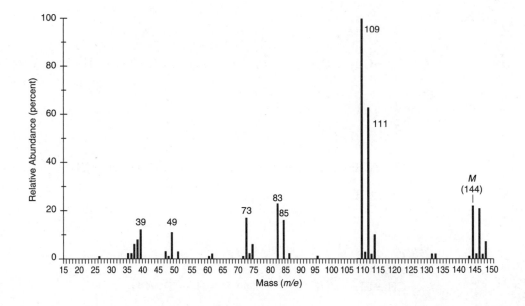

8. The mass spectrum of 3-butyn-2-ol shows a large peak at $m/e = 55$. Draw the structure of the fragment and explain why it is particularly stable.

9. How could the following pairs of isomeric compounds be differentiated by mass spectrometry?

(a)

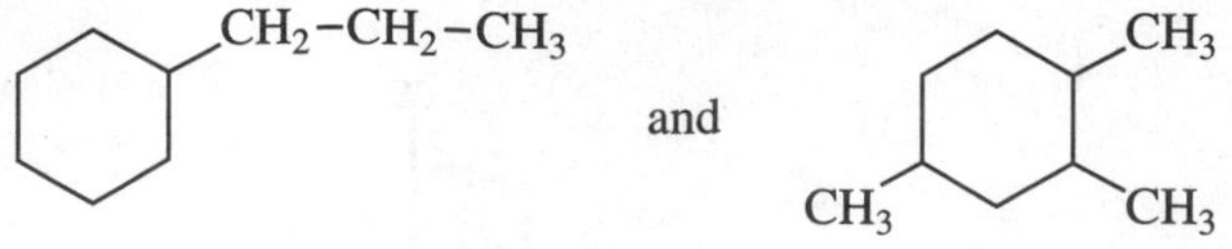

and

(b)

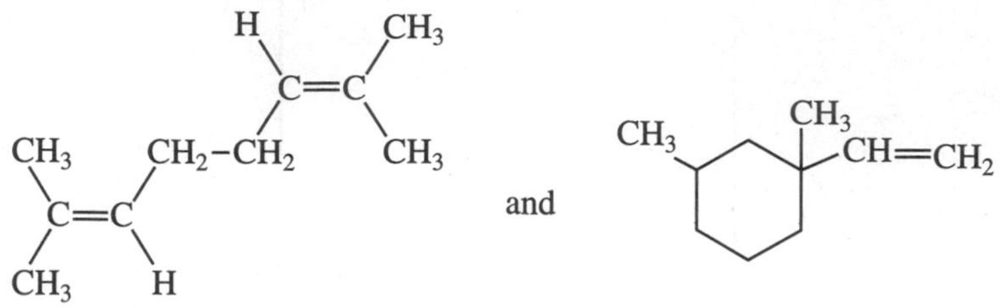

and

(c)

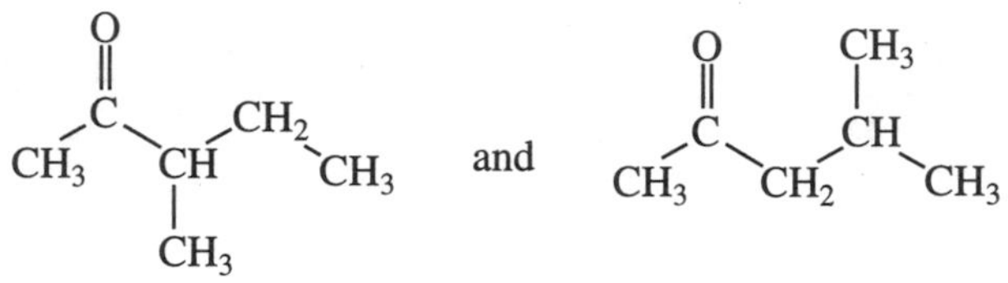

and

(d)

$CH_3-CH_2-CH_2-CH-CH_3$
$\quad\quad\quad\quad\quad\quad\quad |$
$\quad\quad\quad\quad\quad\quad\quad OH$

and

$$CH_3-\overset{\overset{\displaystyle CH_3}{|}}{\underset{\underset{\displaystyle OH}{|}}{C}}-CH_2-CH_3$$

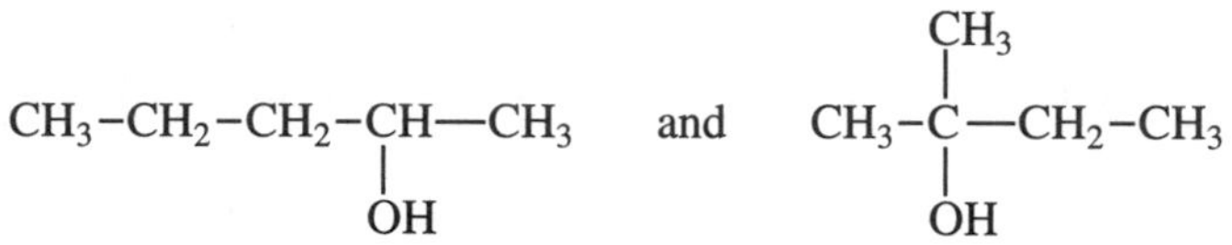

(e) $CH_3-CH_2-CH_2-CH_2-CH_2-Br$ and $CH_3-CH_2-\underset{\underset{\displaystyle Br}{|}}{CH}-CH_2-CH_3$

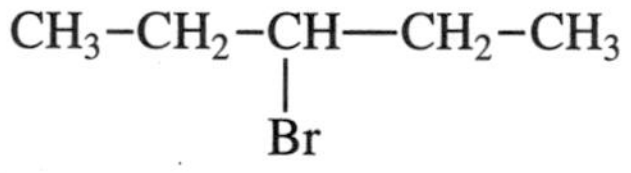

(f)

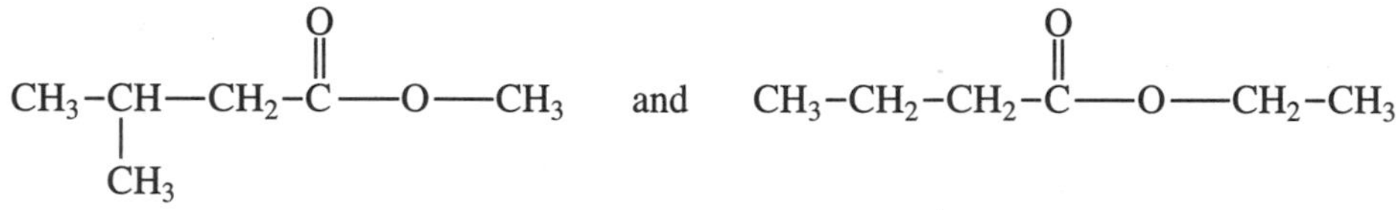

and

(g)

and

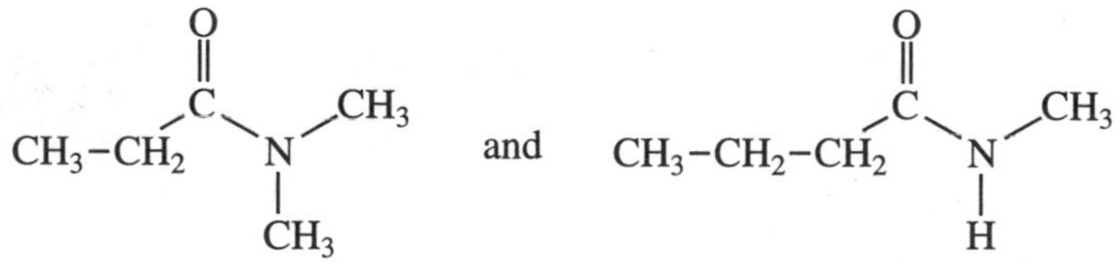

(h) $CH_3-CH_2-CH_2-CH_2-NH_2$ and $CH_3-CH_2-CH_2-NH-CH_3$

(i)

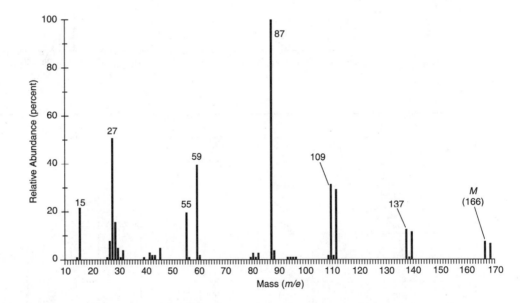

(j)

10. Use the mass spectrum and the additional spectral data provided to deduce the structure of each of the following compounds.

(a) $C_4H_7BrO_2$

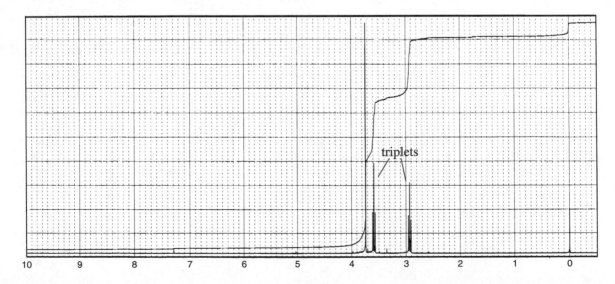

(b) $C_4H_7Cl_2O_2$

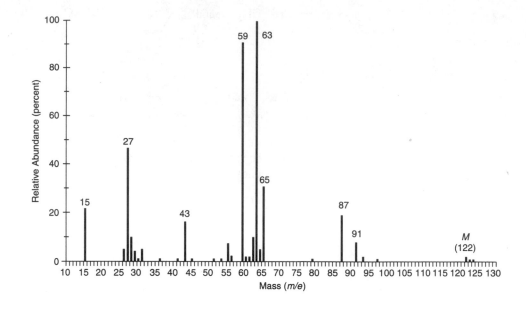

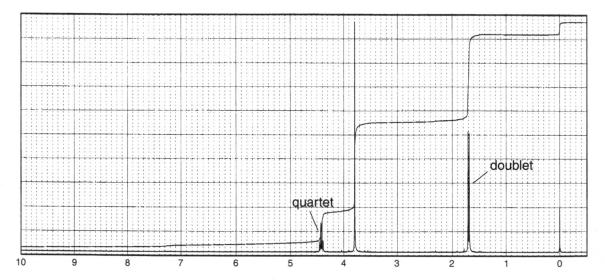

(c) C$_8$H$_6$O$_3$

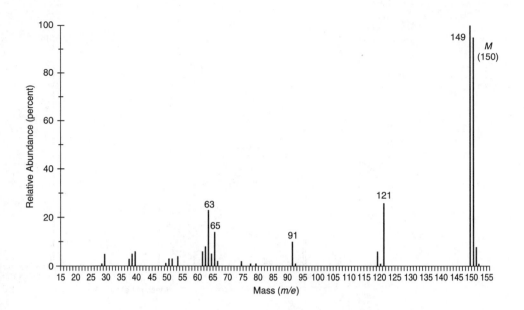

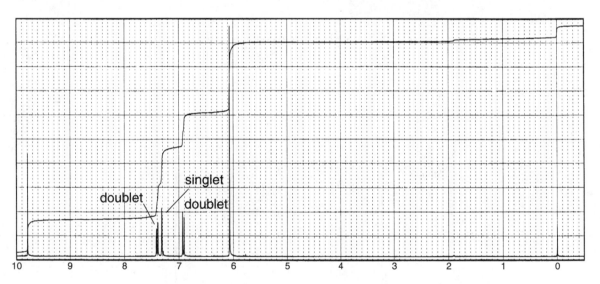

(d) The infrared spectrum lacks any significant peaks above 3000 cm^{-1}.

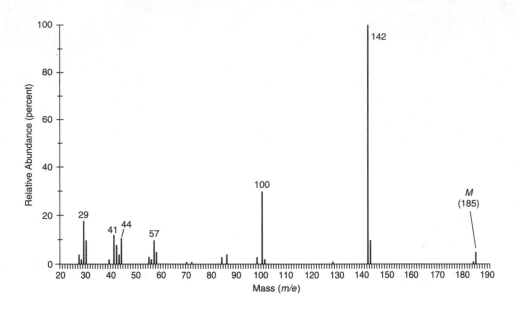

(e) The infrared spectrum contains a single, strong peak at 3280 cm^{-1}.

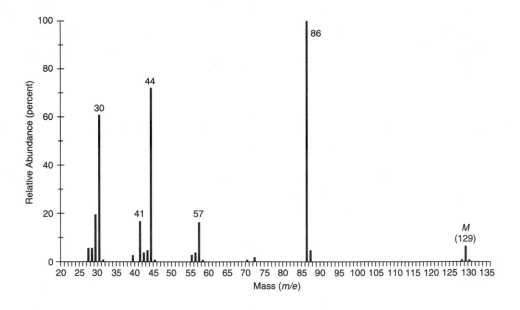

REFERENCES

Beynon, J. H., *Mass Spectrometry and Its Applications to Organic Chemistry*, Elsevier, Amsterdam, 1960.

Beynon, J. H., and A. G. Brenton, *Introduction to Mass Spectrometry*, University of Wales Press, Swansea, 1982.

Biemann, K., *Mass Spectrometry: Organic Chemical Applications*, McGraw–Hill, New York, 1962.

Budzikiewicz, H., C. Djerassi, and D. H. Williams, *Mass Spectrometry of Organic Compounds*, Holden–Day, San Francisco, 1967.

Chapman, J. R., *Computers in Mass Spectrometry*, Academic Press, New York, 1978.

Chapman, J. R., *Practical Organic Mass Spectrometry*, John Wiley, New York, 1985.

Constantin, E., A. Schnell, and M. Thompson, *Mass Spectrometry*, Prentice–Hall, Englewood Cliffs, N.J., 1990.

Dawson, P. H., *Quadrupole Mass Spectrometry*, Elsevier, New York, 1976.

Duckworth, H. E., R. C. Barber, and V. S. Venkatasubramanian, *Mass Spectroscopy,* 2nd ed., Cambridge University Press, Cambridge, England, 1986.

McFadden, W. H., *Techniques of Combined Gas Chromatography/Mass Spectrometry: Applications in Organic Analysis,* Wiley-Interscience, New York, 1989.

McLafferty, F. W., and F. Tureček, *Interpretation of Mass Spectra,* 4th ed., University Science Books, Mill Valley, Calif., 1993.

Pretsch, E., T. Cleve, J. Seibl, and W. Simon, *Tables of Spectral Data for Structure Determination of Organic Compounds,* 2nd ed., Springer-Verlag, Berlin and New York, 1989. Translated from the German by K. Biemann.

Silverstein, R. M., and F. X. Webster, *Spectrometric Identification of Organic Compounds,* 6th ed., John Wiley, New York, 1998.

Selected Web sites

National Institute of Materials and Chemical Research, Tsukuba, Ibaraki, Japan, *Integrated Spectra Data Base System for Organic Compounds (SDBS)* : http://www.aist.go.jp/RIODG/SDBS/menu-e.html

National Institute of Standards and Technology, *NIST Chemistry WebBook*: http://webbook.nist.gov/chemistry/

CHAPTER 9

COMBINED STRUCTURE PROBLEMS

In this chapter you will employ jointly all of the spectroscopic methods we have discussed so far to solve structural problems in organic chemistry. Thirty-four problems are provided to give you practice in applying the principles learned in earlier chapters. The problems involve analysis of the mass spectrum (MS), the infrared (IR) spectrum, and proton and carbon (^{1}H and ^{13}C) NMR. Ultraviolet (UV) spectral data, if provided in the problem, appear in a tabular form rather than as a spectrum. You will notice as you proceed through this chapter that the problems use different "mixes" of spectral information. Thus, you may be provided with a mass spectrum, an infrared spectrum, and a proton NMR spectrum in one problem, and in another you may have available the infrared spectrum and both proton and carbon NMR.

All ^{1}H (proton) NMR spectra were determined at 300 MHz while the ^{13}C (carbon NMR spectra were obtained at 75 MHz. The ^{1}H and ^{13}C were determined in CDCl$_3$ unless otherwise indicated. In some cases, the ^{13}C spectral data have been tabulated, along with the DEPT-135 and DEPT-90 data. Some of the proton NMR spectra have been expanded to show detail. Finally, all infrared spectra on *liquid* samples were obtained neat (with no solvent) on KBr salt plates. The infrared spectra of *solids* have either been melted (cast) onto the salt plate or else determined as a mull (suspension) in Nujol (mineral oil).

The compounds in these problems may contain the following elements: C, H, O, N, S, Cl, Br, and I. In most cases where halogens are present, the mass spectrum should provide you with information as to which halogen atom is present and the number of halogen atoms (p. 403).

There are a number of possible approaches that you may take in solving the problems in this chapter. There are no "right" ways of solving them. In general, however, you should first try to gain an overall impression by looking at the gross features of the spectra provided in the problem. As you do so, you will observe evidence for pieces of the structure. Once you have identified pieces, you can assemble them and test against each of the spectra the validity of the structure you have assembled.

1. **Mass Spectrum.** You should be able to use the mass spectrum to obtain a molecular formula by performing the Rule of Thirteen calculation (p. 7) on the molecular ion peak (*M*) labeled on the spectrum. In most cases, you will need to convert the hydrocarbon formula to one containing a functional group. For example, you may "see" a carbonyl group in the infrared spectrum or ^{13}C spectrum. Make appropriate adjustments to the hydrocarbon formula so that it fits the spectroscopic evidence. When the mass spectrum is not provided in the problem, you will be given the molecular formula. Some of the labeled fragment peaks may provide excellent evidence for the presence of a particular feature in the compound being analyzed.

2. **Infrared Spectrum.** The infrared spectrum provides some idea of the functional group or groups that are present or absent. Look first at the left-hand side of the spectrum to identify functional groups such as O−H, N−H, C≡N, C≡C, C=C, C=O, NO$_2$, and aromatic rings. See Chapter 2, Sections 2.8 and 2.9 (pp. 26–29) for tips on what to look for in the spectrum. Ignore C−H stretching bands during this first "glance" at the spectrum as well as the right-hand side of the spectrum. Determine the type of C=O group you have and also check to see

if there is conjugation with a double bond or aromatic ring. Remember that you can often determine the substitution patterns on alkenes (pp. 39–40) and aromatic rings (pp. 43–45) by using the out-of-plane bending bands (oops). A complete analysis of the infrared spectrum is seldom necessary.

3. **Proton NMR Spectrum.** The proton (^{1}H) NMR spectrum gives information on the numbers and types of hydrogen atoms attached to the carbon skeleton. Chapter 3, Section 3.19 (pp. 138–153) provides information on proton NMR spectra of various functional groups, especially expected chemical shifts values. You will need to determine the integral ratios for the protons by using the integral traces shown. See Chapter 3, Section 3.9 (pp. 118–120) to see how you can obtain the numbers of protons attached to the carbon chain. In most cases, it is not easy to see the splitting patterns of multiplets in the full 300 MHz spectrum. We have, therefore, indicated the multiplicities of peaks as doublet, triplet, quartet, quintet, and sextet on the full spectrum. Singlets are usually easy to see and they have not been labeled. Many problems have been provided with proton expansions. When expansions are provided, Hertz values have been shown so that you can calculate the coupling constants. Often the magnitude of the proton coupling constants will help you to assign structural features to the compound such as the relative position of hydrogen atoms in alkenes (*cis*/*trans* isomers).

4. **Carbon NMR Spectra.** The carbon (^{13}C) NMR spectrum indicates the total number of nonequivalent carbon atoms in the molecule. In some cases, because of symmetry, carbon atoms may have identical chemical shifts. In this latter case, the total number of carbons is less than that found in the molecular formula. Chapter 4 contains important correlation charts that you should review. Figure 4.1 (p. 168) and Table 4.1 (p. 169) show the chemical shift ranges that you should expect for various structural features. Expected ranges for carbonyl groups are shown Figure 4.2 (p. 170). In addition, you may find it useful to calculate approximate ^{13}C chemical shifts values as shown in Appendix 8. Commonly, sp^3 carbon atoms appear to the upfield (right) side of the CDCl$_3$ solvent peak, while the sp^2 carbon atoms in an alkene or in an aromatic ring appear to the left of the solvent peak. Carbon atoms in a C=O group appear furthermost to the left in a carbon spectrum. You should first look on the left-hand side of the carbon spectrum to see if you can identify potential carbonyl groups.

5. **DEPT-135 and DEPT-90 Spectra.** In some cases, the problems list information that can provide valuable information on the types of carbon atoms present in the unknown compound. Review Chapter 4, Section 4.10 (p. 182), A Quick Dip into DEPT) for information on how to determine the presence of CH$_3$, CH$_2$, CH, and C atoms in a carbon spectrum.

6. **Ultraviolet/Visible Spectrum.** The ultraviolet spectrum becomes useful when unsaturation is present in a molecule. See Chapter 7, Section 7.17 (p. 385) for information on how to interpret a UV spectrum.

7. **Determining a Final Structure.** A complete analysis of the information provided in the problems should lead to a unique structure for the unknown compound. Four solved examples are presented first. Note that more than one approach may be taken to the solution of these example problems. Since the problems near the beginning of this chapter are easier, you should attempt them before you move on. Have fun (no kidding)! You may even find that you have as much fun as the authors of this book.

■ **EXAMPLE 1**

Problem

The UV spectrum of this compound shows only end absorption. Determine the structure of the compound.

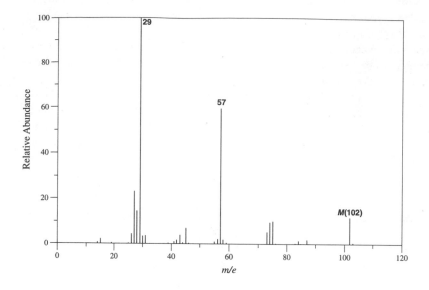

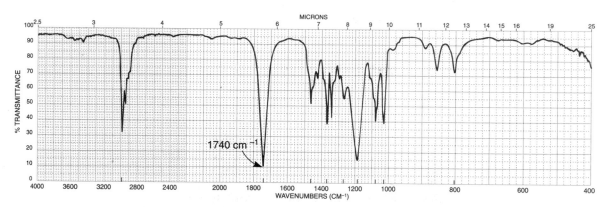

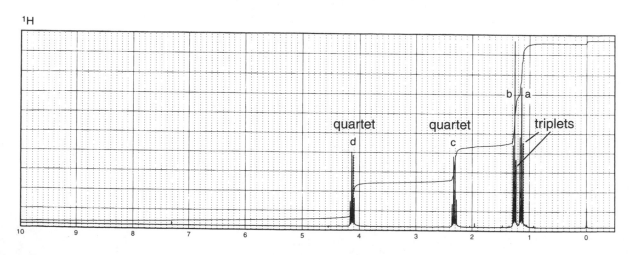

Example 1   **469**

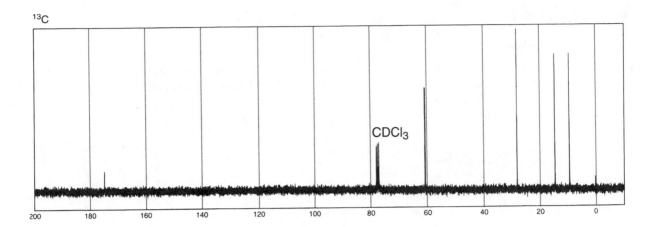

Solution

Notice that this problem does not provide a molecular formula. We need to obtain it from the spectral evidence. The molecular ion peak appears at $m/e = 102$. Using the Rule of Thirteen (p. 7), we can calculate a formula of C_7H_{18} for the peak at 102. The infrared spectrum shows a strong absorption at 1740 cm^{-1}, suggesting that a simple unconjugated ester is present in the compound. The presence of a C—O (strong and broad) at 1200 cm^{-1} confirms the ester. We now know that there are two oxygen atoms in the formula. Returning to the mass spectral evidence, the formula calculated via the Rule of Thirteen was C_7H_{18}. We can modify this formula by converting carbons and hydrogens (one carbon and four hydrogens per oxygen atom) to the two oxygen atoms, yielding the formula $C_5H_{10}O_2$. This is the molecular formula for the compound. We can now calculate the index of hydrogen deficiency for this compound, which equals one, and that corresponds to the unsaturation in the C=O group. The infrared spectrum also shows sp^3 (aliphatic) C—H absorption at less than 3000 cm^{-1}. We conclude that the compound is an aliphatic ester with formula $C_5H_{10}O_2$.

Notice that the ^{13}C NMR shows a total of five peaks, corresponding exactly to the number of carbons in the molecular formula! This is a nice check on our calculation of the formula via the Rule of Thirteen (five carbon atoms). The peak at 174 ppm corresponds to the ester C=O carbon. The peak at 60 ppm is a deshielded carbon atom caused by a neighboring single-bonded oxygen atom. The rest of the carbon atoms are relatively shielded. These three peaks correspond to the remaining part of the carbon chain in the ester.

We could probably derive a couple of possible structures at this point. The ^{1}H NMR spectrum should provide confirmation. Using the integral traces on the spectrum, we should conclude that the peaks shown have the ratio 2:2:3:3 (downfield to upfield). These numbers add up to the 10 total hydrogen atoms in the formula. Now, using the splitting patterns on the peaks, we can determine the structure of the compound. It is ethyl propanoate.

$$\underset{\text{a}}{CH_3}-\underset{\text{c}}{CH_2}-\overset{\displaystyle O}{\overset{\|}{C}}-O-\underset{\text{d}}{CH_2}-\underset{\text{b}}{CH_3}$$

The downfield quartet at 4.1 ppm (**d** protons) results from the neighboring protons on carbon **b**, while the other quartet at 2.4 ppm (**c** protons) results from the protons on carbon **a**. Thus, the proton NMR is consistent with the final structure.

The UV spectrum is uninteresting but supports the identification of structure. Simple esters have weak $n \rightarrow \pi^*$ transitions (205 nm) near the solvent cutoff point. Returning to the mass spectrum, the strong peak at 57 mass units results from an α cleavage of an alkoxy group to yield the acylium ion ($CH_3-CH_2-\overset{+}{C}=O$), which has a mass of 57.

■ EXAMPLE 2

Problem

Determine the structure of a compound with the formula $C_{10}H_{12}O_2$. In addition to the infrared spectrum and proton NMR, the problem includes tabulated data for the normal carbon NMR, DEPT-135, and DEPT-90 spectral data.

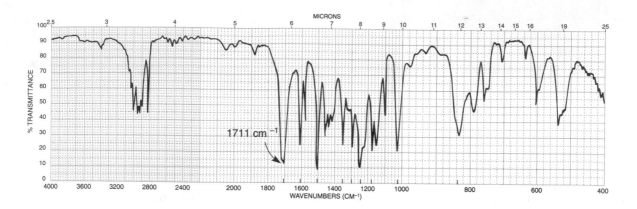

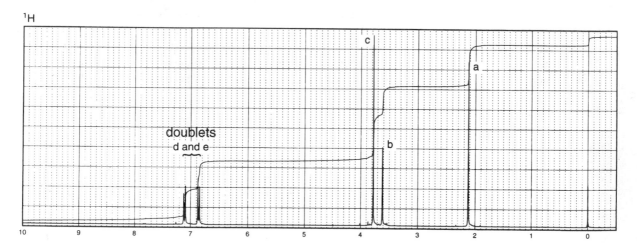

Normal Carbon	DEPT-135	DEPT-90
29 ppm	Positive	No peak
50	Negative	No peak
55	Positive	No peak
114	Positive	Positive
126	No peak	No peak
130	Positive	Positive
159	No peak	No peak
207	No peak	No peak

Example 2 **471**

Solution

We calculate an index of hydrogen deficiency (p. 4) of five. The proton and carbon NMR spectra, as well as the infrared spectrum, suggest an aromatic ring (index = four). The remaining index of one is attributed to a C=O group found in the infrared spectrum at 1711 cm^{-1}. This value for the C=O is close to what you might expect for an unconjugated carbonyl group in a ketone and is too low for an ester. The carbon NMR confirms the ketone C=O; the peak at 207 ppm is typical for a ketone. The carbon NMR spectrum shows only 8 peaks, while 10 are present in the molecular formula. This suggests some symmetry that makes some of the carbon atoms equivalent.

When inspecting the proton NMR spectrum, notice the nice *para* substitution pattern between 6.8 and 7.2 ppm, which appears as a nominal pair of doublets, integrating for two protons in each pair. Also notice in the proton NMR that the upfield portion of the spectrum has protons that integrate for 3:2:3 for a CH$_3$, a CH$_2$, and a CH$_3$, respectively. Also notice that these peaks are unsplit, indicating that there are no neighboring protons. The downfield methyl at 3.8 ppm is next to an oxygen atom, suggesting a methoxy group. The carbon DEPT NMR spectra results confirm the presence of two methyl groups and one methylene group. The methyl group at 55 ppm is deshielded by the presence of an oxygen atom (O−CH$_3$). Keeping in mind the *para*-disubstituted pattern and the singlet peaks in the proton NMR, we derive the following structure for 4-methoxyphenylacetone.

Further confirmation of the *para*-disubstituted ring is obtained from the carbon spectral results. Notice the presence of four peaks in the aromatic region of the carbon NMR spectrum. Two of these peaks (126 and 159 ppm) are *ipso* carbon atoms (no attached protons) that do not show in the DEPT-135 or DEPT-90 spectra. The remaining two peaks at 114 and 130 ppm are assigned to the remaining four carbons (two each equivalent by symmetry). The two carbon atoms **d** show peaks in both of the DEPT experiments, which confirms that they have attached protons (C−H). Likewise, the two carbon atoms **e** have peaks in both DEPT experiments confirming the presence of C−H. The infrared spectrum has a *para* substitution pattern in the out-of-plane region (835 cm^{-1}), which helps to confirm the 1,4-disubstitution on the aromatic ring.

■ EXAMPLE 3

Problem

This compound has the molecular formula $C_9H_{11}NO_2$. Included in this problem are the infrared spectrum, proton NMR with expansions, and carbon NMR spectra data.

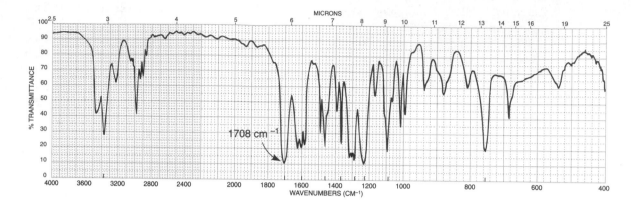

Normal Carbon	DEPT-135	DEPT-90
14 ppm	Positive	No peak
61	Negative	No peak
116	Positive	Positive
119	Positive	Positive
120	Positive	Positive
129	Positive	Positive
131	No peak	No peak
147	No peak	No peak
167	No peak	No peak

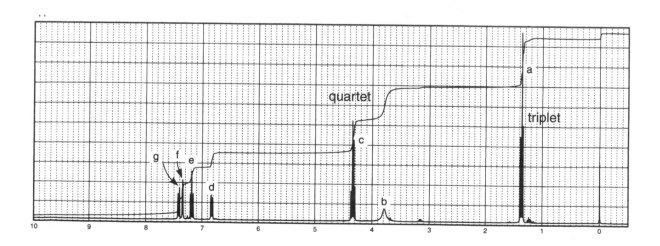

Example 3 473

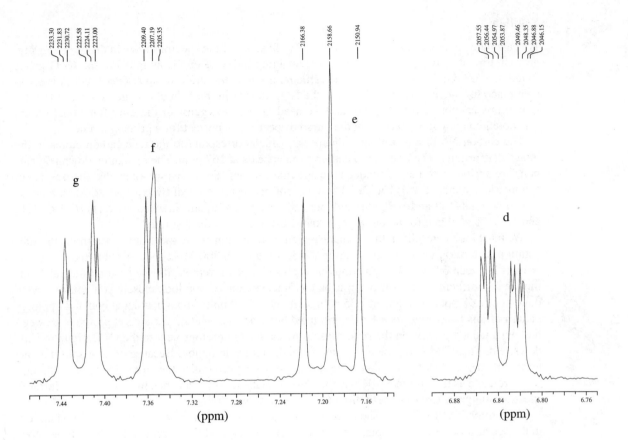

Solution

We calculate an index of hydrogen deficiency of five. All of the spectra shown in this problem suggest an aromatic ring (index = four). The remaining index of one is assigned to the C=O group found at 1708 cm^{-1}. This value for the carbonyl group is too high for an amide. It is in a reasonable place for a conjugated ester. While the NO_2 present in the formula suggests a possible nitro group, this cannot be the case, because we need the two oxygens for the ester functional group. The doublet at about 3400 cm^{-1} in the infrared spectrum is perfect for a primary amine.

The carbon NMR spectrum has nine peaks, which correspond to the nine carbon atoms in the molecular formula. The ester C=O carbon atom appears at 167 ppm. The remaining downfield carbons are attributed to the six unique aromatic ring carbons. From this we know that the ring is not symmetrically substituted. The DEPT results confirm the presence of two carbon atoms with no attached protons (131 and 147 ppm) and four carbon atoms with one attached proton (116, 199, 120, and 129). From this information we now know that the ring is disubstituted.

We must look carefully at the aromatic region in the proton NMR spectrum to determine the substitution pattern on the disubstituted ring. Notice in the full 300 MHz spectrum that there are four protons between 6.8 and 7.5 ppm, and that each set of peaks represents one proton (integrals). Although it is difficult to see the patterns on the full spectrum, if you look closely you will observe a doublet at 7.42 ppm, a singlet at 7.35 ppm, a triplet at 7.19 ppm, and a doublet at bout 6.84 ppm. A pattern of this type suggests a 1,3-disubstituted benzene ring (*meta*). The singlet proton is between the two attached groups on the ring. Anisotropy causes the protons next to the C=O group to shift downfield. The other two protons are upfield in the aromatic region. Because of their splitting patterns and chemical shifts, we should be able to assign the protons in the aromatic region. Although not as reliable as the proton NMR evidence, the aromatic out-of-plane bending bands in the infrared spectrum suggests *meta*-disubstitution: 680, 760, and 880 cm^{-1}.

The proton NMR spectrum shows an ethyl group because of the quartet and triplet found upfield in the spectrum (4.3 and 1.4 ppm, respectively, for the CH_2 and CH_3 groups). Finally, a broad NH_2 peak, integrating for two protons, appears in the proton NMR spectrum at 3.8 ppm. The compound is ethyl 3-aminobenzoate.

Example 3   475

We need to look at the proton expansions provided in the problem to confirm the assignments made for the aromatic protons. The Hertz values shown on the expansions allow us the opportunity to obtain coupling constants that confirms the 1,3-disubstitution pattern. The splittings observed in the expansions can be explained by looking at the coupling constants, ^3J, and ^4J, present in the compound. ^5J couplings are either zero or too small to be observed in the expansions.

7.42 ppm (H_g)	Doublet of triplets (dt) or doublet of doublets of doublets (ddd); $^3J_{eg} = 7.72$ Hz, $^4J_{fg}$ and $^4J_{dg} \approx 1.5$ Hz.
7.35 ppm (H_f)	This proton is located between the two attached groups. The only proton couplings that are observed are small ^4J couplings that result in a closely spaced triplet or more precisely, a doublet of doublets; $^4J_{fg}$ and $^4J_{df} \approx 1.5$ to 2 Hz.
7.19 ppm (H_e)	This proton appears as a widely spaced "triplet." One of the coupling constants, $^3J_{eg} = 7.72$ Hz, was obtained from the pattern at 7.42 ppm. The other coupling constant, $^3J_{de} = 8.09$ Hz, is obtained from the pattern at 6.84 ppm. The pattern appears as a triplet because the coupling constants are nearly equal resulting in an accidental overlap of the center peak in the "triplet." More precisely, we should describe this "triplet" as a doublet of doublets (dd).
6.84 ppm (H_d)	Doublet of doublets of doublets (ddd); $^3J_{de} = 8.09$ Hz, $^4J_{dg} \neq {}^4J_{df}$.

■ EXAMPLE 4

Problem

This compound has the molecular formula $C_5H_7NO_2$. Following are the infrared, proton NMR, and carbon NMR spectra.

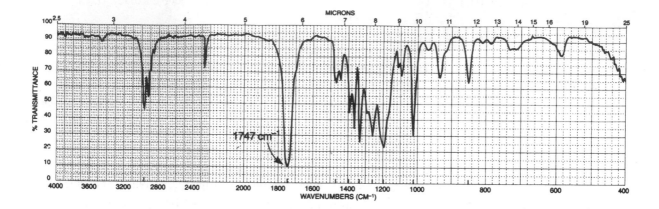

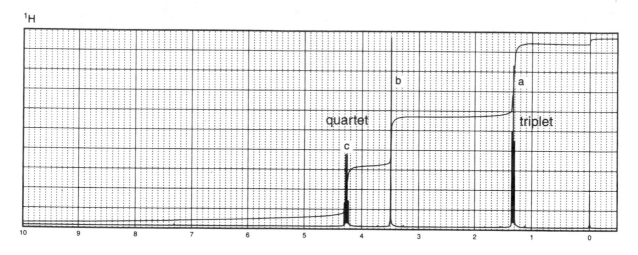

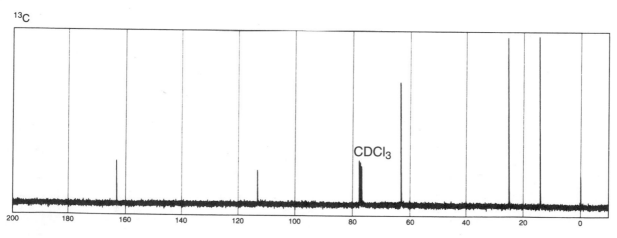

Example 4   **477**

Solution

We calculate an index of hydrogen deficiency of three. A quick glance at the infrared spectrum reveals the source of unsaturation implied by an index of three: a nitrile group at 2260 cm^{-1} (index = two) and a carbonyl group at 1747 cm^{-1} (index = one). The frequency of the carbonyl absorption indicates an unconjugated ester. The appearance of several strong C—O bands near 1200 cm^{-1} confirms the presence of an ester functional group. We can rule out a C≡C bond because they usually absorb at a lower value (2150 cm^{-1}) and have a weaker intensity than compounds that contain C≡N.

The carbon NMR spectrum shows five peaks and thus is consistent with the molecular formula, which contains five carbon atoms. Notice that the carbon atom in the C≡N group has a characteristic value of 113 ppm. In addition, the carbon atom in the ester C=O appears at 163 ppm. One of the remaining carbon atoms (63 ppm) probably lies next to an electronegative oxygen atom. The remaining two carbon atoms, which absorb at 25 and 14 ppm, are attributed to the remaining methylene and methyl carbons. The structure is

$$\underset{\underset{\textbf{b}}{}}{N\!\equiv\!C\!-\!CH_2}\!-\!\overset{\overset{\textstyle O}{\|}}{C}\!-\!O\!-\!\underset{\textbf{c}}{CH_2}\!-\!\underset{\textbf{a}}{CH_3}$$

The proton NMR spectrum shows a classical ethyl pattern: a quartet (2 H) at 4.3 ppm and a triplet (3 H) at 1.3 ppm. The quartet is strongly influenced by the electronegative oxygen atom, which shifts it downfield. There is also a two-proton singlet at 3.5 ppm.

PROBLEMS

*1. The UV spectrum of this compound is determined in 95% ethanol: λ_{max} 290 nm (log ε 1.3).

a)

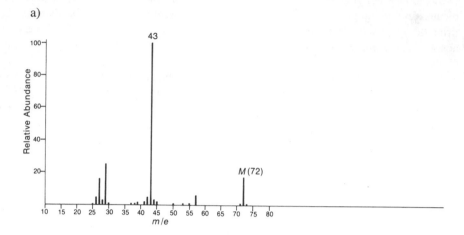

b)

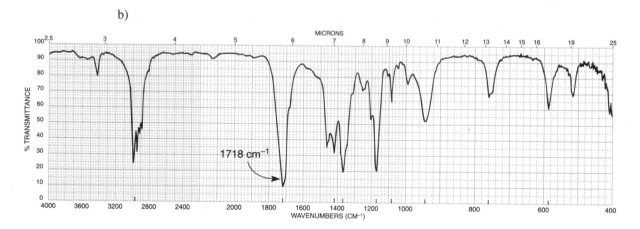

c)

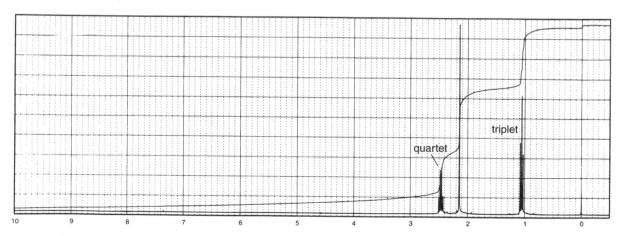

*2. The UV spectrum of this compound shows no maximum above 205 nm. When a drop of aqueous acid is added to the sample, the pattern at 3.6 ppm in the proton NMR spectrum simplifies to a triplet, and the pattern at 3.2 ppm simplifies to a singlet.

a)

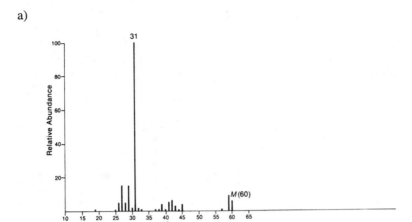

b)

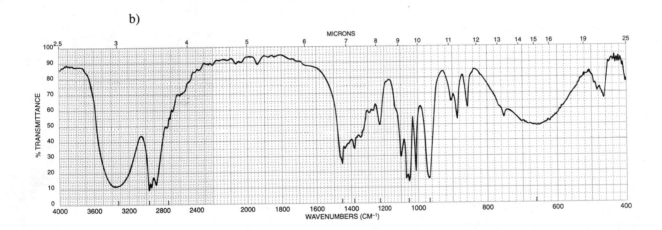

c)

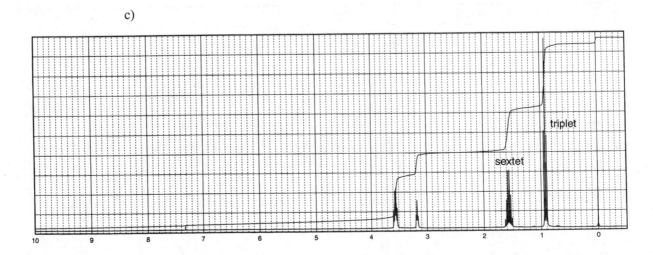

*3. UV spectrum of this compound is determined in 95% ethanol: λ_{max} 280 nm (log ε 1.3).

a)

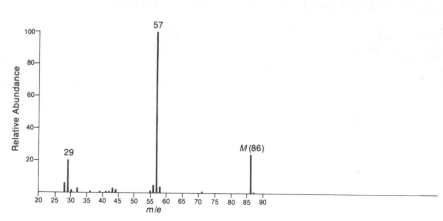

b)

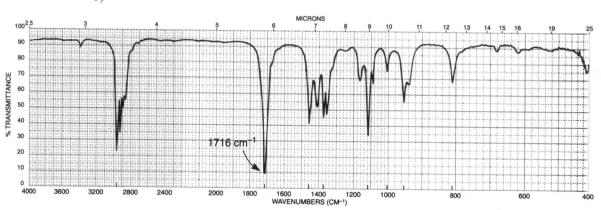

c)

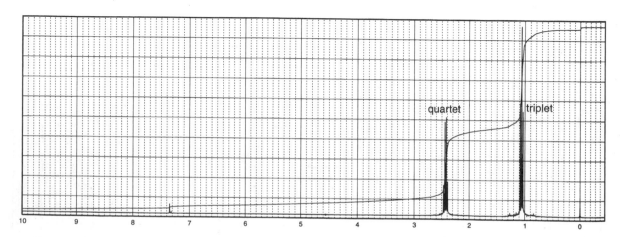

*4. The formula for this compound is $C_6H_{12}O_2$.

a)

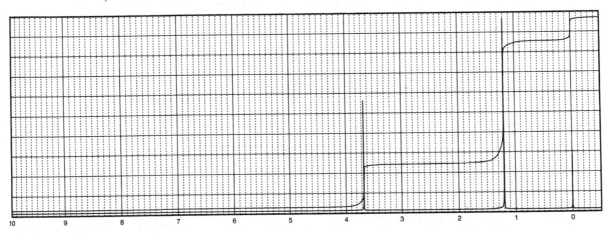

b)

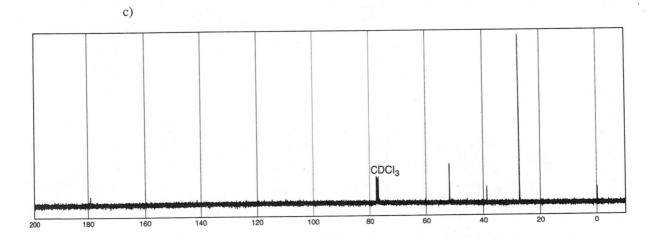

c)

*5. The UV spectrum of this compound is determined in 95% ethanol: strong end absorption and a band with fine structure appearing at λ_{max} 257 nm (log ε 2.4). The IR spectrum was obtained as a Nujol mull. The strong bands at about 2920 and 2860 cm^{-1} from the C—H stretch in Nujol overlap the broad band that extends from 3300 to 2500 cm^{-1}.

a)

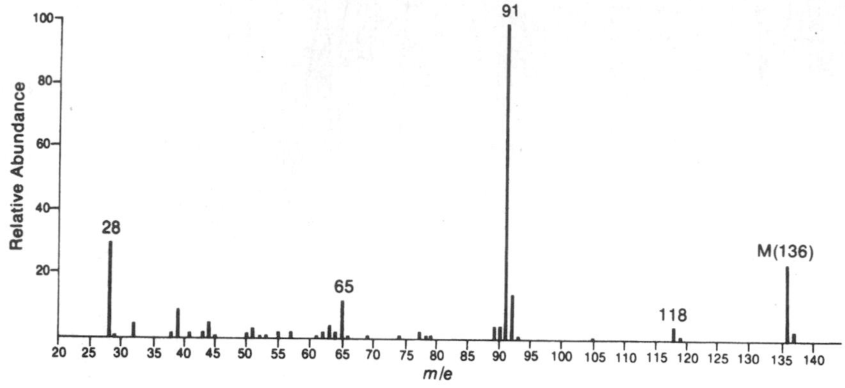

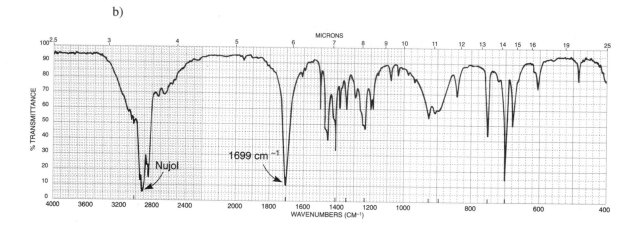

b)

c)

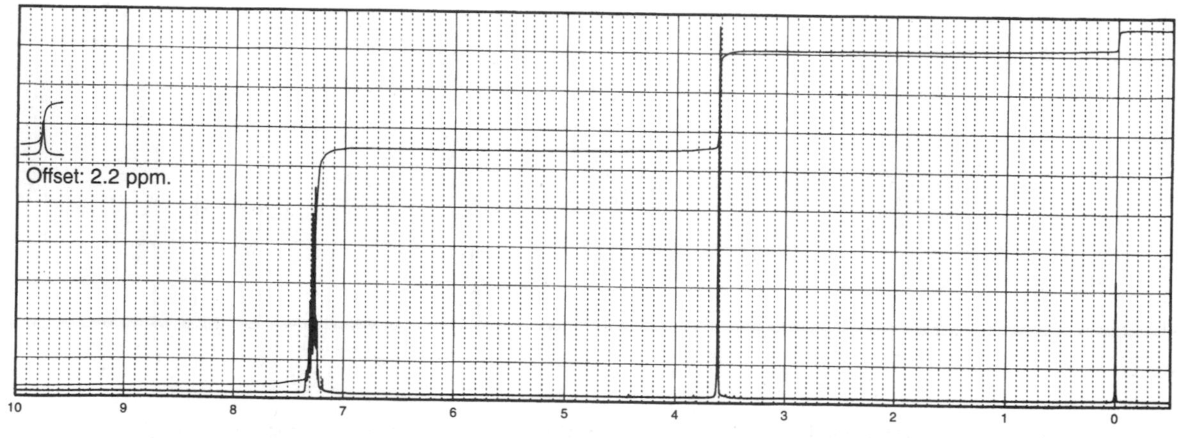

*6. The mass spectrum of this compound shows an intense molecular ion at 172 mass units and an $M + 2$ peak of approximately the same size. The IR spectrum of this solid unknown was obtained in Nujol. The prominent C—H stretching bands centering on about 2900 cm^{-1} are derived from the Nujol and are not part of the unknown. The peak appearing at about 5.3 ppm in the NMR spectrum is solvent dependent. It shifts readily when the concentration is changed.

a)

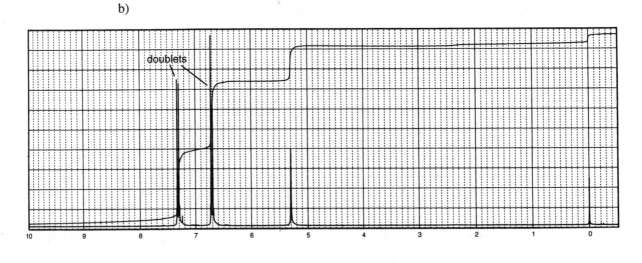

b)

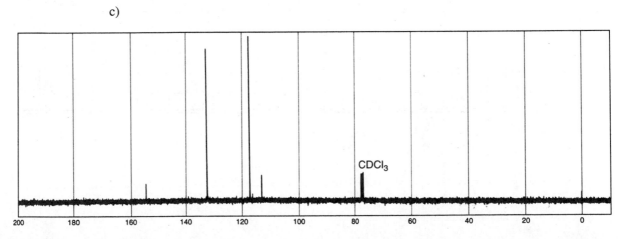

*7. This compound has the molecular formula $C_{11}H_{14}O$.

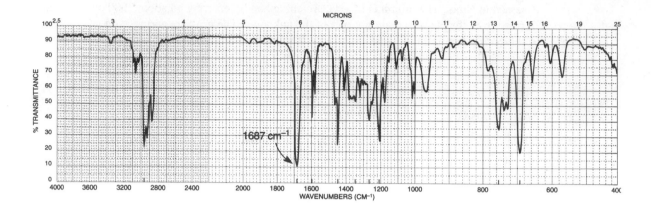

Normal Carbon	DEPT-135	DEPT-90
14 ppm	Positive	No peak
22	Negative	No peak
26	Negative	No peak
38	Negative	No peak
128	Positive	Positive
129	Positive	Positive
133	Positive	Positive
137	No peak	No peak
200	No peak	No peak

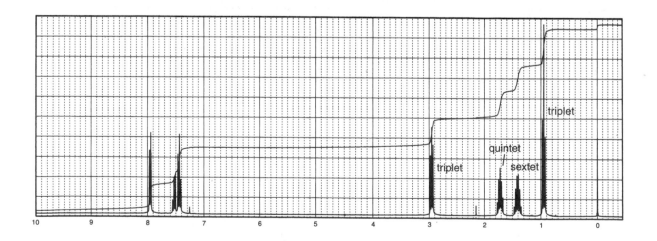

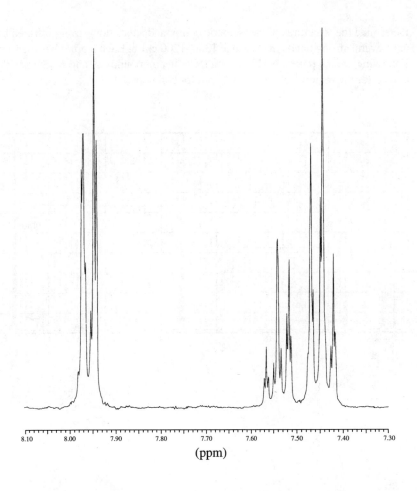

*8. Determine the structures of the isomeric compounds that show strong infrared bands at 1725 cm^{-1} and several strong bands in the range 1300–1200 cm^{-1}. Each isomer has the formula $C_9H_9BrO_2$. Following are the proton NMR spectra for both compounds, **A** and **B**. Expansions have been included for the region from 8.2 to 7.2 ppm for compound **A**.

A.

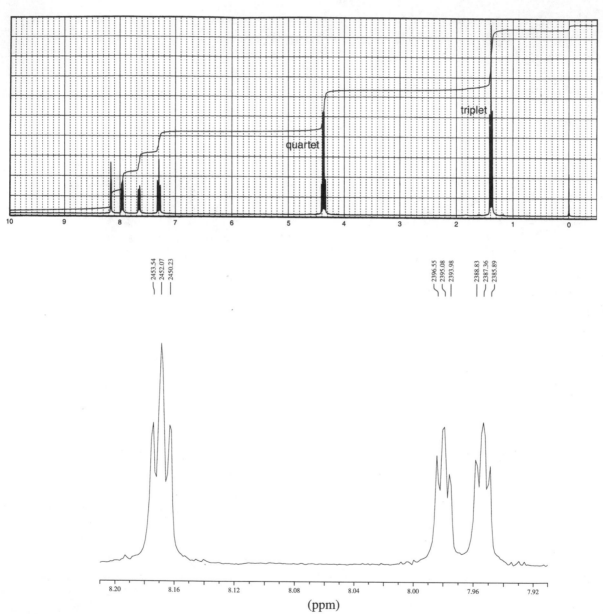

(ppm)

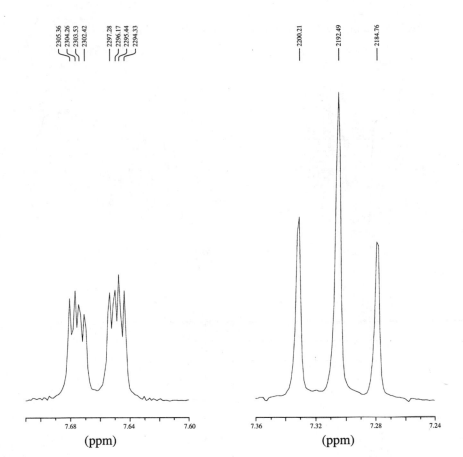

7.68   7.64   7.60
(ppm)

7.36   7.32   7.28   7.24
(ppm)

B.

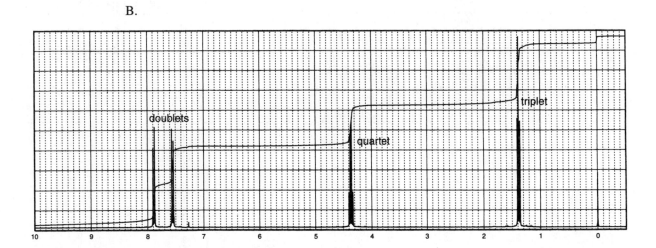

*9. This compound has the molecular formula $C_4H_{11}N$.

a)

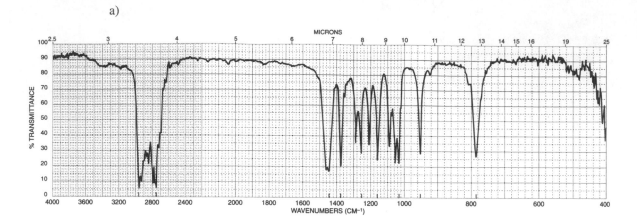

b)

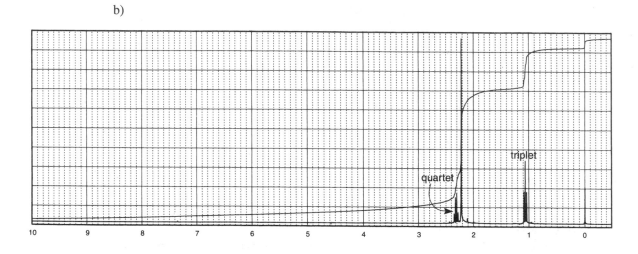

c)

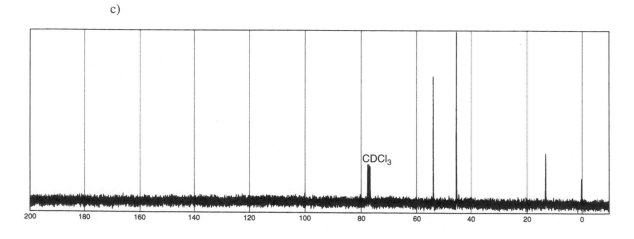

*10. The UV spectrum of this compound is determined in 95% ethanol: λ_{max} 280 nm (log ε 1.3). This compound has the formula $C_5H_{10}O$.

a)

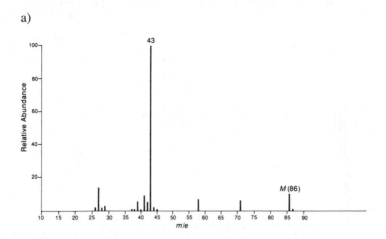

b)

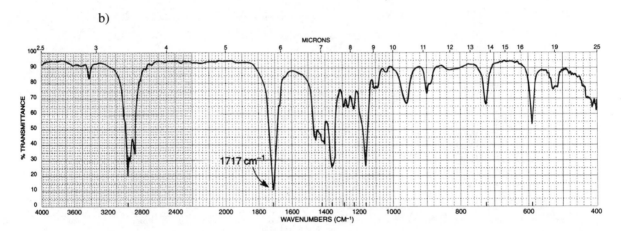

c)

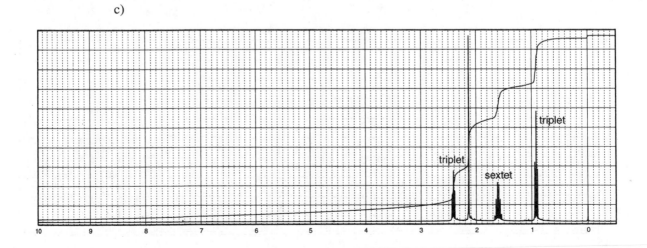

*11. This compound has the formula $C_3H_6O_2$. The UV spectrum of this compound shows no maximum above 205 nm. The carbon NMR spectrum shows peaks at 14, 60, and 161 ppm. The peak at 161 ppm appears as a positive peak in the DEPT-90 spectrum.

a)

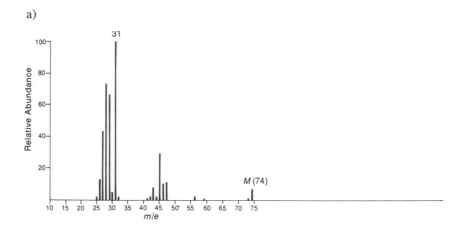

b)

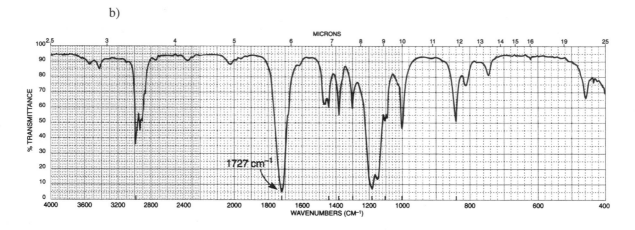

c)

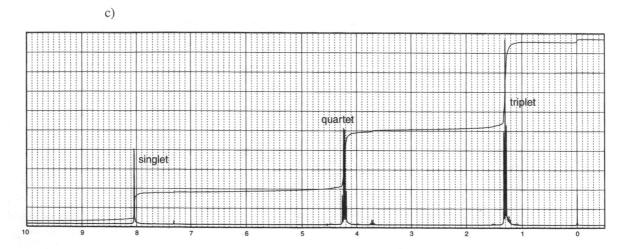

*12. Determine the structures of the isomeric compounds, *A* and *B*, each of which has the formula C_8H_7BrO. The infrared spectrum for compound *A* has a strong absorption band at 1698 cm^{-1} while compound *B* has a strong band at 1688 cm^{-1}. The proton NMR spectrum for compound *A* is shown, along with expansions for the region from 7.7 to 7.2 ppm. The NMR spectrum of compound *B* is also shown.

A.

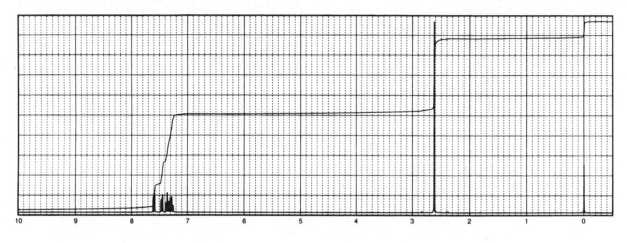

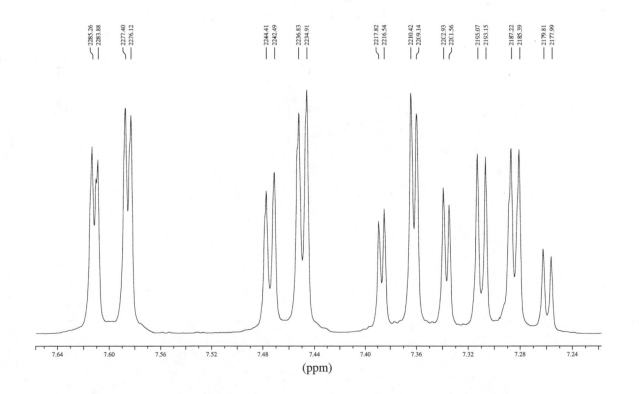

(ppm)

B.

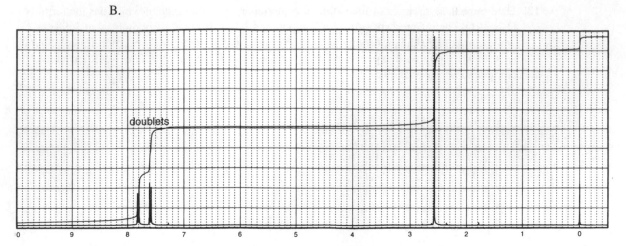

*13. This compound has the formula C_4H_8O. When expanded, the singlet peak at 9.8 ppm in the proton spectrum is actually a triplet. The triplet pattern at 2.4 ppm turns out to be a triplet of doublets when expanded.

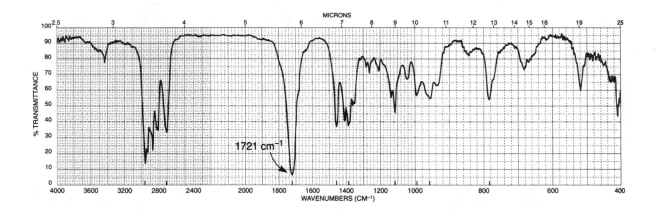

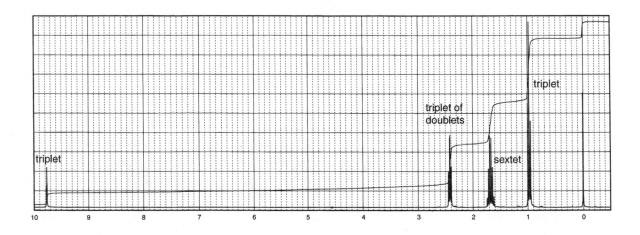

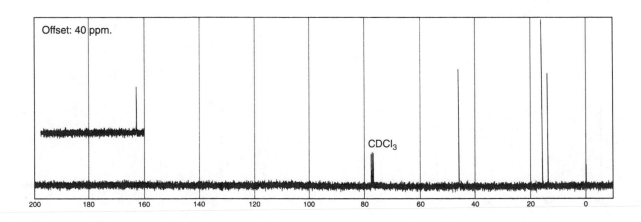

*14. This compound has the formula $C_5H_{12}O$. When a trace of aqueous acid is added to the sample, the proton NMR spectrum resolves into a clean triplet at 3.6 ppm, and the broad peak at 2.2 ppm moves to 4.5 ppm.

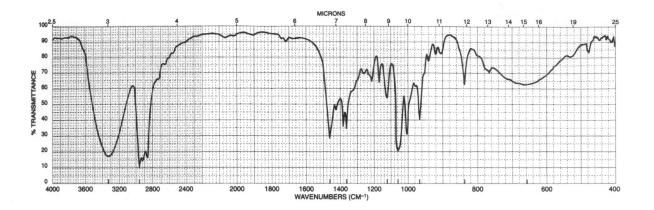

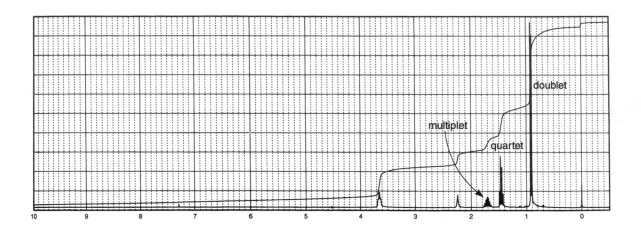

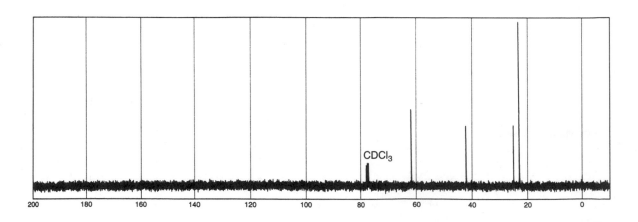

*15. Determine the structures of the isomeric compounds with the formula $C_5H_9BrO_2$. The proton NMR spectra for both compounds follow. The IR spectrum corresponding to the first proton NMR spectrum has strong absorption bands at 1739, 1225, and 1158 cm^{-1}, and that corresponding to the second one has strong bands at 1735, 1237, and 1182 cm^{-1}.

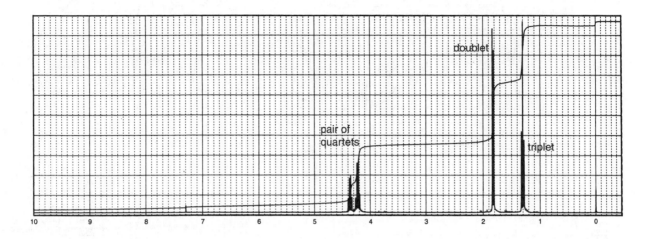

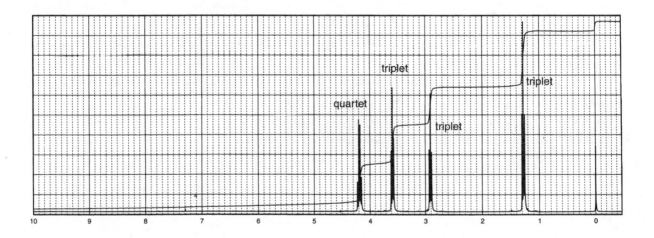

*16. This compound has the molecular formula $C_{10}H_9NO_2$.

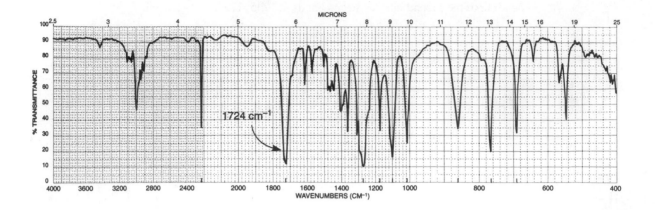

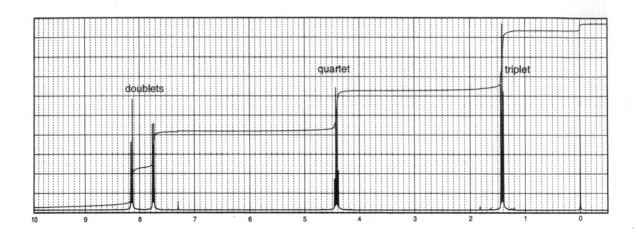

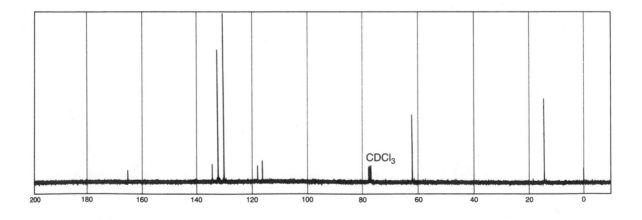

*17. This compound has the formula C_9H_9ClO. The full proton NMR spectrum is shown along with expansions of individual patterns.

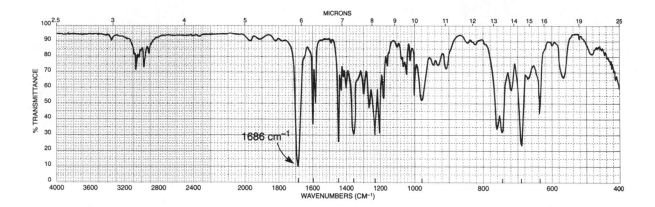

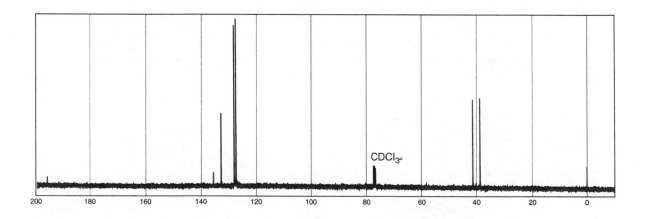

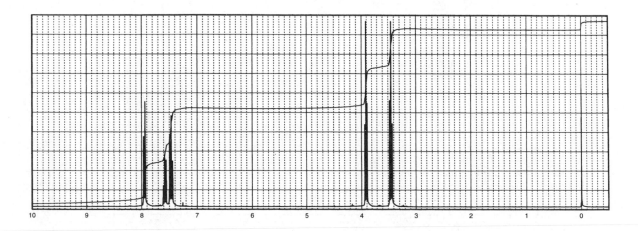

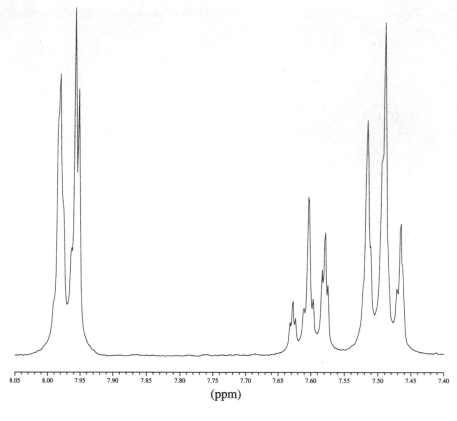

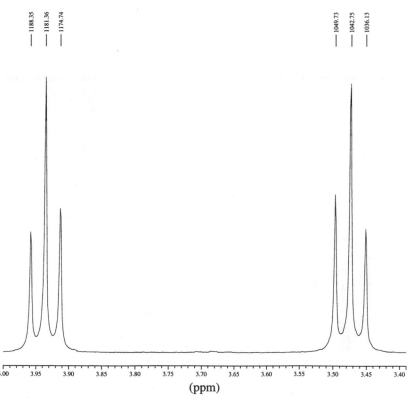

18. The anesthetic procaine (Novocaine) has the formula $C_{13}H_{20}N_2O_2$. In the proton NMR spectrum, each pair of triplets at 2.8 and 4.3 ppm has a coupling constant of 6 Hz. The triplet at 1.1 and the quartet at 2.6 ppm have coupling constants of 7 Hz. The IR spectrum was determined in Nujol. The C—H absorption bands of Nujol at about 2920 cm^{-1} in the IR spectrum obscure the entire C—H stretch region. The carbonyl group appearing at 1669 cm^{-1} in the IR spectrum has an unusually low frequency. Why?

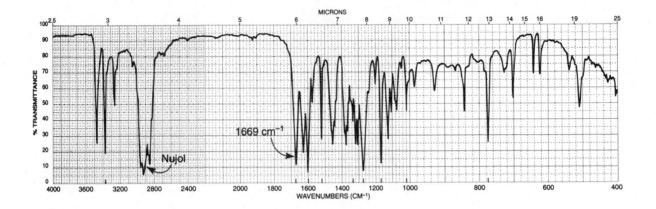

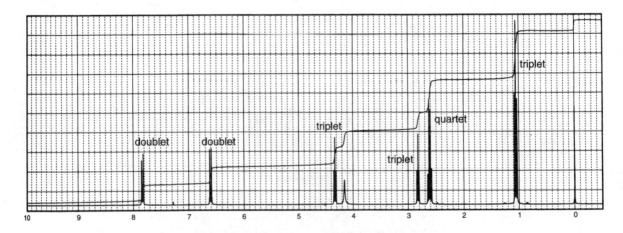

Normal Carbon	DEPT-135	DEPT-90
12 ppm	Positive	No peak
48	Negative	No peak
51	Negative	No peak
63	Negative	No peak
114	Positive	Positive
120	No peak	No peak
132	Positive	Positive
151	No peak	No peak
167	No peak	No peak

19. The UV spectrum of this compound shows no maximum above 250 nm. In the mass spectrum, notice that the patterns for the *M*, *M* + 2, and *M* + 4 peaks have a ratio of 1:2:1 (214, 216, and 218 amn). Draw the structure of the compound and comment on the structures of the mass 135 and 137 fragments.

a)

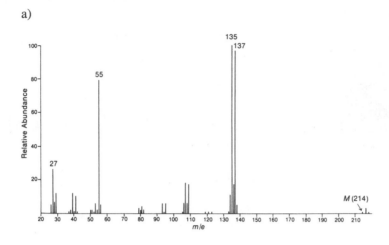

b)

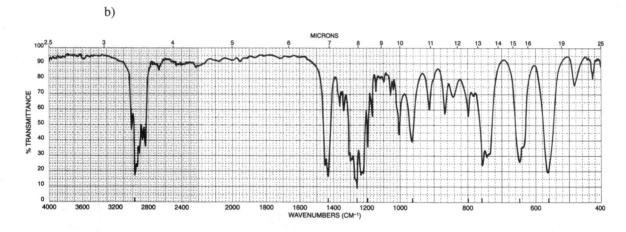

c)

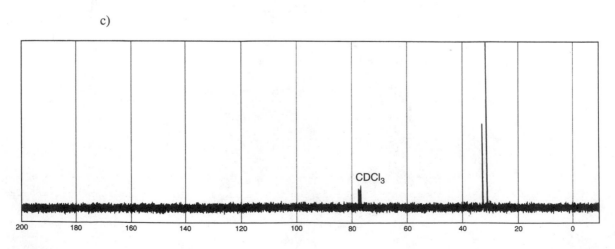

20. The UV spectrum of this compound is determined in 95% ethanol: λ_{max} 225 nm (log ε 4.0) and 270 nm (log ε 2.8). This compound has the formula $C_9H_{12}O_3S$.

a)

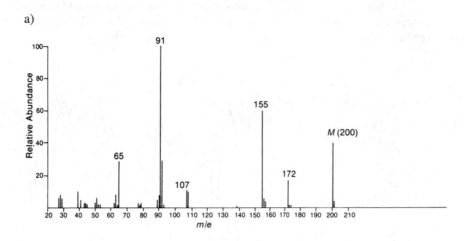

b)

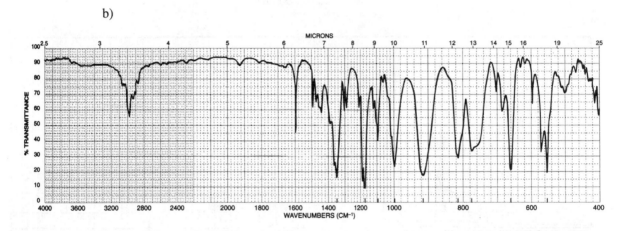

c)

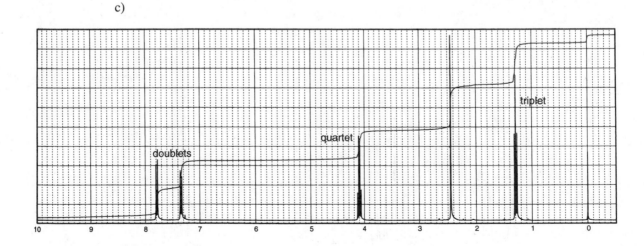

21. This compound has the molecular formula $C_9H_{10}O$. We have supplied you with the IR and proton NMR spectra. The expansions of the interesting sets of peaks centering near 4.3, 6.35, and 6.6 ppm in the proton NMR are provided, as well. Do not attempt to interpret the messy pattern near 7.4 ppm for the aromatic protons. The broad peak at 2.3 ppm (one proton) is solvent and concentration dependent.

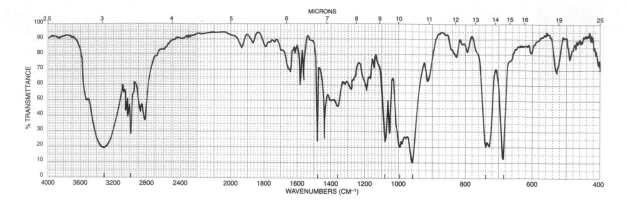

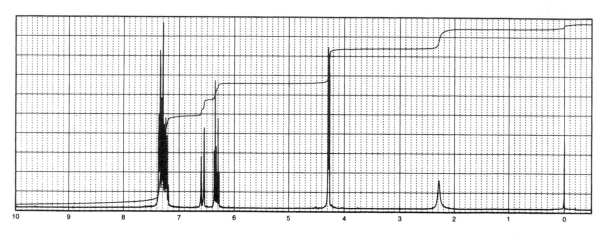

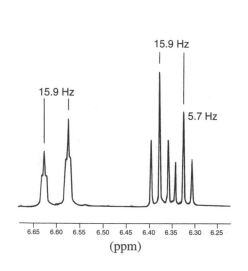

(ppm)

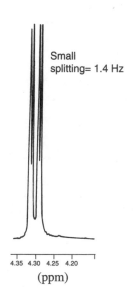

Small
splitting= 1.4 Hz

(ppm)

22. This compound has the formula C_3H_4O. We have supplied you with the IR and proton NMR spectra. Notice that a single peak at 3300 cm^{-1} overlaps the broad peak there. The expansions of the interesting sets of peaks centering near 2.5 and 4.3 ppm in the proton NMR are provided, as well. The peak at 3.25 ppm (one proton) is solvent and concentration dependent.

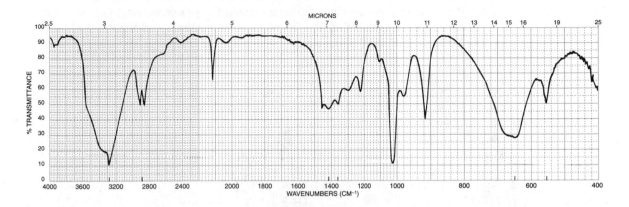

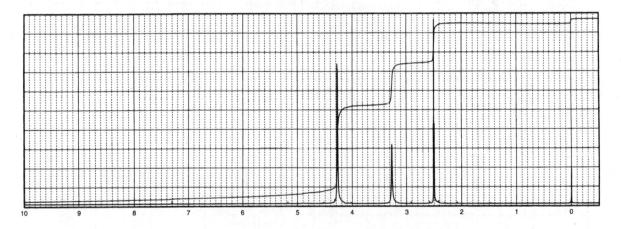

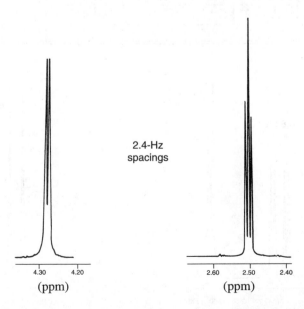

2.4-Hz
spacings

(ppm)   (ppm)

23. This compound has the molecular formula $C_7H_8N_2O_3$. We have supplied you with the IR and proton NMR spectra (run in deuterated DMSO). The expansions of the interesting sets of peaks centering near 7.75, 7.6, and 6.7 ppm in the proton NMR are provided, as well. The peak at 6.45 ppm (two protons) is solvent and concentration dependent. The UV spectrum shows peaks at 204 nm ($\varepsilon\,1.68 \times 10^4$), 260 nm ($\varepsilon\,6.16 \times 10^3$), and 392 nm ($\varepsilon\,1.43 \times 10^4$). The presence of the intense band at 392 nm is an important clue as to the positions of groups on the ring. This band moves to a lower wavelength when acidified. The IR spectrum was determined in Nujol. The C—H bands for Nujol at about 2920 cm^{-1} obscure the C—H bands in the unknown compound.

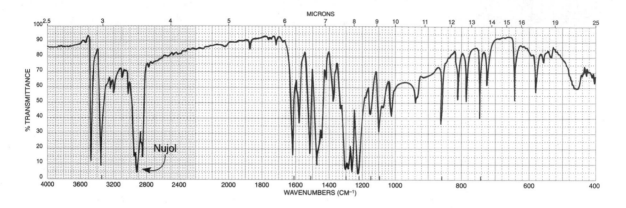

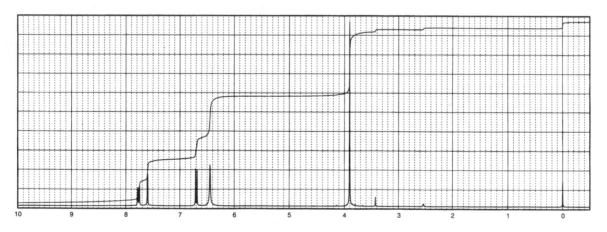

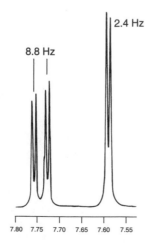

24. This compound has the formula $C_6H_{12}N_2$.

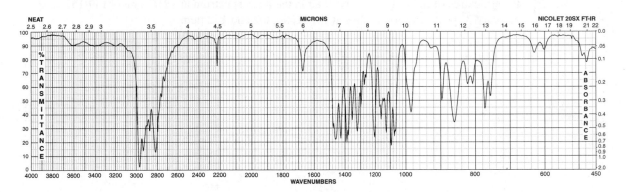

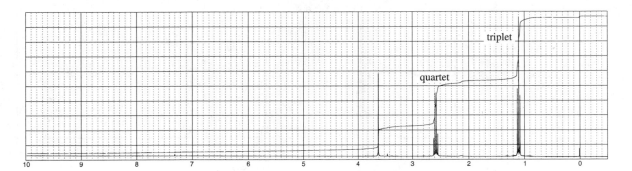

Normal Carbon	DEPT-135	DEPT-90
13 ppm	Positive	No peak
41	Negative	No peak
48	Negative	No peak
213	No peak	No peak

25. This compound has the formula $C_6H_{11}BrO_2$. Determine the structure of this compound. Draw the structures of the fragments observed in the mass spectrum at 121/123 and 149/151. The carbon-13 spectrum shows peaks at 14, 31, 56, 62, and 172 ppm.

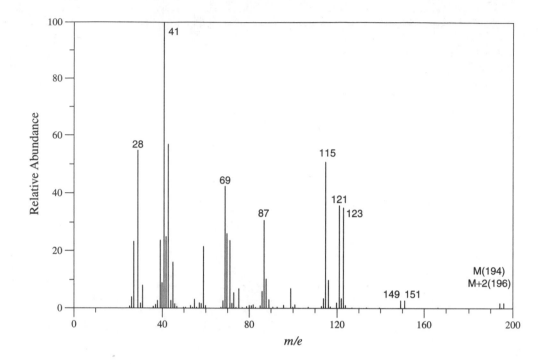

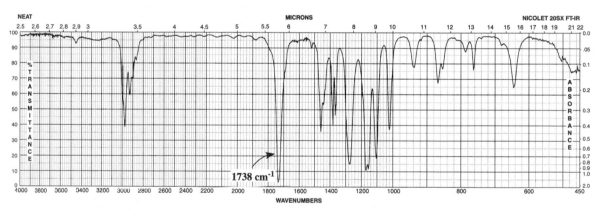

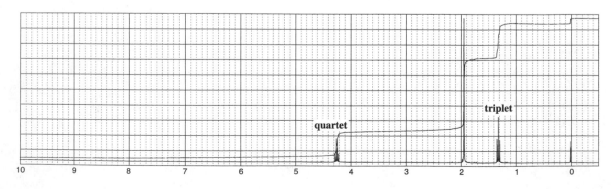

26. This compound has the formula $C_9H_{12}O$. The carbon-13 spectrum shows peaks at 28, 31, 57, 122, 124, 125, and 139 ppm.

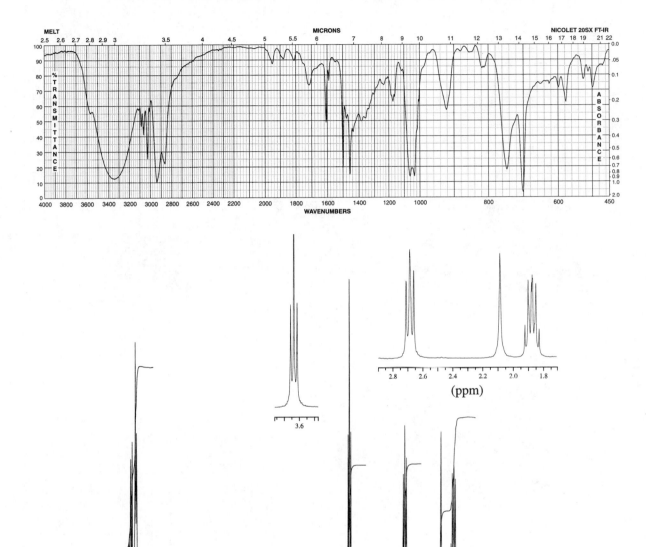

27. This compound has the formula $C_6H_{10}O$. The carbon-13 spectrum shows peaks at 21, 27, 31, 124, 155, and 198 ppm.

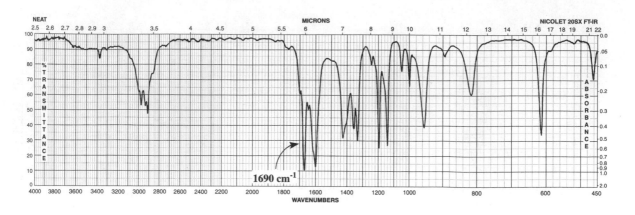

1690 cm⁻¹

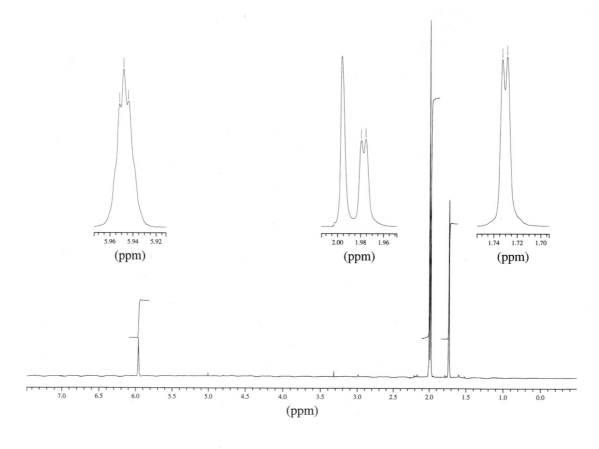

28. This compound has the formula $C_{10}H_{10}O_2$. The carbon-13 spectrum shows peaks at 52, 118, 128, 129, 130, 134, 145, and 167 ppm.

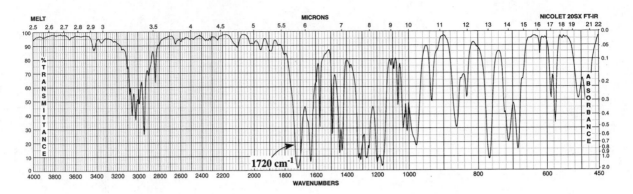

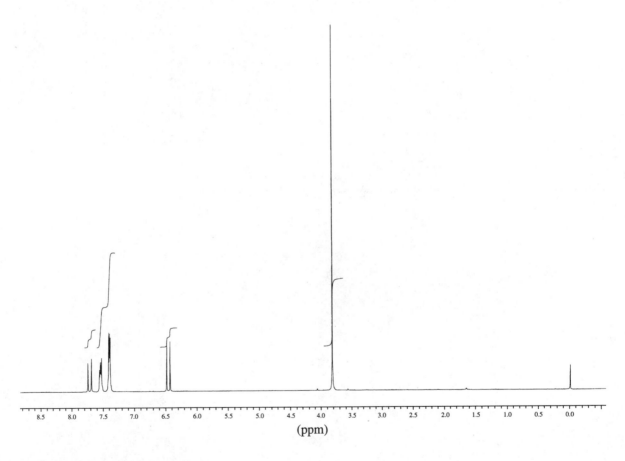

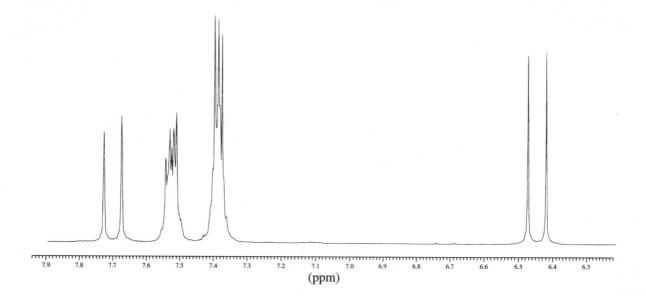

29. This compound has the formula $C_5H_8O_2$. The carbon-13 spectrum shows peaks at 14, 60, 129, 130, and 166 ppm.

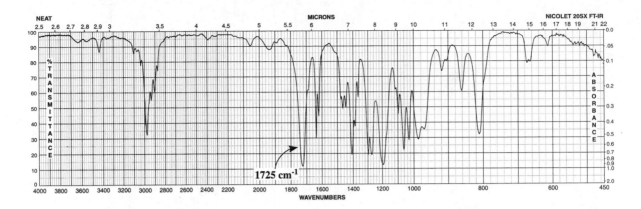

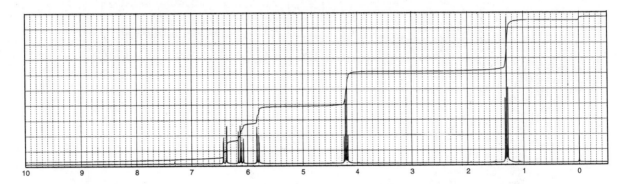

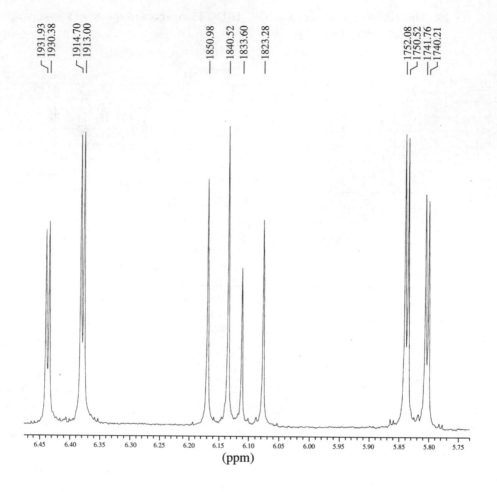

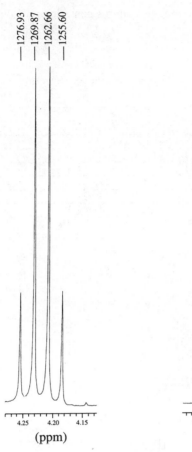

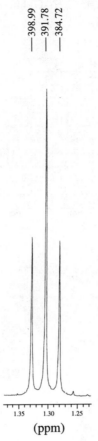

30. This compound has the formula $C_6H_{12}O$. Interpret the patterns centering on 1.3 and 1.58 ppm in the proton NMR spectrum.

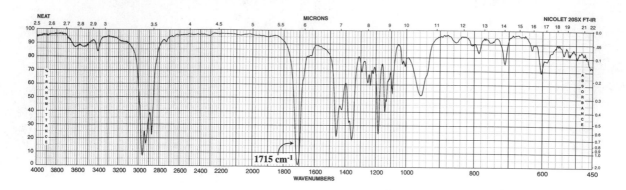

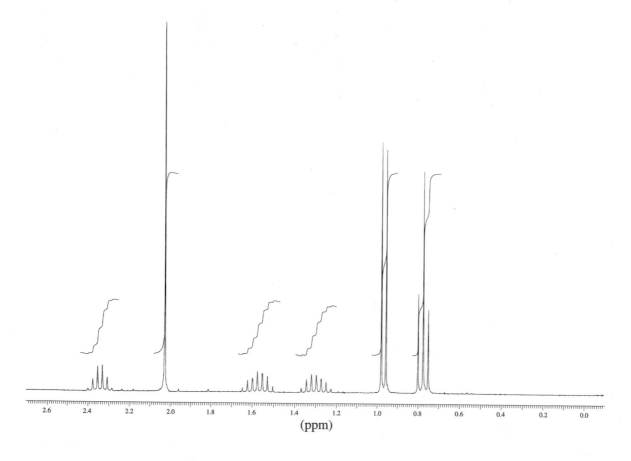

—719.63
—712.78
—705.90
—699.06
—692.18
—685.36

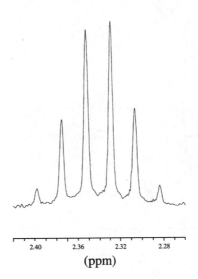

2.40   2.36   2.32   2.28
(ppm)

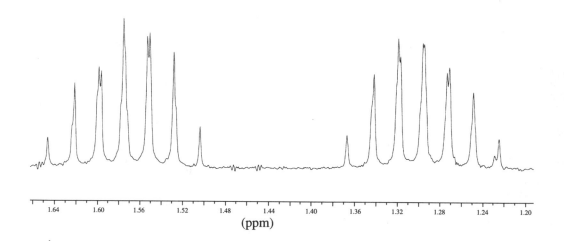

494.05
486.52
479.77
479.06
472.77
466.04
465.36
458.60
451.25
410.29
402.73
395.92
395.33
388.98
382.19
381.54
374.78
367.58

1.64   1.60   1.56   1.52   1.48   1.44   1.40   1.36   1.32   1.28   1.24   1.20

(ppm)

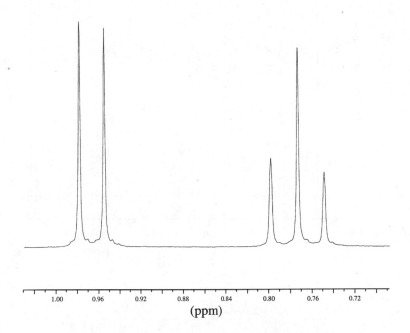

Normal Carbon	DEPT-135	DEPT-90
12 ppm	Positive	No peak
16	Positive	No peak
26	Negative	No peak
28	Positive	No peak
49	Positive	Positive
213	No peak	No peak

31. This compound has the formula $C_9H_{10}O_2$.

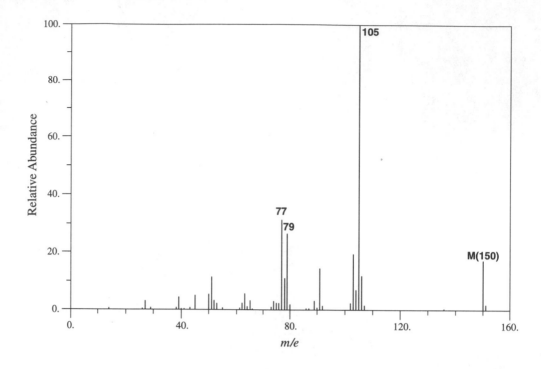

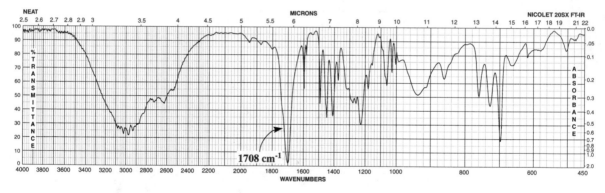

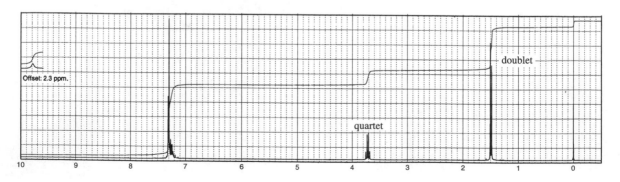

32. This compound has the formula $C_8H_{14}O$.

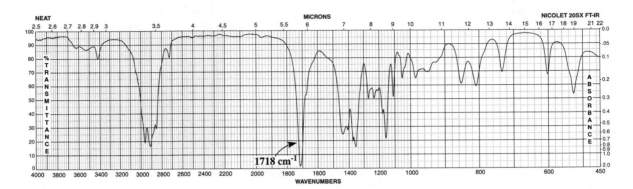

Normal Carbon	DEPT-135	DEPT-90
18 ppm	Positive	No peak
23	Negative	No peak
26	Positive	No peak
30	Positive	No peak
44	Negative	No peak
123	Positive	Positive
133	No peak	No peak
208	No peak	No peak

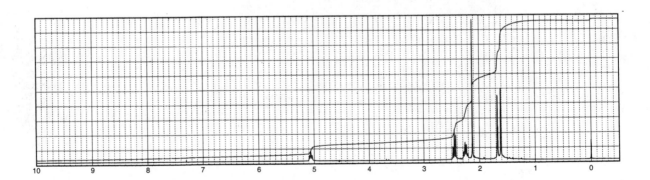

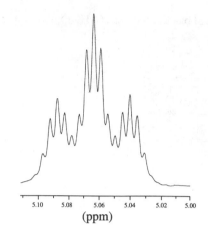

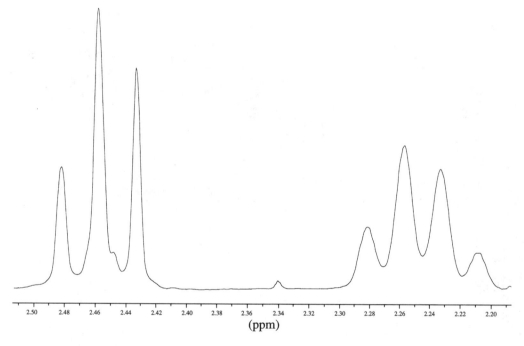

33. This compound has the formula $C_6H_6O_3$. The carbon-13 spectrum shows peaks at 52, 112, 118, 145, 146, and 159 ppm.

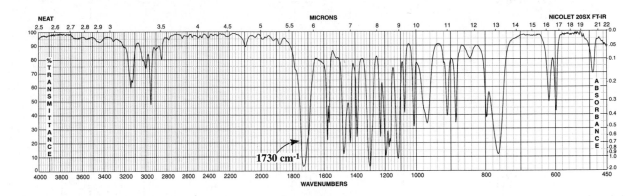

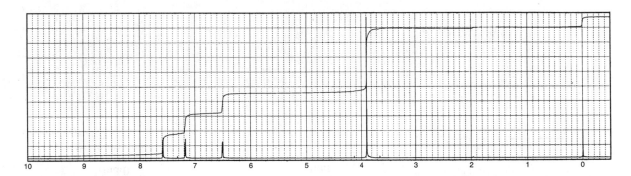

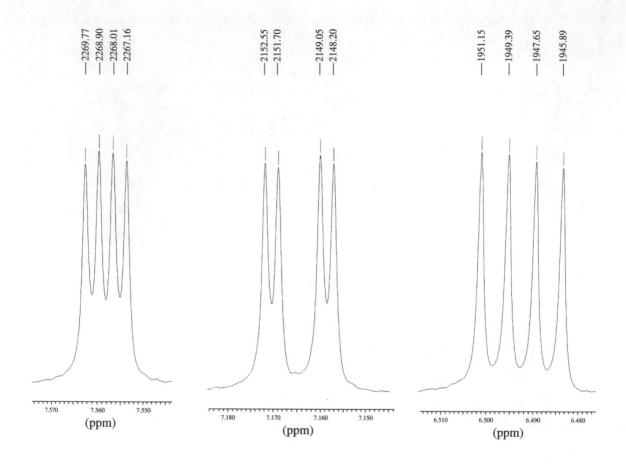

2269.77	2152.55	1951.15
2268.90	2151.70	1949.39
2268.01	2149.05	1947.65
2267.16	2148.20	1945.89

7.570   7.560   7.550
(ppm)

7.180   7.170   7.160   7.150
(ppm)

6.510   6.500   6.490   6.480
(ppm)

34. A compound with the formula $C_9H_8O_3$ shows a strong band at 1661 cm^{-1} in the infrared spectrum. The proton NMR spectrum is shown, but there is a small impurity peak at 3.35 ppm that should be ignored. Expansions are shown for the downfield protons. In addition, the normal carbon-13, DEPT-135, and DEPT-90 spectral results are tabulated.

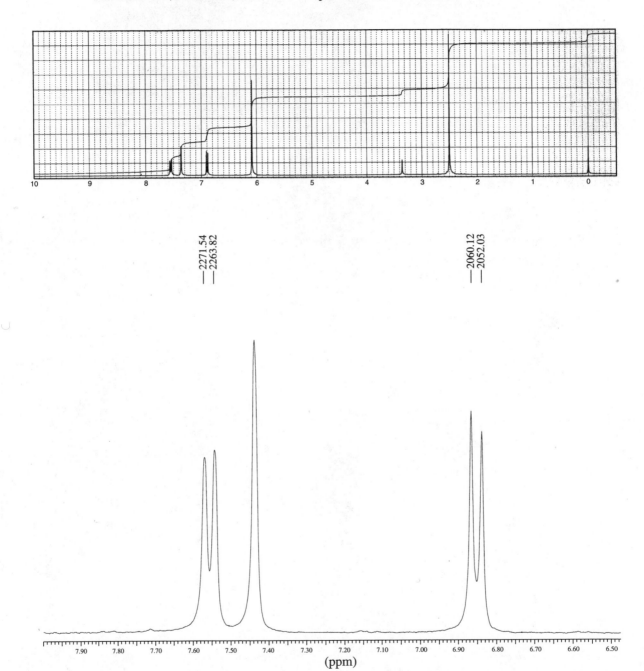

—2271.54
—2263.82

—2060.12
—2052.03

(ppm)

Normal Carbon	DEPT-135	DEPT-90
26 ppm	Positive	No peak
102	Negative	No peak
107	Positive	Positive
108	Positive	Positive
125	Positive	Positive
132	No peak	No peak
148	No peak	No peak
151	No peak	No peak
195	No peak	No peak

SOURCES OF ADDITIONAL PROBLEMS

Books That Contain Combined Spectral Problems

Ault, A., *Problems in Organic Structural Determination*, McGraw-Hill, New York, 1967.

Banks, R. C., E. R. Matjeka, and G. Mercer, *Introductory Problems in Spectroscopy*, Benjamin/Cummings, Menlo Park, CA, 1980.

Davis, R., and C. H. J. Wells, *Spectral Problems in Organic Chemistry*, Chapman and Hall, New York, 1984.

Field, L. D., Sternhell, S., Kalman, J. R., *Organic Structures from Spectra*, 2nd ed., John Wiley, Chichester, England, 1995.

Fuchs, P. L., and C. A. Bunnell, *Carbon-13 NMR Based Organic Spectral Problems*, John Wiley, New York, 1979.

Shapiro, R. H., and C. H. DePuy, *Exercises in Organic Spectroscopy*, 2d ed., Holt, Rinehart & Winston, New York, 1977.

Silverstein, R. M., and F. X. Webster, *Spectrometric Identification of Organic Compounds*, 6th ed., John Wiley, New York, 1998.

Sternhell, S., and J. R. Kalman, *Organic Structures from Spectra*, John Wiley, Chichester, England, 1986.

Tomasi, R. A., *A Spectrum of Spectra*, Sunbelt R&T, Inc., Tulsa, OK, 1992.

Williams, D. H., and I. Fleming, *Spectroscopic Methods in Organic Chemistry*, 4th ed., McGraw-Hill, London, 1987.

Web sites That Have Combined Spectral Problems

The Smith group at the University of Notre Dame has a number of combined problems:
http://www.nd.edu/~smithgrp/structure/workbook.html

The UCLA Department of Chemistry and Biochemistry in connection with Cambridge University Isotopes Laboratories maintains a web site with combined problems. They provide links to other sites with problems.
http://www.chem.ucla.edu/~webnmr/

Web sites That Have Spectral Data

National Institute of Standards and Technology (NIST). This site includes gas phase infrared spectra and mass spectral data. http://webbook.nist.gov/chemistry/

Integrated Spectral Data Base System for Organic Compounds, National Institute of Materials and Chemical Research, Tsukuba, Ibaraki 305-8565, Japan. This database includes infrared, mass spectra, and NMR data (proton and carbon-13).
http://www.aist.go.jp/RIODB/SDBS/menu-e.html

NUCLEAR MAGNETIC RESONANCE SPECTROSCOPY
Part Five: Advanced NMR Techniques

Since the advent of modern, computer-controlled Fourier transform NMR instruments, it has been possible to conduct more sophisticated experiments than those described in preceding chapters. Although a great many specialized experiments can be performed, the following discussion examines only a few of the most important ones.

10.1 PULSE SEQUENCES

Chapter 4, Section 4.5, introduced the concept of **pulse sequences.** In an FT-NMR instrument, the computer that operates the instrument can be programmed to control the timing and duration of the excitation pulse—the radiofrequency pulse used to excite the nuclei from the lower spin state to the upper spin state. Chapter 3, Section 3.7B, discussed the nature of this pulse and the reasons why it can excite all the nuclei in the sample simultaneously. Precise timing can also be applied to any decoupling transmitters that operate during the pulse sequence. As a simple illustration, Figure 10.1 shows the pulse sequence for the acquisition of a simple proton NMR spectrum. The pulse sequence is characterized by an excitation pulse from the transmitter; an **acquisition time,** during which the free-induction decay (FID) pattern is collected in digitized form in the computer; and a **relaxation delay,** during which the nuclei are allowed to relax in order to reestablish the equilibrium populations of the two spin states. After the relaxation delay, a second excitation pulse marks the beginning of another cycle in the sequence.

There are many possible variations on this simple pulse sequence. For example, in Chapter 4 we learned that it is possible to transmit *two* signals into the sample. In ^{13}C NMR spectroscopy, a pulse sequence similar to that shown in Figure 10.1 is transmitted at the absorption frequency of the ^{13}C nuclei. At the same time, a second transmitter, tuned to the frequency of the hydrogen (^1H) nuclei in the sample, transmits a broad band of frequencies to **decouple** the hydrogen nuclei from the ^{13}C nuclei. Figure 10.2 illustrates this type of pulse sequence.

The discussion in Chapter 4, Section 4.5 of methods for determining ^{13}C spectra described how to obtain proton-*coupled* spectra and still retain the benefits of nuclear Overhauser enhancement. In this method, which is called an **NOE-enhanced proton-coupled spectrum** or a **gated decoupling spectrum,** the decoupler is turned on during the interval *before* the pulsing of the ^{13}C nuclei. At the moment the excitation pulse is transmitted, the decoupler is switched off. The decoupler is switched on again during the relaxation delay period. The effect of this pulse sequence is to allow the nuclear Overhauser effect to develop while the decoupler is on. Because the decoupler is switched off during the excitation pulse, spin decoupling of the ^{13}C atoms is not observed (a proton-coupled spectrum is seen). The nuclear Overhauser enhancement decays over a relatively long time period, and, therefore, most of the enhancement is retained as the FID is collected. Once the FID information has been accumulated, the decoupler is switched on again to allow the nuclear Overhauser enhancement to develop before the next excitation pulse. Figure 10.3a shows the pulse sequence for gated decoupling.

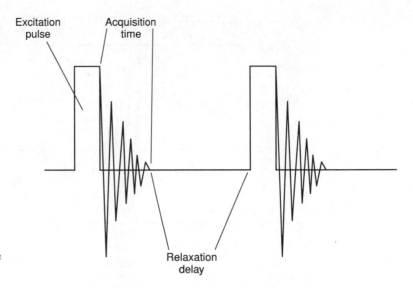

FIGURE 10.1 A simple pulse sequence.

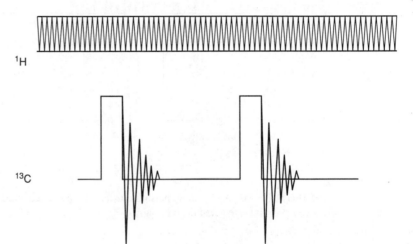

FIGURE 10.2 A proton-decoupled ^{13}C NMR pulse sequence.

The opposite result is obtained if the decoupler is not switched on until *the very instant* the excitation pulse is transmitted. Once the FID data have been collected, the decoupler is switched off until the next excitation pulse. This pulse sequence is called **inverse gated decoupling.** The effect of this pulse sequence is to provide a proton-decoupled spectrum with no NOE enhancement. Because the decoupler is switched off before the excitation pulse, the nuclear Overhauser enhancement is not allowed to develop. Proton decoupling is provided, since the decoupler is switched on during the excitation pulse and the acquisition time. Figure 10.3b shows the pulse sequence for inverse gated decoupling. This technique is used when it is necessary to determine integrals in a ^{13}C spectrum.

The computer that is built into modern FT-NMR instruments is very versatile and enables us to develop more complex and more interesting pulse sequences than the ones shown here. For example, we can transmit second and even third pulses, and we can transmit them along any of the Cartesian axes. The pulses can be transmitted with varying durations, and a variety of times can also be programmed into the sequence. As a result of these pulse programs, nuclei may exchange energy, they may affect each other's relaxation times, or they may encode information about spin coupling from one nucleus to another.

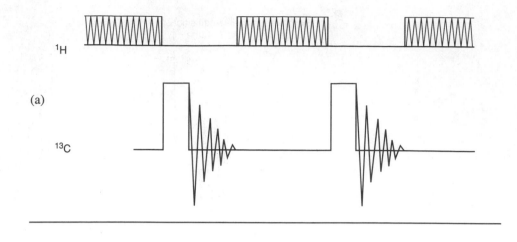

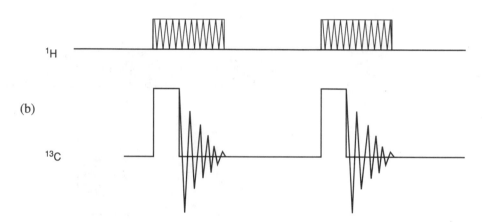

FIGURE 10.3 A simple pulse sequence(a) A pulse sequence for gated decoupling; (b) a pulse sequence for inverse gated decoupling.

We will not describe these complex pulse sequences any further; their description and analysis are beyond the scope of this discussion. Our purpose in describing a few simple pulse sequences in this section is to give you an idea of how a pulse sequence is constructed and how its design may affect the results of an NMR experiment. From this point forward, we shall simply describe the *results* of experiments that utilize some complex sequences and show how the results may be applied to the solution of a molecular structure problem. If you want more detailed information about pulse sequences for the experiments described in the following sections, consult one of the works listed in the references at the end of this chapter.

10.2 PULSE WIDTHS, SPINS, AND MAGNETIZATION VECTORS

To gain some appreciation for the advanced techniques that this chapter describes, you must spend some time learning what happens to a magnetic nucleus when it receives a pulse of radiofrequency energy. The nuclei that concern us in this discussion, ^1H and ^{13}C, are magnetic. They have a finite spin, and a spinning charged particle generates a magnetic field. That means that each individual nucleus behaves as a tiny magnet. The nuclear magnetic moment, or the strength of the magnet, each nucleus can be illustrated as a vector (Fig. 10.4a). When the magnetic nuclei are placed in a large, strong magnetic field, they tend to align themselves with the strong field, much as a compass needle aligns itself with the earth's magnetic field. Figure 10.4b shows this alignment. In the following discussion, it would be very inconvenient to continue describing the behavior of each individual nucleus. We can simplify the discussion by considering that the magnetic field vectors for each nucleus give rise to a resultant vector called the **nuclear magnetization vector** or the **bulk-magnetization vector.** Figure 10.4b also shows this vector. The individual nuclear magnetic vectors precess about the principal magnetic field axis (Z). They have random precessional motions that are not in phase; vector addition produces the resultant, the nuclear (bulk) magnetization vector, which is aligned along the Z axis. We can more easily describe an effect that involves the individual magnetic nuclei by examining the behavior of the nuclear magnetization vector.

The small arrows in Figure 10.4 represent the individual magnetic moments. In this picture we are viewing the orientations of the magnetic moment vectors from a stationary position, as if we were standing on the laboratory floor watching the nuclei precess inside the magnetic field. This view, or **frame of reference,** is thus known as the **laboratory frame** or the **stationary frame.** We can simplify the study of the magnetic moment vectors by imagining a set of coordinate axes that rotates in the same direction and at the same speed as the average nuclear magnetic moment precesses. This reference frame is called the **rotating frame,** and it rotates about the Z axis. We can visualize these vectors more easily by considering them within the rotating frame, much as we can visualize the complex motions of objects on the earth more easily by first observing their motions from the earth, alone, even though the earth is spinning on its axis, rotating about the sun, and moving through the solar system. We can label the axes of the rotating frames X', Y', and Z' (identical with Z). In this rotating frame, the microscopic magnetic moments are stationary (are not rotating) because the reference frame and the microscopic moments are rotating at the same speed and in the same direction.

Because the small microscopic moments (vectors) from each nucleus add together, what our instrument sees is the *net* or *bulk* magnetization vector for the whole sample. We will refer to this bulk magnetization vector in the discussions that follow.

In a Fourier transform NMR instrument, the radiofrequency is transmitted into the sample in a **pulse** of very short duration—typically on the order of 1 to 10 microseconds (μsec); during this time, the

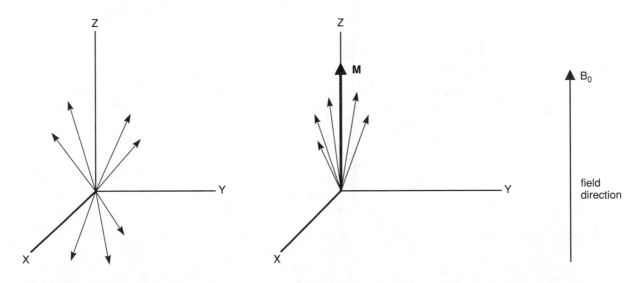

(a) A collection of magnetic nuclei, showing the individual magnetic moments.

(b) Magnetic nuclei aligned in an external magnetic field; (M) represents the bulk magnetization vector.

FIGURE 10.4 Nuclear magnetization (laboratory frame).

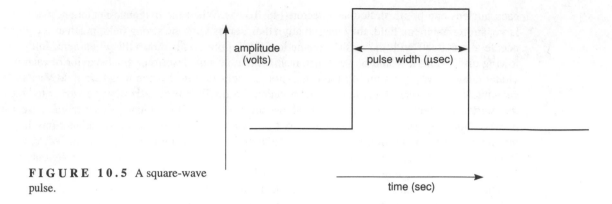

FIGURE 10.5 A square-wave pulse.

radiofrequency transmitter is suddenly turned on and, after about 10 μsec, suddenly turned off again. The pulse can be applied along either the X' or the Y' axis and in either the positive or the negative direction. The shape of the pulse, expressed as a plot of DC voltage versus time, looks like Figure 10.5.

When we apply this pulse to the sample, the magnetization vector of every magnetic nucleus begins to precess about the axis of the new pulse.[1] If the pulse is applied along the X' axis, the magnetization vectors all begin to tip in the same direction at the same time. The vectors tip to greater or smaller extents depending on the duration of the pulse. In a typical experiment, the duration of the pulse is selected to cause a specific **tip angle** of the bulk magnetization vector (the resultant vector of all of the individual vectors), and a pulse duration (known as a **pulse width**) is selected to result in a 90° rotation of the bulk magnetization vector. Such a pulse is known as a **90° pulse.** Figure 10.6 shows its effect along the X' axis. At the same time, if the duration of the pulse were twice as long, the bulk magnetization vector would tip to an angle of 180° (it would point straight down in Fig. 10.6). A pulse of this duration is called a **180° pulse.**

What happens to the magnetization vector following a 90° pulse? At the end of the pulse, the B_0 field is still present, and the nuclei continue to precess about it. If we focus for the moment on the

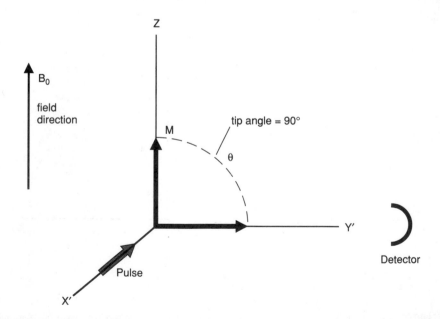

FIGURE 10.6 The effect of a 90° pulse (**M** is the bulk magnetization vector for the sample).

---

[1] Recall from Chapter 3 that if the duration of the pulse is short, the pulse has an uncertain frequency. The range of the uncertainty is sufficiently wide that all of the magnetic nuclei absorb energy from the pulse.

nuclei whose precessional frequencies exactly match the frequency of the rotating frame, we expect the magnetization vector to remain directed along the Y' axis (see Fig. 10.6).

In the **laboratory frame,** the Y' component corresponds to a magnetization vector rotating in the XY plane. The magnetization vector rotates in the XY plane because the individual nuclear magnetization vectors are precessing about Z (the principal field axis). Before the pulse, individual nuclei have random precessional motions and are not in phase. The pulse causes **phase coherence** to develop, so that all of the vectors precess in phase (see Fig. 10.7). Because all of the individual vectors precess about the Z axis, **M**, the resultant of all of these vectors, also rotates in the XY plane.

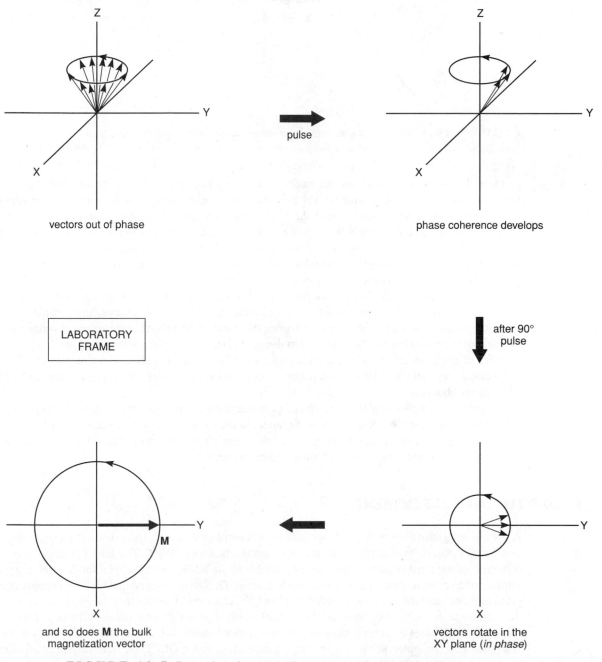

FIGURE 10.7 Precession of magnetization vectors in the XY plane.

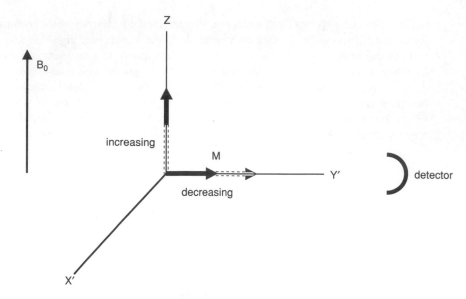

F I G U R E   1 0 . 8   Decay of the magnetization vector components as a function of time.

Once the pulse stops, however, the excited nuclei begin to relax (lose excitation energy and invert their individual nuclear spins). Over time, such **relaxation processes** diminish the magnitude of the bulk magnetization vector along the Y' axis and increase it along the Z axis, as illustrated in Figure 10.8. These changes in bulk magnetization occur as a result of both spin inversion to restore the Boltzmann distribution (spin–lattice relaxation) *and* loss of phase coherence (spin–spin relaxation). If we wait long enough, eventually the bulk magnetization returns to its equilibrium value, and the bulk magnetization vector points along the Z axis.

A receiver coil is situated in the XY plane, where it senses the rotating magnetization. As the Y' component grows smaller, the oscillating voltage from the receiver coil decays, and it reaches zero when the magnetization has been restored along the Z axis. The record of the receiver voltage as a function of time is called the **free-induction decay** (**FID**), because the nuclei are allowed to precess "freely" in the absence of an X-axis field. Figure 3.15 in Chapter 3 shows an example of a free-induction decay pattern. When such a pattern is analyzed via a Fourier transform, the typical NMR spectrum is obtained.

To understand how some of the advanced experiments work, it is helpful to develop an appreciation of what an excitation pulse does to the nuclei in the sample and how the magnetization of the sample nuclei behaves during the course of a pulsed experiment. At this point, we shall turn our attention to three of the most important advanced experiments.

10.3 THE DEPT EXPERIMENT

A very useful pulse sequence in ^{13}C spectroscopy is employed in the experiment called **Distortionless Enhancement by Polarization Transfer,** better known as **DEPT.** The DEPT method has become one of the most important techniques available to the NMR spectroscopist for determining the number of hydrogens attached to a given carbon atom. The pulse sequence involves a complex program of pulses and delay times in both the ^1H and ^{13}C channels. The result of this pulse sequence is that carbon atoms with one, two, and three attached hydrogens exhibit different **phases** as they are recorded. The phases of these hydrogens will also depend on the duration of the delays that are programmed into the pulse sequence. In one experiment, called a DEPT-45, only carbon atoms that bear an attached hydrogen will produce a peak. With a slightly different delay, a second experiment (called a DEPT-90) shows peaks only for those carbon atoms that are part of a **methine** (CH) group. With an even longer delay, a DEPT-135 spectrum is obtained. In a DEPT-135 spectrum, methine and methyl carbons give rise to positive peaks, whereas methylene carbons appear as inverse peaks. Section 10.4 will develop the reasons why carbon atoms with different numbers of attached hydrogens behave differently in this type of experiment. Quaternary carbons, which have no attached hydrogens, are not recorded in a DEPT experiment.

There are several variations on the DEPT experiment. In one form, separate spectra are traced on a single sheet of paper. On one spectrum, only the methyl carbons are shown; on the second spectrum, only the methylene carbons are traced; on the third spectrum, only the methine carbons appear; and on the fourth trace, all carbon atoms that bear hydrogen atoms are shown. In another variation on this experiment, the peaks due to methyl, methylene, and methine carbons are all traced on the same line, with the methyl and methine carbons appearing as positive peaks and the methylene carbons appearing as negative peaks.

In many instances, a DEPT spectrum makes spectral assignments easier than does a proton-decoupled ^{13}C spectrum. Figure 10.9 is the DEPT-135 spectrum of **isopentyl acetate.**

$$\overset{6}{CH_3}-\overset{\overset{\displaystyle O}{\|}}{\underset{5}{C}}-O-\overset{4}{CH_2}-\overset{3}{CH_2}-\underset{\underset{1}{\overset{|}{CH_3}}}{\overset{2}{CH}}-\overset{1}{CH_3}$$

The two methyl carbons (numbered **1**) can be seen as the tallest peak (at 22.3 ppm), while the methyl group on the acetyl function (numbered **6**) is a shorter peak at 20.8 ppm. The methine carbon (**2**) is a still smaller peak at 24.9 ppm. The methylene carbons produce the inverted peaks: carbon **3** appears at 37.1 ppm, and carbon **4** appears at 63.0 ppm. Carbon **4** is deshielded since it is near the electronegative oxygen atom. The carbonyl carbon (**5**) does not appear in the DEPT spectrum since it has no attached hydrogen atoms.

ISOPENTYL ACETATE IN $CDCl_3$—DEPT-135

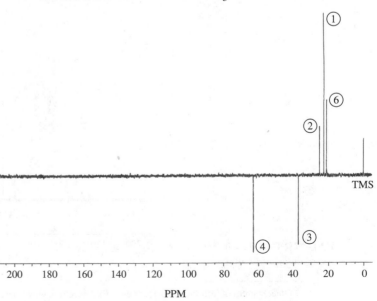

FIGURE 10.9 DEPT-135 spectrum of isopentyl acetate.

Clearly, the DEPT technique is a very useful adjunct to ^{13}C NMR spectroscopy. Whereas the peaks in a typical ^{13}C NMR spectrum appear as singlets, the results of the DEPT experiment can tell us whether a given peak arises from a carbon on a methyl group, a methylene group, or a methine group. By comparing the results of the DEPT spectrum with the original ^{13}C NMR spectrum, we can also identify the peaks that must arise from quaternary carbons. Quaternary carbons, which bear no hydrogens, appear in the ^{13}C NMR spectrum but are missing in the DEPT spectrum.

Another example that demonstrates some of the power of the DEPT technique is the terpenoid alcohol **citronellol.**

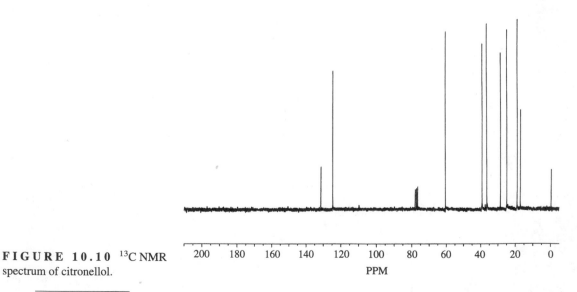

Figure 10.10 is the proton-decoupled ^{13}C NMR spectrum of citronellol. We can easily assign certain features of the spectrum to particular carbon atoms of the molecule by examining the chemical shifts and intensities. For example, the peak at 131 ppm is assigned to carbon 7, while the taller peak at 125 ppm must arise from carbon 6, which has an attached hydrogen. The pattern appearing between 15 and 65 ppm, however, is much more complex and thus more difficult to interpret.

The DEPT spectrum of citronellol (Fig. 10.11) makes the specific assignment of individual carbon atoms much easier.[2] Our earlier assignment of the peak at 125 ppm to carbon 6 is confirmed, because that peak appears positive in the DEPT-135 spectrum. Notice that the peak at 131 ppm is missing in the DEPT spectrum, since carbon 7 has no attached hydrogens. The peak at 61 ppm is negative in the DEPT-135 spectrum, indicating that it is due to a methylene group. Combining that information with our knowledge of the deshielding effects of electronegative elements enables us to assign this peak to carbon 1.

FIGURE 10.10 ^{13}C NMR spectrum of citronellol.

200   180   160   140   120   100   80   60   40   20   0

PPM

[2] Other sources of information besides the DEPT spectrum were consulted in making these assignments (see the references and problem 4).

CITRONELLOL DEPT

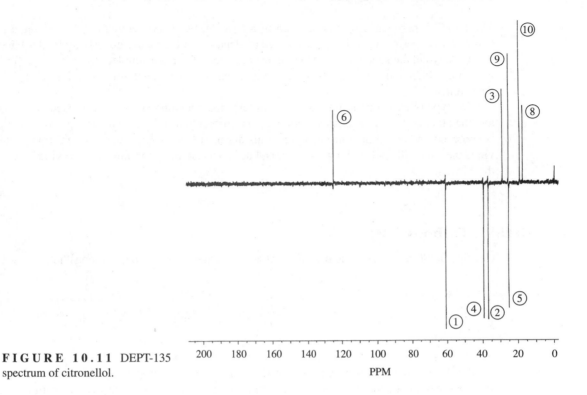

FIGURE 10.11 DEPT-135 spectrum of citronellol.

Shifting our attention to the upfield portion of the spectrum, we can identify the three methyl carbons, since they appear at the highest values of magnetic field and give positive peaks in the DEPT-135 spectrum. We can assign the peak at 17 ppm to carbon 8 and the peak at 19 ppm to carbon 10 (see footnote 2).

The most interesting feature of the spectrum of citronellol appears at 25 ppm. When we look carefully at the DEPT-135 spectrum, we see that this peak is actually *two* peaks, appearing by coincidence at the same value of chemical shift. The DEPT-135 spectrum shows clearly that one of the peaks is positive (corresponding to the methyl carbon at C9) and the other is negative (corresponding to the methylene carbon at C5).

We can assign the remaining peaks in the spectrum by noting that only one positive peak remains in the DEPT-135 spectrum. This peak must correspond to the methine position at C3 (30 ppm). The two remaining negative peaks (at 37 and 40 ppm) are assigned to the methylene carbons at C4 and C2. Without additional information, it is not possible to make a more specific assignment of these two carbons (see problem 4).

These examples should give you an idea of the capabilities of the DEPT technique. It is an excellent means of distinguishing among methyl, methylene, methine, and quaternary carbons in a ^{13}C NMR spectrum.

Certain NMR instruments are also programmed to record the results of a DEPT experiment directly onto a proton-decoupled ^{13}C NMR spectrum. In this variation, called a ^{13}C NMR spectrum with multiplicity analysis, each of the singlet peaks of a proton-decoupled spectrum is labeled according to whether that peak would appear as a singlet, doublet, triplet, or quartet if proton coupling were considered.

◾ 10.4 DETERMINING THE NUMBER OF ATTACHED HYDROGENS

The DEPT experiment is a modification of a basic NMR experiment called the **attached proton test (APT)** experiment. Although a detailed explanation of the theory underlying the DEPT experiment is beyond the scope of this book, an examination of a much simpler experiment (APT) should give you sufficient insight into the DEPT experiment so that you will understand how its results are determined.

This type of experiment uses two transmitters, one operating at the proton resonance frequency and the other at the ^{13}C resonance frequency. The proton transmitter serves as a proton decoupler; it is switched on and off at appropriate intervals during the pulse sequence. The ^{13}C transmitter provides the usual 90° pulse along the X' axis, but it can also be programmed to provide pulses along the Y' axis.

A. Methine Carbons (CH)

Consider a ^{13}C atom with one proton attached to it, where J is the coupling constant.

$$-\overset{|}{\underset{|}{C}}-H \qquad {}^{1}J_{CH}$$

After a 90° pulse, the bulk magnetization vector, **M**, is directed along the Y' axis. The result of this simple experiment should be a single line, since there is only one vector that is rotating at exactly the same frequency as the Larmor precessional frequency.

In this case, however, the attached hydrogen splits this resonance into a doublet. The resonance does not occur exactly at the Larmor frequency; rather, coupling to the proton produces *two* vectors. One of the vectors rotates $J/2$ Hz *faster* than the Larmor frequency, and the other vector rotates $J/2$ Hz *slower* than the Larmor frequency. One vector results from a coupling to the proton with its magnetic moment aligned with the magnetic field, and the other vector results from a coupling to the proton with its magnetic moment aligned against the magnetic field. The two vectors are separated in the rotating frame.

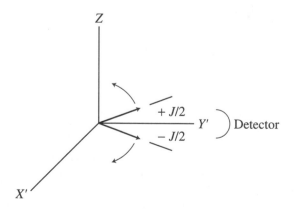

The vectors are moving relative to the rotating frame at a speed of $J/2$ revolutions per second but in opposite directions. The time required for one revolution is therefore the inverse of this speed, or $2/J$ seconds per revolution. At time $\frac{1}{4}(2/J)$, the vectors have made one-fourth of a revolution and are opposite each other along the X' axis. At this point no signal is detected by the receiver, because there is no component of magnetization along the Y' axis (the resultant of these two vectors is zero).

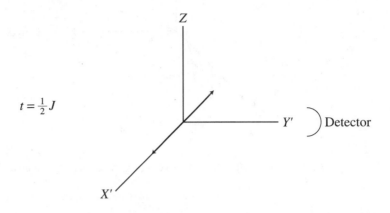

At time $\frac{1}{2}(2/J) = 1/J$, the vectors have realigned along the Y' axis but in the negative direction. An inverted peak would be produced if we collected signal at that time. So, if $t = 1/J$, a methine carbon should show an inverted peak.

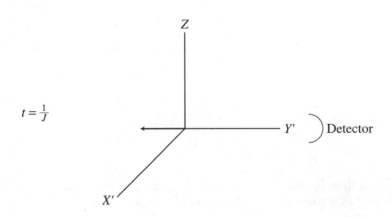

B. Methylene Carbons (CH₂)

If we examine the fate of a ^{13}C atom with *two* attached protons, we find different behavior.

$$H-\overset{|}{\underset{|}{C}}-H \qquad {}^{1}J_{CH}$$

In this case there are *three* vectors for the ^{13}C nucleus, because the two attached protons split the ^{13}C resonance into a triplet. One of these vectors remains stationary in the rotating frame, while the other two move apart with a speed of J revolutions per second.

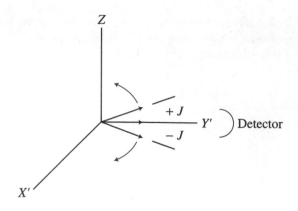

At time $\frac{1}{2}(1/J)$, the two moving vectors have realigned along the negative Y' axis,

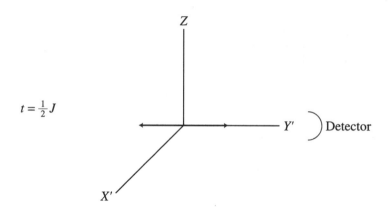

and at time $1/J$, they have realigned along the positive Y axis. The vectors thus produce a normal peak if they are detected at time $t = 1/J$. Therefore, a methylene carbon should show a normal (positive) peak.

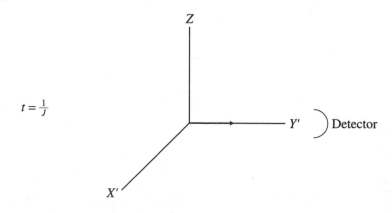

C.   Methyl Carbons (CH₃)

In the case of a methyl carbon,

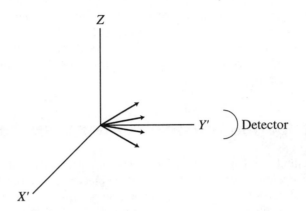

there should be *four* vectors, corresponding to the four possible spin states of a collection of three hydrogen nuclei.

An analysis of the precessional frequencies of these vectors shows that after time $t = 1/J$, the methyl carbon should also show an inverted peak.

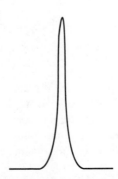

D.   Quaternary Carbons (C)

An unprotonated carbon simply shows one magnetization vector, which precesses at the Larmor frequency (i.e., it always points along the Y' axis). A normal peak is recorded at time $t = 1/J$.

E.   The Final Result

In this type of experiment, we should see a normal peak for every quaternary carbon and methylene carbon and an inverted peak for every methine carbon and methyl carbon. We can thus tell whether the number of hydrogens attached to the carbon is *even* or *odd*.

In the form of this experiment known as the DEPT experiment, the pulse sequence is more complex than those described in the preceding paragraphs. By varying the pulse widths and delay times, it is possible to obtain separate spectra for methyl, methylene, and methine carbons. In the normal manner of presenting DEPT spectra (e.g., a DEPT-135 spectrum), the trace that combines the spectra of all these types of ^{13}C atoms is inverted from the presentation described for the attached proton test (APT). Therefore, in the spectra presented in Figures 10.9 and 10.11, carbon atoms that bear odd numbers of hydrogens appear as positive peaks, carbon atoms that bear even numbers of hydrogens appear as negative peaks, and unprotonated carbon atoms do not appear.

In the DEPT experiment, results similar to those described here for the APT experiment are obtained. A variety of pulse angles and delay times are incorporated into the pulse sequence. The result of the DEPT experiment is that methyl, methylene, methine, and quaternary carbons can be distinguished from one another.

10.5 INTRODUCTION TO TWO-DIMENSIONAL SPECTROSCOPIC METHODS

The methods we have described to this point are examples of one-dimensional experiments. In a **one-dimensional experiment,** the signal is presented as a function of a single parameter, usually the chemical shift. In a **two-dimensional experiment,** there are two coordinate axes. Generally these axes also represent ranges of chemical shifts. The signal is presented as a function of each of these chemical shift ranges. The data are plotted as a grid; one axis represents one chemical shift range, the second axis represents the second chemical shift range, and the third dimension constitutes the magnitude (intensity) of the observed signal. The result is a form of contour plot where contour lines correspond to signal intensity.

In a normal pulsed NMR experiment, the 90° **excitation pulse** is followed immediately by a **data acquisition phase** in which the FID is recorded and the data are stored in the computer. In experiments that use complex pulse sequences, such as DEPT, a **preparation phase** is included before data acquisition. During the preparation phase, the nuclear magnetization vectors are allowed to precess, and information may be exchanged between magnetic nuclei. In other words, a given nucleus may become encoded with information about the spin state of another nucleus which may be nearby.

Of the many types of two-dimensional experiments, two find the most frequent application. One of these is **H—H Correlation Spectroscopy,** better known by its acronym, **COSY.** In a COSY experiment, the chemical shift range of the proton spectrum is plotted on both axes. The second important technique is **Heteronuclear Correlation Spectroscopy,** better known as the **HETCOR** technique. In a HETCOR experiment, the chemical shift range of the proton spectrum is plotted on one axis, while the chemical shift range of the ^{13}C spectrum for the same sample is plotted on the second axis.

10.6 THE COSY TECHNIQUE

When we obtain the splitting patterns for a particular proton and interpret it in terms of the numbers of protons located on adjacent carbons, we are using only one of the ways in which NMR spectroscopy can be applied to a structure proof problem. We may also know that a certain proton has two equivalent protons nearby that are coupled with a *J* value of 4 Hz, another nearby proton coupled with a *J* value of 10 Hz, and three others nearby that are coupled by 2 Hz. This gives a very rich pattern for the proton we are observing, but we can interpret it, with a little effort, by using a tree diagram. Selective spin decoupling may be used to collapse or sharpen portions of the spectrum in order to obtain more direct information about the nature of coupling patterns. However, each of these methods can become tedious and very difficult with complex spectra. What is needed is a simple, unbiased, and convenient method for relating coupled nuclei.

A. An Overview of the COSY Experiment

The pulse sequence for a ^1H COSY experiment contains a variable delay time, t_1, as well as an acquisition time, t_2. The experiment is repeated with different values of t_1, and the data collected during t_2 are stored in the computer. The value of t_1 is increased by regular, small intervals for each experiment, so that the data that are collected consist of a series of FID patterns collected during t_2, each with a different value of t_1.

To identify which protons couple to each other, the coupling interaction is allowed to take place during t_1. During the same period, the individual nuclear magnetization vectors spread as a result of spin-coupling interactions. These interactions modify the signal that is observed during t_2. Unfortunately, the mechanism of the interaction of spins in a COSY experiment is too complex to be described completely in a simple manner. A pictorial description must suffice.

Consider a system in which two protons are coupled to each other.

$$-\underset{\underset{H_a}{|}}{\overset{|}{C}}-\underset{\underset{H_x}{|}}{\overset{|}{C}}-$$

An initial relaxation delay and a pulse **prepare** the spin system by rotating the bulk magnetization vectors of the nuclei by 90°. At this point, the system can be described mathematically as a sum of terms, each containing the spin of only one of the two protons. The spins then **evolve** during the variable delay period (called t_1). In other words, they precess under the influences of both chemical shift and mutual spin—spin coupling. This precession modifies the signal that we finally observe during the acquisition time (t_2). In addition, mutual coupling of the spins has the mathematical effect of converting some of the single-spin terms to **products,** which contain the magnetization components of *both* nuclei. The product terms are the ones we will find most useful in analyzing the COSY spectrum.

Following the evolution period, a second 90° pulse is introduced; this constitutes the next essential part of the sequence, the **mixing period** (which we have not discussed previously). The mixing pulse has the effect of distributing the magnetization among the various spin states of the coupled nuclei. Magnetization that has been encoded by chemical shift during t_1 can be detected at another chemical shift during t_2. The mathematical description of the system is too complex to be treated here. Rather, we can say that two important types of terms arise in the treatment. The first type of term, which does not contain much information that is useful to us, results in the appearance of diagonal peaks in the two-dimensional plot. The more interesting result of the pulse sequences comes from the terms that contain the precessional frequencies of *both* coupled nuclei. The magnetization represented by these terms has been modulated (or "labeled") by the chemical shift of one nucleus during t_1 and, after the mixing pulse, by the precession of the other nucleus during t_2. The resulting off-diagonal peaks (**cross peaks**) show the correlations of pairs of nuclei by means of their spin–spin coupling. When the data are subjected to a Fourier transform, the resulting spectrum plot shows the chemical shift of the first proton plotted along one axis (f_1) and the chemical shift of the second proton plotted along the other axis (f_2). The existence of the off-diagonal peak that corresponds to the chemical shifts of both protons is *proof* of spin coupling between the two protons. If there had been no coupling, their magnetizations would not have given rise to off-diagonal peaks. In the COSY spectrum of a complete molecule, the pulses are transmitted with short duration and high power so that all possible off-diagonal peaks are generated. The result is a complete description of the coupling partners in a molecule.

Since each axis spans the entire chemical shift range, something on the order of a thousand individual FID patterns, each incremented in t_1, must be recorded. With instruments operating at a high spectrometer frequency (high-field instruments), even more FID patterns must be collected. As a result, a typical COSY experiment may require about a half hour to be completed. Furthermore, since each FID pattern must be stored in a separate memory block in the computer, this type of experiment requires a computer with a large available memory. Nevertheless, most modern instruments are capable of performing COSY experiments routinely.

B.   How to Read COSY Spectra

2-Nitropropane. To see what type of information a COSY spectrum may provide, we shall consider several examples of increasing complexity. The first is the COSY spectrum of 2-nitropropane. In this simple molecule, we expect to observe coupling between the protons on the two methyl groups and the proton at the methine position.

$$CH_3-CH-CH_3$$
$$|$$
$$NO_2$$

Figure 10.12 is the COSY spectrum of 2-nitropropane. The first thing to note about the spectrum is that the proton NMR spectrum of the compound being studied is plotted along both the horizontal and vertical axes, and each axis is calibrated according to the chemical shift values (in parts per million). The COSY spectrum shows distinct spots on a diagonal, extending from the upper right corner of the spectrum down to the lower left corner. By extending vertical and horizontal lines from each spot on the diagonal, you can easily see that each spot on the diagonal corresponds with the same peak on each coordinate axis. The diagonal peaks serve only as reference points. The important peaks in the spectrum are the **off-diagonal peaks.** In the spectrum of 2-nitropropane, we can extend a horizontal line from the spot at 1.56 ppm (which is labeled **A** and corresponds to the methyl protons). This horizontal line eventually encounters an off-diagonal spot (at the upper left of the COSY spectrum) that corresponds to the methine proton peak at 4.66 ppm (labeled **B**). A vertical line drawn from this off-diagonal spot intersects the spot on the diagonal that corresponds to the methine proton (**B**). The presence of this off-diagonal spot, which correlates the methyl proton spot and the methine proton spot, confirms that the methyl protons are coupled to the methine protons, as we would have expected. A similar result would have been obtained by drawing a vertical line from the 1.56-ppm spot (**A**) and a horizontal line from the 4.66-ppm spot (**B**). The two lines would have intersected at the second off-diagonal spot (at the lower right of the COSY spectrum). The vertical and horizontal lines described in this analysis are drawn on the COSY spectrum in Figure 10.12.

Isopentyl Acetate. In practice, we would not require a COSY spectrum to fully interpret the NMR spectrum of 2-nitropropane. The preceding analysis illustrated how to interpret a COSY spectrum, using a simple, easy-to-understand example. A more interesting example is the COSY spectrum of **isopentyl acetate** (Fig. 10.13).

$$\overset{6}{CH_3}-\overset{O}{\underset{5}{\overset{\|}{C}}}-O-\overset{4}{CH_2}-\overset{3}{CH_2}-\overset{2}{CH}-\overset{1}{CH_3}$$
$$\underset{\underset{1}{CH_3}}{|}$$

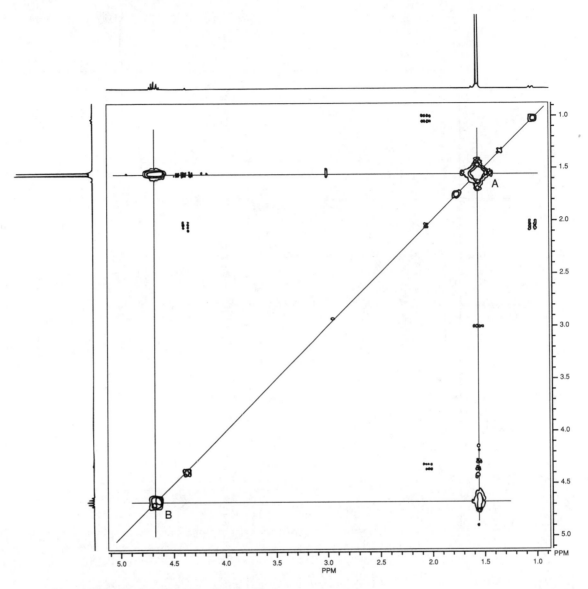

FIGURE 10.12 COSY spectrum of 2-nitropropane.

Again we see coordinate axes; the proton spectrum of isopentyl acetate is plotted along each axis. The COSY spectrum shows a distinct set of spots on a diagonal, with each spot corresponding to the same peak on each coordinate axis. Lines have been drawn to help you identify the correlations. In the COSY spectrum of isopentyl acetate, we see that the protons of the two equivalent methyl groups (**1**) correlate with the methine proton (**2**). We can also see correlation between the two methylene groups (**3** and **4**) and between the methine proton (**2**) and the neighboring methylene (**3**). The methyl group of the acetate moiety (**6**) does not show off-diagonal peaks, because the acetyl methyl protons are not coupled to other protons in the molecule.

You may have noticed that each of the COSY spectra shown in this section contains additional spots besides the ones examined in our discussion. Often these "extra" spots have much lower intensities than the principal spots on the plot. The COSY method can sometimes detect interactions between nuclei over ranges that extend beyond three bonds. Besides this long-range coupling, nuclei

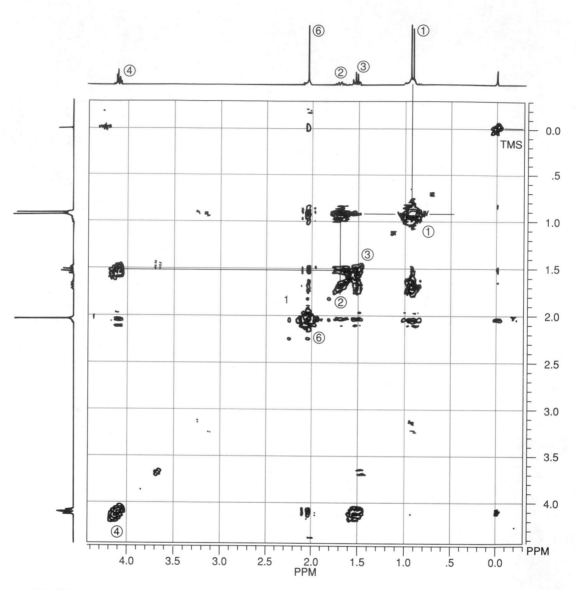

FIGURE 10.13 COSY spectrum of isopentyl acetate.

that are several atoms apart but that are close together *spatially* also may produce off-diagonal peaks. We learn to ignore these minor peaks in our interpretation of COSY spectra. In some variations of the method, however, spectroscopists make use of such long-range interactions to produce two-dimensional NMR spectra that specifically record this type of information.

Citronellol. The COSY spectrum of citronellol (see the structural formula on p. 534) is a third example. The spectrum (Fig. 10.14) is rather complex in appearance. Nevertheless, we can identify certain important coupling interactions. Again, lines have been drawn to help you identify the correlations. The proton on C6 is clearly coupled to the protons on C5. Closer examination of the spectrum also reveals that the proton on C6 is coupled through allylic (four-bond) coupling to the two methyl groups at C8 and C9. The protons on C1 are coupled to two nonequivalent protons on C2 (at 1.4 and 1.6 ppm). They are nonequivalent, owing to the presence of a stereocenter in the molecule at C3. The splitting of the methyl protons at C10 by the methine proton at C3 can also be seen, al-

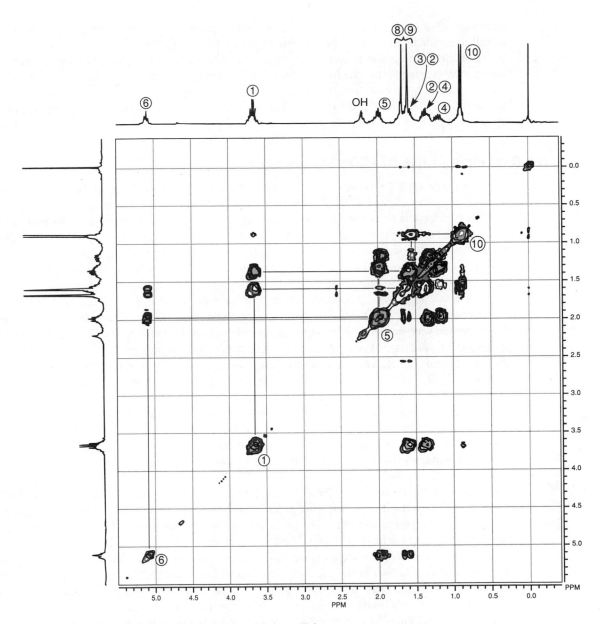

FIGURE 10.14 COSY spectrum of citronellol.

though the C3 spot on the diagonal line is obscured by other spots that are superimposed on it. However, from the COSY spectrum we can determine that the methine proton at C3 must occur at the same chemical shift as one of the C8 or C9 methyl groups (1.6 ppm). Thus, a great deal of useful information can be obtained even from a complicated COSY pattern.

10.7 THE HETCOR TECHNIQUE

Protons and carbon atoms interact in two very important ways. First, they both have magnetic properties, and they can induce relaxation in one another. Second, the two types of nuclei can be spin-coupled to each other. This latter interaction can be very useful, since directly bonded protons and carbons have a J value that is at least a power of 10 larger than nuclei related by two-bond or three-bond couplings. This marked distinction between orders of coupling provides us with a sensitive way of identifying carbons and protons that are directly bonded to one another.

To obtain a correlation between carbons and attached protons in a two-dimensional experiment, we must be able to plot the chemical shifts of the ^{13}C atoms along one axis and the chemical shifts of the protons along the other axis. A spot of intensity in this type of two-dimensional spectrum would indicate the existence of a C—H bond. The heteronuclear chemical shift correlation (HETCOR) experiment is designed to provide the desired spectrum.

A.   An Overview of the HETCOR Experiment

As we did in the COSY experiment, we want to allow the magnetization vectors of the protons to precess according to different rates, as dictated by their chemical shifts. Therefore, we apply a 90° pulse to the protons, then include an evolution time (t_1). This pulse tips the bulk magnetization vector into the $X'Y'$ plane. During the evolution period, the proton spins precess at a rate determined by their chemical shifts and the coupling to other nuclei (both protons and carbons). Protons bound to ^{13}C atoms experience not only their own chemical shifts during t_1 but also homonuclear spin coupling and heteronuclear spin coupling to the attached ^{13}C atoms. It is the interaction between ^1H nuclei and ^{13}C nuclei that produces the correlation that interests us. After the evolution time, we apply simultaneous 90° pulses to both the protons and the carbons. These pulses transfer magnetization from protons to carbons. Since the carbon magnetization was "labeled" by the proton precession frequencies during t_1, the ^{13}C signals that are detected during t_2 are modulated by the chemical shifts of the coupled protons. The ^{13}C magnetization can then be detected in t_2 to identify a particular carbon carrying each type of proton modulation.

A HETCOR experiment, like all two-dimensional experiments, describes the environment of the nuclei during t_1. Because of the manner in which the HETCOR pulse sequence has been constructed, the only interactions that are responsible for modulating the proton spin states are the proton chemical shifts and homonuclear couplings. Each ^{13}C atom may have one or more peaks appearing on the f_2 axis that correspond to its chemical shift. The proton chemical shift modulation causes the two-dimensional intensity of the proton signal to appear at an f_1 value that corresponds to the proton chemical shift. Further proton modulations of much smaller frequency arise from homonuclear (H—H) couplings. These provide fine structure on the peaks along the f_1 axis. We can interpret the fine structure exactly as we would in a normal proton spectrum, but in this case we understand that the proton chemical shift value belongs to a proton that is attached to a specific ^{13}C nucleus that appears at its own carbon chemical shift value.

We can thus assign carbon atoms on the basis of known proton chemical shifts, or we can assign protons on the basis of known carbon chemical shifts. For example, we might have a crowded proton spectrum but a carbon spectrum that is well resolved (or vice versa). This approach makes the HETCOR experiment particularly useful in the interpretation of the spectra of large, complex molecules. An even more powerful technique is to use results from both the HETCOR and COSY experiments together.

B.   How to Read HETCOR Spectra

2-Nitropropane. Figure 10.15 is an example of a simple HETCOR plot. In this case, the sample substance is 2-nitropropane.

$$CH_3-CH-CH_3$$
$$|$$
$$NO_2$$

It is common practice to plot the proton spectrum of the compound being studied along one axis and the carbon spectrum along the other axis. Each spot of intensity on the two-dimensional plot indicates a carbon atom that bears the corresponding protons. In Figure 10.15, you should be able to see a peak corresponding to the methyl carbons, which appear at 21 ppm in the carbon spectrum (horizontal

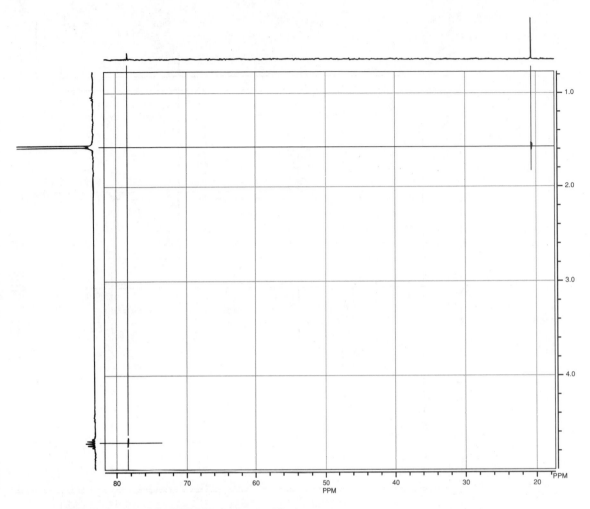

FIGURE 10.15 HETCOR spectrum of 2-nitropropane.

axis), and a peak at 79 ppm corresponding to the methine carbon. On the vertical axis, you should also be able to find the doublet for the methyl protons at 1.56 ppm (proton spectrum) and a septet for the methine proton at 4.66 ppm. If you drew a vertical line from the methyl peak of the carbon spectrum (21 ppm) and a horizontal line from the methyl peak of the proton spectrum (1.56 ppm), the two lines would intersect at the exact point on the two-dimensional plot where a spot is marked. This spot indicates that the protons at 1.56 ppm and the carbons at 21 ppm represent the same position of the molecule. That is, the hydrogens are attached to the indicated carbon. In the same way, the spot in the lower left corner of the HETCOR plot correlates with the carbon peak at 79 ppm and the proton septet at 4.66 ppm, indicating that these two absorptions represent the same position in the molecule.

Isopentyl Acetate. A second, more complex example is isopentyl acetate. Figure 10.16 is the HETCOR plot for this substance.

$$\underset{6}{CH_3}-\overset{\overset{\displaystyle O}{\parallel}}{\underset{5}{C}}-O-\underset{4}{CH_2}-\underset{3}{CH_2}-\underset{\underset{\displaystyle \underset{1}{CH_3}}{|}}{\underset{2}{CH}}-\underset{1}{CH_3}$$

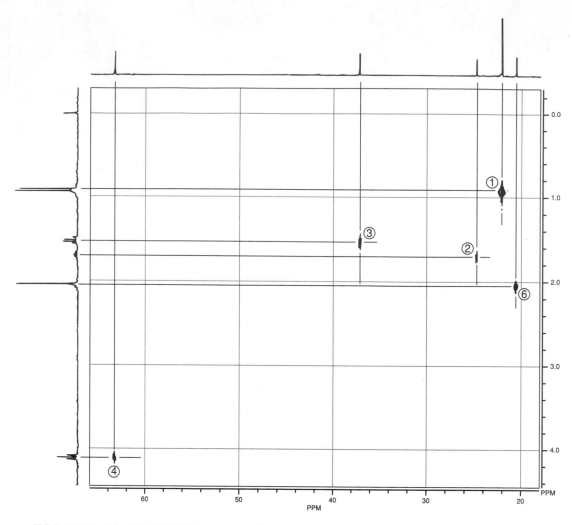

F I G U R E   1 0 . 1 6   HETCOR spectrum of isopentyl acetate.

Each spot on the HETCOR plot has been labeled with a number and lines have been drawn to help you see the correlations between proton peaks and carbon peaks. The carbon peak at 23 ppm and the proton doublet at 0.92 ppm correspond to the methyl groups (**1**); the carbon peak at 25 ppm and the proton multiplet at 1.69 ppm correspond to the methine position (**2**); and the carbon peak at 37 ppm and the proton quartet at 1.52 ppm correspond to the methylene group (**3**). The other methylene group (**4**) is deshielded by the nearby oxygen atom. Therefore, a spot on the HETCOR plot for this group appears at 63 ppm on the carbon axis and 4.10 ppm on the proton axis. It is interesting that the methyl group of the acetyl function (**6**) appears downfield of the methyl groups of the isopentyl group (**1**) in the proton spectrum (2.04 ppm). We expect this chemical shift, since the methyl protons should be deshielded by the anisotropic nature of the carbonyl group. In the carbon spectrum, however, the carbon peak appears *upfield* of the methyl carbons of the isopentyl group. A spot on the HETCOR plot that correlates these two peaks confirms that assignment.

4-Methyl-2-Pentanol. Figure 10.17 is a final example that illustrates some of the power of the HETCOR technique for 4-methyl-2-pentanol. Lines have been drawn on the spectrum to help you find the correlations.

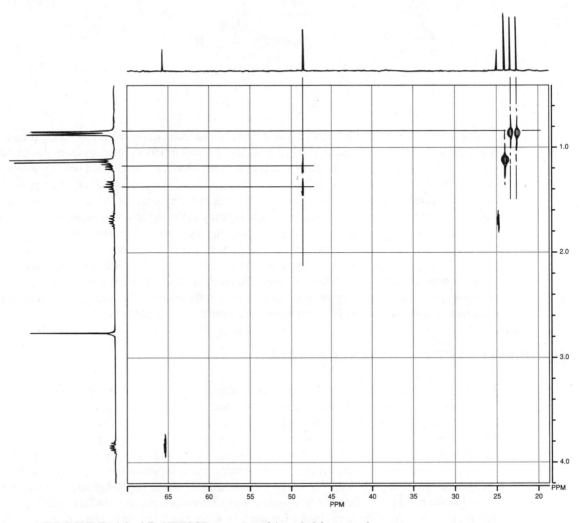

FIGURE 10.17 HETCOR spectrum of 4-methyl-2-pentanol.

$$
\begin{array}{c}
\overset{5}{CH_3} \\
| \\
\overset{1}{CH_3}-\overset{2}{CH}-\overset{3}{CH_2}-\overset{}{\underset{4}{CH}}-\overset{6}{CH_3} \\
| \\
OH
\end{array}
$$

This molecule has a stereocenter at carbon 2. An examination of the HETCOR plot for 4-methyl-2-pentanol reveals *two* spots that correspond to the two methylene protons on carbon 3. At 48 ppm on the carbon axis, two contour spots appear, one at about 1.20 ppm on the proton axis and the other at about 1.40 ppm. The HETCOR plot is telling us that there are two non-equivalent protons attached to carbon 3. If we examine a Newman projection of this molecule, we find that the presence of the stereocenter makes the two methylene protons (**a** and **b**) non-equivalent (see Sections 5.14 and 5.15). As a result, they appear at different values of chemical shift.

The *carbon* spectrum also reveals the effect of a stereocenter in the molecule. In the proton spectrum, the apparent doublet (actually it is a *pair* of doublets) at 0.91 ppm arises from the six protons on the methyl groups, which are labeled **5** and **6** in the preceding structure. Looking across to the right on the HETCOR plot, you will find two contour spots, one corresponding to 22 ppm and the other corresponding to 23 ppm. These two carbon peaks arise because the two methyl groups are also not quite equivalent; the distance from one methyl group to the oxygen atom is not quite the same as that from the other methyl group, when we consider the most likely conformation for the molecule.

A great many advanced techniques can be applied to complex molecules. We have introduced only a few of the most important ones here. As computers become faster and more powerful, as chemists evolve their understanding of what different pulse sequences can achieve, and as scientists write more sophisticated computer programs to control those pulse sequences and treat data, it will become possible to apply NMR spectroscopy to increasingly complex systems.

10.8 MAGNETIC RESONANCE IMAGING

The principles that govern the NMR experiments described throughout this textbook have begun to find application in the field of medicine. A very important diagnostic tool in medicine is a technique known as **magnetic resonance imaging (MRI).** In the space of only a few years, MRI has found wide use in the diagnosis of injuries and other forms of abnormality. It is quite common for sports fans to hear of a football star who has sustained a knee injury and had it examined via an MRI scan.

Typical magnetic resonance imaging instruments use a superconducting magnet with a field strength on the order of 1 Tesla. The magnet is constructed with a very large inner cavity so that an entire human body can fit inside. A transmitter–receiver coil (known as a **surface coil**) is positioned outside the body, near the area being examined. In most cases, the ^{1}H nucleus is the one studied, since it is found in the water molecules that are present in and around living tissue. In a manner somewhat analogous to that used for an X-ray-based CAT (computerized axial tomography) scan, a series of planar images is collected and stored in the computer. The planar images can be obtained from various angles. When the data have been collected, the computer processes the results and generates a three-dimensional picture of the proton density in the region of the body under study.

The ^{1}H nuclei of water molecules that are not bound within living cells have a relaxation time different from the nuclei of water molecules bound within tissue. Water molecules that appear in a highly ordered state have relaxation times shorter than water molecules that appear in a more random state. The degree of ordering of water molecules within tissues is greater than that of water molecules that are part of the fluid flowing within the body. Furthermore, the degree of ordering of water molecules may be different in different types of tissue, especially in diseased tissue as compared with normal tissue. Specific pulse sequences detect these differences in relaxation time for the protons of water molecules in the tissue being examined. When the results of the scans are processed, the image that is produced shows different densities of signals, depending upon the degree to which the water molecules are in an ordered state. As a result, the "picture" that we see shows the various types of tissue clearly. The radiologist can then examine the image to determine whether any abnormality exists.

As a brief illustration of the type of information that can be obtained from MRI, consider the image in Figure 10.18. This is a view of the patient's skull from the spinal column, looking toward the top of the patient's head. The light-colored areas represent the locations of soft tissues of the

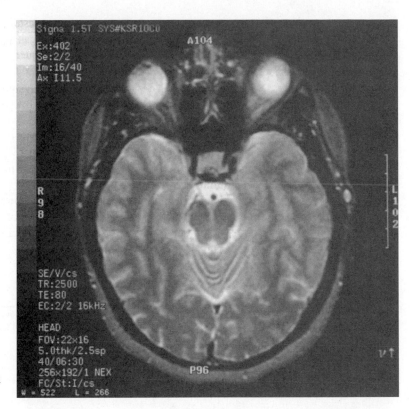

FIGURE 10.18 MRI scan of a skull, showing soft tissues of the brain and eyes.

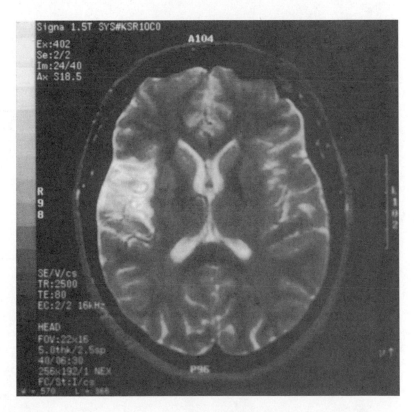

FIGURE 10.19 MRI scan of a skull, showing the presence of an infarct.

brain. Because bone does not contain a very high concentration of water molecules, the MRI image provides only a somewhat dim view of the bones of the spinal column. The two bulbous features at the top of the image are the person's eyes.

Figure 10.19 is another MRI scan of the same patient shown in Figure 10.18. On the left side of the image is an area that appears brightly white. This patient has suffered an **infarct**, an area of dying tissue resulting from an obstruction of the blood vessels supplying that part of the brain. In other words, the patient has had a stroke, and the MRI scan has clearly shown exactly where this lesion has occurred. The physician can use very specific information of this type to develop a plan for treatment.

The MRI method is not limited to the study of water molecules. Pulse sequences designed to study the distribution of lipids are also used.

The MRI technique has several advantages over conventional X-ray or CAT scan techniques; it is better suited for studies of abnormalities of soft tissue or of metabolic disorders. Furthermore, unlike other diagnostic techniques, MRI is noninvasive and painless, and it does not require the patient to be exposed to large doses of X-rays or radioisotopes.

PROBLEMS

*1. Using the following set of DEPT-135, COSY, and HETCOR spectra, provide a complete assignment of all protons and carbons for C_4H_9Cl.

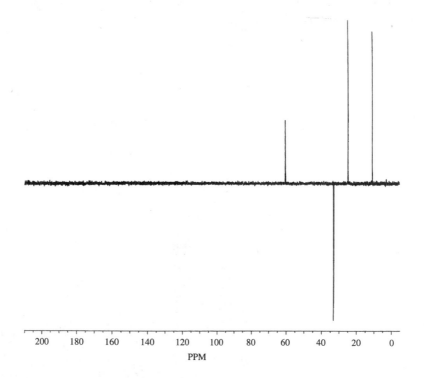

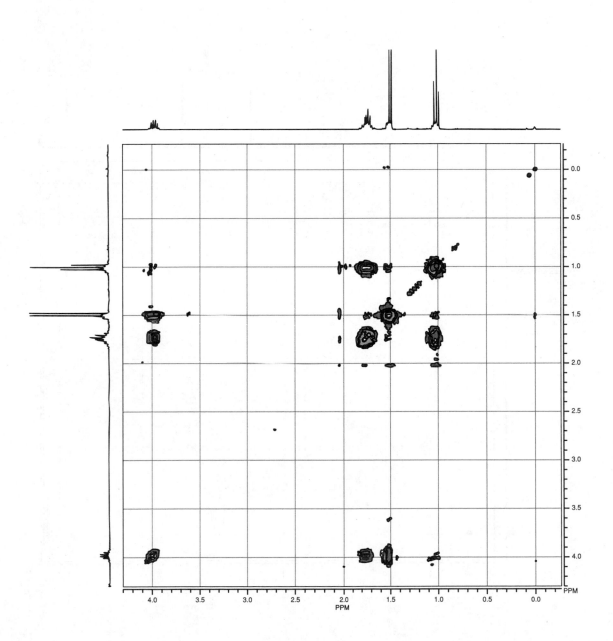

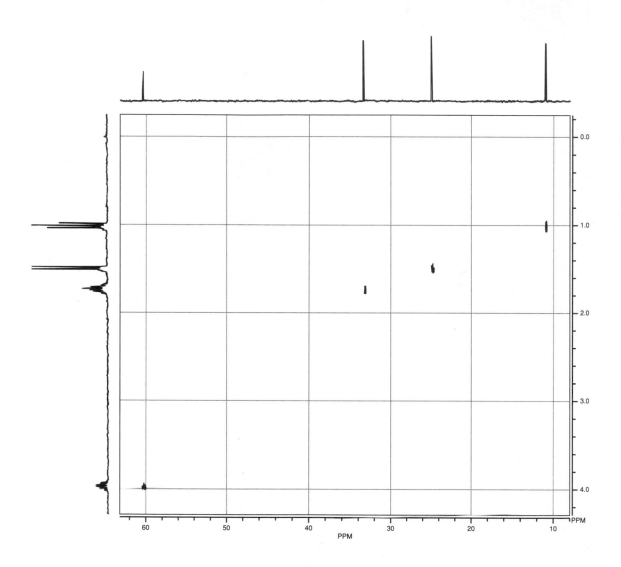

2. Determine the structure for a compound with formula $C_{13}H_{12}O_2$. The infrared spectrum shows a strong band at 1680 cm^{-1}. The normal carbon spectrum is shown in a stacked plot along with the DEPT-135 and DEPT-90 spectra. The proton spectrum and expansions are provided in the problem along the COSY spectrum. Assign all of the protons and carbons for this compound.

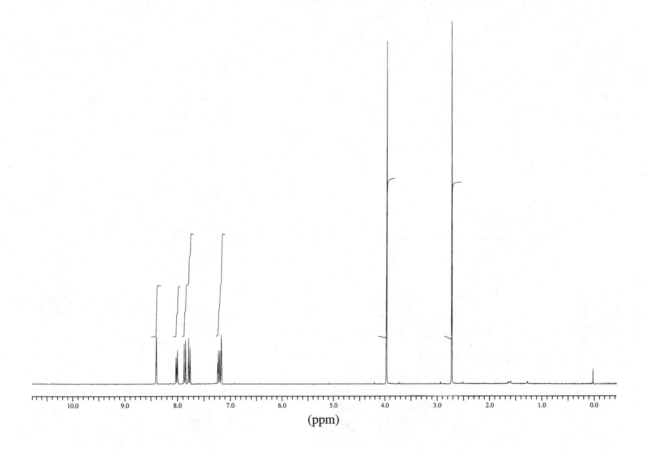

(ppm)

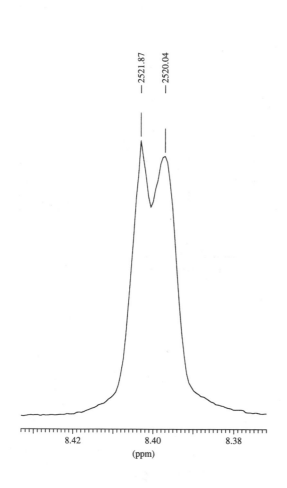

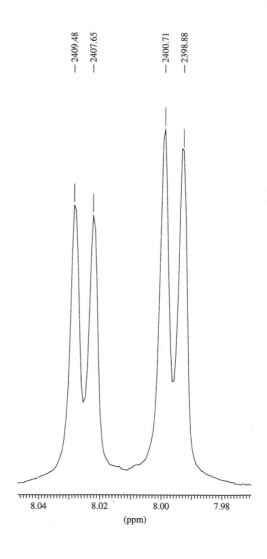

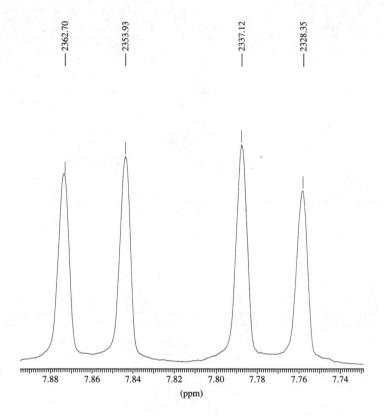

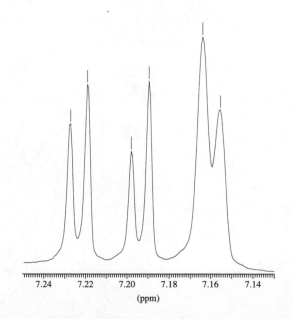

DEPT-90

DEPT-135

197.8181

159.6834

137.2069
132.5229
131.0402
129.9955
127.7265
127.0189
124.5814
119.6503

105.6657

77.4156
77.0000
76.5732

55.3435

26.4869

Normal
carbon

CDCl$_3$

(ppm)

CHCl$_3$

CHCl$_3$

(ppm)

(ppm)

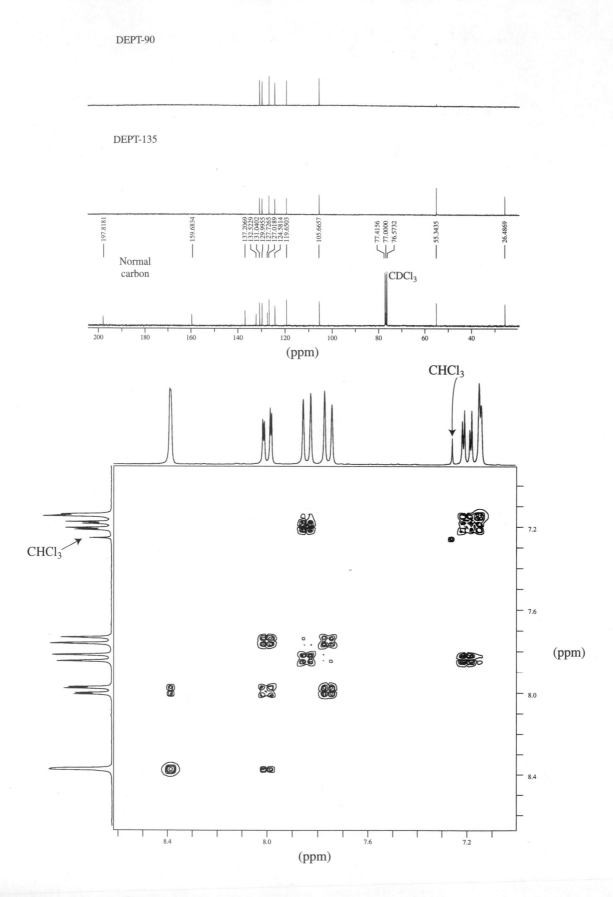

*3. Assign each of the peaks in the following DEPT spectrum of $C_6H_{14}O$. *Note*: There is more than one possible answer.

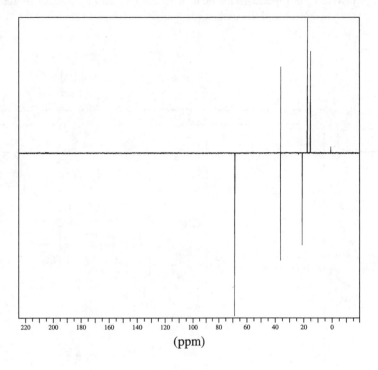

(ppm)

*4. The following HETCOR spectrum is for citronellol. Use the structural formula on p. 534, as well as the DEPT-135 spectrum (Fig. 10.11) and the COSY spectrum (Fig. 10.14), to provide a complete assignment of all carbons and hydrogens in the molecule, especially the assignment of the carbon resonances at C2 and C4.

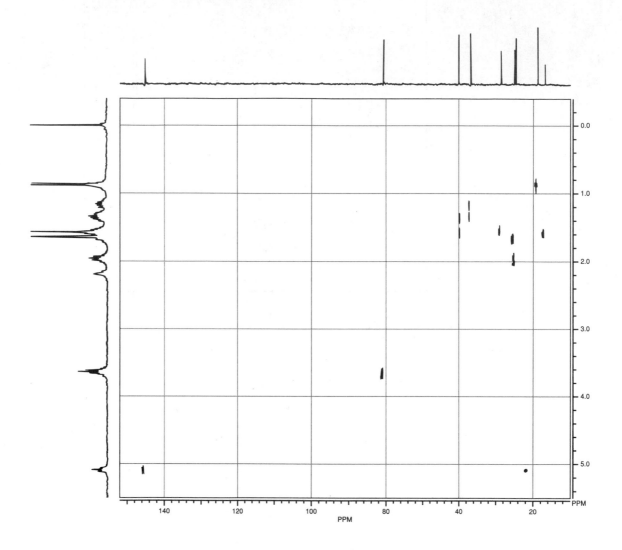

*5. Geraniol has the structure

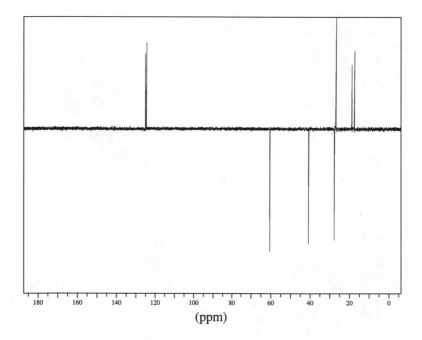

Use the DEPT-135, COSY, and HETCOR spectra to provide a complete assignment of all protons and carbons in geraniol. (*Hint*: The assignments you determined in Problem 4 may help you here.)

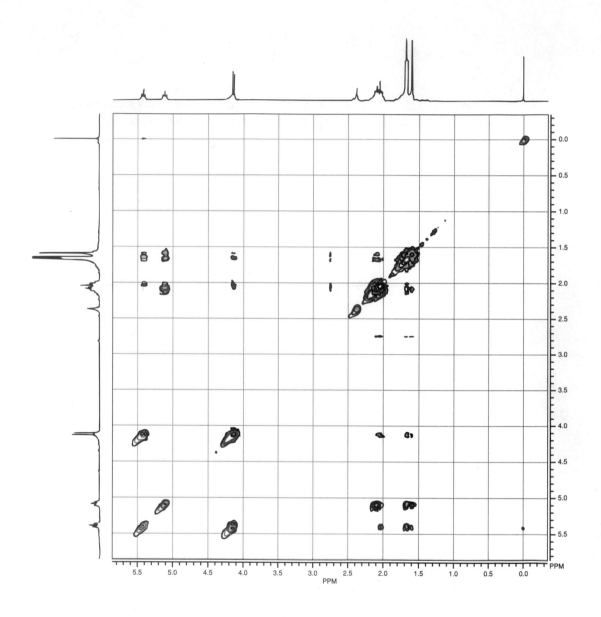

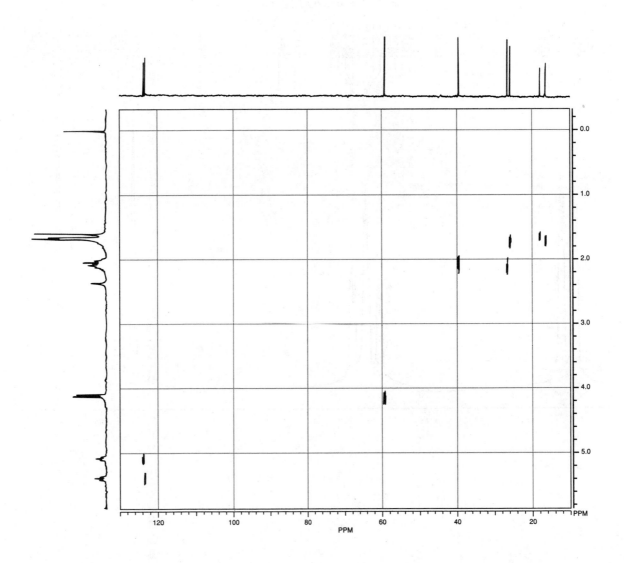

*6. The following set of spectra includes an expansion of the aromatic region of the proton NMR spectrum of **methyl salicylate,** as well as a HETCOR spectrum. Provide a complete assignment of all aromatic protons and unsubstituted ring carbons in methyl salicylate. (*Hint*: Consider the resonance effects of the substituents to determine relative chemical shifts of the aromatic hydrogens. Also try calculating the expected chemical shifts using the data provided in Appendix 6.)

METHYL SALICYLATE EXPANSION

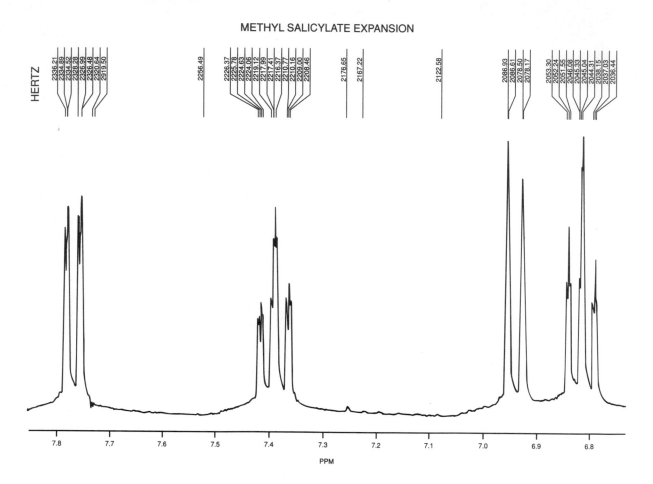

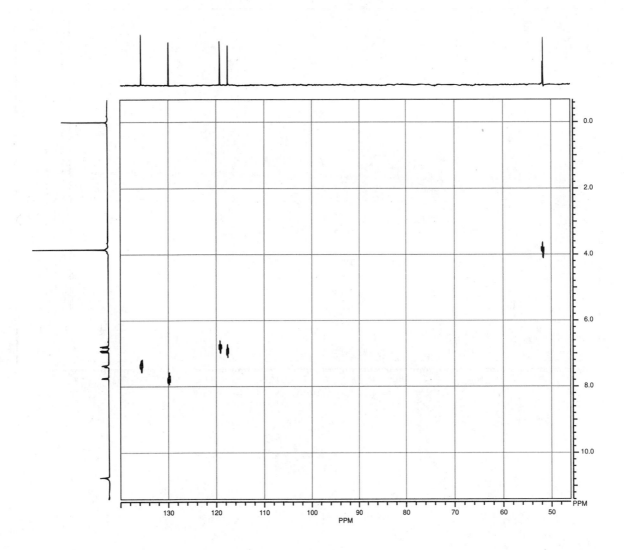

7. Determine the structure for a compound with formula $C_6H_{12}O_2$. The infrared spectrum shows a strong and broad band from 3400 to 2400 cm^{-1} and also at 1710 cm^{-1}. The proton NMR spectrum and expansions are shown, but a peak appearing at 12.0 ppm is not shown in the full spectrum. Fully interpret the proton NMR spectrum, especially the patterns between 2.1 and 2.4 ppm. A HETCOR spectrum is provided in this problem. Comment on the carbon peaks appearing at 29 and 41 ppm in the HETCOR spectrum. Assign all of the protons and carbons for this compound.

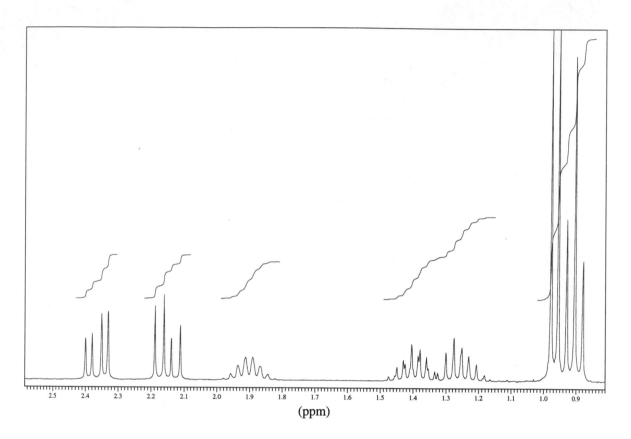

(ppm)

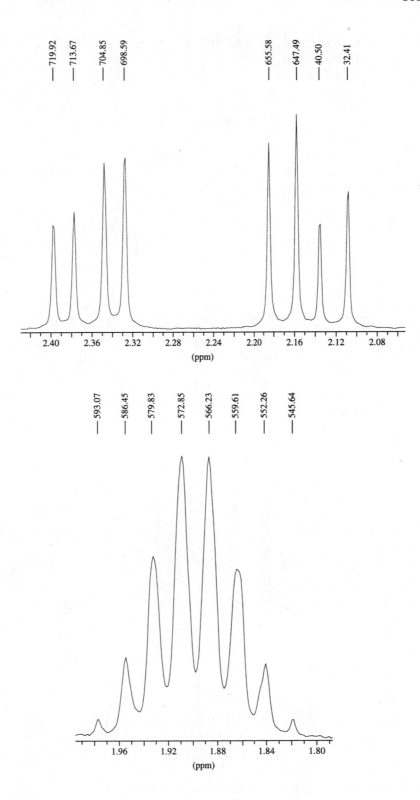

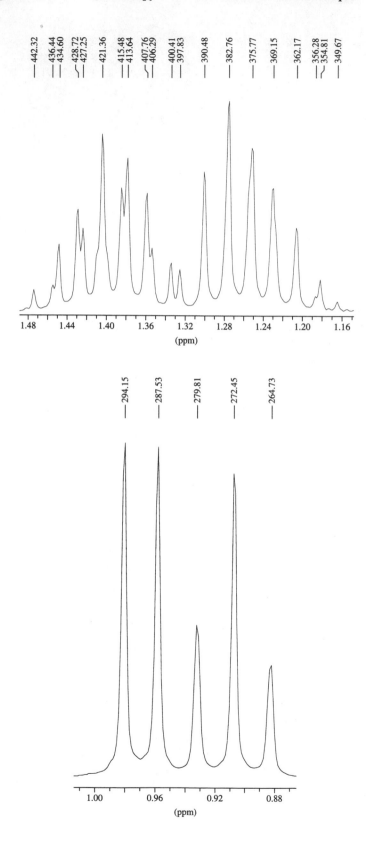

HETCOR

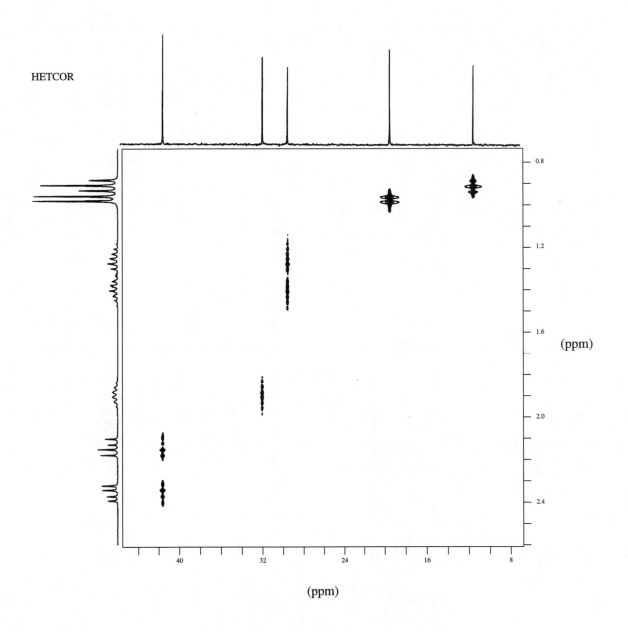

(ppm)

(ppm)

8. Determine the structure for a compound with formula $C_{11}H_{16}O$. This compound is isolated from jasmine flowers. The infrared spectrum shows strong bands at 1700 and 1648 cm^{-1}. The proton NMR spectrum, with expansions, along with the HETCOR, COSY, and DEPT spectra are provided in this problem. The DEPT-90 spectrum is not shown, but it has peaks at 125 and 132 ppm. This compound is synthesized from 2,5-hexanedione by monoalkylation with (*Z*)-1-chloro-2-pentene, followed by aldol condensation. Assign all of the protons and carbons for this compound.

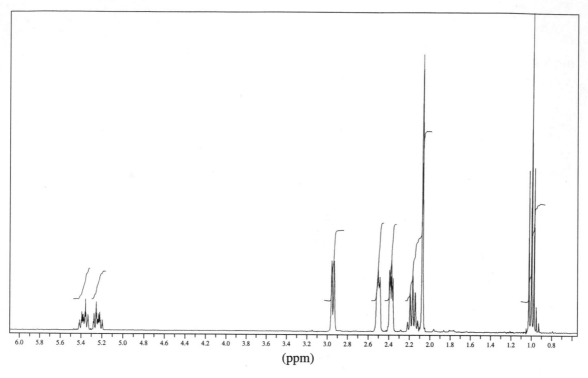

(ppm)

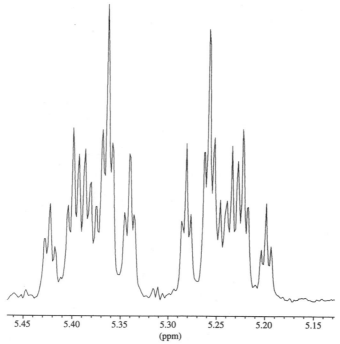

(ppm)

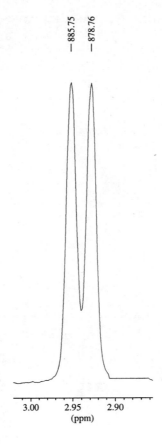

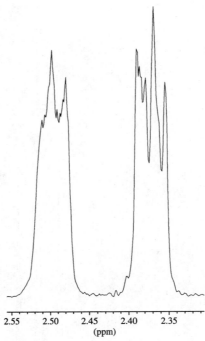

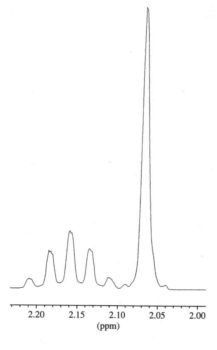

2.20   2.15   2.10   2.05   2.00
(ppm)

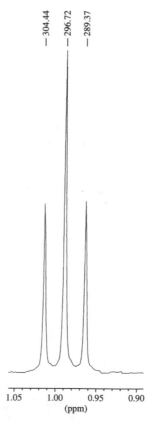

1.05   1.00   0.95   0.90
(ppm)

HETCOR

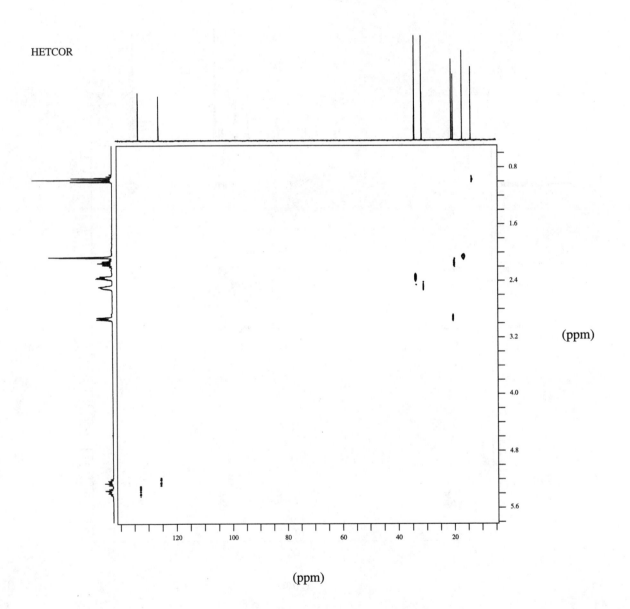

(ppm)

(ppm)

COSY

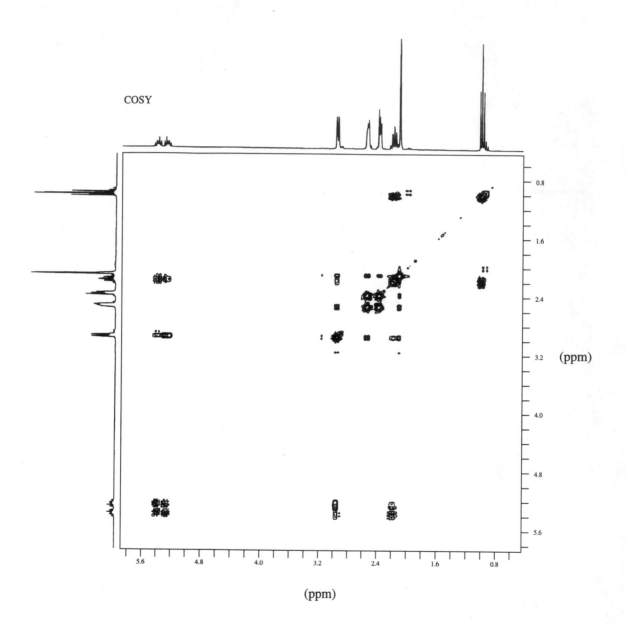

(ppm)

(ppm)

Normal Carbon

209.0915

170.4213

139.3743
132.3202
125.0565

77.4751
77.0557
76.6364

34.2669
31.6459

21.1472
20.5631
17.2832
14.1680

CDCl₃

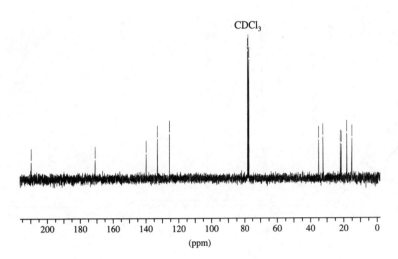

| 200 | 180 | 160 | 140 | 120 | 100 | 80 | 60 | 40 | 20 | 0 |

(ppm)

DEPT-135

132.3202
125.0565

34.2669
31.6459

21.1472
20.5631
17.2981
14.1680

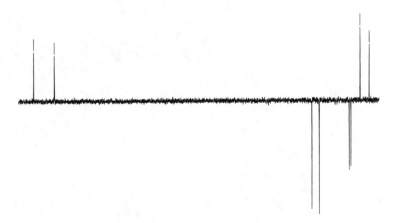

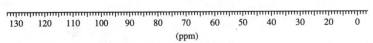

| 130 | 120 | 110 | 100 | 90 | 80 | 70 | 60 | 50 | 40 | 30 | 20 | 0 |

(ppm)

9. Determine the structure for a compound with formula $C_{10}H_8O_3$. The infrared spectrum shows strong bands at 1720 and 1620 cm^{-1}. In addition, the infrared spectrum has bands at 1580, 1560, 1508, 1464, and 1125 cm^{-1}. The proton NMR spectrum, with expansions, along with the COSY and DEPT spectra are provided in this problem. Assign all of the protons and carbons for this compound.

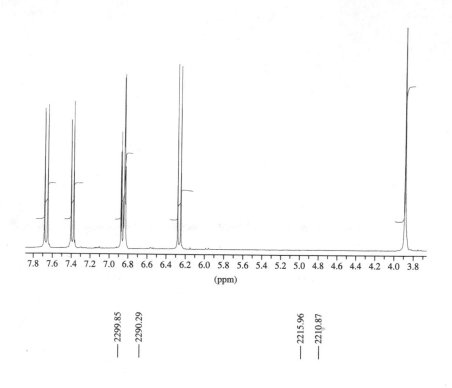

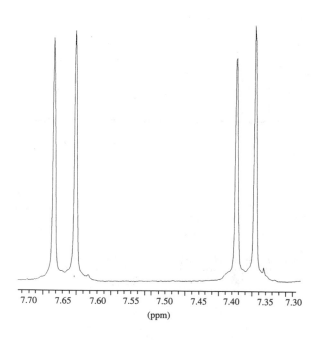

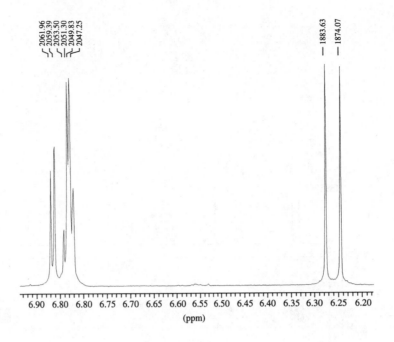

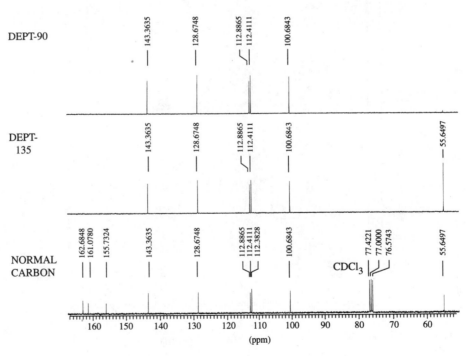

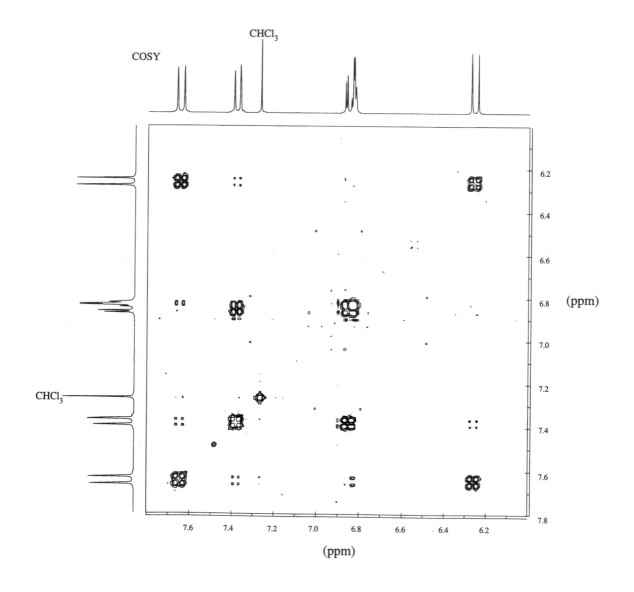

REFERENCES

Becker, E. D., *High Resolution NMR: Theory and Chemical Applications*, 3rd ed., Academic Press, San Diego, CA, 2000.

Croasmun, W. R., and R. M. K. Carlson, editors, *Two-Dimensional NMR Spectroscopy*, VCH Publishers, New York, 1994.

Derome, A. E., *Modern NMR Techniques for Chemistry Research*, Pergamon Press, Oxford, England, 1987.

Friebolin, H., *Basic One- and Two-Dimensional NMR Spectroscopy*, 3rd rev. ed., Wiley-VCH, Weinheim, Germany, 1998.

Sanders, J. K. M., and B. K. Hunter, *Modern NMR Spectroscopy: A Guide for Chemists*, 2nd ed., Oxford University Press, Oxford, England, 1993.

Schraml, J., and J. M. Bellama, *Two-Dimensional NMR Spectroscopy*, Wiley, New York, 1988.

Silverstein, R. M., and F. X. Webster, *Spectrometric Identification of Organic Compounds*, 6th ed., Wiley, New York, 1998 (Chapter 6).

Another valuable source of information about advanced NMR methods is a series of articles which appeared in *Journal of Chemical Education* under the general title "The Fourier Transform in Chemistry." The specific volume and page citations are as follows.

Volume 66 (1989), p. A213
Volume 66 (1989), p. A243
Volume 67 (1990), p. A93
Volume 67 (1990), p. A100
Volume 67 (1990), p. A125

Selected Web sites

Webspectra: Problems in NMR and IR Spectroscopy (C. A. Merlic, Project Director): http://www.chem.ucla.edu/~webnmr

The Basics of NMR (Joseph P. Hornek, Ph. D.): http://www.cis.rit.edu/htbooks/nmr

ANSWERS TO SELECTED PROBLEMS

CHAPTER 1

1. (a) 90.50% carbon; 9.50% hydrogen   (b) C_4H_5

2. 32.0% carbon; 5.4% hydrogen; 62.8% chlorine; $C_3H_6Cl_2$

3. $C_2H_5NO_2$

4. 180.2 = molecular mass. Molecular formula is $C_9H_8O_4$.

5. Equivalent weight = 52.3

6. (a) 6   (b) 1   (c) 3   (d) 6   (e) 12

7. The index of hydrogen deficiency = 1. There cannot be a triple bond, since the presence of a triple bond would require an index of hydrogen deficiency of at least 2.

8. (a) 59.96% carbon; 5.75% hydrogen; 34.29% oxygen   (b) $C_7H_8O_3$ (c) $C_{21}H_{24}O_9$
   (d) A maximum of two aromatic (benzenoid) rings

9. (a) $C_8H_8O_2$ (b) $C_8H_{12}N_2$ (c) $C_7H_8N_2O$ (d) $C_5H_{12}O_4$

10. Molecular formula = $C_8H_{10}N_4O_2$
    Index of hydrogen deficiency = 6

11. Molecular formula = $C_{21}H_{30}O_2$
    Index of hydrogen deficiency = 7

CHAPTER 2

1. (a) Propargyl chloride (3-chloropropyne)   (b) *p*-Cymene (4-isopropyltoluene)
   (c) *m*-Toluidine (3-methylaniline)   (d) *o*-Cresol (2-methylphenol)
   (e) *N*-Ethylaniline   (f) 2-Chlorotoluene
   (g) 2-Chloropropanoic acid   (h) 3-Methyl-1-butanol
   (i) 5-Hexen-2-one   (j) 1,2,3,4-Tetrahydronaphthalene
   (k) 3-(Dimethylamino)propanenitrile   (l) 1,2-Epoxybutane

2. Citronellal

3. *trans*-Cinnamaldehyde (*trans*-3-phenyl-2-propenal)

4. Upper spectrum, *trans*-3-hexen-1-ol; Lower spectrum, *cis*-3-hexen-1-ol

5. (a) Structure B (ethyl cinnamate)   (b) Structure C (cyclobutanone)
   (c) Structure D (2-ethylaniline)   (d) Structure A (propiophenone)
   (e) Structure D (butanoic anhydride)

6. Poly(acrylonitrile-styrene); poly(methyl methacrylate); polyamide (nylon)

C H A P T E R   3

1. (a) $-1, 0, +1$ (b) $-\frac{1}{2}, +\frac{1}{2}$ (c) $-\frac{5}{2}, -\frac{3}{2}, -\frac{1}{2}, +\frac{1}{2}, +\frac{3}{2}, +\frac{5}{2}$ (d) $-\frac{1}{2}, +\frac{1}{2}$

2. 128 Hz/60 MHz = 2.13 ppm

3. (a) 180 Hz   (b) 1.50 ppm

4. See Figures 3.22 and 3.23. The methyl protons are in a shielding region. Acetonitrile shows similar anisotropic behavior to acetylene.

5. *o*-Hydroxyacetophenone is intramolecularly hydrogen bonded. The proton is deshielded (12.05 ppm). Changing concentration does not alter the extent of hydrogen bonding. Phenol is intermolecularly hydrogen bonded. The extent of hydrogen bonding depends upon concentration.

6. The methyl groups are in a shielding region of the double bonds. See Figure 3.23.

7. The carbonyl group deshields the *ortho* protons owing to anisotropy.

8. The methyl groups are in the shielding region of the double-bonded system. See Figure 3.24.

9. The spectrum will be similar to that in Figure 3.25, with some differences in chemical shifts. Spin arrangements: H_A will be identical to the pattern in Figure 3.32 (triplet); H_B will see one adjacent proton and will appear as a doublet ($+\frac{1}{2}$ and $-\frac{1}{2}$).

10. The isopropyl group will appear as a septet for the α-H (methine). From Pascal's triangle, the intensities are 1:6:15:20:15:6:1. The CH_3 groups will be a doublet.

11. Downfield doublet, area = 2, for the protons on carbon 1 and carbon 3; upfield triplet, area = 1, for the proton on carbon 2.

12. $X-CH_2-CH_2-Y$, where $X \neq Y$.

13. Upfield triplet for the C-3 protons, area = 3; intermediate sextet for the C-2 protons, area = 2; and downfield triplet for the C-1 protons, area = 2

14. Ethyl acetate (ethyl ethanoate)

15. Isopropylbenzene

16. 2-Bromobutanoic acid

17. (a) Propyl acetate   (b) Isopropyl acetate

18. 1,3-Dibromopropane

19. 2,2-Dimethoxypropane

20. (a) Isobutyl propanoate   (b) *t*-Butyl propanoate   (c) Butyl propanoate

21. (a) 3-Chloropropanoic acid   (b) 2-Chloropropanoic acid

22. (a) 2-Phenylbutane (*sec*-butylbenzene)   (b) 1-Phenylbutane (butylbenzene)

23. 2-Phenylethylamine

CHAPTER 4

1. Methyl acetate

2. (c) 7 peaks   (d) 3 peaks
   (e) 5 peaks   (f) 10 peaks
   (g) 10 peaks   (h) 4 peaks
   (i) 5 peaks   (j) 6 peaks
   (k) 8 peaks

3. (a) 2-Methyl-2-propanol   (b) 2-Butanol   (c) 2-Methyl-1-propanol

4. Methyl methacrylate (methyl 2-methyl-2-propenoate)

5. (a) 2-Bromo-2-methylpropane   (b) 2-Bromobutane   (c) 1-Bromobutane
   (d) 1-Bromo-2-methylpropane

6. (a) 4-Heptanone   (b) 2,4-Dimethyl-3-pentanone   (c) 4,4-Dimethyl-2-pentanone

17. 2,3-Dimethyl-2-butene. A primary cation rearranges to a tertiary cation via a hydride shift. E1 elimination forms the tetrasubstituted alkene.

18. (a) Three equal-sized peaks for ^{13}C coupling to a single D atom; quintet for ^{13}C coupling to two D atoms.

    (b) Fluoromethane: doublet for ^{13}C coupling to a single F atom ($^{1}J > 180$ Hz).
    Trifluoromethane: quartet for ^{13}C coupling to three F atoms ($^{1}J > 180$ Hz).
    1,1-Difluoro-2-chloroethane: triplet for carbon-1 coupling to two F atoms ($^{1}J > 180$ Hz); triplet for carbon-2 coupling to two F atoms ($^{2}J \approx 40$ Hz).
    1,1,1-trifluoro-2-chloroethane: quartet for carbon-1 coupling to three F atoms ($^{1}J > 180$ Hz); quartet for carbon-2 coupling to three F atoms ($^{2}J \approx 40$ Hz).

19. C1 = 128.5 + 9.3 = 137.8 ppm; C2 = 128.5 + 0.7 = 129.2 ppm; C3 = 128.5 − 0.1 = 128.4 ppm; C4 = 128.5 − 2.9 = 125.6 ppm.

20. All carbons are numbered according to IUPAC rules. The following information is given: the name of the compound, the number of the table used (T2–T7, Appendix 8), and, where needed, the name of the reference compound used (from T1, Appendix 8). If actual values are known, they are given in parentheses.

    (a) Methyl vinyl ether, T4 (actual: 153.2, 84.2 ppm)
        C1 = 123.3 + 29.4 = 152.7   C2 = 123.3 − 38.9 = 84.4

    (b) Cyclopentanol, T3-cyclopentane (actual: 73.3, 35.0, 23.4 ppm)
        C1 = 25.6 + 41 = 66.6   C2 = 25.6 + 8 = 33.6   C3 = 25.6 − 5 = 20.6

    (c) 2-Pentene, T5 or T6 (actual: 123.2, 132.7 ppm)
        C2 = 123.3 + 10.6 − 7.9 − 1.8 = 124.2   C3 = 123.3 + 10.6 + 7.2 − 7.9 = 133.2
        Using Table A5.4:
        C2 = 123.3 + 10.6 − 9.7 = 124.2   C3 = 123.3 + 15.5 − 7.9 = 130.9

(d) *ortho*-Xylene, T7
   C1,C2 = 128.5 + 9.3 + 0.7 = 138.5
   C3,C6 = 128.5 + 0.7 − 0.1 = 129.1
   C4,C5 = 128.5 − 0.1 − 2.9 = 125.5
   meta-Xylene, T7 (actual: 137.6, 130.0, 126.2, 128.2 ppm)
   C1,C3 = 128.5 + 9.3 − 0.1 = 137.7
   C2 = 128.5 + 0.7 + 0.7 = 129.9
   C4,C6 = 128.5 + 0.7 − 2.9 = 126.3
   C5 = 128.5 − 0.1 − 0.1 = 128.3
   para-Xylene, T7
   C1,C4 = 128.5 + 9.3 − 2.9 = 134.9
   C2,C3,C5,C6 = 128.5 + 0.7 − 0.1 = 129.1

(e) 3-Pentanol, T3-pentane (actual: 9.8, 29.7, 73.8 ppm)
   C1,C5 = 13.9 − 5 = 8.9   C2,C4 = 22.8 + 8 = 30.8   C3 = 34.7 + 41 = 75.7

(f) 2-Methylbutanoic acid, T3-butane
   C1 = 13.4 + 2 = 15.4   C2 = 25.2 + 16 = 41.2   C3 = 25.2 + 2 = 27.2   C4 = 13.4 − 2 + 11.4

(g) 1-Phenyl-1-propene, T4 or T6
   C1 = 123.3 + 12.5 − 7.9 = 127.9   C2 = 123.3 + 10.6 − 11 = 122.9

(h) 2,2-Dimethylbutane, T3 or T2 (actual: 29.1, 30.6, 36.9, 8.9 ppm)
   Using Table A8.3: C1 = 13.4 + 8 + 8 = 29.4   C2 = 25.2 + 6 + 6 = 37.2
   C3 = 25.2 + 8 + 8 = 41.2   C4 = 13.4 − 2 − 2 = 9.4
   Using Table A8.2:
   C1 = −2.3 + [9.1(1) + 9.4(3) − 2.5(1)] + [(−3.4)] = 29.1
   C2 = −2.3 + [9.1(4) + 9.4(1)] + [3(−1.5) + (−8.4)] = 30.6
   C3 = −2.3 + [9.1(2) + 9.4(3)] + [(0) + (−7.5)] = 36.6
   C4 = −2.3 + [9.1(1) + 9.4(1) − 2.5(3)] + [(0)] = 8.7

(i) 2,3-Dimethyl-2-pentenoic acid, T6
   C2 = 123.3 + 4 + 10.6 − 7.9 − 7.9 − 1.8 = 120.3
   C3 = 123.3 + 10.6 + 10.6 + 7.2 + 9 − 7.9 = 152.8

(j) 4-Octene, T5, and assume *trans* geometry
   C4,C5 = 123.3 + [10.6 + 7.2 − 1.5] − [7.9 + 1.8 − 1.5] = 131.4
   To estimate *cis*, correct as follows: 131.4 − 1.1 = 130.3

(k) 4-Aminobenzoic acid, T7
   C1 = 128.5 + 2.9 − 9.5 = 121.9   C2 = 128.5 + 1.3 + 1.3 = 131.1
   C3 = 128.5 − 12.4 + 0.4 = 116.5   C4 = 128.5 + 19.2 + 4.3 = 152.0

(l) 1-Pentyne, T3-propane
   C3 = 15.8 + 4.5 = 20.3   C4 = 16.3 + 5.4 = 21.7   C5 = 15.8 − 3.5 = 12.3

(m) Methyl 2-methylpropanoate, T3-propane
   C2 = 16.3 + 17 = 33.3   C3 = 15.8 + 2 = 17.8

(n) 2-Pentanone, T3-propane
   C3 = 15.8 + 30 = 45.8   C4 = 16.3 + 1 = 17.3   C5 = 15.8 − 2 = 13.8

(o) Bromocyclohexane, T3-cyclohexane
   C1 = 26.9 + 25 = 51.9   C2 = 26.9 + 10 = 36.9   C3 = 26.9 − 3 = 23.9
   C4 = 26.9 (no correction)

(p) 2-Methylpropanoic acid, T3-propane
   C1 = 15.8 + 2 = 17.8   C2 = 16.3 + 16 = 32.3

(q) 4-Nitroaniline, T7 (actual: 155.1, 112.8, 126.3, 136.9 ppm)
   C1 = 128.5 + 19.2 + 6.0 = 153.7   C2 = 128.5 − 12.4 + 0.9 = 117.0
   C3 = 128.5 + 1.3 − 5.3 = 124.5   C4 = 128.5 + 19.6 − 9.5 = 138.6
   2-Nitroaniline, T7
   C1 = 128.5 + 19.2 − 5.3 = 142.4   C2 = 128.5 − 12.4 + 19.6 = 135.7
   C3 = 128.5 − 5.3 + 1.3 = 124.5   C4 = 128.5 + 0.9 − 9.5 = 119.9
   C5 = 128.5 + 1.3 + 0.9 = 130.7   C6 = 128.5 − 12.4 + 0.9 = 117.0

(r) 1,3-Pentadiene, T4
   C3 = 123.3 + 13.6 − 7.9 = 129.0   C4 = 123.3 + 10.6 − 7 = 126.9

(s) Cyclohexene, T5 (actual: 127.3 ppm)
   C1,C2 = 123.3 + [10.6 + 7.2 − 1.5] − [7.9 + 1.8 − 1.5] + [−1.1] = 130.3

(t) 4-Methyl-2-pentene, T5, and assume *trans*
   C2 = 123.3 + [10.6(1)] − [7.9(1) + 1.8(2)] = 122.4
   C3 = 123.3 + [10.6(1) + 7.2(2)] − [7.9(1)] + 2.3 = 142.7

CHAPTER 5

---

1. Refer to Section 5.8 for instructions on measuring coupling constants using the Hertz values that are printed above the expansions of the proton spectra.

   (a) Vinyl acetate (Fig. 5.17): all vinyl protons are doublets of doublets.

   H_a = 4.57 ppm, $^3J_{ac}$ = 6.25 Hz and $^2J_{ab}$ = 1.47 Hz.

   H_b = 4.88 ppm. The coupling constants are not consistent; $^3J_{bc}$ =13.98 or 14.34 Hz from the spacing of the peaks. $^2J_{ab}$ = 1.48 or 1.84 Hz. It is often the case that the coupling constants are not consistent (see Section 5.8). More consistent coupling constants can be obtained from analysis of proton H_c.

   H_c = 7.27 ppm, $^3J_{bc}$ = 13.97 Hz and $^3J_{ac}$ = 6.25 Hz from the spacing of the peaks.

   Summary of coupling constants from the analysis of the spectrum: $^3J_{ac}$ =6.25 Hz, $^3J_{bc}$ = 13.97 Hz and $^2J_{ab}$ = 1.47 Hz. They can be rounded off to: 6.3, 14.0 and 1.5 Hz, respectively.

   (b) *trans*-Crotonic acid (Fig. 5.21).

   H_a = 1.92 ppm (methyl group at C-4). It appears as a doublet of doublets (dd) because it shows both ^3J and ^4J couplings; $^3J_{ac}$ = 6.9 Hz and $^4J_{ab}$ allylic = 1.6 Hz.

   H_b = 5.86 ppm (vinyl proton at C-2). It appears as a doublet of quartets (dq); $^3J_{bc}$ trans = 15.6 Hz and $^4J_{ab}$ allylic = 1.6 Hz.

   H_c = 7.10 ppm (vinyl proton at C-3). It appears as a doublet of quartets (dq), with some partial overlap of the quartets; $^3J_{bc}$ trans = 15.6 Hz and $^3J_{ac}$ = 6.9 Hz. Notice that H_c is shifted further downfield than H_b because of the resonance effect of the carboxyl group and also a through-space deshielding by the oxygen atom in the carbonyl group.

$H_d = 12.2$ ppm (singlet, acid proton on carboxyl group).

(c) 2-Nitrophenol (Fig. 5.49). H_a and H_b are shielded by the electron releasing effect of the hydroxyl group caused by the non-bonded electrons on the oxygen atom being involved in resonance. They can be differentiated by their appearance: H_a is a triplet with some fine structure and H_b is a doublet with fine structure. H_d is deshielded by the electron withdrawing effect and by the anisotropy of the nitro group. Notice that the pattern is a doublet with some fine structure. H_c is assigned by a process of elimination. It lacks any of the above effects that shields or deshields that proton. It appears as a triplet with some fine structure.

$H_a = 7.00$ ppm (ddd); $^3J_{ac} \cong {}^3J_{ad} = 8.5$ Hz and $^4J_{ab} = 1.5$ Hz. H_a could also be described as a triplet of doublets (td) since $^3J_{ac}$ and $^3J_{ad}$ are nearly equal.

$H_b = 7.16$ ppm (dd); $^3J_{bc} = 8.5$ Hz and $^4J_{ab} = 1.5$ Hz.

$H_c = 7.60$ ppm (ddd or td); $^3J_{ac} \cong {}^3J_{bc} = 8.5$ Hz and $^4J_{cd} = 1.5$ Hz.

$H_d = 8.12$ ppm (dd); $^3J_{ad} = 8.5$ Hz and $^4J_{cd} = 1.5$ Hz; $^5J_{bd} = 0$.

The OH group is not shown in the spectrum.

(d) 3-Nitrobenzoic acid (Fig. 5.50). H_d is significantly deshielded by the anisotropy of both the nitro and carboxyl groups and appears furthest downfield. It appears as a narrowly spaced triplet. This proton only shows ^4J couplings. H_b is *ortho* to a carboxyl group while H_c is *ortho* to a nitro group. Both protons are deshielded, but the nitro group shifts a proton further downfield than for a proton next to a carboxyl group (see Appendix 6). Both H_b and H_c are doublets with fine structure consistent with their positions on the aromatic ring. H_a is relatively shielded and appears upfield as a widely spaced triplet. This proton does not experience any anisotropy effect because of its distance away from the attached groups. H_a has only ^3J couplings ($^5J_{ad} = 0$).

$H_a = 7.72$ ppm (dd); $^3J_{ac} = 8.1$ Hz and $^3J_{ab} = 7.7$ Hz (these values come from analysis of H_b and H_c, below). Since the coupling constants are similar, the pattern appears as an accidental triplet.

$H_b = 8.45$ ppm (ddd or dt); $^3J_{ab} = 7.7$ Hz; $^4J_{bd} \cong {}^4J_{bc} = 1.5$ Hz. The pattern is an accidental doublet of triplets.

$H_c = 8.50$ ppm (ddd); $^3J_{ac} = 8.1$ Hz and $^4J_{cd} \neq {}^4J_{bc}$.

$H_d = 8.96$ ppm (dd). The pattern appears to be a narrowly spaced triplet, but is actually an accidental triplet since $^4J_{bd} \neq {}^4J_{cd}$.

The carboxyl proton is not shown in the spectrum.

(e) Furfuryl alcohol (Fig. 5.51). The chemical shift values and coupling constants for a furanoid ring are given in Appendix 4 and 5.

$H_a = 6.24$ ppm (doublet of quartets); $^3J_{ab} = 3.2$ Hz and $^4J_{ac} = 0.9$ Hz. The quartet pattern results from a nearly equal ^4J coupling of H_a to the two methylene protons in the CH_2OH group and the ^4J coupling of H_a to H_c (n + 1 rule, three protons plus one equals four, a quartet).

$H_b = 6.31$ ppm (dd); $^3J_{ab} = 3.2$ Hz and $^3J_{bc} = 1.9$ Hz.

$H_c = 7.36$ ppm (dd); $^3J_{bc} = 1.9$ Hz and $^4J_{ac} = 0.9$ Hz.

The CH_2 and OH groups are not shown in the spectrum.

(f) 2-Methylpyridine (Fig. 5.52). Typical chemical shift values and coupling constants for a pyridine ring are given in Appendix 4 and 5.

H_a = 7.08 ppm (dd); $^3J_{ac}$ = 7.4 Hz and $^3J_{ad}$ = 4.8 Hz.

H_b = 7.14 ppm (d); $^3J_{bc}$ = 7.7 Hz and $^4J_{ab} \cong 0$ Hz.

H_c = 7.56 ppm (ddd or td). This pattern is a likely accidental triplet of doublets because $^3J_{ac} \cong {}^3J_{bc}$ and $^4J_{cd}$ = 1.8 Hz.

H_d = 8.49 ppm ("doublet"). Because of the broadened peaks in this pattern, it is impossible to extract the coupling constants. We expect a doublet of doublets, but $^4J_{cd}$ is not resolved from $^3J_{ad}$. The adjacent nitrogen atom may be responsible for the broadened peaks.

2. (a) J_{ab} = 0 Hz   (b) J_{ab} ~ 10 Hz   (c) J_{ab} = 0 Hz   (d) J_{ab} ~ 1 Hz
   (e) J_{ab} = 0 Hz   (f) J_{ab} ~ 10 Hz   (g) J_{ab} = 0 Hz   (h) J_{ab} = 0 Hz
   (i) J_{ab} ~ 10 Hz; J_{ac} ~ 16 Hz; J_{bc} ~ 1 Hz

3.

H_a = 2.80 ppm (singlet, CH_3).

H_b = 5.98 ppm (doublet); $^3J_{bd}$ = 9.9 Hz and $^2J_{bc}$ = 0 Hz.

H_c = 6.23 ppm (doublet); $^3J_{cd}$ = 16.6 Hz and $^2J_{bc}$ = 0 Hz.

H_d = 6.61 ppm (doublet of doublets); $^3J_{cd}$ = 16.6 Hz and $^3J_{bd}$ = 9.9 Hz.

4.

H_a = 0.88 ppm (triplet, CH_3); $^3J_{ac}$ = 7.4 Hz.

H_c = 2.36 ppm (quartet, CH_2; $^3J_{ac}$ = 7.4 Hz.

H_b = 1.70 ppm (doublet of doublets, CH_3); $^3J_{be}$ = 6.8 Hz and $^4J_{bd}$ = 1.6 Hz.

H_d = 5.92 ppm (doublet of quartets, vinyl proton). The quartets are narrowly spaced, suggesting a four bond coupling, ^4J; $^3J_{de}$ = 15.7 Hz and $^4J_{bd}$ = 1.6 Hz.

H_e = 6.66 ppm (doublet of quartets, vinyl proton). The quartets are widely spaced, suggesting a three bond coupling, ^3J; $^3J_{de}$ = 15.7 Hz and $^3J_{be}$ = 6.8 Hz. H_e appears further downfield than H_d (see the answer to problem 1b for an explanation).

5.

H_a = 0.96 ppm (triplet, CH_3); $^3J_{ab}$ = 7.4 Hz.

H_d = 6.78 ppm (doublet of triplets, vinyl proton). The triplets are widely spaced suggesting a three bond coupling, ^3J; $^3J_{cd}$ = 15.4 Hz and $^3J_{bd}$ = 6.3 Hz. H_d appears further downfield than H_c (see the answer to problem 1b for an explanation).

H_b = 2.21 ppm (quartet of doublets of doublets, CH_2) resembles a quintet with fine structure. $^3J_{ab}$ = 7.4 Hz and $^3J_{bd}$ = 6.3 Hz are derived from the H_a and H_d patterns while $^4J_{bc}$ = 1.5 Hz is obtained from the H_b pattern (left hand doublet at 2.26 ppm) or from the H_c pattern.

H_c = 5.95 ppm (doublet of doublets of triplets, vinyl proton). The triplets are narrowly spaced, suggesting a four bond coupling, ^4J; $^3J_{cd}$ = 15.4 Hz, $^3J_{ce}$ = 7.7 Hz and $^4J_{bc}$ = 1.5 Hz.

H_e = 9.35 ppm (doublet, aldehyde proton); $^3J_{ce}$ = 7.7 Hz.

6. Structure **A** would show allylic coupling. The C−H bond orbital is parallel to the π system of the double bond leading to more overlap. A stronger coupling of the two protons results.

14. 3-Bromoacetophenone. The aromatic region of the proton spectrum shows one singlet, two doublets and one triplet consistent with a 1,3-disubstituted (*meta*) pattern. Each carbon atom in the aromatic ring is unique leading to the observed six peaks in the carbon spectrum. The downfield peak at near 197 ppm is consistent with a ketone C=O. The integral value (3H) in the proton spectrum and the chemical shift value (2.6 ppm) indicates that a methyl group is present. The most likely possibility is that there is an acetyl group attached to the aromatic ring. A bromine atom is the other substituent on the ring.

15. Valeraldehyde (pentanal). The aldehyde peak on carbon 1 appears at 9.8 ppm. It is split into a triplet by the two methylene protons on carbon 2 (^3J = 1.9 Hz). Aldehyde protons often have smaller three-bond (vicinal) coupling constants than typically found. The pattern at 2.4 ppm (triplet of doublets) is formed from coupling with the two protons on carbon 3 (^3J = 7.4 Hz) and with the single aldehyde proton on carbon 1 (^3J = 1.9 Hz).

16. The DEPT spectral results indicate that the peak at 15 ppm is a CH_3 group; 40 and 63 ppm peaks are CH_2 groups; 115 and 130 ppm peaks are CH groups; 125 and 158 ppm peaks are quaternary (*ipsi* carbons). The 179 ppm peak in the carbon spectrum is a C=O group at a value typical for esters and carboxylic acids. A carboxylic acid is indicated since a broad peak appears at 12.5 ppm in the proton spectrum. The value for the chemical shift of the methylene carbon peak at 63 ppm indicates an attached oxygen atom. Confirmation of this is seen in the proton spectrum (4 ppm, a quartet), leading to the conclusion that the compound has an ethoxy group (triplet at 1.4 ppm for the CH_3 group). A *para* disubstituted aromatic ring is indicated with the carbon spectrum (two C−H and two C with no protons). This substitution pattern is also indicated in the proton spectrum (two doublets at 6.8 and 7.2 ppm). The remaining methylene group at 40 ppm in the carbon spectrum is a singlet in the proton spectrum indicating no adjacent protons. The compound is 4-ethoxyphenylacetic acid.

22. (a) In the *proton* NMR, one fluorine atom splits the CH_2 ($^2J_{HF}$) into a doublet. This doublet is shifted downfield because of the influence of the electronegative fluorine atom. The CH_3 group is too far away from the fluorine atom and thus appears upfield as a singlet.

(b) Now the operating frequency of the NMR is changed so that only *fluorine* atoms are observed. The fluorine NMR would show a triplet for the single fluorine atom because of the two adjacent protons ($n + 1$ Rule). This would be the only pattern observed in the spectrum. Thus, we do not see protons directly in a fluorine spectrum because the spectrometer is operating at a different frequency. We do see, however, the *influence* of the protons on the fluorine spectrum. The J values would be the same as those obtained from the *proton* NMR.

23. The aromatic proton spectral data indicates a 1,3-disubstituted (*meta* substituted) ring. One attached substituent is a methyl group (2.35 ppm, integrating for 3H). Since the ring is disubstituted, the remaining substituent would be an oxygen atom attached to the remaining two carbon atoms with one proton and four fluorine atoms in the "ethoxy" group. This substituent would most likely be a 1,1,2,2-tetrafluoroethoxy group. The most interesting pattern is the widely spaced triplet of triplets centering on 5.85 ppm; $^2J_{HF} = 53.1$ Hz for the proton on carbon 2 of the ethoxy group coupled to two adjacent fluorine atoms (two bond, ^2J) and $^3J_{HF} = 2.9$ Hz for this same proton on carbon 2 coupled to the remaining two fluorine atoms on carbon 1 (three bond, ^3J) from this proton. The compound is 1-methyl-3-(1,1,2,2-tetrafluoroethoxy)benzene.

25. In the *proton* NMR, the attached deuterium, which has a spin = 1, splits the methylene protons into a triplet (equal intensity for each peak, a 1:1:1 pattern). The methyl group is too far removed from deuterium to have any influence, and it will be a singlet. Now change the frequency of the NMR to a value where only *deuterium* undergoes resonance. Deuterium will see two adjacent protons on the methylene group, splitting it into a triplet (1:2:1 pattern). No other peaks will be observed since, at this NMR frequency, the only atom observed is deuterium. Compare the results to the answers in Problem 22.

26. Two singlets will appear in the proton NMR spectrum: a downfield CH_2 and an upfield CH_3 group. Compare this result to the answer in problem 22a.

27. Phosphorus has a spin of $\frac{1}{2}$. The two methoxy groups, appearing at about 3.7 ppm in the proton NMR, are split into a doublet by the phosphorus atom ($^3J_{HP} \cong 8$ Hz). Since there are two equivalent methoxy groups, the protons integrate for 6H. The methyl group directly attached to the same phosphorus atom appears at about 1.5 ppm (integrates for 3H). This group is split by phosphorus into a doublet ($^2J_{HP} \cong 13$ Hz). Phosphorus coupling constants are provided in Appendix 5.

30. (a) δ_H ppm $= 0.23 + 1.70 = 1.93$ ppm
    (b) δ_H ppm (α to two C=O groups) $= 0.23 + 1.70 + 1.55 = 3.48$ ppm
    δ_H ppm (α to one C=O group) $= 0.23 + 1.70 + 0.47 = 2.40$ ppm
    (c) δ_H ppm $= 0.23 + 2.53 + 1.55 = 4.31$ ppm
    (d) δ_H ppm $= 0.23 + 1.44 + 0.47 = 2.14$ ppm
    (e) δ_H ppm $= 0.23 + 2.53 + 2.53 + 0.47 = 5.76$ ppm
    (f) δ_H ppm $= 0.23 + 2.56 + 1.32 = 4.11$ ppm

31. (a) δ_H ppm (*cis* to COOCH$_3$) $= 5.25 + 1.15 - 0.29 = 6.11$ ppm
    δ_H ppm (*trans* to COOCH$_3$) $= 5.25 + 0.56 - 0.26 = 5.55$ ppm
    (b) δ_H ppm (*cis* to CH$_3$) $= 5.25 + 0.84 - 0.26 = 5.83$ ppm
    δ_{II} ppm (*cis* to COOCH$_3$) $= 5.25 + 1.15 + 0.44 = 6.84$ ppm
    (c) δ_H ppm (*cis* to C$_6$H$_5$) $= 5.25 + 0.37 = 5.62$ ppm
    δ_H ppm (*gem* to C$_6$H$_5$) $= 5.25 + 1.35 = 6.60$ ppm
    δ_H ppm (*trans* to C$_6$H$_5$) $= 5.25 - 0.10 = 5.15$ ppm
    (d) δ_H ppm (*cis* to C$_6$H$_5$) $= 5.25 + 0.37 + 1.10 = 6.72$ ppm
    δ_H ppm (*cis* to COCH$_3$) $= 5.25 + 1.13 + 1.35 = 7.73$ ppm
    (e) δ_H ppm (*cis* to CH$_3$) $= 5.25 + 0.67 - 0.26 = 5.66$ ppm
    δ_H ppm (*cis* to CH$_2$OH) $= 5.25 - 0.02 + 0.44 = 5.67$ ppm
    (f) δ_H ppm $= 5.25 + 1.10 - 0.26 - 0.29 = 5.80$ ppm

32. In the answers provided here, numbering begins with the group attached at the top of the ring.

   (a) δ_H (proton 2 and 6) $= 7.27 - 0.14 + 0.26 = 7.39$ ppm
   δ_H (proton 3 and 5) $= 7.27 - 0.06 + 0.95 = 8.16$ ppm

   (b) δ_H (proton 2) $= 7.27 - 0.48 + 0.95 = 7.74$ ppm
   δ_H (proton 4) $= 7.27 - 0.44 + 0.95 = 7.78$ ppm
   δ_H (proton 5) $= 7.27 - 0.09 + 0.26 = 7.44$ ppm
   δ_H (proton 6) $= 7.27 - 0.48 + 0.38 = 7.17$ ppm

   (c) δ_H (proton 3) $= 7.27 - 0.09 + 0.95 = 8.13$ ppm
   δ_H (proton 4) $= 7.27 - 0.44 + 0.26 = 7.09$ ppm
   δ_H (proton 5) $= 7.27 - 0.09 + 0.38 = 7.56$ ppm
   δ_H (proton 6) $= 7.27 - 0.48 + 0.26 = 7.05$ ppm

   (d) δ_H (proton 2 and 6) $= 7.27 + 0.71 - 0.25 = 7.73$ ppm
   δ_H (proton 3 and 5) $= 7.27 + 0.10 - 0.75 = 6.62$ ppm

   (e) δ_H (proton 3) $= 7.27 + 0.10 - 0.75 = 6.62$ ppm
   δ_H (proton 4) $= 7.27 + 0.21 - 0.25 = 7.23$ ppm
   δ_H (proton 5) $= 7.27 + 0.10 - 0.65 = 6.72$ ppm
   δ_H (proton 6) $= 7.27 + 0.71 - 0.25 = 7.73$ ppm

   (f) δ_H (proton 2 and 6) $= 7.27 + 0.71 - 0.02 = 7.96$ ppm
   δ_H (proton 3 and 5) $= 7.27 + 0.10 + 0.03 = 7.40$ ppm

   (g) δ_H (proton 3) $= 7.27 + 0.18 + 0.03 + 0.38 = 7.86$ ppm
   δ_H (proton 4) $= 7.27 + 0.28 - 0.02 + 0.26 = 7.79$ ppm
   δ_H (proton 5) $= 7.27 + 0.18 - 0.09 + 0.95 = 8.31$ ppm

   (h) δ_H (proton 2) $= 7.27 + 0.85 + 0.95 - 0.02 = 9.05$ ppm
   δ_H (proton 5) $= 7.27 + 0.18 + 0.26 + 0.03 = 7.74$ ppm
   δ_H (proton 6) $= 7.27 + 0.85 + 0.38 - 0.02 = 8.48$ ppm

   (i) δ_H (proton 2 and 6) $= 7.27 - 0.56 - 0.02 = 6.69$ ppm
   δ_H (proton 3 and 5) $= 7.27 - 0.12 + 0.03 = 7.18$ ppm

C H A P T E R   6

1. The methylene group is a quartet of doublets. Draw a tree diagram where the quartet has spacings of 7 Hz. This represents the ^3J (three bond coupling) to the CH_3 group from the methylene protons. Now split each leg of the quartet into doublets (5 Hz). This represents the ^3J (three bond coupling) of the methylene protons to the O—H group. The pattern can also be interpreted as a doublet of quartets, where the doublet (5 Hz) is constructed first, followed by splitting each leg of the doublet into quartets (7 Hz spacings).

2. 2-Methyl-3-buten-2-ol. $H_a = 1.3$ ppm; $H_b = 1.9$ ppm; $H_c = 5.0$ ppm (doublet of doublets, $^3J_{ce} = 10.7$ Hz (*cis*) and $^2J_{cd} = 0.9$ Hz (*geminal*)); $H_d = 5.2$ ppm (doublet of doublets, $^3J_{de} = 17.4$ Hz (*trans*) and $^2J_{cd} = 0.9$ Hz (*geminal*)); $H_e = 6.0$ ppm (doublet of doublets, $^3J_{de} = 17.4$ Hz and $^3J_{ce} = 10.7$ Hz.

3. 2-Bromophenol. The unexpanded spectrum shows two doublets and two triplets, consistent with a 1,2-disubstituted (*ortho*) pattern. Each shows fine structure in the expansions (^4J). Assignments can be made by assuming that the two upfield protons (shielded) are *ortho* and *para* with respect to the electron-releasing OH group. The other two patterns can be assigned by a process of elimination.

4. The two structures shown here are the ones that can be derived from 2-methylphenol. The infrared spectrum shows a significantly shifted conjugated carbonyl group which suggests that the OH group is releasing electrons and providing single bond character to the C=O group, consistent with 4-hydroxy-3-methylacetophenone (the other compound would not have as significant a shift in the C=O). The 3136 cm^{-1} peak is an OH group, also seen in the NMR spectrum as a solvent-dependent peak. Both structures shown would be expected to show a singlet and two doublets in the aromatic region of the NMR spectrum. The positions of the downfield singlet and doublet in the spectrum fit the calculated values from Appendix 6 for 4-hydroxy-3-methylacetophenone more closely than for 3-hydroxy-4-methylacetophenone (calculated values are shown on each structure). The other doublet appearing at 6.9 ppm is a reasonable fit to the calculated value of 6.79 ppm. It is interesting to note that the two *ortho* protons in 3-hydroxy-4-methylacetophenone are deshielded by the C=O group and shielded by the OH group leading to little shift from the base value of 7.27 (Appendix 6). In conclusion, the NMR spectrum and calculated values best fit 4-hydroxy-3-methylacetophenone.

4-hydroxy-3-methylacetophenone   3-hydroxy-4-methylacetophenone

5. All of the compounds would have a singlet and two doublets in the aromatic portion of the NMR spectrum. When comparing the calculated values to the observed chemical shifts, it is important to compare the *relative* positions of each proton (positions of doublet, singlet, and doublet). Don't be concerned with slight differences (about ± 0.10 Hz) in the calculated *vs* observed values. The calculated values for third compound fits the observed spectral data better than the first two.

			observed
6.51 s	6.59 d	6.48 d	6.57 d
6.62 d	6.84 d	6.51 s	6.64 s
6.95 d	6.87 s	6.95 d	6.97 d

6. 3-Methyl-3-buten-1-ol. The DEPT spectral results show a CH$_3$ group at 22 ppm, two CH$_2$ groups at 41 and 60 ppm. The peaks at 112 ppm (CH$_2$) and 142 ppm (C with no attached H) are part of a vinyl group. The peaks at 4.78 and 4.86 ppm in the proton spectrum are the

protons on the terminal double bond. The 4.78 ppm pattern (fine structure) shows long range coupling (^{4}J) to the methyl and methylene groups. The methylene group at 2.29 ppm is broadened because of non-resolved ^{4}J coupling.

9. 4-Butylaniline

10. 2,6-Dibromoaniline

12. 2,4-Dichloroaniline. The broad peak at about 4 ppm is assigned to the $-NH_2$ group. The doublet at 7.23 ppm is assigned to the proton on carbon 3 (it appears as a near singlet in the upper trace). Proton 3 is coupled, long range, to the proton on carbon 5 ($^4J = 2.3$ Hz). The doublet of doublets centering on 7.02 ppm is assigned to the proton on carbon 5. It is coupled to the proton on carbon 6 ($^3J = 8.6$ Hz) and also to proton 3 ($^4J = 2.3$ Hz). Finally, the doublet at 6.65 ppm is assigned to the proton on carbon 6 ($^3J = 8.6$ Hz), which arises from coupling to the proton on carbon 5. There is no sign of ^5J coupling in this compound.

13. Alanine

21. Rapid equilibration at room temperature between chair conformations leads to one peak. As one lowers the temperature, the interconversion is slowed down until, at temperatures below $-66.7°C$, peaks due to the axial and equatorial hydrogens are observed. Axial and equatorial hydrogens have different chemical shifts under these conditions.

22. The *t*-butyl-substituted rings are conformationally locked. The hydrogen at C4 has different chemical shifts, depending upon whether it is axial or equatorial. 4-Bromocyclohexanes are conformationally mobile. No difference between axial and equatorial hydrogens is observed until the rate of chair–chair interconversion is decreased by lowering the temperature.

CHAPTER 7

1. (a) $\varepsilon = 13,000$ (b) $I_0/I = 1.26$

2. (a) 2,4-Dichlorobenzoic acid or 3,4-dichlorobenzoic acid   (b) 4,5-Dimethyl-4-hexen-3-one
   (c) 2-Methyl-1-cyclohexenecarboxaldehyde

3. (a) Calculated: 215 nm   observed: 213 nm
   (b) Calculated: 249 nm   observed: 249 nm
   (c) Calculated: 214 nm   observed: 218 nm
   (d) Calculated: 356 nm   observed: 348 nm
   (e) Calculated: 244 nm   observed: 245 nm
   (f) Calculated: 303 nm   observed: 306 nm
   (g) Calculated: 249 nm   observed: 245 nm
   (h) Calculated: 281 nm   observed: 278 nm
   (i) Calculated: 275 nm   observed: 274 nm
   (j) Calculated: 349 nm   observed: 348 nm

4. 166 nm: $n \to \sigma^*$
   189 nm: $\pi \to \pi^*$
   279 nm: $n \to \pi^*$

5. Each absorption is due to $n \to \sigma^*$ transitions. As one goes from the *chloro* to the *bromo* to the *iodo* group, the electronegativity of the halogens decreases. The orbitals interact to different degrees, and the energies of the n and the σ^* states differ.

6. (a) $\sigma \to \sigma^*$, $\sigma \to \pi^*$, $\pi \to \pi^*$, and $\pi \to \sigma^*$

   (b) $\sigma \to \sigma^*$, $\sigma \to \pi^*$, $\pi \to \pi^*$, $\pi \to \sigma^*$, $n \to \sigma^*$, and $n \to \pi^*$

   (c) $\sigma \to \sigma^*$ and $n \to \sigma^*$

   (d) $\sigma \to \sigma^*$, $\sigma \to \pi^*$, $\pi \to \pi^*$, $\pi \to \sigma^*$, $n \to \sigma^*$, and $n \to \pi^*$

   (e) $\sigma \to \sigma^*$ and $n \to \sigma^*$

   (f) $\sigma \to \sigma^*$

CHAPTER 8

1. $C_{43}H_{50}N_4O_6$

2. $C_{34}H_{44}O_{13}$

3. $C_{12}H_{10}O$

4. C_6H_{12}

5. C_7H_9N

6. C_3H_7Cl

7. (a) Methylcyclohexane
   (b) 2-Methyl-1-pentene
   (c) 2-Methyl-2-hexanol
   (d) Ethyl isobutyl ether
   (e) 2-Methylpropanal
   (f) 3-Methyl-2-heptanone
   (g) Ethyl octanoate
   (h) 2-Methylpropanoic acid
   (i) 4-Methylbenzoic acid
   (j) Butylamine
   (k) 2-Propanethiol
   (l) Nitroethane
   (m) Propanenitrile
   (n) Iodoethane
   (o) Chlorobenzene
   (p) 1-Bromobutane
   (q) Bromobenzene
   (r) 1,1-Dichloroethane
   (s) 1,2,3-Trichloro-1-propene

CHAPTER 9

1. 2-Butanone

2. 1-Propanol

3. 3-Pentanone

4. Methyl trimethylacetate (methyl 2,2-dimethylpropanoate)

5. Phenylacetic acid

6. 4-Bromophenol

7. Valerophenone (1-phenyl-1-pentanone)

8. Ethyl 3-bromobenzoate; ethyl 4-bromobenzoate

9. *N,N*-dimethylethylamine

10. 2-Pentanone

11. Ethyl formate

12. 2-Bromoacetophenone; 4-bromoacetophenone

13. Butyraldehyde (butanal)

14. 3-Methyl-1-butanol

15. Ethyl 2-bromopropionate (ethyl 2-bromopropanoate);
    ethyl 3-bromopropionate (ethyl 3-bromopropanoate)

16. Ethyl 4-cyanobenzoate

17. 3-Chloropropiophenone (3-chloro-1-phenyl-1-propanone)

C H A P T E R   1 0

1.

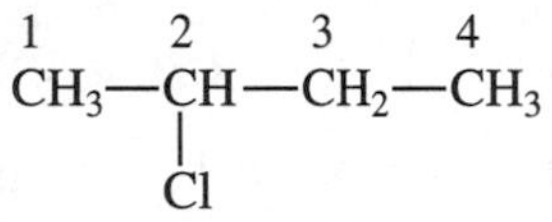

$$\overset{1}{CH_3}-\overset{2}{CH}-\overset{3}{CH_2}-\overset{4}{CH_3}$$
$$|$$
$$Cl$$

Proton #1: 1.5 ppm   Carbon #1: 24 ppm
Proton #2: 4.0 ppm   Carbon #2: 60 ppm
Proton #3: 1.7 ppm   Carbon #3: 33 ppm (inverted peak indicates CH_2)
Proton #4: 1.0 ppm   Carbon #4: 11 ppm

3.

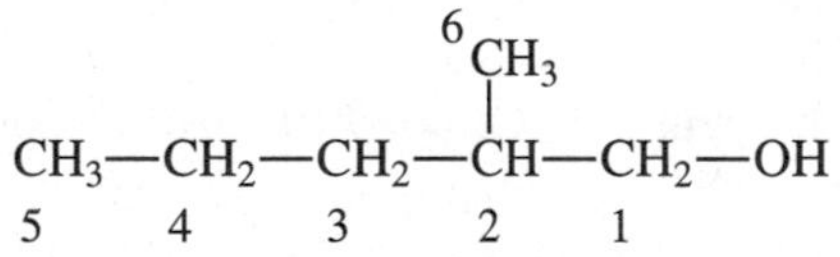

$$\overset{6}{CH_3}$$
$$|$$
$$\overset{}{CH_3}-\overset{}{CH_2}-\overset{}{CH_2}-\overset{}{CH}-\overset{}{CH_2}-OH$$
$$\;\;5\quad\;\;4\quad\;\;3\quad\;\;2\quad\;\;1$$

Carbon #1: 68 ppm
Carbon #2: 35.2 ppm
Carbon #3: 35.3 ppm
Carbon #4: 20 ppm
Carbon #5: 14 ppm
Carbon #6: 16 ppm

3-Methyl-1-pentanol and 4-methyl-1-pentanol would be expected to give similar DEPT spectra. They are also acceptable answers based on the information provided.

4.

Proton #1: 3.8 ppm   Carbon #1: 61 ppm
Proton #2: 1.4 and 1.6 ppm   Carbon #2: 40 ppm
Proton #3: 1.6 ppm   Carbon #3: 30 ppm
Proton #4: 1.2 and 1.3 ppm   Carbon #4: 37 ppm
Proton #5: 2.0 ppm   Carbon #5: 25 ppm
Proton #6: 5.2 ppm   Carbon #6: 125 ppm
Proton #7: —   Carbon #7: 131 ppm
Proton #8: 1.6 ppm   Carbon #8: 17 ppm
Proton #9: 1.7 ppm   Carbon #9: 25 ppm
Proton #10: 0.9 ppm   Carbon #10: 19 ppm

5. Proton #1: 4.1 ppm   Carbon #1: 59 ppm
    Proton #2: 5.4 ppm   Carbon #2: 124 ppm
    Proton #3: —   Carbon #3: —
    Proton #4: 2.1 ppm   Carbon #4: 39 ppm
    Proton #5: 2.2 ppm   Carbon #5: 26 ppm
    Proton #6: 5.1 ppm   Carbon #6: 124.5 ppm
    Proton #7: —   Carbon #7: —
    Proton #8: 1.6 ppm   Carbon #8: 18 ppm
    Proton #9: 1.7 ppm   Carbon #9: 16 or 25 ppm
    Proton #10: 1.7 ppm   Carbon #10: 16 or 25 ppm

6.

Proton #3: 6.95 ppm   Carbon #3: 117 ppm
Proton #4: 7.40 ppm   Carbon #4: 136 ppm
Proton #5: 6.82 ppm   Carbon #5: 119 ppm
Proton #6: 7.75 ppm   Carbon #6: 130 ppm
$J_{3,4} = 8$ Hz $J_{3,5} = 1$ Hz $J_{3,6} \sim 0$ Hz
$J_{4,5} = 7$ Hz $J_{4,6} = 2$ Hz $J_{5,6} = 8$ Hz

APPENDICES

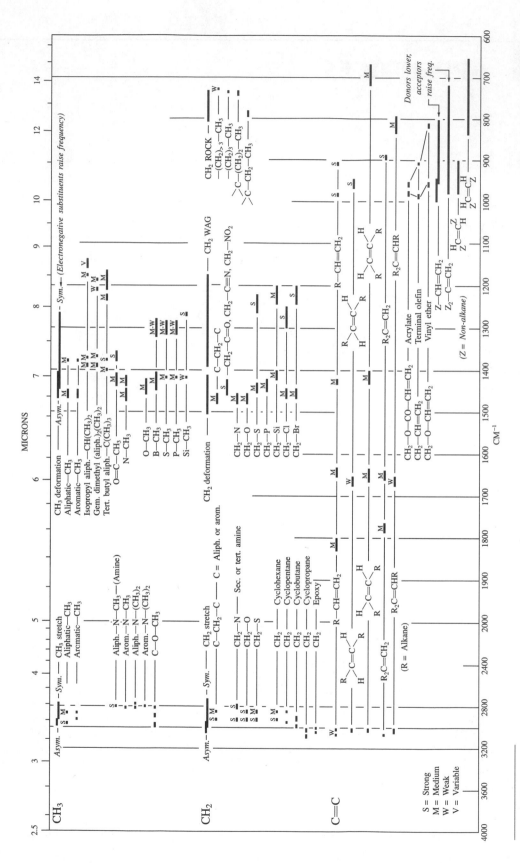

Colthup spectra–structure correlation charts for infrared frequencies in the 4000–600 cm^{-1} region from Lin-Vien, D., N. B. Colthup, W. G. Fateley, and J. G. Grasselli, *The Handbook of Infrared and Raman Characteristic Frequencies of Organic Molecules*, Academic Press, New York, 1991.

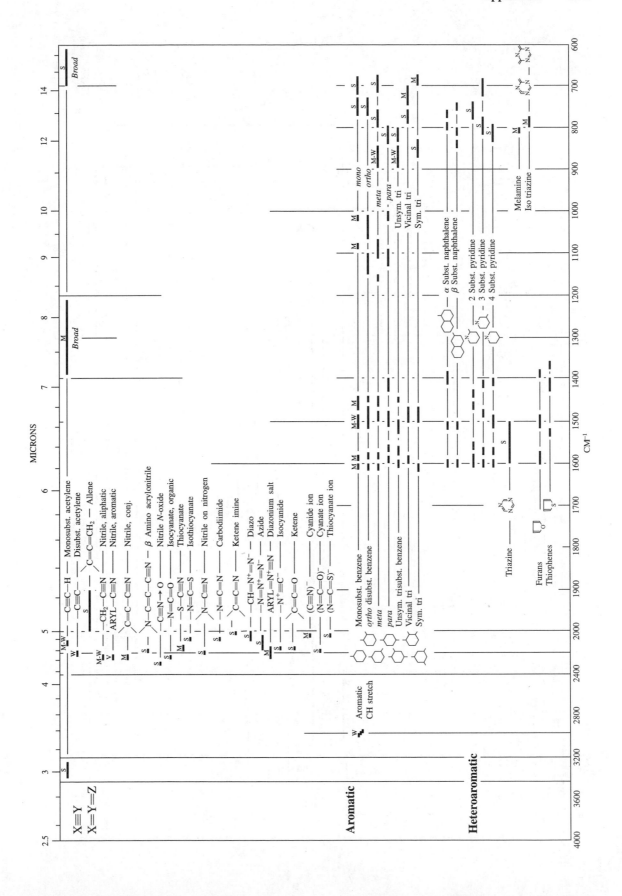

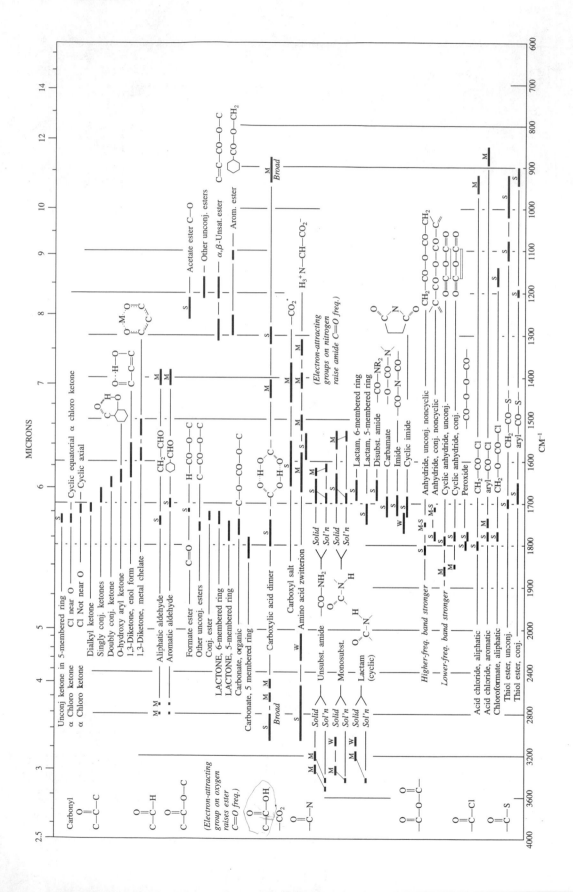

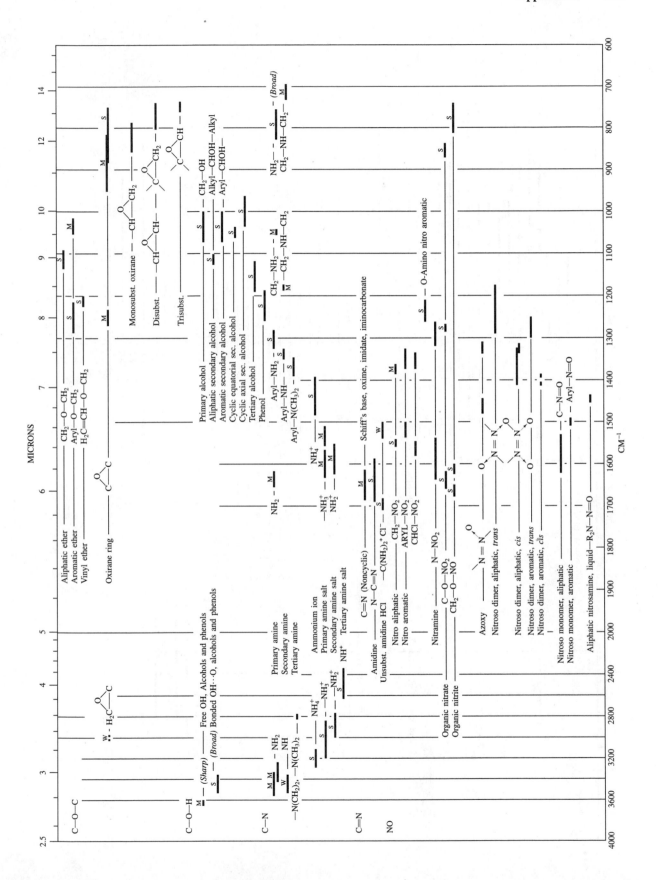

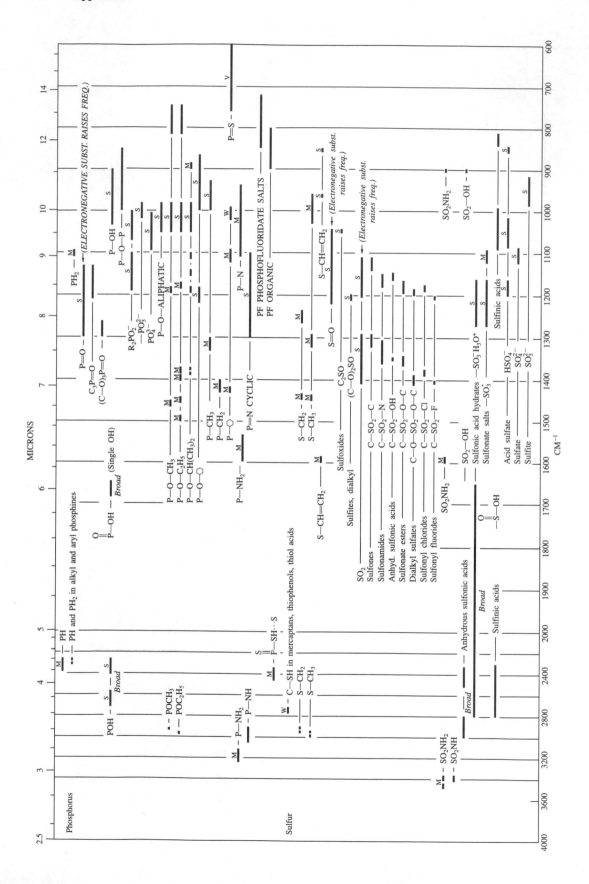

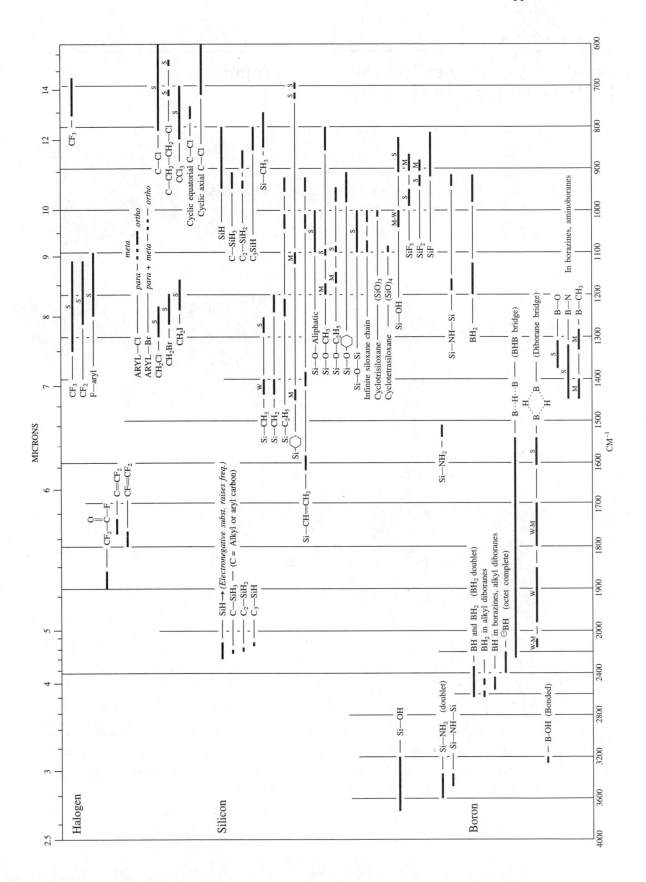

APPENDIX 2

Approximate ^{1}H Chemical Shift Ranges (ppm) for Selected Types of Protons^a

R–CH$_3$	0.7 – 1.3		R–N–C–H	2.2 – 2.9
R–CH$_2$–R	1.2 – 1.4		R–S–C–H	2.0 – 3.0
R$_3$CH	1.4 – 1.7			
R–C=C–C–H	1.6 – 2.6		I–C–H	2.0 – 4.0
R–C–C–H, H–C–C–H (O)	2.1 – 2.4		Br–C–H	2.7 – 4.1
RO–C–C–H, HO–C–C–H (O)	2.1 – 2.5		Cl–C–H	3.1 – 4.1
N≡C–C–H	2.1 – 3.0		R–S–O–C–H (O, O)	ca. 3.0
⬡–C–H	2.3 – 2.7		RO–C–H, HO–C–H	3.2 – 3.8
R–C≡C–H	1.7 – 2.7		R–C–O–C–H (O)	3.5 – 4.8
R–S–H	var	1.0 – 4.0^b	O$_2$N–C–H	4.1 – 4.3
R–N–H	var	0.5 – 4.0^b	F–C–H	4.2 – 4.8
R–O–H	var	0.5 – 5.0^b		
⬡–O–H	var	4.0 – 7.0^b	R–C=C–H	4.5 – 6.5
⬡–N–H	var	3.0 – 5.0^b	⬡–H	6.5 – 8.0
R–C–N–H (O)	var	5.0 – 9.0^b	R–C–H (O)	9.0 – 10.0
			R–C–OH (O)	11.0 – 12.0

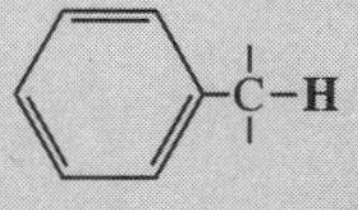

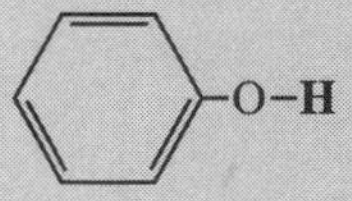

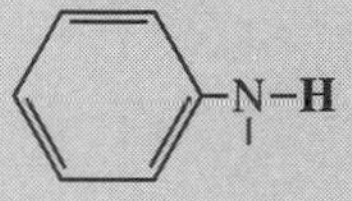

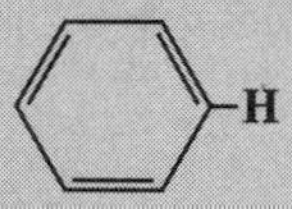

^{a} For those hydrogens shown as $-\overset{|}{\underset{|}{C}}-H$, if that hydrogen is part of a methyl group (CH$_3$) the shift is generally at the low end of the range given, if the hydrogen is in a methylene group ($-CH_2-$) the shift is intermediate, and if the hydrogen is in a methine group ($-CH-$) the shift is typically at the high end of the range given.

^{b} The chemical shift of these groups is variable, depending not only on the chemical environment in the molecule, but also on concentration, temperature, and solvent.

A P P E N D I X   3

Some Representative ¹H Chemical Shift Values[a] for Various Types of Protons[b]

[a]Chemical shift values refer to the boldface protons **H**, not to regular H.

[b]Adapted with permission from Landgrebe, J. A., *Theory and Practice in the Organic Laboratory*, 4th ed., Brooks/Cole Publishing, Pacific Grove, CA, 1993.

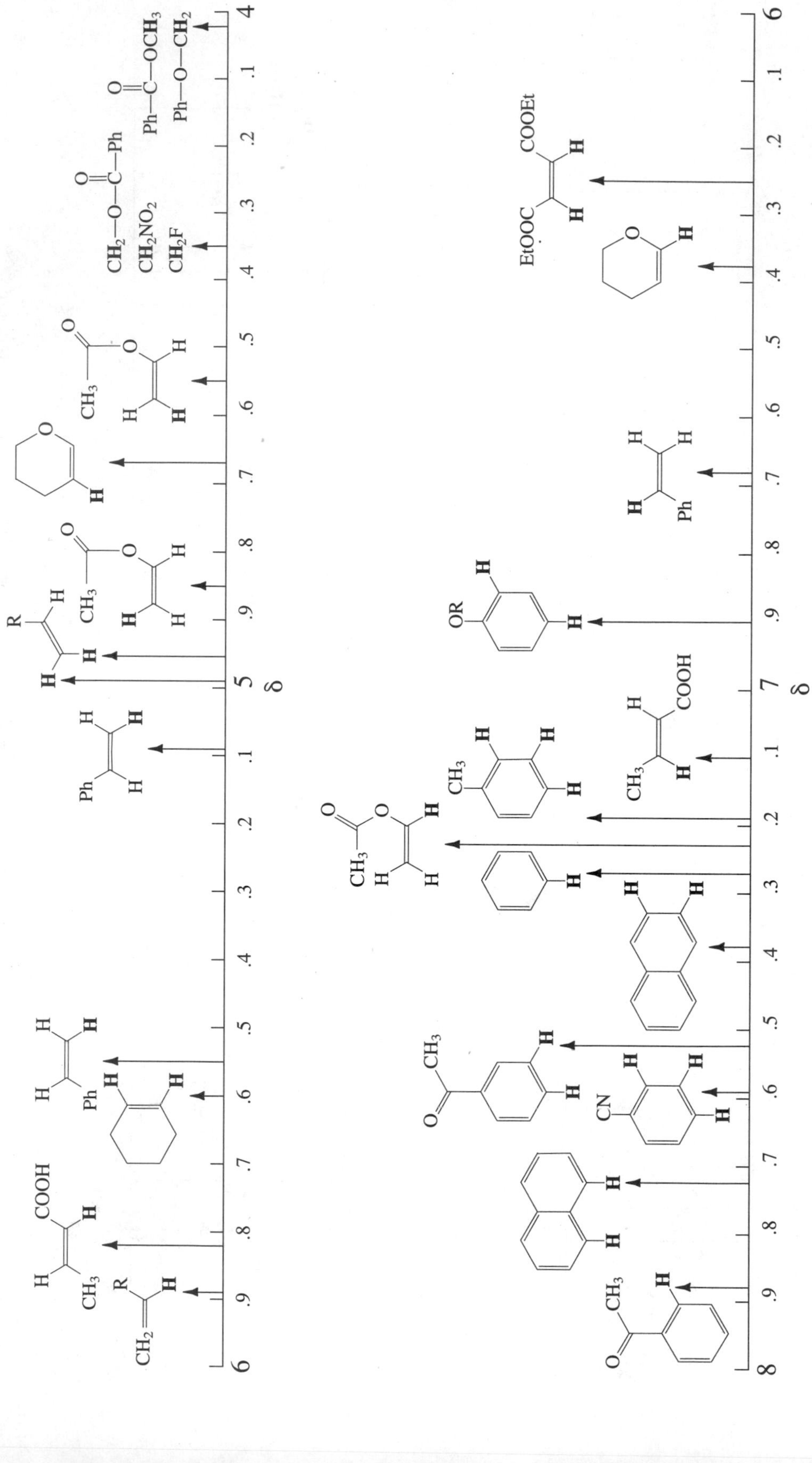

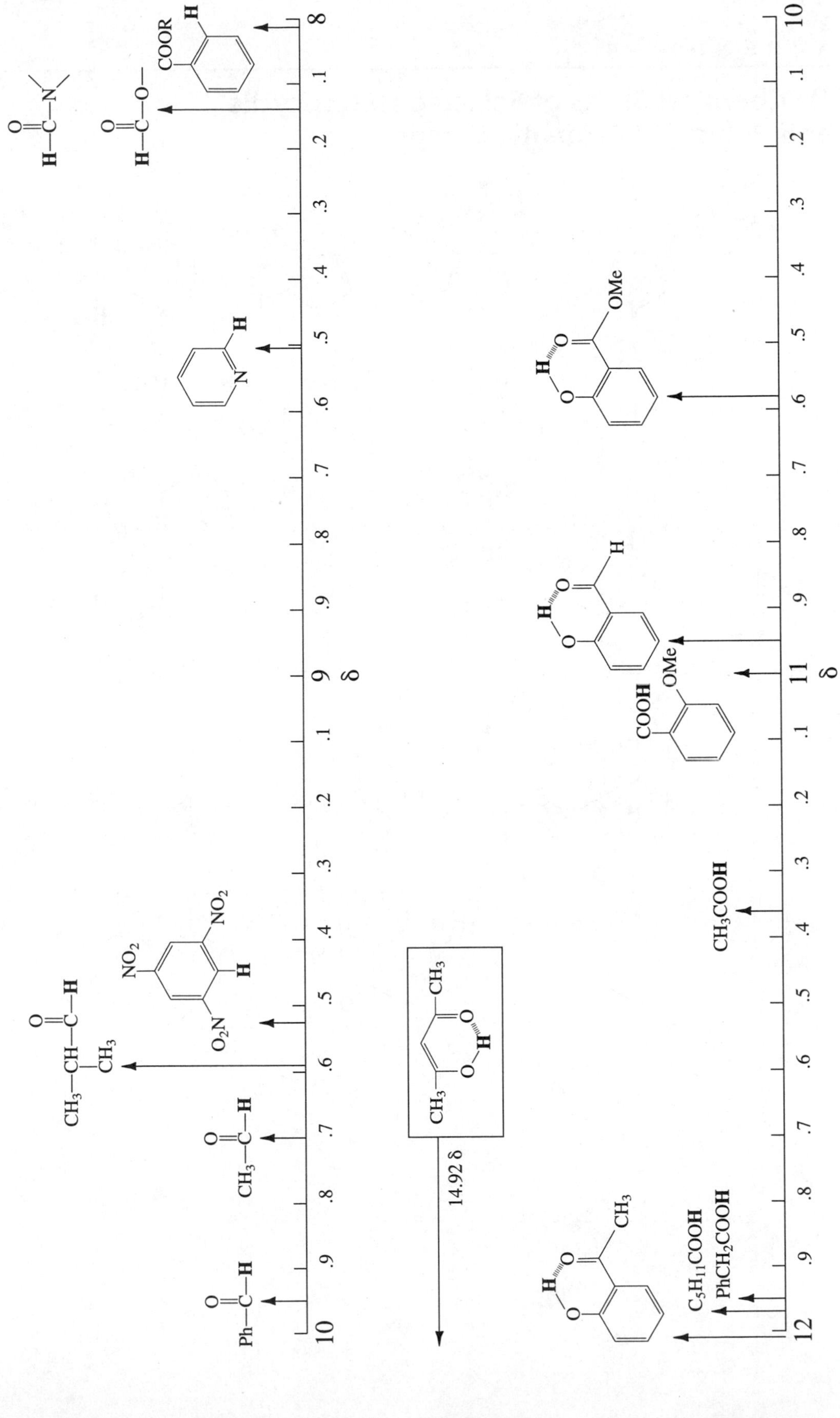

Ph—C—H
CH3—C—H
CH3—CH—C—H
CH3
NO2
O2N
H
H—C—N
H—C—O—
COOR
H
N
14.92 δ
CH3
CH3
O
H
OMe
COOH
OMe
H
CH3COOH
C5H11COOH
PhCH2COOH
CH3
10 .9 .8 .7 .6 .5 .4 .3 .2 .1 9 δ .9 .8 .7 .6 .5 .4 .3 .2 .1 8
12 .9 .8 .7 .6 .5 .4 .3 .2 .1 11 δ .9 .8 .7 .6 .5 .4 .3 .2 .1 10

APPENDIX 4

^{1}H Chemical Shifts of Selected Heterocyclic and Polycyclic Aromatic Compounds

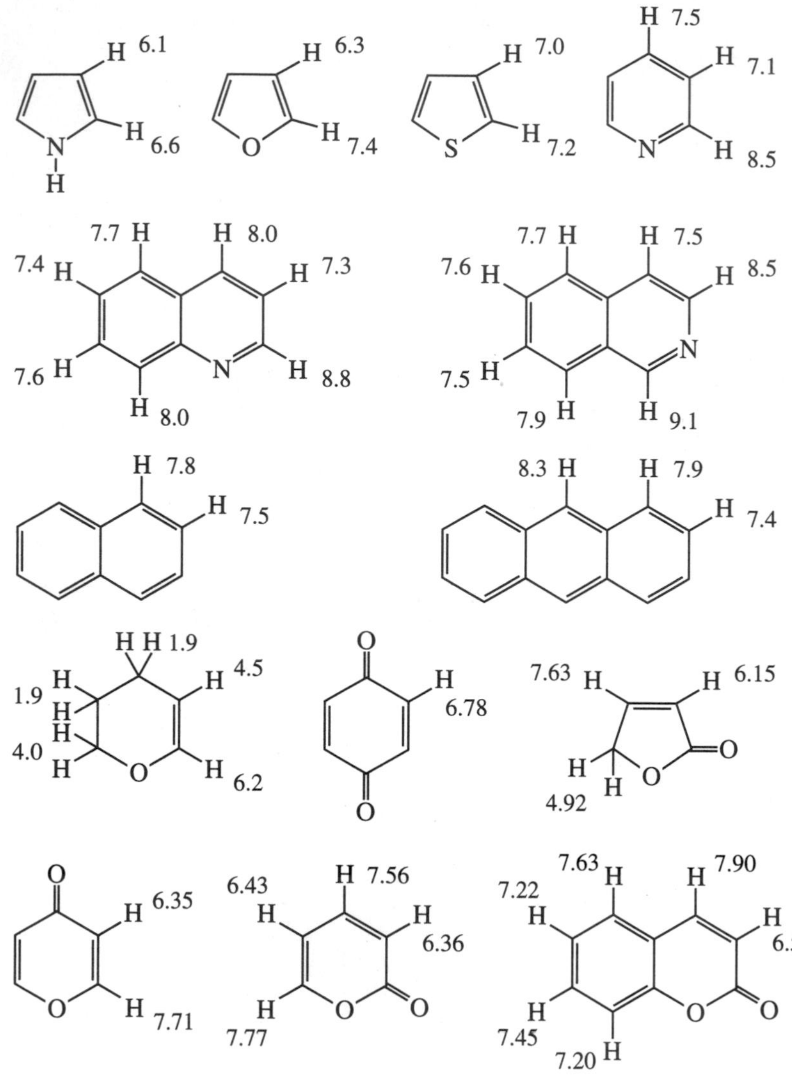

H 6.1 · H 6.6 (pyrrole, N–H)

H 6.3 · H 7.4 (furan, O)

H 7.0 · H 7.2 (thiophene, S)

H 7.5 · H 7.1 · H 8.5 (pyridine, N)

7.7 H · 7.4 H · 7.6 H · H 8.0 · H 7.3 · H 8.8 · H 8.0 (quinoline, N)

7.7 H · 7.6 H · 7.5 H · H 7.5 · H 8.5 · 7.9 H · H 9.1 (isoquinoline, N)

H 7.8 · H 7.5 (naphthalene)

8.3 H · H 7.9 · H 7.4 (anthracene)

H H 1.9 · 1.9 H · H · H · 4.0 H · H 4.5 · H 6.2 (O)

O · H 6.78 · O (benzoquinone)

7.63 H · H 6.15 · H · H 4.92 · O (furanone)

O · H 6.35 · H 7.71 (O, pyranone)

6.43 H · H 7.56 · H 6.36 · H · 7.77 · O · O (pyranone)

7.22 H · 7.63 H · H 7.90 · H · 7.45 H · 7.20 H · H 6.5 · O · O (coumarin)

APPENDIX 5

Typical Proton Coupling Constants

ALKANES AND SUBSTITUTED ALKANES

Type			Typical Value (Hz)	Range (Hz)	
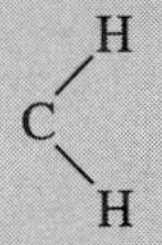	^2J	geminal	12	12–15	(For a 109° H—C—H angle)
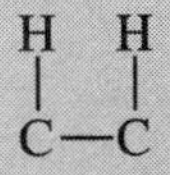	^3J	vicinal	7	6–8	(Depends on HCCH dihedral angle)
	^3J ^3J ^3J	a,a a,e e,e	10 5 3	8–14 0–7 0–5	In conformationally rigid systems (in systems undergoing inversion, all $J \approx 7$–8 Hz)
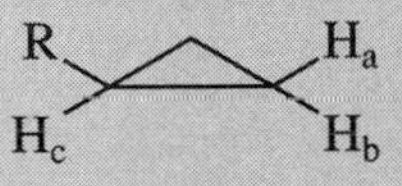	^3J ^3J ^2J	cis (H_bH_c) trans (H_aH_c) gem (H_aH_b)	9 6 6	6–12 4–8 3–9	
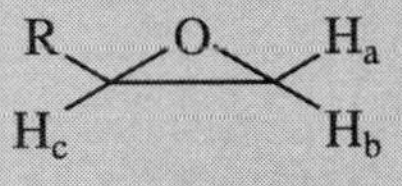	^3J ^3J ^2J	cis (H_bH_c) trans (H_aH_c) gem (H_aH_b)	4 2.5 6	2–5 1–3 4–6	
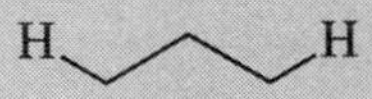	^4J		0	0–7	(W-configuration obligatory—strained systems have the larger values)

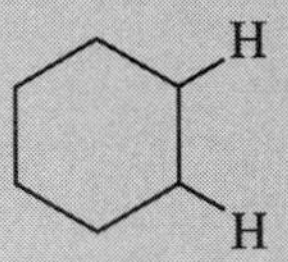

ALKENES AND CYCLOALKENES (^2J AND ^3J)

Type		Typical Value (Hz)	Range (Hz)	Type		Typical Value (Hz)	Range (Hz)
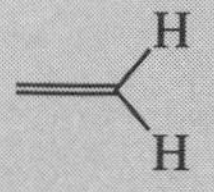	^2J gem	<1	0–5	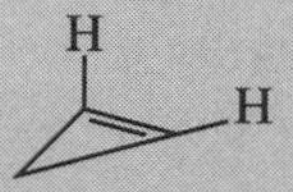	^3J	2	0–2
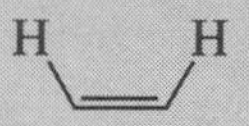	^3J cis	10	6–15	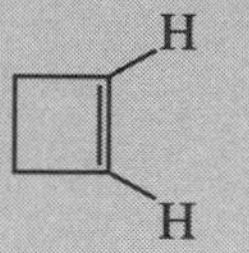	^3J	4	2–4
	^3J trans	16	11–18		^3J	6	5–7
	^3J	5	4–10		^3J	10	8–11
	^3J	10	9–13				

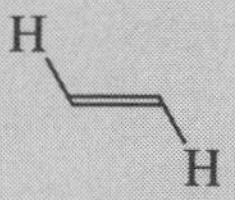

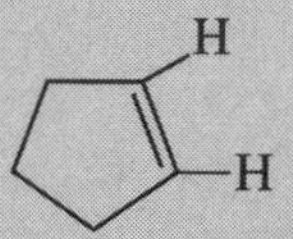

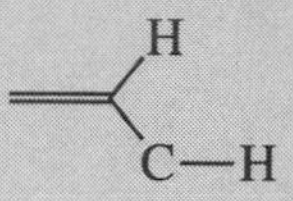

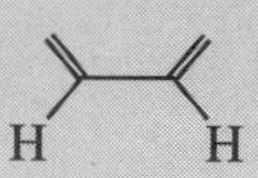

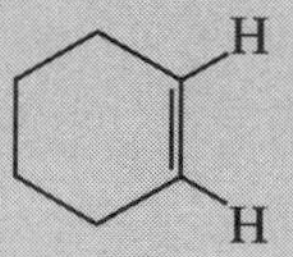

ALKENES AND ALKYNES (^4J AND ^5J)

Type		Typical Value (Hz)	Range (Hz)	Type		Typical Value (Hz)	Range (Hz)
H–C=C–C–H Allylic	^4J (cis or trans)	1	0–3	H–C≡C–C–H Allylic	^4J	2	2–3
H–C–C=C–C–H Homoallylic	^5J	0	0–1.5	H–C–C≡C–C–H Homoallylic	^5J	2	2–3

AROMATICS AND HETEROCYCLES

Type		Typical Value (Hz)	Range (Hz)	Type		Range (Hz)
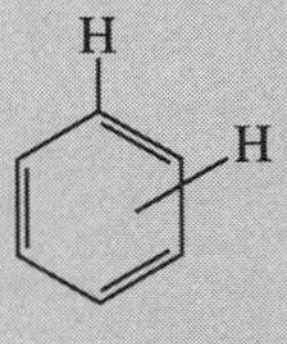	^3J ortho	8	6–10		^3J $\alpha\beta$	4.6–5.8
	^4J meta	3	1–4		^4J $\alpha\beta'$	1.0–1.5
	^5J para	<1	0–2		^4J $\alpha\alpha'$	2.1–3.3
					^3J $\beta\beta'$	3.0–4.2
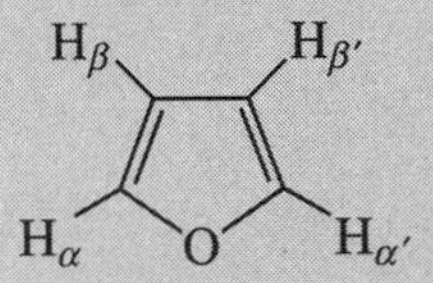	^3J $\alpha\beta$		1.6–2.0		^3J $\alpha\beta$	4.9–5.7
	^4J $\alpha\beta'$		0.3–0.8		^4J $\alpha\gamma$	1.6–2.0
	^4J $\alpha\alpha'$		1.3–1.8		^5J $\alpha\beta'$	0.7–1.1
	^3J $\beta\beta'$		3.2–3.8		^4J $\alpha\alpha'$	0.2–0.5
					^3J $\beta\gamma$	7.2–8.5
					^4J $\beta\beta'$	1.4–1.9
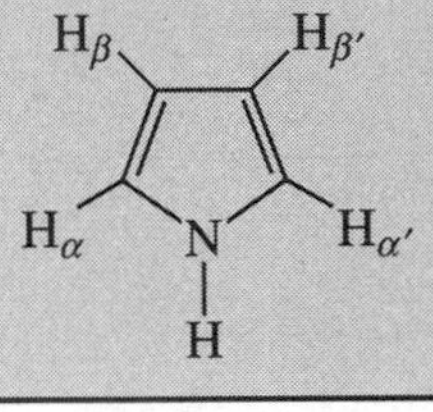	^3J $\alpha\beta$		2.0–2.6			
	^4J $\alpha\beta'$		1.0–1.5			
	^4J $\alpha\alpha'$		1.8–2.3			
	^3J $\beta\beta'$		2.8–4.0			

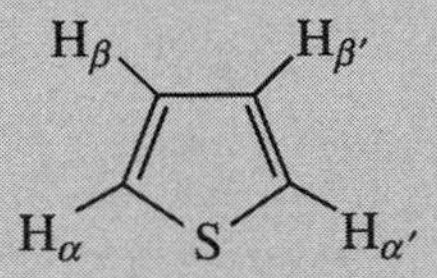

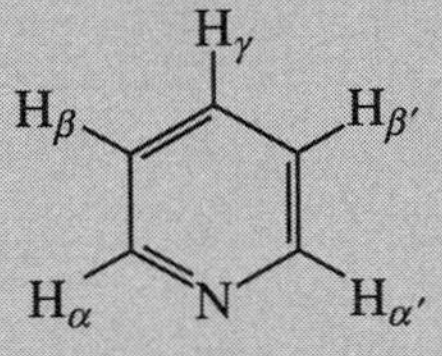

ALCOHOLS

Type		Typical Value (Hz)	Range (Hz)
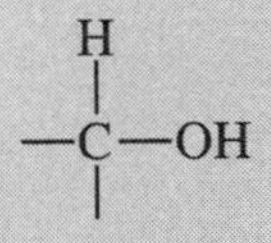	^3J	5	4–10
(No exchange occurring)			

ALDEHYDES

Type		Typical Value (Hz)	Range (Hz)
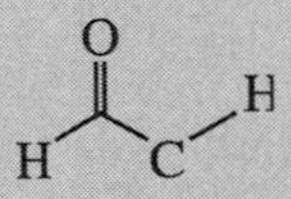	^3J	2	1–3
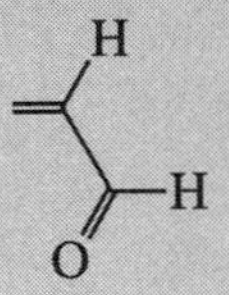	^3J	6	5–8

PROTON–OTHER NUCLEUS COUPLING CONSTANTS

Type	Typical Value (Hz)	Type	Typical Value (Hz)	Type	Typical Value (Hz)
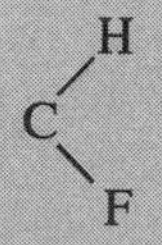 C(H)(F) ^2J	44–81	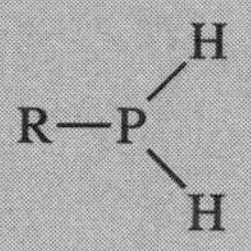 R—P(H)(H) ^1J	~190	N—H	~52
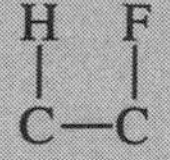 H—C—C—F ^3J	3–25 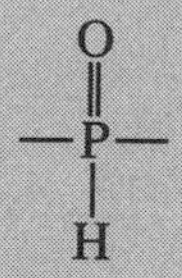	—P(=O)(H)— ^1J	~650	H—C—N—H	0
C(D)(H) ^2J	~2	—C(H)—P(=O)— ^2J	~13		
 H—C—C—D ^3J	<1 (Leads only to peak broadening)	—C(H)—C—P(=O)— ^3J	~17		
		—C(H)—O—P(=O)— ^3J	~8		

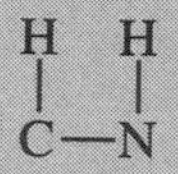

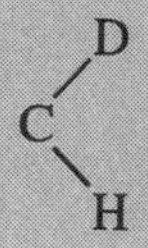

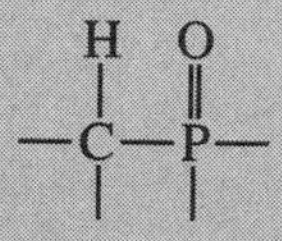

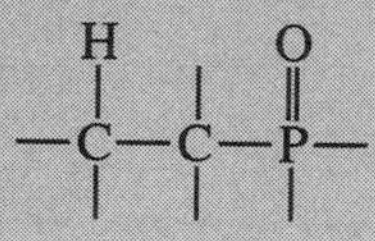

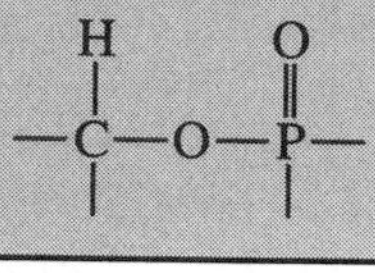

APPENDIX 6

Calculation of Proton (^{1}H) Chemical Shifts

TABLE A6.1
^{1}H CHEMICAL-SHIFT CALCULATIONS FOR DISUBSTITUTED METHYLENE COMPOUNDS

$$X-CH_2-X \quad \text{or} \quad X-CH_2-Y \qquad \boxed{\delta_H \text{ ppm} = 0.23 + \Sigma \text{ constants}}$$

Substituents	Constants	Substituents	Constants
Alkanes, alkenes, alkynes, aromatics		Bonded to oxygen	
$-R$	0.47	$-OH$	2.56
$\diagdown C = C \diagup$	1.32	$-OR$	2.36
$-C\equiv C-$	1.44	$-OCOR$	3.13
$-C_6H_5$	1.85	$-OC_6H_5$	3.23
Bonded to nitrogen and sulfur		Bonded to halogen	
$-NR_2$	1.57	$-F$	4.00
$-NHCOR$	2.27	$-Cl$	2.53
$-NO_2$	3.80	$-Br$	2.33
$-SR$	1.64	$-I$	1.82
Ketones		Derivatives of carboxylic acids	
$-COR$	1.70	$-COOR$	1.55
$-COC_6H_5$	1.84	$-CONR_2$	1.59
		$-C\equiv N$	1.70

Example Calculations

The formula allows you to calculate the *approximate* chemical-shift values for protons (^{1}H) based on methane (0.23 ppm). Although it is possible to calculate chemical shifts for any proton (methyl, methylene, or methine), agreement with actual experimental values is best with *disubstituted* compounds of the type $X-CH_2-Y$ or $X-CH_2-X$.

$Cl-\mathbf{CH_2}-Cl$	$\delta_H = 0.23 + 2.53 + 2.53 = 5.29$ ppm; actual = 5.30 ppm
$C_6H_5-\mathbf{CH_2}-O-\overset{\displaystyle \parallel}{\underset{\displaystyle O}{C}}-CH_3$	$\delta_H = 0.23 + 1.85 + 3.13 = 5.21$ ppm; actual = 5.10 ppm
$C_6H_5-\mathbf{CH_2}-\overset{\displaystyle \parallel}{\underset{\displaystyle O}{C}}-O-CH_3$	$\delta_H = 0.23 + 1.85 + 1.55 = 3.63$ ppm; actual = 3.60 ppm
$CH_3-CH_2-\mathbf{CH_2}-NO_2$	$\delta_H = 0.23 + 3.80 + 0.47 = 4.50$ ppm; actual = 4.38 ppm

TABLE A6.2
¹H CHEMICAL-SHIFT CALCULATIONS FOR SUBSTITUTED ALKENES

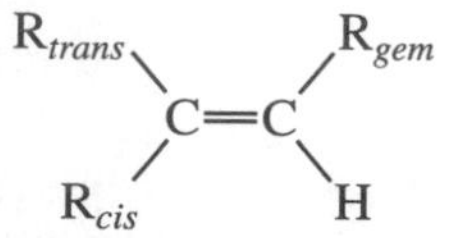

$$\delta_H \text{ ppm} = 5.25 + \delta_{gem} + \delta_{cis} + \delta_{trans}$$

Substituents (−R)	δ_{gem}	δ_{cis}	δ_{trans}
Saturated carbon groups			
Alkyl	0.44	−0.26	−0.29
−CH_2−O−	0.67	−0.02	−0.07
Aromatic groups			
−C_6H_5	1.35	0.37	−0.10
Carbonyl, acid derivatives, and nitrile			
COR	1.10	1.13	0.81
−COOH	1.00	1.35	0.74
−COOR	0.84	1.15	0.56
−C≡N	0.23	0.78	0.58
Oxygen groups			
−OR	1.18	−1.06	−1.28
−OCOR	2.09	−0.40	−0.67
Nitrogen groups			
−NR_2	0.69	−1.19	−1.31
−NO_2	1.87	1.30	0.62
Halogen groups			
−Cl	1.00	0.19	0.03
−Br	1.04	0.40	0.55

Example Calculations

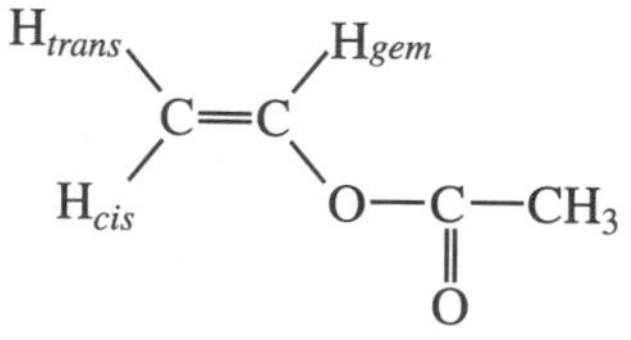

H_{gem} = 5.25 + 2.09 = 7.34 ppm; actual = 7.25 ppm

H_{cis} = 5.25 − 0.40 = 4.85 ppm; actual = 4.85 ppm

H_{trans} = 5.25 − 0.67 = 4.58 ppm; actual = 4.55 ppm

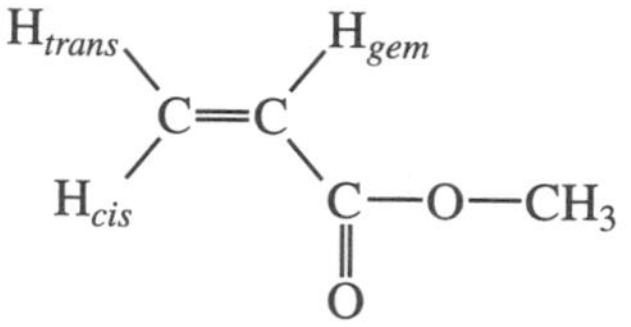

H_{gem} = 5.25 + 0.84 = 6.09 ppm; actual = 6.14 ppm

H_{cis} = 5.25 + 1.15 = 6.40 ppm; actual = 6.42 ppm

H_{trans} = 5.25 + 0.56 = 5.81 ppm; actual = 5.82 ppm

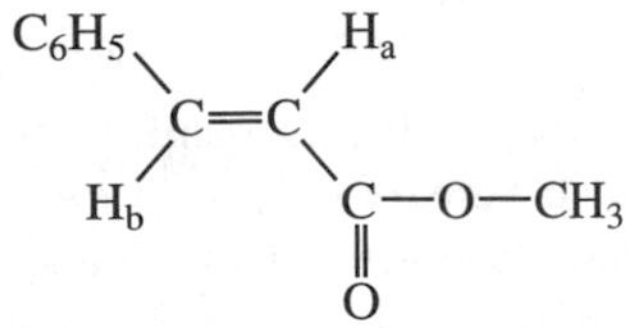

$$H_a \begin{cases} \delta_{gem} \text{ for } -COOR = 0.84 \\ \delta_{cis} \text{ for } -C_6H_5 = 0.37 \\ H_a = 5.25 + 0.84 + 0.37 = 6.46 \text{ ppm}; \\ \qquad\qquad\qquad \text{actual} = 6.43 \text{ ppm} \end{cases}$$

$$H_b \begin{cases} \delta_{gem} \text{ for } -C_6H_5 = 1.35 \\ \delta_{cis} \text{ for } -COOR = 1.15 \\ H_b = 5.25 + 1.35 + 1.15 = 7.75 \text{ ppm}; \\ \qquad\qquad\qquad \text{actual} = 7.69 \text{ ppm} \end{cases}$$

TABLE A6.3
^{1}H CHEMICAL-SHIFT CALCULATIONS FOR SUBSTITUTED BENZENE RINGS

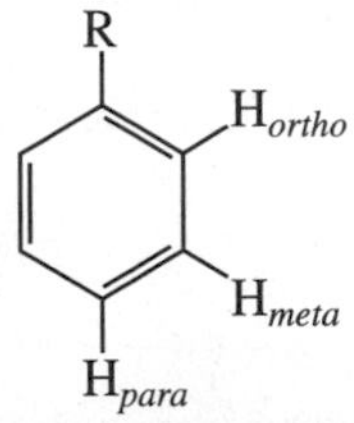

$$\delta_H \text{ ppm} = 7.27 + \Sigma\delta$$

Substituents (−R)	δ_{ortho}	δ_{meta}	δ_{para}
Saturated carbon groups			
Alkyl	−0.14	−0.06	−0.17
−CH$_2$OH	−0.07	−0.07	−0.07
Aldehydes and ketones			
−CHO	0.56	0.22	0.29
−COR	0.62	0.14	0.21
Carboxylic acids and derivatives			
−COOH	0.85	0.18	0.27
−COOR	0.71	0.10	0.21
−C≡N	0.36	0.18	0.28
Oxygen groups			
−OH	−0.56	−0.12	−0.45
−OCH$_3$	−0.48	−0.09	−0.44
−OCOCH$_3$	−0.25	0.03	−0.13
Nitrogen groups			
−NH$_2$	−0.75	−0.25	−0.65
−NO$_2$	0.95	0.26	0.38
Halogen groups			
−Cl	0.03	−0.02	−0.09
−Br	0.18	−0.08	−0.04

Example Calculations

The formula allows you to calculate the *approximate* chemical-shift values for protons (^1H) on a benzene ring. Although the values given in the table are for *monosubstituted benzenes*, it is possible to estimate chemical shifts for disubstituted and trisubstituted compounds by adding values from the table. The calculations for *meta-* and *para*-disubstituted benzenes often agree closely with actual values. More significant deviations from the experimental values are expected with *ortho*-disubstituted and trisubstituted benzenes. With these types of compounds, steric interactions cause groups such as carbonyl and nitro to turn out of the plane of the ring and thereby lose conjugation. Calculated values are often lower than the actual chemical shifts for *ortho*-disubstituted and trisubstituted benzenes.

$H_{ortho} = 7.27 + 0.71 = 7.98$ ppm; actual $= 8.03$ ppm

$H_{meta} = 7.27 + 0.10 = 7.37$ ppm; actual $= 7.42$ ppm

$H_{para} = 7.27 + 0.21 = 7.48$ ppm; actual $= 7.53$ ppm

H_a $\begin{cases} \delta_{ortho} \text{ for } -Cl = 0.03 \\ \delta_{meta} \text{ for } -NO_2 = 0.26 \\ H_a = 7.27 + 0.03 + 0.26 = 7.56 \text{ ppm; actual} = 7.50 \text{ ppm} \end{cases}$

H_b $\begin{cases} \delta_{meta} \text{ for } -Cl = -0.02 \\ \delta_{ortho} \text{ for } -NO_2 = 0.95 \\ H_b = 7.27 - 0.02 + 0.95 = 8.20 \text{ ppm; actual} = 8.20 \text{ ppm} \end{cases}$

H_a $\begin{cases} \delta_{meta} \text{ for } -Cl = -0.02 \\ \delta_{meta} \text{ for } -NO_2 = 0.26 \\ H_a = 7.27 - 0.02 + 0.26 = 7.51 \text{ ppm; actual} = 7.51 \text{ ppm} \end{cases}$

H_b $\begin{cases} \delta_{ortho} \text{ for } -Cl = 0.03 \\ \delta_{para} \text{ for } -NO_2 = 0.38 \\ H_b = 7.27 + 0.03 + 0.38 = 7.68 \text{ ppm; actual} = 7.69 \text{ ppm} \end{cases}$

H_c $\begin{cases} \delta_{para} \text{ for } -Cl = -0.09 \\ \delta_{ortho} \text{ for } -NO_2 = 0.95 \\ H_c = 7.27 - 0.09 + 0.95 = 8.13 \text{ ppm; actual} = 8.12 \text{ ppm} \end{cases}$

H_d $\begin{cases} \delta_{ortho} \text{ for } -Cl = 0.03 \\ \delta_{ortho} \text{ for } -NO_2 = 0.95 \\ H_d = 7.27 + 0.03 + 0.95 = 8.25 \text{ ppm; actual} = 8.21 \text{ ppm} \end{cases}$

APPENDIX 7

Approximate ^{13}C Chemical-Shift Values (ppm) for Selected Types of Carbon

Types of Carbon	Range (ppm)	Types of Carbon	Range (ppm)
R—CH$_3$	8–30	C≡C	65–90
R$_2$CH$_2$	15–55	C=C	100–150
R$_3$CH	20–60	C≡N	110–140
C—I	0–40	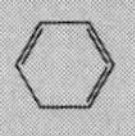 ⬡	110–175
C—Br	25–65	$\underset{\text{R—C—OR, R—C—OH}}{\overset{\text{O}\quad\quad\text{O}}{\parallel\quad\quad\parallel}}$	155–185
C—N	30–65	$\overset{\text{O}}{\underset{\text{R—C—NH}_2}{\parallel}}$	155–185
C—Cl	35–80	$\overset{\text{O}}{\underset{\text{R—C—Cl}}{\parallel}}$	160–170
C—O	40–80	$\underset{\text{R—C—R, R—C—H}}{\overset{\text{O}\quad\quad\text{O}}{\parallel\quad\quad\parallel}}$	185–220

APPENDIX 8

Calculation of ^{13}C Chemical Shifts

TABLE A8.1
^{13}C CHEMICAL SHIFTS OF SELECTED HYDROCARBONS (PPM)

Compound	Formula	C1	C2	C3	C4	C5
Methane	CH_4	−2.3				
Ethane	CH_3CH_3	5.7				
Propane	$CH_3CH_2CH_3$	15.8	16.3			
Butane	$CH_3CH_2CH_2CH_3$	13.4	25.2			
Pentane	$CH_3CH_2CH_2CH_2CH_3$	13.9	22.8	34.7		
Hexane	$CH_3(CH_2)_4CH_3$	14.1	23.1	32.2		
Heptane	$CH_3(CH_2)_5CH_3$	14.1	23.2	32.6	29.7	
Octane	$CH_3(CH_2)_6CH_3$	14.2	23.2	32.6	29.9	
Nonane	$CH_3(CH_2)_7CH_3$	14.2	23.3	32.6	30.0	30.3
Decane	$CH_3(CH_2)_8CH_3$	14.2	23.2	32.6	31.1	30.5
2-Methylpropane		24.5	25.4			
2-Methylbutane		22.2	31.1	32.0	11.7	
2-Methylpentane		22.7	28.0	42.0	20.9	14.3
2,2-Dimethylpropane		31.7	28.1			
2,2-Dimethylbutane		29.1	30.6	36.9	8.9	
2,3-Dimethylbutane		19.5	34.4			
Ethylene	$CH_2{=}CH_2$	123.3				
Cyclopropane		−3.0				
Cyclobutane		22.4				
Cyclopentane		25.6				
Cyclohexane		26.9				
Cycloheptane		28.4				
Cyclooctane		26.9				
Cyclononane		26.1				
Cyclodecane		25.3				
Benzene		128.5				

TABLE A8.2
^{13}C CHEMICAL-SHIFT CALCULATIONS FOR LINEAR AND BRANCHED ALKANES

$$\delta_C = -2.3 + 9.1\alpha + 9.4\beta - 2.5\gamma + 0.3\delta + 0.1\varepsilon + \Sigma \text{ (steric corrections) ppm}$$

α, β, γ, δ, and ε are the numbers of carbon atoms in the α, β, γ, δ, and ε positions relative to the carbon atom being observed.

$$\cdots\cdots C_\varepsilon - C_\delta - C_\gamma - C_\beta - C_\alpha - \underset{\uparrow}{C} - C_\alpha - C_\beta - C_\gamma - C_\delta - C_\varepsilon \cdots\cdots$$

Steric corrections are derived from the following table (use all that apply, even if they apply more than once).

	Steric Corrections (ppm)			
Carbon Atom Observed	*Type of Carbons Attached*			
	Primary	**Secondary**	**Tertiary**	**Quaternary**
Primary	0	0	−1.1	−3.4
Secondary	0	0	−2.5	−7.5
Tertiary	0	−3.7	−9.5	−15.0
Quaternary	−1.5	−8.4	−15.0	−25.0

Example

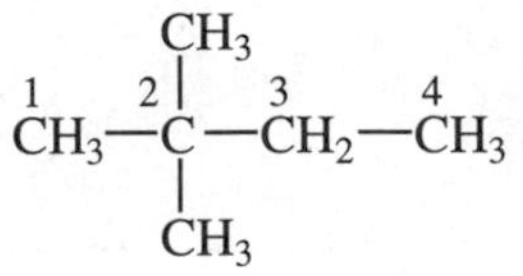

$$\underset{\underset{\displaystyle CH_3}{|}}{\overset{\overset{\displaystyle CH_3}{|}}{\underset{1}{CH_3} - \underset{2}{C} - \underset{3}{CH_2} - \underset{4}{CH_3}}}$$

2,2-Dimethylbutane

Actual values: C1 29.1 ppm
C2 30.6 ppm
C3 36.9 ppm
C4 8.9 ppm

C1 = −2.3 + 9.1(1) + 9.4(3) − 2.5(1) + 0.3(0) + 0.1(0) **+ [1(−3.4)]** = 29.1 ppm
   Steric correction (boldface) = primary with 1 adjacent quaternary

C2 = −2.3 + 9.1(4) + 9.4(1) − 2.5(0) + 0.3(0) + 0.1(0) **+ [3(−1.5)] + [1(−8.4)]** = 30.6 ppm
   Steric corrections = quaternary/3 adj. primary, and quaternary/1 adj. secondary

C3 = −2.3 + 9.1(2) + 9.4(3) − 2.5(0) + 0.3(0) + 0.1(0) **+ [1(0)] + [1(−7.5)]** = 36.6 ppm
   Steric corrections = secondary/1 adj. primary, and secondary/1 adj. quaternary

C4 = −2.3 + 9.1(1) + 9.4(1) − 2.5(3) + 0.3(0) + 0.1(0) **+ [1(0)]** = 8.7 ppm
   Steric correction = primary/1 adj. secondary

TABLE A8.3
^{13}C SUBSTITUENT INCREMENTS FOR ALKANES AND CYCLOALKANES (PPM)[a]

Substituent Y	Terminal: $Y-C_\alpha-C_\beta-C_\gamma$			Internal: $C_\gamma-C_\beta-\overset{Y}{\underset{\vert}{C_\alpha}}-C_\beta-C_\gamma$		
	α	β	γ	α	β	γ
—D	−0.4	−0.1	0			
—CH_3	9	10	−2	6	8	−2
—$CH=CH_2$	20	6	−0.5			−0.5
—$C\equiv CH$	4.5	5.4	−3.5			−3.5
—C_6H_5	23	9	−2	17	7	−2
—CHO	31	0	−2			
—$COCH_3$	30	1	−2	24	1	−2
—COOH	21	3	−2	16	2	−2
—COOR	20	3	−2	17	2	−2
—$CONH_2$	22	2.5	−0.5			−0.5
—CN	4	3	−3	1	3	−3
—NH_2	29	11	−5	24	10	−5
—NHR	37	8	−4	31	6	−4
—NR_2	42	6	−3			−3
—NO_2	63	4		57	4	
—OH	48	10	−5	41	8	−5
—OR	58	8	−4	51	5	−4
—$OCOCH_3$	51	6	−3	45	5	−3
—F	68	9	−4	63	6	−4
—Cl	31	11	−4	32	10	−4
—Br	20	11	−3	25	10	−3
—I	−6	11	−1	4	12	−1

[a]Add these increments to the values given in Table A8.1.

Example 1

$$\overset{1}{CH_3}-\overset{2}{CH}-\overset{3}{CH_2}-\overset{4}{CH_3} \quad \textbf{2-Butanol}$$
$$\underset{OH}{\mid}$$

Using the values for butane listed in Table A8.1 and the internal substituent corrections from Table A8.3, we calculate:

				Actual value
C1 = 13.4 +	8 = 21.4 ppm			22.6 ppm
C2 = 25.2 +	41 = 66.2 ppm			68.7 ppm
C3 = 25.2 +	8 = 33.2 ppm			32.0 ppm
C4 = 13.4 + (−5) =	8.4 ppm			9.9 ppm

Example 2

$$HO-\overset{1}{CH_2}-\overset{2}{CH_2}-\overset{3}{CH_2}-\overset{4}{CH_3} \quad \textbf{1-Butanol}$$

Using the values for butane listed in Table A8.1 and the terminal substituent corrections from Table A8.3, we calculate:

		Actual value
C1 = 13.4 + 48 = 61.4 ppm		61.4 ppm
C2 = 25.2 + 10 = 35.2 ppm		35.0 ppm
C3 = 25.2 + (−5) = 20.2 ppm		19.1 ppm
C4 = 13.4 = 13.4 ppm		13.6 ppm

Example 3

$$Br-\overset{1}{CH_2}-\overset{2}{CH_2}-\overset{3}{CH_3} \quad \textbf{1-Bromopropane}$$

Using the values for propane listed in Table A8.1 and the terminal substituent corrections from Table A8.3, we calculate:

		Actual value
C1 = 15.8 + 20 = 35.8 ppm		35.7 ppm
C2 = 16.3 + 11 = 27.3 ppm		26.8 ppm
C3 = 15.8 + (−3) = 12.8 ppm		13.2 ppm

TABLE A8.4
^{13}C SUBSTITUENT INCREMENTS FOR ALKENES (PPM)[a,b]

$$\overset{1}{Y-C}=\overset{2}{C}-X$$
$$\uparrow$$

Substituent	Y	X
—H	0	0
—CH$_3$	10.6	−7.9
—CH$_2$CH$_3$	15.5	−9.7
—CH=CH$_2$	13.6	−7
—C$_6$H$_5$	12.5	−11
—COOH	4.2	8.9
—NO$_2$	22.3	−0.9
—OCH$_3$	29.4	−38.9
—OCOCH$_3$	18.4	−26.7
—F	24.9	−34.3
—Cl	2.6	−6.1
—Br	−7.9	−1.4
—I	−38.1	7.0

[a]*Corrections for C1; add these increments to the base value of ethylene (123.3 ppm).*

[b]*Calculate C1 as shown in the diagram. Redefine C2 as C1 when estimating values for C2.*

Example 1

$$Br-\overset{1}{C}H=\overset{2}{C}H-\overset{3}{C}H_3 \quad \textbf{1-Bromopropene}$$

		Actual values	
		cis	*trans*
C1 = 123.3 + (−7.9) + (−7.9) = 107.5 ppm		108.9	104.7 ppm
C2 = 123.3 + 10.6 + (−1.4) = 132.5 ppm		129.4	132.7 ppm

Example 2

$$HOO\overset{1}{C}-\overset{2}{C}H=\overset{3}{C}H-\overset{4}{C}H_3 \quad \textbf{Crotonic acid}$$

	Actual value (trans)
C2 = 123.3 + 4.2 + (−7.9) = 119.6 ppm	122.0 ppm
C3 = 123.3 + 10.6 + 8.9 = 142.8 ppm	147.0 ppm

TABLE A8.5
^{13}C CHEMICAL-SHIFT CALCULATIONS FOR LINEAR AND BRANCHED ALKENES^a

$$\delta_{C1} = 123.3 + [10.6\alpha + 7.2\beta - 1.5\gamma] - [7.9\alpha' + 1.8\beta' - 1.5\gamma'] + \Sigma \text{ (steric corrections)}$$

α, β, γ and α', β', γ' are the numbers of carbon atoms in those same positions relative to C1:

$$C\gamma—C\beta—C\alpha—\overset{1}{C}=\overset{2}{C}—C\alpha'—C\beta'—C\gamma'$$

Steric corrections are applied as follows (use all that apply):

Cα and Cα' are *trans* (*E*-configuration)	0
Cα and Cα' are *cis* (*Z*-configuration)	−1.1
Two alkyl substituents at C1 (two Cα)	−4.8
Two alkyl substituents at C2 (two Cα')	+2.5
Two or three alkyl substituents at Cβ	+2.3

^{a}*Calculate C1 as shown in the diagram. Redefine C2 as C1 when calculating values for C2.*

Example 1

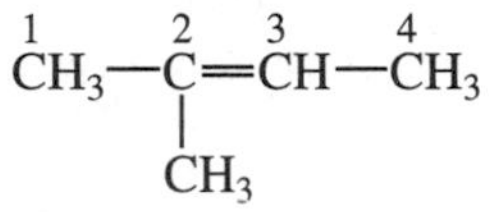

$$\overset{1}{C}H_3—\overset{2}{C}=\overset{3}{C}H—\overset{4}{C}H_3 \quad \textbf{2-Methyl-2-butene}$$
$$|$$
$$CH_3$$

Actual value

C2 = 123.3 + [10.6(2)] − [7.9(1)] + [(−4.8) + (−1.1)] = 130.7 ppm   131.4 ppm
C3 = 123.3 + [10.6(1)] − [7.9(2)] + [(+2.5) + (−1.1)] = 119.5 ppm   118.7 ppm

Example 2

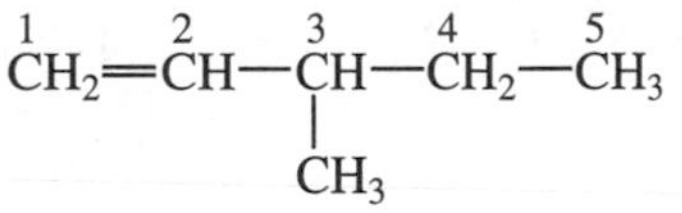

$$\overset{1}{C}H_2=\overset{2}{C}H—\overset{3}{C}H—\overset{4}{C}H_2—\overset{5}{C}H_3 \quad \textbf{3-Methyl-1-pentene}$$
$$|$$
$$CH_3$$

Actual value

C1 = 123.3 + [0] − [7.9(1) + 1.8(2) − 1.5(1)]   = 113.3 ppm   112.9 ppm
C2 = 123.3 + [10.6(1) + 7.2(2) − 1.5(1)] − [0] + [(+2.3)] = 149.1 ppm   144.9 ppm

Example 3

$$\overset{1}{C}H_3—\overset{2}{C}H=\overset{3}{C}H—\overset{4}{C}H_3 \quad \textbf{2-Butene}$$

Actual value

C2 (*cis* isomer)   = C3 = 123.3 + [10.6(1)] − [7.9(1)] + [(−1.1)] = 124.9 ppm   124.6 ppm
C2 (*trans* isomer) =   C3 = 123.3 + [10.6(1)] − [7.9(1)] + [0]   = 126.0 ppm   126.0 ppm

TABLE A8.6
^{13}C SUBSTITUENT INCREMENTS FOR ALKENE (VINYL) CARBONS[a,b]

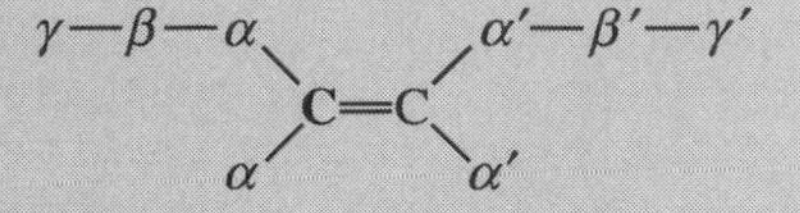

Substituent	α	β	γ	α'	β'	γ'
Carbon	10.6	7.2	−1.5	−7.9	−1.8	−1.5
—C$_6$H$_5$	12			−11		
—OR	29	2		−39	−1	
—OCOR	18			−27		
—COR	15			6		
—COOH	4			9		
—CN	−16			15		
—Cl	3	−1		−6	2	
—Br	−8	0		−1	2	
—I	−38			7		

[a]*In the upper chains, if a group is in the β or γ position, the preceding atoms (α and/or β) are assumed to be carbon atoms. Add these increments to the base value of ethylene (123.3 ppm).*

[b]*Calculate C1 as shown in the diagram. Redefine C2 as C1 when estimating values for C2.*

Example 1

$$\overset{1}{\text{Br}-\text{CH}}=\overset{2}{\text{CH}}-\overset{3}{\text{CH}_3} \quad \textbf{1-Bromopropene}$$

	Actual values	
	cis	*trans*
C1 = 123.3 − 8 − 7.9 = 107.4 ppm	108.9	104.7 ppm
C2 = 123.3 + 10.6 − 1 = 132.9 ppm	129.4	132.7 ppm

Example 2

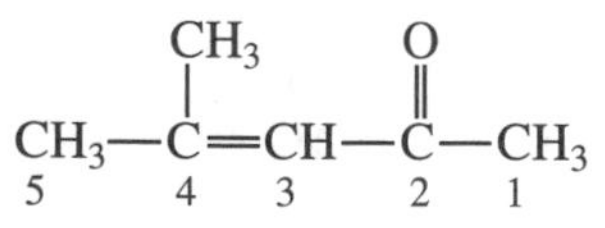

$$\underset{5}{\text{CH}_3}-\underset{4}{\overset{\overset{\text{CH}_3}{|}}{\text{C}}}=\underset{3}{\text{CH}}-\underset{2}{\overset{\overset{\text{O}}{\|}}{\text{C}}}-\underset{1}{\text{CH}_3} \quad \textbf{Mesityl oxide}$$

	Actual value
C3 = 123.3 + 15 − 7.9 − 7.9 = 122.5 ppm	124.3 ppm
C4 = 123.3 + 10.6 + 10.6 + 6 = 150.5 ppm	154.6 ppm

TABLE A8.7
^{13}C SUBSTITUENT INCREMENTS FOR BENZENE RINGS (PPM)[a]

Substituent Y	α (ipso)	o (ortho)	m (meta)	p (para)
$-CH_3$	9.3	0.7	−0.1	−2.9
$-CH_2CH_3$	15.6	−0.5	0	−2.6
$-CH(CH_2)_2$	20.1	−2.0	0	−2.5
$-C(CH_3)_3$	22.2	−3.4	−0.4	−3.1
$-CH=CH_2$	9.1	−2.4	0.2	−0.5
$-C\equiv CH$	−5.8	6.9	0.1	0.4
$-C_6H_5$	12.1	−1.8	−0.1	−1.6
$-CHO$	8.2	1.2	0.6	5.8
$-COCH_3$	7.8	−0.4	−0.4	2.8
$-COC_6H_5$	9.1	1.5	−0.2	3.8
$-COOH$	2.9	1.3	0.4	4.3
$-COOCH_3$	2.0	1.2	−0.1	4.8
$-CN$	−16.0	3.6	0.6	4.3
$-NH_2$	19.2	−12.4	1.3	−9.5
$-N(CH_3)_2$	22.4	−15.7	0.8	−11.8
$-NHCOCH_3$	11.1	−9.9	0.2	−5.6
$-NO_2$	19.6	−5.3	0.9	6.0
$-OH$	26.6	−12.7	1.6	−7.3
$-OCH_3$	31.4	−14.4	1.0	−7.7
$-OCOCH_3$	22.4	−7.1	−0.4	−3.2
$-F$	35.1	−14.3	0.9	−4.5
$-Cl$	6.4	0.2	1.0	−2.0
$-Br$	−5.4	3.4	2.2	−1.0
$-I$	−32.2	9.9	2.6	−7.3

[a]Add these increments to the base value for benzene-ring carbons (128.5 ppm).

Example 1

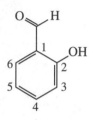

Mesitylene

$$C1,C3,C5 = 128.5 + 9.3 - 0.1 - 0.1 = 137.6 \text{ ppm}$$
$$C2,C4,C6 = 128.5 + 0.7 + 0.7 - 2.9 = 127.0 \text{ ppm}$$

Observed
137.4 ppm
127.1 ppm

Example 2

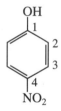

Salicylaldehyde

$$C1 = 128.5 + 8.2 - 12.7 = 124.0 \text{ ppm}$$
$$C2 = 128.5 + 26.6 + 1.2 = 156.3 \text{ ppm}$$
$$C3 = 128.5 - 12.7 + 0.6 = 116.4 \text{ ppm}$$
$$C4 = 128.5 + 1.6 + 5.8 = 135.9 \text{ ppm}$$
$$C5 = 128.5 - 7.3 + 0.6 = 121.8 \text{ ppm}$$
$$C6 = 128.5 + 1.2 + 1.6 = 131.3 \text{ ppm}$$

Observed
121.0 ppm
161.4 ppm
117.4 ppm
136.6 ppm
119.6 ppm
133.6 ppm

Example 3

4-Nitrophenol

$$C1 = 128.5 + 26.6 + 6.0 = 161.1 \text{ ppm}$$
$$C2 = 128.5 - 12.7 + 0.9 = 116.7 \text{ ppm}$$
$$C3 = 128.5 + 1.6 - 5.3 = 124.8 \text{ ppm}$$
$$C4 = 128.5 + 19.6 + 7.3 = 140.8 \text{ ppm}$$

Observed
161.5 ppm
115.9 ppm
126.4 ppm
141.7 ppm

APPENDIX 9

^{13}C Coupling Constants

^{13}C–proton coupling constants (^1J)
  sp^3 ^{13}C—H   115–125 Hz
  sp^2 ^{13}C—H   150–170 Hz
  sp ^{13}C—H   250–270 Hz

^{13}C–proton coupling constants (^2J)
  ^{13}C—C—H   0–60 Hz

^{13}C–deuterium coupling constants (^1J)
  ^{13}C—D   20–30 Hz

^{13}C–fluorine coupling constants (^1J)
  ^{13}C—F   165–370 Hz

^{13}C–fluorine coupling constants (^2J)
  ^{13}C—C—F   18–45 Hz

^{13}C–phosphorus coupling constants (^1J)
  ^{13}C—P   48–56 Hz

^{13}C–phosphorus coupling constants (^2J)
  ^{13}C—C—P   4–6 Hz

Example

$(CH_3-CH_2)_4P^+X^-$

^{13}C–phosphorus coupling constants (^1J)
  ^{13}C—P   143 Hz

^{13}C–phosphorus coupling constants (^2J and ^3J)
  ^{13}C—O—P   6–7 Hz
  ^{13}C—C—O—P   6–7 Hz

Example

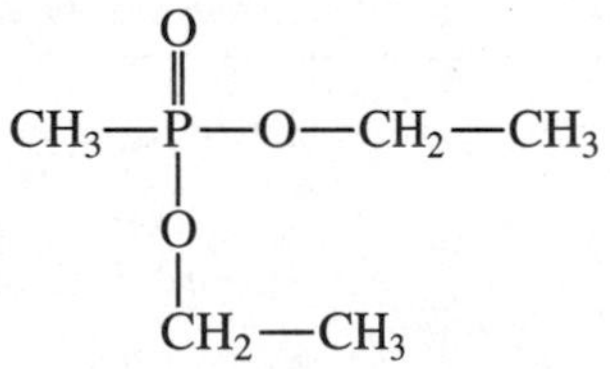

$$CH_3-\underset{\underset{CH_2-CH_3}{\overset{O}{|}}}{\overset{\overset{O}{\|}}{P}}-O-CH_2-CH_3$$

APPENDIX 10

^{1}H and ^{13}C Chemical Shifts for Common NMR Solvents

TABLE A10.1
^{1}H CHEMICAL-SHIFT VALUES (PPM) FOR SOME COMMON NMR SOLVENTS

Solvent	Deuterated Form	Chemical Shift (Multiplicity)[a]
Acetone	Acetone-d_6	2.05 (5)
Acetonitrile	Acetonitrile-d_3	1.93 (5)
Benzene	Benzene-d_6	7.15 (broad)
Carbon tetrachloride	—	—
Chloroform	Chloroform-d	7.25 (1)
Dimethylsulfoxide	Dimethylsulfoxide-d_6	2.49 (5)
Water	Deuterium oxide	4.82 (1)
Methanol	Methanol-d_4	4.84 (1) hydroxyl
		3.30 (5) methyl
Methylene chloride	Methylene chloride-d_2	5.32 (3)

[a]Where multiplets apply, the center peak is given and the number of lines is indicated in parentheses. No proton peak should be observed in the completely deuterated solvents listed. However, multiplets will arise from coupling of a proton with deuterium because the solvents are not 100% isotopically pure. For example, acetone-d_6 has a trace of acetone-d_5 in it, while chloroform-d has some chloroform present.

TABLE A10.2
^{13}C CHEMICAL-SHIFT VALUES FOR SOME COMMON NMR SOLVENTS (PPM)

Solvent	Deuterated Form	Chemical Shift (Multiplicity)[a]
Acetone	Acetone-d_6	206.0 (1) carbonyl
		24.8 (7) methyl
Benzene	Benzene-d_6	128.0 (3)
Chloroform	Chloroform-d	77.0 (3)
Dimethylsulfoxide	Dimethylsulfoxide-d_6	39.5 (7)
Dioxane	Dioxane-d_8	66.5 (5)
Methanol	Methanol-d_4	49.0 (7)
Methylene chloride	Methylene chloride-d_2	54.0 (5)

[a]Where multiplets apply, the center peak is given and the number of lines is indicated in parentheses. These multiplets arise from the coupling of carbon with the deuterium.

APPENDIX 11

Tables of Precise Masses and Isotopic Abundance Ratios for Molecular Ions under Mass 100 Containing Carbon, Hydrogen, Nitrogen, and Oxygen[a]

	Precise Mass	M + 1	M + 2
16			
CH_4	16.0313	1.15	
17			
NH_3	17.0266	0.43	
18			
H_2O	18.0106	0.07	0.20
26			
C_2H_2	26.0157	2.19	0.01
27			
CHN	27.0109	1.48	
28			
N_2	28.0062	0.76	
CO	27.9949	1.12	
C_2H_4	28.0313	2.23	0.01
29			
CH_3N	29.0266	1.51	
30			
CH_2O	30.0106	1.15	0.20
C_2H_6	30.0470	2.26	0.01
31			
CH_5N	31.0422	1.54	
32			
O_2	31.9898	0.08	0.40
N_2H_4	32.0375	0.83	
CH_4O	32.0262	1.18	0.20
40			
C_3H_4	40.0313	3.31	0.04
41			
C_2H_3N	41.0266	2.59	0.02
42			
CH_2N_2	42.0218	1.88	0.01
C_2H_2O	42.0106	2.23	0.21
C_3H_6	42.0470	3.34	0.04

[a]Adapted with permission from Beynon, J. H., Mass Spectrometry and Its Application to Organic Chemistry, Elsevier, Amsterdam, 1960. The precise masses are calculated on the basis of the most abundant isotope of carbon having a mass of 12.0000.

	Precise Mass	M + 1	M + 2
43			
CH_3N_2	43.0297	1.89	0.01
C_2H_5N	43.0422	2.62	0.02
44			
N_2O	44.0011	0.80	0.20
CO_2	43.9898	1.16	0.40
CH_4N_2	44.0375	1.91	0.01
C_2H_4O	44.0262	2.26	0.21
C_3H_8	44.0626	3.37	0.04
45			
CH_3NO	45.0215	1.55	0.21
C_2H_7N	45.0579	2.66	0.02
46			
NO_2	45.9929	0.46	0.40
CH_2O_2	46.0054	1.19	0.40
CH_4NO	46.0293	1.57	0.21
CH_6N_2	46.0532	1.94	0.01
C_2H_6O	46.0419	2.30	0.22
47			
CH_5NO	47.0371	1.58	0.21
48			
O_3	47.9847	0.12	0.60
CH_4O_2	48.0211	1.22	0.40
52			
C_4H_4	52.0313	4.39	0.07
53			
C_3H_3N	53.0266	3.67	0.05
54			
$C_2H_2N_2$	54.0218	2.96	0.03
C_3H_2O	54.0106	3.31	0.24
C_4H_6	54.0470	4.42	0.07
55			
C_2HNO	55.0058	2.60	0.22
C_3H_5N	55.0422	3.70	0.05
56			
$C_2H_4N_2$	56.0375	2.99	0.03
C_3H_4O	56.0262	3.35	0.24
C_4H_8	56.0626	4.45	0.08
57			
CH_3N_3	57.0328	2.27	0.02
C_2H_3NO	57.0215	2.63	0.22
C_3H_7N	57.0579	3.74	0.05

	Precise Mass	$M + 1$	$M + 2$
58			
CH_2N_2O	58.0167	1.92	0.21
$C_2H_2O_2$	58.0054	2.27	0.42
$C_2H_6N_2$	58.0532	3.02	0.03
C_3H_6O	58.0419	3.38	0.24
C_4H_{10}	58.0783	4.48	0.08
59			
$CHNO_2$	59.0007	1.56	0.41
CH_5N_3	59.0484	2.31	0.02
C_2H_5NO	59.0371	2.66	0.22
C_3H_9N	59.0736	3.77	0.05
60			
CH_4N_2O	60.0324	1.95	0.21
$C_2H_4O_2$	60.0211	2.30	0.04
$C_2H_8N_2$	60.0688	3.05	0.03
C_3H_8O	60.0575	3.41	0.24
61			
CH_3NO_2	61.0164	1.59	0.41
CH_7N_3	61.0641		
C_2H_7NO	61.0528	2.69	0.22
62			
CH_2O_3	62.0003	1.23	0.60
CH_6N_2O	62.0480	1.98	0.21
$C_2H_6O_2$	62.0368	2.34	0.42
63			
CH_5NO_2	63.0320	1.62	0.41
64			
CH_4O_3	64.0160	1.26	0.60
66			
C_5H_6	66.0470	5.50	0.12
67			
C_4H_5N	67.0422	4.78	0.09
68			
$C_3H_4N_2$	68.0375	4.07	0.06
C_4H_4O	68.0262	4.43	0.28
C_5H_8	68.0626	5.53	0.12
69			
$C_2H_3N_3$	69.0328	3.35	0.04
C_3H_3NO	69.0215	3.71	0.25
C_4H_7N	69.0579	4.82	0.09
70			
$C_2H_2N_2O$	70.0167	3.00	0.23
$C_3H_2O_2$	70.0054	3.35	0.44
$C_3H_6N_2$	70.0532	4.10	0.07
C_4H_6O	70.0419	4.46	0.28
C_5H_{10}	70.0783	5.56	0.13

	Precise Mass	M + 1	M + 2
71			
C_2HNO_2	71.0007	2.64	0.42
$C_2H_5N_3$	71.0484	3.39	0.04
C_3H_5NO	71.0371	3.74	0.25
C_4H_9N	71.0736	4.85	0.09
72			
$C_2H_4N_2O$	72.0324	3.03	0.23
$C_3H_4O_2$	72.0211	3.38	0.44
$C_3H_8N_2$	72.0688	4.13	0.07
C_4H_8O	72.0575	4.49	0.28
C_5H_{12}	72.0939	5.60	0.13
73			
$C_2H_3NO_2$	73.0164	2.67	0.42
$C_2H_7N_3$	73.0641	3.42	0.04
C_3H_7NO	73.0528	3.77	0.25
$C_4H_{11}N$	73.0892	4.88	0.10
74			
$C_2H_2O_3$	74.0003	2.31	0.62
$C_2H_6N_2O$	74.0480	3.06	0.23
$C_3H_6O_2$	74.0368	3.42	0.44
$C_3H_{10}N_2$	74.0845	4.17	0.07
$C_4H_{10}O$	74.0732	4.52	0.28
75			
$CHNO_3$	74.9956	1.60	0.61
$C_2H_5NO_2$	75.0320	2.70	0.43
$C_2H_9N_3$	75.0798	3.45	0.05
C_3H_9NO	75.0684	3.81	0.25
76			
$C_2H_4O_3$	76.0160	2.34	0.62
$C_2H_8N_2O$	76.0637	3.09	0.24
$C_3H_8O_2$	76.0524	3.45	0.44
77			
CH_3NO_3	77.0113	1.63	0.61
$C_2H_7NO_2$	77.0477	2.73	0.43
78			
$C_2H_6O_3$	78.0317	2.38	0.62
C_6H_6	78.0470	6.58	0.18
79			
CH_5NO_3	79.0269	1.66	0.61
C_5H_5N	79.0422	5.87	0.14
80			
C_6H_8	80.0626	6.61	0.18
81			
C_5H_7N	81.0579	5.90	0.14

	Precise Mass	M + 1	M + 2
82			
$C_4H_6N_2$	82.0532	4.18	0.11
C_5H_6O	82.0419	5.54	0.32
C_6H_{10}	82.0783	6.64	0.19
83			
$C_3H_5N_3$	83.0484	4.47	0.08
C_4H_5NO	83.0371	4.82	0.29
C_5H_9N	83.0736	5.93	0.15
84			
$C_3H_4N_2O$	84.0324	4.11	0.27
$C_4H_4O_2$	84.0211	4.47	0.48
$C_4H_8N_2$	84.0688	5.21	0.11
C_5H_8O	84.0575	5.57	0.33
C_6H_{12}	84.0939	6.68	0.19
85			
$C_3H_3NO_2$	85.0164	3.75	0.45
$C_3H_7N_3$	85.0641	4.50	0.08
C_4H_7NO	85.0528	4.86	0.29
$C_5H_{11}N$	85.0892	5.96	0.15
86			
$C_3H_2O_3$	86.0003	3.39	0.64
$C_3H_6N_2O$	86.0480	4.14	0.27
$C_4H_6O_2$	86.0368	4.50	0.48
$C_4H_{10}N_2$	86.0845	5.25	0.11
$C_5H_{10}O$	86.0732	5.60	0.33
C_6H_{14}	86.1096	6.71	0.19
87			
C_2HNO_3	86.9956	2.68	0.62
$C_3H_5NO_2$	87.0320	3.78	0.45
$C_3H_9N_3$	87.0798	4.53	0.08
C_4H_9NO	87.0684	4.89	0.30
$C_5H_{13}N$	87.1049	5.99	0.15
88			
$C_3H_4O_3$	88.0160	3.42	0.64
$C_3H_8N_2O$	88.0637	4.17	0.27
$C_4H_8O_2$	88.0524	4.53	0.48
$C_4H_{12}N_2$	88.1001	5.28	0.11
$C_5H_{12}O$	88.0888	5.63	0.33
89			
$C_2H_3NO_3$	89.0113	2.71	0.63
$C_3H_7NO_2$	89.0477	3.81	0.46
$C_3H_{11}N_3$	89.0954	4.56	0.84
$C_4H_{11}NO$	89.0841	4.92	0.30
90			
$C_3H_6O_3$	90.0317	3.46	0.64
$C_3H_{10}N_2O$	90.0794	4.20	0.27
$C_4H_{10}O_2$	90.0681	4.56	0.48

	Precise Mass	M + 1	M + 2
91			
$C_2H_5NO_3$	91.0269	2.74	0.63
$C_2H_9N_3O$	91.0746	3.49	0.25
$C_3H_9NO_2$	91.0634	3.85	0.46
92			
$C_3H_8O_3$	92.0473	3.49	0.64
C_7H_8	92.0626	7.69	0.26
93			
$C_2H_7NO_3$	93.0426	2.77	0.63
C_6H_7N	93.0579	6.98	0.21
94			
$C_5H_6N_2$	94.0532	6.26	0.17
C_6H_6O	94.0419	6.62	0.38
C_7H_{10}	94.0783	7.72	0.26
95			
$C_4H_5N_3$	95.0484	5.55	0.13
C_5H_5NO	95.0371	5.90	0.34
C_6H_9N	95.0736	7.01	0.21
96			
$C_4H_4N_2O$	96.0324	5.19	0.31
$C_5H_4O_2$	96.0211	5.55	0.53
$C_5H_8N_2$	96.0688	6.29	0.17
C_6H_8O	96.0575	6.65	0.39
C_7H_{12}	96.0939	7.76	0.26
97			
$C_4H_3NO_2$	97.0164	4.83	0.49
$C_4H_7N_3$	97.0641	5.58	0.13
C_5H_7NO	97.0528	5.94	0.35
$C_6H_{11}N$	97.0892	7.04	0.21
98			
$C_4H_6N_2O$	98.0480	5.22	0.31
$C_5H_6O_2$	98.0368	5.58	0.53
$C_5H_{10}N_2$	98.0845	6.33	0.17
$C_6H_{10}O$	98.0732	6.68	0.39
C_7H_{14}	98.1096	7.79	0.26
99			
$C_4H_5NO_2$	99.0320	4.86	0.50
$C_4H_9N_3$	99.0798	5.61	0.13
C_5H_9NO	99.0684	5.97	0.35
$C_6H_{13}N$	99.1049	7.07	0.21
100			
$C_4H_8N_2O$	100.0637	5.25	0.31
$C_5H_8O_2$	100.0524	5.61	0.53
$C_5H_{12}N_2$	100.1001	6.36	0.17
$C_6H_{12}O$	100.0888	6.72	0.39
C_7H_{16}	100.1253	7.82	0.26

APPENDIX 12

Common Fragment Ions under Mass 105[a]

m/e	Ions	m/e	Ions
14	CH_2	44	$CH_2CH=O + H$
15	CH_3		CH_3CHNH_2
16	O		CO_2
17	OH		$NH_2C=O$
18	H_2O		$(CH_3)_2N$
	NH_4	45	CH_3CHOH
19	F		CH_2CH_2OH
	H_3O		CH_2OCH_3
26	$C\equiv N$		$\overset{O}{\overset{\|}{C}}-OH$
27	C_2H_3		
28	C_2H_4		$CH_3CH-O + H$
	CO	46	NO_2
	N_2 (air)	47	CH_2SH
	$CH=NH$		CH_3S
29	C_2H_5	48	$CH_3S + H$
	CHO	49	CH_2Cl
30	CH_2NH_2	51	CHF_2
	NO		C_4H_3
31	CH_2OH	53	C_4H_5
	OCH_3	54	$CH_2CH_2C\equiv N$
32	O_2 (air)	55	C_4H_7
33	SH		$CH_2=CHC=O$
	CH_2F	56	C_4H_8
34	H_2S	57	C_4H_9
35	Cl		$C_2H_5C=O$
36	HCl	58	$CH_3-\underset{CH_2}{\overset{\|}{C}}=O + H$
39	C_3H_3		
40	$CH_2C\equiv N$		
41	C_3H_5		$C_2H_5CHNH_2$
	$CH_2C=H + H$		$(CH_3)_2NCH_2$
	C_2H_2NH		$C_2H_5NHCH_2$
42	C_3H_6		C_2H_2S
43	C_3H_7		
	$CH_3C=O$		
	C_2H_5N		

[a]Adapted with permission from Silverstein, R. M. and F. X. Webster, Spectrometric Identification of Organic Compounds, 6th ed., John Wiley & Sons, New York, 1998.

m/e	Ions	m/e	Ions
59	$(CH_3)_2COH$	74	$CH_2{-}\overset{\displaystyle O}{\overset{\|}{C}}{-}OCH_3 + H$
	$CH_2OC_2H_5$		
	$\overset{\displaystyle O}{\overset{\|}{C}}{-}OCH_3$	75	$\overset{\displaystyle O}{\overset{\|}{C}}{-}OC_2H_5 + 2H$
	$NH_2C{=}O \quad + H$ $\quad\quad\underset{CH_2}{\|}$		$CH_2SC_2H_5$
			$(CH_3)_2CSH$
	CH_3OCHCH_3		$(CH_3O)_2CH$
	CH_3CHCH_2OH	77	C_6H_5
60	$CH_2C{=}O \quad + H$ $\quad\quad\underset{OH}{\|}$	78	$C_6H_5 + H$
		79	$C_6H_5 + 2H$
	CH_2ONO		Br
61	$\overset{\displaystyle O}{\overset{\|}{C}}{-}OCH_3 + 2H$	80	$CH_3SS + H$
	CH_2CH_2SH	81	C_6H_9
	CH_2SCH_3		
65	 (or C_5H_5)		
		82	$CH_2CH_2CH_2CH_2C{\equiv}N$
66	 (or C_5H_6)		CCl_2
			C_6H_{10}
67	C_5H_7	83	C_6H_{11}
68	$CH_2CH_2CH_2C{\equiv}N$		$CHCl_2$
69	C_5H_9	85	C_6H_{13}
	CF_3		$C_4H_9C{=}O$
	$CH_3CH{=}CHC{=}O$		$CClF_2$
	$CH_2{=}C(CH_3)C{=}O$	86	$C_3H_7\overset{\displaystyle O}{\overset{\|}{C}}{-}CH_2 + H$
70	C_5H_{10}		
71	C_5H_{11}		$C_4H_9CHNH_2$ and isomers
	$C_3H_7C{=}O$	87	C_3H_7CO
72	$C_2H_5\overset{\displaystyle O}{\overset{\|}{C}}{-}CH_2$		Homologs of 73
			$CH_2CH_2COCH_3$
	$C_3H_7CHNH_2$	88	$CH_2{-}\overset{\displaystyle O}{\overset{\|}{C}}{-}OC_2H_5 + H$
	$(CH_3)N{=}C{=}O$		
	$C_2H_5NHCHCH_3$ and isomers		
73	Homologs of 59		

m/e	Ions	m/e	Ions
89	$\overset{O}{\overset{\|}{C}}-OC_3H_7 + 2H$ (phenyl)—C	94	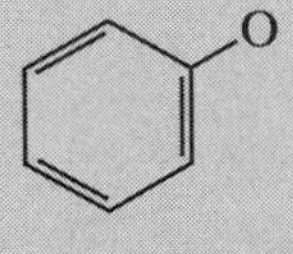 (cyclohexadienyl)—O + H
90	CH_3CHONO_2 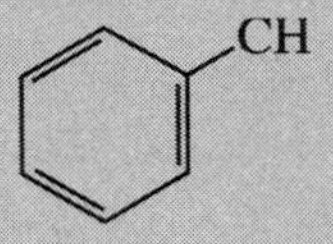 (phenyl)—CH	96	$CH_2CH_2CH_2CH_2CH_2C\equiv N$
	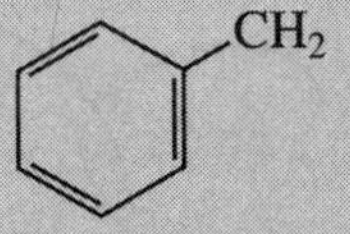 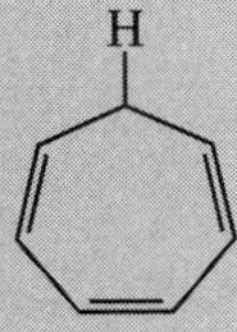 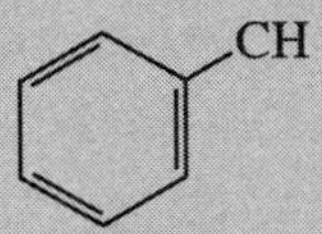 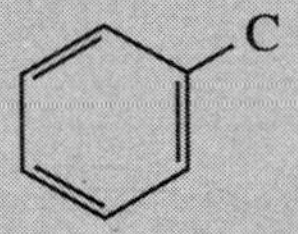 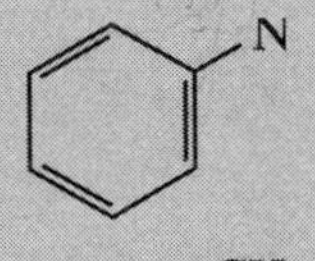	97	C_7H_{13}
	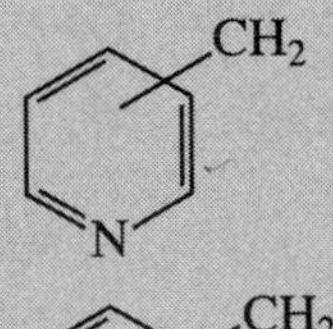 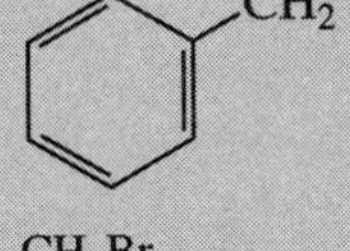	99	C_7H_{15} $C_6H_{11}O$
91	(phenyl)—CH_2 or (tropylium H) (phenyl)—CH + H (phenyl)—C + 2H $(CH_2)_4Cl$ (phenyl)—N	100	$\overset{O}{\overset{\|}{C_4H_9C}}-CH_2 + H$ $C_5H_{11}CHNH_2$
92	(pyridyl)—CH_2 (phenyl)—CH_2 + H	101	$\overset{O}{\overset{\|}{C}}-OC_4H_9$
		102	$CH_2\overset{O}{\overset{\|}{C}}-OC_3H_7 + H$
		103	$\overset{O}{\overset{\|}{C}}-OC_4H_9 + 2H$ $C_5H_{11}S$ $CH(OCH_2CH_3)_2$
		104	$C_2H_5CHONO_2$
93	CH_2Br (methylphenol)—OH C_7H_9 (phenyl)—O	105	(phenyl)—C=O (phenyl)—CH_2CH_2 (phenyl)—$CHCH_3$

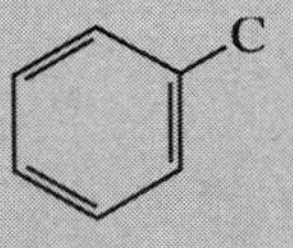

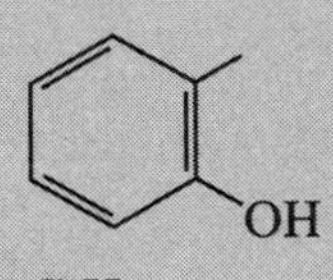

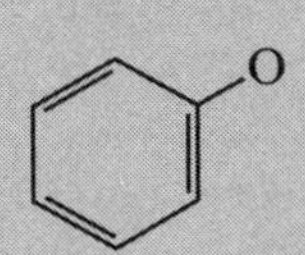

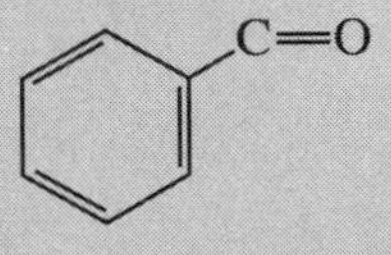

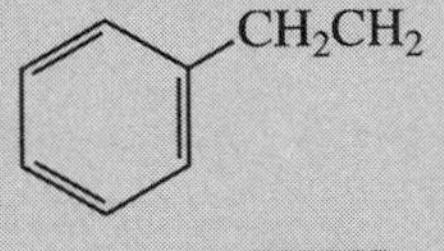

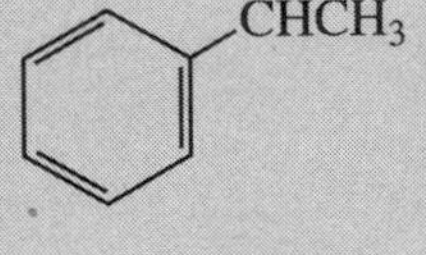

APPENDIX 13

A Handy-Dandy Guide to Mass Spectral Fragmentation Patterns

Alkanes

Good M^+

14-amu fragments

Alkenes

Distinct M^+

Loss of 15, 29, 43, and so on

Cycloalkanes

Strong M^+

Loss of $CH_2=CH_2$ $M-28$

Loss of alkyl

Aromatics

Strong M^+

$C_7H_7^+$ $m/e = 91$, weak $m/e = 65$ $(C_5H_5^+)$

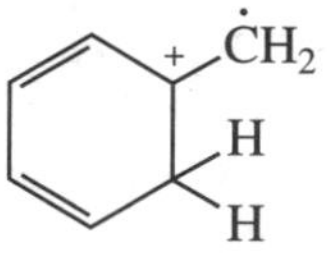

$m/e = 92$ Transfer of *gamma* hydrogens

Halides

Cl and Br doublets (M^+ and $M+2$)

$m/e = 49$ or 51 $CH_2=Cl^+$

$m/e = 93$ or 95 $CH_2=Br^+$

$M-36$ Loss of HCl

$m/e = 91$ or 93

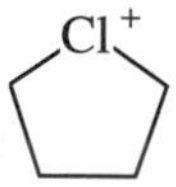

$m/e = 135$ or 137

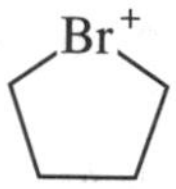

$M-79$ $(M-81)$ Loss of Br·

$M-127$ Loss of I·

Alcohols

M^+ weak or absent

Loss of alkyl

$CH_2=OH^+$ $m/e = 31$

$RCH=OH^+$ $m/e = 45, 59, 73, \ldots$

$R_2C=OH^+$ $m/e = 59, 73, 87, \ldots$

$M-18$ Loss of H_2O

$M-46$ Loss of $H_2O + CH_2=CH_2$

Phenols

Strong M^+
Strong $M - 1$ Loss of H·
$M - 28$ Loss of CO

Ethers

M^+ stronger than alcohols
Loss of alkyl
Loss of OR′ $M - 31, M - 45, M - 59$, and so on
$CH_2=OR'^+$ $m/e = 45, 59, 73, \ldots$

Amines

M^+ weak or absent
Nitrogen Rule
$m/e = 30$ $CH_2=NH_2^+$ (base peak)
Loss of alkyl

Aldehydes

Weak M^+
$M - 29$ Loss of HCO
$M - 43$ Loss of $CH_2=CHO$
$m/e = 44$

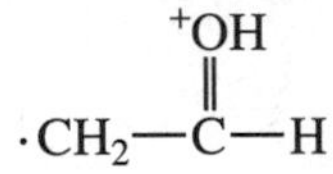

$$·CH_2-\overset{\overset{+OH}{\|}}{C}-H \qquad \text{Transfer of } gamma \text{ hydrogens}$$

or 58, 72, 86, . . .

Aromatic Aldehydes

Strong M^+
$M - 1$ Loss of H·
$M - 29$ Loss of H· and CO

Ketones

M^+ intense
$M - 15, M - 29, M - 43, \ldots$ Loss of alkyl group
$m/e = 43$ CH_3CO^+
$m/e = 58, 72, 86, \ldots$ Transfer of *gamma* hydrogens
$m/e = 55$ $^+CH_2-CH=C=O$ Base peak for cyclic ketones

$m/e = 83$ $C\equiv O^+$ in cyclohexanone

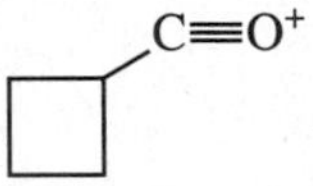

$m/e = 42$ in cyclohexanone

$m/e = 105$ in aryl ketones

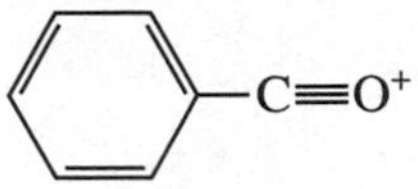

$m/e = 120$ Transfer of *gamma* hydrogens

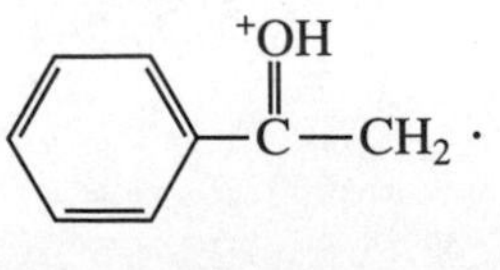

Carboxylic Acids

M^+ weak but observable
$M - 17$ Loss of OH
$M - 45$ Loss of COOH
$m/e = 45$ $^+$COOH

$m/e = 60$ $^+$OH   Transfer of *gamma* hydrogens
                    ‖
          HO—C—CH$_2$ ·

Aromatic Acids

M^+ large
$M - 17$ Loss of OH
$M - 45$ Loss of COOH
$M - 18$ *Ortho* effect

Methyl Esters

M^+ weak but observable
$M - 31$ Loss of OCH$_3$
$m/e = 59$ $^+$COOCH$_3$

$m/e = 74$ $^+$OH   Transfer of *gamma* hydrogens
                    ‖
        CH$_3$O—C—CH$_2$

Higher Esters

M^+ weaker than for RCOOCH$_3$
Same pattern as in methyl esters
$M - 45, M - 59, M - 73$ Loss of OR
$m/e = 73, 87, 101$ $^+$COOR

$m/e = 88, 102, 116$ $^+$OH   Transfer of *gamma* hydrogens
                              ‖
                  RO—C—CH$_2$ ·

$m/e = 28, 42, 56, 70$ *Beta* hydrogens on alkyl group

$m/e = 61, 75, 89$ $^+$OH   Long alkyl chain
                            ‖
                  R—C—OH

$m/e = 108$ Loss of CH$_2$=C=O   Benzyl or acetate ester

$m/e = 105$ ⬡—C≡O$^+$

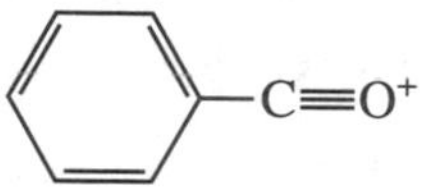

$m/e = 77$ 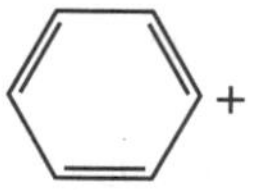   ⬡$^+$ weak

$M - 32, M - 46, M - 60$ *ortho* effect—loss of ROH

APPENDIX 14

Index of Spectra

Infrared Spectra

Acetophenone, 57
Acetyl chloride, 70
Anisole, 49
Benzaldehyde, 55
Benzenesulfonamide, 81
Benzenesulfonyl chloride, 81
Benzenethiol, 79
Benzoic acid, 61
Benzonitrile, 76
Benzoyl chloride, 70
Benzyl isocyanate, 76
2-Butanol, 46
Butylamine, 73
Butyronitrile, 75
Carbon tetrachloride, 83
Chloroform, 84
para-Cresol, 46
Crotonaldehyde, 55
Cyclohexane, 31
Cyclohexene, 32
Cyclopentanone, 58
Decane, 30
Dibutyl ether, 49
Dibutylamine, 73
meta-Diethylbenzene, 42
ortho-Diethylbenzene, 42
para-Diethylbenzene, 42
Ethyl 3-aminobenzoate, 472
Ethyl butyrate, 63
Ethyl cyanoacetate, 476
Ethyl propionate, 468
1-Hexanol, 45
1-Hexene, 32
Isobutyric acid, 61
Leucine, 79
Mesityl oxide, 57
4-Methoxyphenylacetone, 470
Methyl benzoate, 64
3-Methyl-2-butanone, 25
Methyl methacrylate, 63
Methyl salicylate, 64
Methyl *p*-toluenesulfonate, 81
N-Methylacetamide, 69
N-Methylaniline, 74

Mineral oil, 30
Nitrobenzene, 77
1-Nitrohexane, 77
Nonanal, 55
Nujol, 30
1-Octyne, 33
4-Octyne, 34
2,4-Pentanedione, 58
cis-2-Pentene, 32
trans-2-Pentene, 33
Propionamide, 68
Propionic anhydride, 72
Styrene, 43
Toluene, 41
Tributylamine, 73
Vinyl acetate, 64

Mass Spectra

Acetophenone, 428
p-Anisic acid, 433
Benzaldehyde, 425
Benzene, 414
Benzonitrile, 439
Benzyl alcohol, 421
Bicyclo[2.2.1]heptane, 410
1-Bromo-2-chloroethane, 445
1-Bromohexane, 441
Butane, 406
1-Butanol, 417
2-Butanol, 418
2-Butanone, 426
1-Butene, 411
cis-2-Butene, 411
Butyraldehyde, 424
Butyric acid, 432
1-Chloro-2-methylbenzene, 447
2-Chloroheptane, 442
Cyclohexanol, 420
Cyclohexanone, 427
Cyclopentane, 409
Dibromomethane, 444
Dichloromethane, 444
Diethylamine, 434
Diisopropyl ether, 422
Dopamine, 395

^{1}H NMR Spectra

^{13}C NMR Spectra

COSY Spectra

DEPT Spectra

HETCOR Spectra

NOE Difference Spectra

Ultraviolet-Visible Spectra

INDEX

I-1